Mathematical Methods for Engineers and Scientists 1

Kwong-Tin Tang

Mathematical Methods for Engineers and Scientists 1

Complex Analysis and Linear Algebra

Second Edition

Springer

Kwong-Tin Tang
Department of Physics
Pacific Lutheran University
Tacoma, WA, USA

ISBN 978-3-031-05680-2   ISBN 978-3-031-05678-9 (eBook)
https://doi.org/10.1007/978-3-031-05678-9

This Springer imprint is published by the registered company Springer Nature Switzerland AG
The registered company address is: Gewerbestrasse 11, 6330 Cham, Switzerland

In loving memory of my parents

Preface to Second Edition

There are two more chapters in this edition.

The one on conformal mapping is by readers' request. Hopefully this means that the subject is important and my leisurely approach is appealing. I must confess that I taught these materials only once before. They are mostly my personal notes.

There were three chapters on determinants and matrices. Together they form the body of linear algebra. However, I found it easy to get lost in the details of calculations. So I wrote a summary of the computation aspects and gave a simple geometrical description of linear algebra in terms of vector space. These are the contents of Chap. 5. Because of the time line they were written, there are some overlaps with the following chapters. Since pedagogically a certain amount of repetition is not necessarily bad, I did not remove them.

I want to thank all those who pointed out the mistakes of the first edition. Especially I want to thank my colleagues Professor Richard Louie who gave me a list of errors, and Prof. Chang-li Yiu who wrote part of the vector space. I must also thank my wife, Pauline, who has taken up all house chores. Once again I want to thank Mattew Hacker for his help with computer problems.

May this set of books be an acceptable offering in the eyes of my Lord Jesus Christ.

Tacoma, Washington   Kwong-Tin Tang
March 2022

Preface to the First Edition

For thirty-some years, I have taught two "Mathematical Physics" courses. One of them was previously named "Engineering Analysis". There are several textbooks of unquestionable merit for such courses, but I could not find one that fitted our needs. It seemed to me that students might have an easier time if some changes were made in these books. I ended up using class notes. Actually I felt the same about my own notes, so they got changed again and again. Throughout the years, many students and colleagues have urged me to publish them. I resisted until now, because the topics were not new and I was not sure that my way of presenting them was really that much better than others. In recent years, some former students came back to tell me that they still found my notes useful and looked at them from time to time. The fact that they always singled out these courses, among many others I have taught, made me think that besides being kind, they might even mean it. Perhaps it is worthwhile to share these notes with a wider audience.

It took far more work than expected to transcribe the lecture notes into printed pages. The notes were written in an abbreviated way without much explanation between any two equations, because I was supposed to supply the missing links in person. How much detail I would go into depended on the reaction of the students. Now without them in front of me, I had to decide the appropriate amount of derivation to be included. I chose to err on the side of too much detail rather than too little. As a result, the derivation does not look very elegant, but I also hope it does not leave any gap in students' comprehension.

Precisely stated and elegantly proved theorems looked great to me when I was a young faculty member. But in later years, I found that elegance in the eyes of the teacher might be stumbling blocks for students. Now I am convinced that before the student can use a mathematical theorem with confidence, he must first develop an intuitive feeling. The most effective way to do that is to follow a sufficient number of examples.

This book is written for students who want to learn but need a firm hand-holding. I hope they will find the book readable and easy to learn from. Learning, as always, has to be done by the student herself or himself. No one can acquire mathematical skill without doing problems, the more the better. However, realistically students have a finite amount of time. They will be overwhelmed if problems are too numerous, and frustrated if problems are too difficult. A common practice in textbooks is to list a large number of problems and let the instructor to choose a few for assignments. It seems to me that is not a confidence building strategy. A self-learning person would not know what to choose. Therefore a moderate number of not overly difficult problems, with answers, are selected at the end of each chapter. Hopefully after the student has successfully solved all of them, he will be encouraged to seek more challenging ones. There are plenty of problems in other books. Of course, an instructor can always assign more problems at levels suitable to the class.

On certain topics, I went farther than most other similar books, not in the sense of esoteric sophistication, but in making sure that the student can carry out the actual calculation. For example, the diagonalization of a degenerate Hermitian matrix is of considerable importance in many fields. Yet to make it clear in a succinct way is not easy. I used several pages to give a detailed explanation of a specific example.

Professor I.I. Rabi used to say "All textbooks are written with the principle of least astonishment". Well, there is a good reason for that. After all, textbooks are supposed to explain away the mysteries and make the profound obvious. This book is no exception. Nevertheless, I still hope the reader will find something in this book exciting.

This volume consists of three chapters on complex analysis and three chapters on theory of matrices. In subsequent volumes, we will discuss vector and tensor analysis, ordinary differential equations and Laplace transforms, Fourier analysis and partial differential equations. Students are supposed to have already completed two or three semesters of calculus and a year of college physics.

This book is dedicated to my students. I want to thank my A students and B students, their diligence and enthusiasm have made teaching enjoyable and worthwhile. I want to thank my C students and D students, their difficulties and mistakes made me search for better explanations.

I want to thank Brad Oraw for drawing many figures in this book, and Matthew Hacker for helping me to typeset the manuscript.

I want to express my deepest gratitude to Professor S.H. Patil, Indian Institute of Technology, Bombay. He has read the entire manuscript and provided many excellent suggestions. He has also checked the equations and the problems and corrected numerous errors. Without his help and encouragement, I doubt there is this book.

The responsibility for remaining errors is, of course, entirely mine. I will greatly appreciate if they are brought to my attention.

Tacoma, Washington   K.T Tang
October 2005

Contents

Part I
Complex Analysis

Chapter 1
Complex Numbers

The most compact equation in all of mathematics is surely

$$e^{i\pi} + 1 = 0. \tag{1.1}$$

In this one equation, the five fundamental constants coming from four major branches of classical mathematics: –arithmetic $(0, 1)$, algebra (i), geometry (π), and analysis (e),—are connected by the three most important mathematic operations:—addition, multiplication, and exponentiation,—into two nonvanshing terms.

The reader is probably aware that (1.1) is but one of the consequences of the miraculous Euler formula (discovered around 1740 by Leonhard Euler)

$$e^{i\theta} = \cos\theta + i\sin\theta. \tag{1.2}$$

When $\theta = \pi$, $\cos\pi = -1$, and $\sin\pi = 0$, it follows that $e^{i\pi} = -1$.

Much of the computations involving complex numbers are based on the Euler formula. To provide a proper setting for the discussion of this formula, we will first present a sketch of our number system and some historic background. This will also give us a framework to review some of the basic mathematics operations.

1.1 Our Number System

Anyone who encounters for the first time these equations cannot help but be intrigued by the strange properties of the numbers such as e and i. But strange is relative, with sufficient familiarity, the strange object of yesterday becomes the common thing of today. For example, nowadays no one will be bothered by the negative numbers, but for a long time negative numbers were regarded as "strange" or "absurd". For two thousand years, mathematics thrived without negative numbers. The Greeks did

© The Author(s), under exclusive license to Springer Nature Switzerland AG 2022
K.-T. Tang, *Mathematical Methods for Engineers and Scientists 1*,
https://doi.org/10.1007/978-3-031-05678-9_1

not recognize negative numbers and did not need them. Their main interest was geometry, for the description of which positive numbers are entirely sufficient. Even after Hindu mathematician Brahmagupta "invented" zero around 628, and negative numbers were interpreted as a loss instead of a gain in financial matters, medieval Europe mostly ignored them.

Indeed, so long as one regards subtraction as an act of "taken away", negative numbers are absurd. One cannot take away, say, three apples from two.

Only after the development of the axiomatic algebra, the full acceptance of negative numbers into our number system was made possible. It is also within the framework of axiomatic algebra, irrational numbers and complex numbers are seen to be natural parts of our number system.

By axiomatic method, we mean the step by step development of a subject from a small set of definitions and a chain of logical consequences derived from them. This method had long been followed in geometry, ever since the Greeks established it as a rigorous mathematical discipline.

1.1.1   Addition and Multiplication of Integers

We start with the assumption that we know what integers are, what zero is, and how to count. Although mathematicians could go even further back and describe the theory of sets in order to derive the properties of integers, we are not going in that direction.

We put the integers on a line with increasing order as in the following diagram.

$$
\begin{array}{cccccccc}
0 & 1 & 2 & 3 & 4 & 5 & 6 & 7 \cdots \\
 & & \downarrow & & & \uparrow & & \\
2 & & -- & -- & & -- & -- &
\end{array}
$$

If we start with a certain integer a, and we count successively one unit b times to the right, the number we arrive at we call $a + b$, and that defines *addition* of integers. For example, starting at 2, and going up 3 units, we arrive at 5. So 5 is equal to $2 + 3$.

Once we have defined addition, then we can consider this: if we start with nothing and add a to it, b times in succession, we call the result *multiplication* of integers; we call it b *times* a.

Now as a consequence of these definitions it can be easily shown that these operations satisfy certain simple rules concerning the order in which the computations can proceed. They are the familiar commutative, associative, and distributive laws:

$$
\begin{array}{lll}
a + b = b + a & \textit{Commutative Law of Addition} & \\
a + (b + c) = (a + b) + c & \textit{Associative Law of Addition} & \\
ab = ba & \textit{Commutative Law of Multiplication} & (1.3) \\
(ab)c = a(bc) & \textit{Associative Law of Multiplication} & \\
a(b + c) = ab + ac & \textit{Distributive Law} &
\end{array}
$$

These rules characterize the **elementary algebra**. We say elementary algebra because there is a branch of mathematics called modern algebra in which some of the rules such as $ab = ba$ are abandoned, but we shall not discuss that.

Among the integers, 1 and 0 have special properties:

$$a + 0 = a$$
$$a \cdot 1 = a.$$

So 0 is the additive identity and 1 is the multiplicative identity. Furthermore,

$$0 \cdot a = 0,$$

and if $ab = 0$, either a or/and b is zero.

Now we can also have a succession of multiplications: if we start with 1 and multiply by a, b times in succession, we call that *raising to power*: a^b. It follows from this definition that

$$(ab)^c = a^c b^c$$
$$a^b a^c = a^{(b+c)}$$
$$(a^b)^c = a^{(bc)}.$$

These results are well known and we shall not belabor them.

1.1.2 Inverse Operations

In addition to the direct operation of addition, multiplication, and raising to a power, we have also the inverse operations, which are defined as follows. Let us assume a and c are given, and that we wish to find what values of b satisfy such equations as $a + b = c$, $ab = c$, $b^a = c$.

If $a + b = c$, b is defined as $c - a$, which is called *subtraction*. The operation called division is also clear: if $ab = c$, then $b = c/a$ defines division—a solution of the equation $ab = c$ "backwards".

Now if we have a power $b^a = c$ and we ask ourselves, " What is b?", it is called ath root of c: $b = \sqrt[a]{c}$. For instance, if we ask ourselves the following question, "What integer, raised to third power, equals 8?", then the answer is cube root of 8; it is 2. The direct and inverse operations are summarized as follows:

Operation		*Inverse Operation*	
(a) *addition* :	$a + b = c$	(a') *subtraction* :	$b = c - a$
(b) *multiplication* :	$ab = c$	(b') *division* :	$b = c/a$
(c) *power* :	$b^a = c$	(c') *root* :	$b = \sqrt[a]{c}$

Insoluble Problems

When we try to solve simple algebraic equations using these definitions, we soon discover some insoluble problems, such as the following. Suppose we try to solve the equation $b = 3 - 5$. That means, according to our definition of subtraction, that we must find a number which, when added to 5, gives 3. And of course there is no such number, because we consider only positive integers; this is an insoluble problem.

1.1.3   Negative Numbers

In the grand design of algebra, the way to overcome this difficulty is to broaden the number system through **abstraction** and **generalization**. We abstract the original definitions of addition and multiplication from the rules and integers. We assume the rules to be true in general on a wider class of numbers, even though they are originally derived on a smaller class. Thus, rather using the integers to symbolically define the rules, we use the rules as the definition of the symbols, which then represent a more general kind of number. As an example, by working with the rules alone we can show that $3 - 5 = 0 - 2$. In fact we can show that one can make all subtractions, provided we define a whole set of new numbers: $0 - 1$, $0 - 2$, $0 - 3$, $0 - 4$, and so on (abbreviated as -1, -2, -3, -4, ...), called the negative numbers.

So we have increased the range of objects over which the rules work, but the meaning of the symbols is different. One cannot say, for instance, that -2 times 5 really means to add 5 together successively -2 times. That means nothing. But we require the negative numbers to obey all the rules.

For example, we can use the rules to show that -3 times -5 is equal to 15. Let $x = -3(-5)$, this is equivalent to $x + 3(-5) = 0$, or $x + 3(0 - 5) = 0$. By the rules, we can write this equation as

$$x + 0 - 15 = (x + 0) - 15 = x - 15 = 0.$$

Thus, $x = 15$. Therefore negative a times negative b is equal to positive ab,

$$(-a)(-b) = ab.$$

An interesting problem comes up in taking powers. Suppose we wish to discover what $a^{(3-5)}$ means. We know that $3 - 5$ is a solution of the problem, $(3 - 5) + 5 = 3$. Therefore

$$a^{(3-5)+5} = a^3.$$

Since

$$a^{(3-5)+5} = a^{(3-5)}a^5 = a^3,$$

it follows that
$$a^{(3-5)} = a^3/a^5.$$

Thus, in general

$$a^{n-m} = \frac{a^n}{a^m}.$$

If $n = m$, we have

$$a^0 = 1.$$

In addition, we found out what it means to raise a negative power. Since

$$3 - 5 = -2, \quad a^3/a^5 = \frac{1}{a^2},$$

So

$$a^{-2} = \frac{1}{a^2}.$$

If our number system consists of only positive and negative integers, then $1/a^2$ is a meaningless symbol, because if a is a positive or negative integer, the square of it is greater than 1, and we do not know what we mean by 1 divided by a number greater than 1! So this is another insoluble problem.

1.1.4 Fractional Numbers

The great plan is to continue the process of generalization; whenever we find another problem that we cannot solve we extend our realm of numbers. Consider division: we cannot find a number which is an integer, even a negative integer, which is equal to the result of dividing 3 by 5. So we simply say that 3/5 is another number, called fraction number. With the fraction number defined as a/b where a and b are integers and $b \neq 0$, we can talk about multiplying and adding fractions. For example, if $A = a/b$ and $B = c/b$, then by definition $bA = a$, $bB = c$, so $b(A + B) = a + c$. Thus, $A + B = (a + c)/b$. Therefore

$$\frac{a}{b} + \frac{c}{b} = \frac{a+c}{b}.$$

Similarly, we can show

$$\frac{a}{b} \times \frac{c}{d} = \frac{ac}{bd}, \quad \frac{a}{b} + \frac{c}{d} = \frac{ad+cb}{bd}.$$

It can also be readily shown that fractional numbers satisfy the rules defined in (1.3). For example, to prove the commutative law of multiplication, we can start with

$$\frac{a}{b} \times \frac{c}{d} = \frac{ac}{bd}, \quad \frac{c}{d} \times \frac{a}{b} = \frac{ca}{db}.$$

Since a, b, c, d are integers, so $ac = ca$ and $bd = db$. Therefore $\frac{ac}{bd} = \frac{ca}{db}$. It follows that

$$\frac{a}{b} \times \frac{c}{d} = \frac{c}{d} \times \frac{a}{b}.$$

Take another example of powers: what is $a^{3/5}$? We know only that $(3/5)5 = 3$, since that was the definition of $3/5$. So we know also that

$$(a^{(3/5)})^5 = a^{(3/5)(5)} = a^3$$

Then by the definition of roots we find that

$$a^{(3/5)} = \sqrt[5]{a^3}.$$

In this way we can define what we mean by putting fractions in the various symbols. It is a remarkable fact that all the rules still work for positive and negative integers, as well as for fractions!

Historically, the positive integers and their ratios (the fractions) were embraced by the ancients as natural numbers. These natural numbers together with their negative counter parts are known as **rational numbers** in our present day language.

The Greeks, under the influence of the teaching of Pythagoras, elevated fractional numbers to the central pillar of their mathematical and philosophical system. They believed that fractional numbers are prime cause behind everything in the world, from the laws of musical harmony to the motion of planets. So it was quite a shock when they found that there are numbers that cannot be expressed as a fraction.

1.1.5   *Irrational Numbers*

The first evidence of the existence of the irrational number (a number that is not a rational number) came from finding the length of the diagonal of a unit square. If the length of the diagonal is x, then by Pythagorean theorem $x^2 = 1^2 + 1^2 = 2$. Therefore $x = \sqrt{2}$. When people assumed this number is equal to some fraction, say m/n where m and n have no common factors, they found this assumption leads to a contradiction.

The argument goes as follows. If $\sqrt{2} = m/n$, then $2 = m^2/n^2$, or $2n^2 = m^2$. This means m^2 is an even integer. Furthermore, m itself must also be an even integer, since the square of an odd number is always odd. Thus $m = 2k$ for some integer k.

It follows that $2n^2 = (2k)^2$, or $n^2 = 2k^2$. But this means n is also an even integer. Therefore, m and n have a common factor of 2, contrary to the assumption that they have no common factors. Thus $\sqrt{2}$ cannot be a fraction.

This was shocking to the Greeks, not only because of philosophical arguments, but also because mathematically, fractions form a **dense set** of numbers. By this we mean that between any two fractions, no matter how close, we can always squeeze in another. For example,

$$\frac{1}{100} = \frac{2}{200} > \frac{2}{201} > \frac{2}{202} = \frac{1}{101}.$$

So we find $\frac{2}{201}$ between $\frac{1}{100}$ and $\frac{1}{101}$. Now between $\frac{1}{100}$ and $\frac{2}{201}$, we can squeeze in $\frac{4}{401}$, since

$$\frac{1}{100} = \frac{4}{400} > \frac{4}{401} > \frac{4}{402} = \frac{2}{201}.$$

This process can go on ad infinitum. So it seems only natural to conclude—as the Greeks did—that fractional numbers are continuously distributed on the number line. However, the discovery of irrational numbers showed that fractions, despite of their density, leave "holes" along the number line.

To bring the irrational numbers into our number system is in fact quite the most difficult step in the processes of generalization. A fully satisfactory theory of irrational numbers was not given until 1872 by Richard Dedekind (1831–1916), who made a careful analysis of continuity and ordering. To make the set of real numbers a continuum, we need the irrational numbers to fill the "holes" left by the rational numbers on the number line. A real number is any number that can be written as a decimal. There are three types of decimals: terminating, nonterminating but repeating, and nonterminating and nonrepeating. The first two types represent rational numbers, such as $\frac{1}{4} = 0.25$; $\frac{2}{3} = 0.666\cdots$. The third type represents irrational numbers, like $\sqrt{2} = 1.4142135\cdots$.

From a practical point of view, we can always approximate an irrational number by truncating the unending decimal. If higher accuracy is needed, we simply take more decimal places. Since any decimal when stopped somewhere is rational, this means that an irrational number can be represented by a sequence of rational numbers with progressively increasing accuracy. This is good enough for us to perform mathematical operations with irrational numbers.

1.1.6 Imaginary Numbers

We go on in the process of generalization. Are there any other insoluble equations? Yes, there are. For example, it is impossible to solve this equation: $x^2 = -1$. The square of no rational, of no irrational, of nothing that we have discovered so far, is equal to -1. So again we have to generalize our numbers to still a wider class.

This time we extend our number system to include the solution of this equation and introduce the symbol i for $\sqrt{-1}$ (engineers call it j to avoid confusion with current). Of course someone could call it $-i$ since it is just as good a solution. The only property of i is that $i^2 = -1$. Certainly, $x = -i$ also satisfies the equation $x^2 + 1 = 0$. Therefore it must be true that any equation we can write is equally valid if the sign of i is changed everywhere. This is called taking the **complex conjugate**.

We can make up numbers by adding successively $i's$, and multiplying $i's$ by numbers, and adding other numbers and so on, according to all our rules. In this way we find that numbers can all be written as $a + ib$, where a and b are real numbers, i.e., the numbers we have defined up until now. The number i is called the **unit imaginary number**. Any real multiple of i is called **pure imaginary**. The most general number is of course of the form $a + ib$ and is called a **complex number**. Things do not get any worse if we add and multiply two such numbers. For example:

$$(a + bi) + (c + di) = (a + c) + (b + d)i. \tag{1.4}$$

In accordance with the distributive law, the multiplication of two complex number is defined as

$$(a + bi)(c + di) = ac + a(di) + (bi)c + (bi)(di)$$
$$= ac + (ad)i + (bc)i + (bd)ii = (ac - bd) + (ad + bc)i, \tag{1.5}$$

since $ii = i^2 = -1$. Therefore all the numbers have this mathematical form.

It is customary to use a single letter, z, to denote a complex number $z = a + bi$. Its real and imaginary parts are written as $\text{Re}(z)$ and $\text{Im}(z)$, respectively. With this notation, $Re(z) = a$, $Im(z) = b$.

The equation $z_1 = z_2$ holds if and only if

$$Re(z_1) = Re(z_2) \quad and \quad Im(z_1) = Im(z_2).$$

Thus any equation involving complex numbers can be interpreted as a pair of real equations.

The complex conjugate of the number $z = a + bi$ is usually denoted as either z^*, or $\bar{z}$, and is given by $z^* = a - bi$. An important relation is that the product of a complex number and its complex conjugate is a real number

$$zz^* = (a + bi)(a - bi) = a^2 + b^2.$$

With this relation, the division of two complex numbers can also be written as the sum of a real part and an imaginary part

$$\frac{a + bi}{c + di} = \frac{a + bi}{c + di}\frac{c - di}{c - di} = \frac{ac + bd}{c^2 + d^2} + \frac{bc - ad}{c^2 + d^2}i.$$

Example 1.1.1 Express the following in the form of $a + bi$,

$$(a)\ (6+2i) - (1+3i), \qquad (b)\ (2-3i)(1+i),$$

$$(c)\ \left(\frac{1}{2-3i}\right)\left(\frac{1}{1+i}\right).$$

Solution 1.1.1

$$(a)\ (6+2i) - (1+3i) = (6-1) + i(2-3) = 5 - i.$$

$$(b)\ (2-3i)(1+i) = 2(1+i) - 3i(1+i) = 2 + 2i - 3i - 3i^2$$
$$= (2+3) + i(2-3) = 5 - i.$$

$$(c)\ \left(\frac{1}{2-3i}\right)\left(\frac{1}{1+i}\right) = \frac{1}{(2-3i)(1+i)} = \frac{1}{5-i}$$
$$= \frac{5+i}{(5-i)(5+i)} = \frac{5+i}{5^2-i^2} = \frac{5}{26} + \frac{1}{26}i.$$

Historically, Italian mathematician Girolamo Cardano was credited as the first to consider the square root of a negative number in 1545 in connection with solving quadratic equations. But after introducing the imaginary numbers, he immediately dismissed them as "useless". He had a good reason to think that way. At Cardano's time, mathematics was still synonymous with geometry. Thus the quadratic equation $x^2 = mx + c$ was thought as a vehicle to find the intersection points of the parabola $y = x^2$ and the line $y = mx + c$. For an equation such as $x^2 = -1$, the horizontal line $y = -1$ will obviously not intersect the parabola $y = x^2$ which is always positive. The absence of the intersection was thought as the reason of the occurrence of the imaginary numbers.

It was the cubic equation that forced complex numbers to be taken seriously. For a cubic curve $y = x^3$, the values of y go from $-\infty$ to $+\infty$. A line will always hit the curve at least once. In 1572, Rafael Bombeli considered the equation

$$x^3 = 15x + 4,$$

which clearly has a solution of $x = 4$. Yet at the time, it was known that this kind of equation could be solved by the following formal procedure. Let $x = a + b$, then

$$x^3 = (a+b)^3 = a^3 + 3ab(a+b) + b^3,$$

which can be written as

$$x^3 = 3abx + (a^3 + b^3).$$

The problem will be solved, if we can find a set of values a and b satisfying the conditions

$$3ab = 15 \quad \text{and} \quad a^3 + b^3 = 4.$$

Since $a^3 b^3 = 5^3$ and $b^3 = 4 - a^3$, we have

$$a^3 (4 - a^3) = 5^3,$$

which is a quadratic equation in a^3,

$$\left(a^3\right)^2 - 4a^3 + 125 = 0.$$

The solution of such an equation was known for thousands of years,

$$a^3 = \frac{1}{2} \left(4 \pm \sqrt{16 - 500}\right) = 2 \pm 11i.$$

It follows that

$$b^3 = 4 - a^3 = 2 \mp 11i.$$

Therefore

$$x = a + b = (2 + 11i)^{1/3} + (2 - 11i)^{1/3}.$$

Clearly, the interpretation that the appearance of imaginary numbers signifies no solution of the geometric problem is not valid. In order to have the solution come out to equal 4, Bombeli assumed

$$(2 + 11i)^{1/3} = 2 + bi; \quad (2 - 11i)^{1/3} = 2 - bi.$$

To justify this assumption, he had to use the rules of addition and multiplication of complex numbers. With the rules listed in (1.4) and (1.5), it can be readily shown that

$$(2 + bi)^3 = 8 + 3 (4) (bi) + 3(2) (bi)^2 + (bi)^3$$
$$= \left(8 - 6b^2\right) + \left(12b - b^3\right) i.$$

With $b = \pm 1$, he obtained

$$(2 \pm i)^3 = 2 \pm 11i,$$

and

$$x = (2 + 11i)^{1/3} + (2 - 11i)^{1/3} = 2 + i + 2 - i = 4.$$

Thus he established that problems with real coefficients required complex arithmetic for solutions.

Despite Bombelli's work, complex numbers were greeted with suspicion, even hostility for almost 250 years. Not until the beginning of the nineteenth century, complex numbers were fully embraced as members of our number system. The acceptance of complex numbers was largely due to the work and reputation of K.F. Gauss.

Karl Friedrich Gauss (1777–1855) of Germany was given the title of "the prince of mathematics" by his contemporaries as a tribute to his great achievements in almost every branch of mathematics. At the age of 22, Gauss in his doctoral dissertation gave the first rigorous proof of what we now call the Fundamental Theorem of Algebra. It says that a polynomial of degree n always has exactly n complex roots. This shows that complex numbers are not only necessary to solve a general algebraic equation, but they are also sufficient. In other words, with the invention of i, every algebraic equation can be solved. This is a fantastic fact. It is certainly not self-evident. In fact, the process by which our number system is developed would make us think that we will have to keep on inventing new numbers to solve yet unsoluble equations. It is a miracle that this is not the case. With the last invention of i, our number system is complete. Therefore a number, no matter how complicated it looks, can always be reduced to the form of $a + bi$, where a and b are real numbers.

1.2 Logarithm

1.2.1 Napier's Idea of Logarithm

Rarely a new idea was embraced so quickly by the entire scientific community with such enthusiasm as the invention of logarithm. Although it was merely a device to simplify computation, its impact on scientific developments could not be overstated.

Before seventeenth-century scientists had to spend much of their time doing numerical calculations. The Scoltish baron, John Napier (1550–1617) thought to relieve this burden as he wrote: "Seeing there is nothing that is so troublesome to mathematical practice than multiplications, divisions, square and cubical extractions of great numbers,.......I began therefore in my mind by what certain and ready art I might remove those hinderance". His idea was this: if we could write any number as a power of some given, fixed number b (later to be called base), then the multiplication of numbers would be equivalent to the addition of their exponents. He called the power logarithm.

In modern notation, this works as follows. If

$$b^{x_1} = N_1; \quad b^{x_2} = N_2$$

then by definition

$$x_1 = \log_b N_1; \quad x_2 = \log_b N_2.$$

Obviously,

$$x_1 + x_2 = \log_b N_1 + \log_b N_2.$$

Since

$$b^{x_1+x_2} = b^{x_1} b^{x_2} = N_1 N_2,$$

again by definition

$$x_1 + x_2 = \log_b N_1 N_2.$$

Therefore

$$\log_b N_1 N_2 = \log_b N_1 + \log_b N_2.$$

Suppose we have a table, in which N and $\log_b N$ (the power x) are listed side by side. To multiply two numbers N_1 and N_2, you first look up $\log_b N_1$ and $\log_b N_2$ in the table. You then add the two numbers. Next, find the number in the body of the table that matches the sum, and read backward to get the product $N_1 N_2$.

Similarly we can show

$$\log_b \frac{N_1}{N_2} = \log_b N_1 - \log_b N_2,$$

$$\log_b N^n = n \log_b N, \quad \log_b N^{1/n} = \frac{\log_b N}{n}.$$

Thus, division of numbers would be equivalent to subtraction of their exponents, raising a number to nth power would be equivalent to multiplying the exponent by n, and finding the nth root of a number would be equivalent to dividing the exponent by n. In this way the drudgery of computations is greatly reduced.

Now the question is, with what base b should we compute. Actually it makes no difference what base is used, as long as it is not exactly equal to 1. We can use the same principle all the time. Besides, if we are using logarithms to any particular base, we can find logarithms to any other base merely by multiplying a factor, equivalent to a change of scale. For example, if we know the logarithm of all numbers with base b, we can find the logarithm of N with base a. First if $a = b^x$, then by definition, $x = \log_b a$, therefore

$$a = b^{\log_b a}. \tag{1.6}$$

To find $\log_a N$, first let $y = \log_a N$. By definition $a^y = N$. With a given by (1.6), we have

$$\left(b^{\log_b a}\right)^y = b^{y \log_b a} = N.$$

Again by definition, (or take logarithm of both sides of the equation)

$$y \log_b a = \log_b N.$$

Thus

$$y = \frac{1}{\log_b a} \log_b N.$$

Since $y = \log_a N$, it follows

$$\log_a N = \frac{1}{\log_b a} \log_b N.$$

This is known as change of base. Having a table of logarithm with base b will enable us to calculate the logarithm to any other base.

In any case, the key is, of course, to have a table. Napier chose a number slightly less than one as the base and spent 20 years to calculate the table. He published his table in 1614. His invention was quickly adopted by scientists all across Europe and even in far away China. Among them was the astronomer Johannes Kepler, who used the table with great success in his calculations of the planetary orbits. These calculations became the foundation of Newton's classical dynamics and his law of gravitation.

1.2.2 Briggs' Common Logarithm

Henry Briggs (1561–1631), a professor of geometry in London, was so impressed by Napier's table, he went to Scotland to meet the great inventor in person. Briggs suggested that a table of base 10 would be more convenient. Napier readily agreed. Briggs undertook the task of additional computations. He published his table in 1624. For 350 years, the logarithmic table and the slide rule (constructed with the principle of logarithm) were indispensable tools of every scientist and engineer.

The logarithm in Briggs' table is now known as the common logarithm. In modern notation, if we write $x = \log N$ without specifying the base, it is understood that the base is 10, and $10^x = N$.

Today logarithmic tables are replaced by hand-held calculators, but logarithmic function remains central to mathematical sciences.

It is interesting to see how logarithms were first calculated. In addition to historic interests, it will help us to gain some insights into our number system.

Table 1.1 Successive square roots of ten

x (log N)	10^x (N)	$(10^x - 1)/x$
1	10.0	9.00
$\frac{1}{2} = 0.5$	3.16228	4.32
$(\frac{1}{2})^2 = 0.25$	1.77828	3.113
$(\frac{1}{2})^3 = 0.125$	1.33352	2.668
$(\frac{1}{2})^4 = 0.0625$	1.15478	2.476
$(\frac{1}{2})^5 = 0.03125$	1.074607	2.3874
$(\frac{1}{2})^6 = 0.015625$	1.036633	2.3445
$(\frac{1}{2})^7 = 0.0078125$	1.018152	2.3234
$(\frac{1}{2})^8 = 0.00390625$	1.0090350	2.3130
$(\frac{1}{2})^9 = 0.001953125$	1.0045073	2.3077
$(\frac{1}{2})^{10} = 0.00097656$	1.0022511	2.3051
$(\frac{1}{2})^{11} = 0.00048828$	1.0011249	2.3038
$(\frac{1}{2})^{12} = 0.00024414$	1.0005623	2.3032
$(\frac{1}{2})^{13} = 0.00012207$	1.000281117	2.3029
$(\frac{1}{2})^{14} = 0.000061035$	1.000140548	2.3027
$(\frac{1}{2})^{15} = 0.0000305175$	1.000070272	2.3027
$(\frac{1}{2})^{16} = 0.0000152587$	1.000035135	2.3026
$(\frac{1}{2})^{17} = 0.0000076294$	1.0000175675	2.3026

Since a simple process for taking square roots was known, Briggs computed successive square roots of 10. A sample of the results is shown in Table 1.1. The powers (x) of 10 are given in the first column and the results, 10^x, are given in the second column. For example, the second row is the square root of 10, that is $10^{1/2} = \sqrt{10} = 3.16228$. The third row is the square root of the square root of 10, $(10^{1/2})^{1/2} = 10^{1/4} = 1.77828$. So on and so forth, we get a series of successive square roots of 10. With a hand-held calculator, you can readily verify these results.

In the table we noticed that when 10 is raised to a very small power, we get 1 plus a small number. Furthermore, the small numbers that are added to 1 begins to look as though we are merely dividing by 2 each time we take a square root. In other words, it looks that when x is very small, $10^x - 1$ is proportional to x. To find the proportionality constant, we list $(10^x - 1)/x$ in column 3. At the top of the table, these ratios are not equal, but as they come down, they get closer and closer to a constant value. To the accuracy of 5 significant digits, the proportional constant is equal to 2.3026. So we find that when s is very small

$$10^s = 1 + 2.3026 \, s. \tag{1.7}$$

Briggs computed successively 27 square roots of 10, and used (1.7) to obtain another 27 squares roots.

Since $10^x = N$ means $x = \log N$, the first column in Table 1.1 is also the logarithm of the corresponding number in the second column. For example, the second row is the square root of 10, that is $10^{1/2} = 3.16228$. Then by definition, we know

$$\log(3.16228) = 0.5.$$

If we want to know the logarithm of a particular number N, and N is not exactly the same as one of the entries in the second column, we have to break up N as a product of a series of numbers which are entries of the table. For example, suppose we want to know the logarithm of 1.2. Here is what we do. Let $N = 1.2$, and we are going to find a series of n_i in column 2 such that

$$N = n_1 n_2 n_3 \cdots.$$

Since all n_i are greater than one, so $n_i < N$. The number in column 2 closest to 1.2 satisfying this condition is 1.15478, So we choose $n_1 = 1.15478$, and we have

$$\frac{N}{n_1} = \frac{1.2}{1.15478} = 1.039159 = n_2 n_3 \cdots$$

The number smaller than and closest to 1.039159 is 1.036633. So we choose $n_2 = 1.036633$, thus

$$\frac{N}{n_1 n_2} = \frac{1.039159}{1.036633} = 1.0024367.$$

With $n_3 = 1.0022511$, we have

$$\frac{N}{n_1 n_2 n_3} = \frac{1.0024367}{1.0022511} = 1.0001852.$$

The plan is to continue this way until the right-hand side is equal to one. But most likely, sooner or later, the right-hand side will fall beyond the table and is still not exactly equal to one. In our particular case, we can go down a couple of more steps. But for the purpose of illustration, let us stop here. So

$$N = n_1 n_2 n_3 (1 + \Delta n)$$

where $\Delta n = 0.0001852$. Now

$$\log N = \log n_1 + \log n_2 + \log n_3 + \log(1 + \Delta n).$$

The terms on the right-hand side, except the last one, can be read from the table. For the last term, we will make use of (1.7). By definition, if s is very small, (1.7) can be written as

$$s = \log(1 + 2.3026s).$$

Let $\Delta n = 2.3026s$, so $s = \frac{\Delta n}{2.3026} = \frac{0.0001852}{2.3026} = 8.04 \times 10^{-5}$. It follows

$$\log(1 + \Delta n) = \log[1 + 2.3026\,(8.04 \times 10^{-5})] = 8.04 \times 10^{-5}.$$

With $\log n_1 = 0.0625$, $\log n_2 = 0.015625$, $\log n_3 = 0.0009765$ from the table, we arrived at

$$\log(1.2) = 0.0625 + 0.015625 + 0.0009765 + 0.0000804 = 0.0791819.$$

The value of $\log(1.2)$ should be 0.0791812. Clearly if we have a larger table we can have as many accurate digits as we want. In this way Briggs calculated the logarithms to 16 decimal places and reduced them to 14 when he published his table, so there were no rounding errors. With minor revisions, Briggs' table remained the basis for all subsequent logarithmic tables for the next 300 years.

1.3   A Peculiar Number Called e

1.3.1   The Unique Property of e

Equation (1.7) expresses a very interesting property of our number system. If we let $t = 2.3026$ s, then for a very small t, (1.7) becomes

$$10^{\frac{t}{2.3026}} = 1 + t. \tag{1.8}$$

To simplify the writing, let us denote

$$10^{\frac{1}{2.3026}} = e. \tag{1.9}$$

Thus (1.8) says that e raised to a very small power is equal to one plus the small power

$$e^t = 1 + t \quad for \ t \to 0. \tag{1.10}$$

Because of this, we find the derivative of e^x is equal to itself.

Recall the definition of the derivative of a function:

$$\frac{df(x)}{dx} = \lim_{\Delta x \to 0} \frac{f(x + \Delta x) - f(x)}{\Delta x}.$$

So

$$\frac{de^x}{dx} = \lim_{\Delta x \to 0} \frac{e^{x+\Delta x} - e^x}{\Delta x} = \lim_{\Delta x \to 0} \frac{e^x(e^{\Delta x} - 1)}{\Delta x}.$$

Now Δx approaches zero as a limit, certainly it is very small, so we can write (1.10) as

$$e^{\Delta x} = 1 + \Delta x.$$

Thus

$$\frac{e^x(e^{\Delta x} - 1)}{\Delta x} = \frac{e^x(1 + \Delta x - 1)}{\Delta x} = e^x.$$

Therefore

$$\frac{de^x}{dx} = e^x. \tag{1.11}$$

The function e^x (or written as $\exp(x)$) is generally called the natural exponential function, or simply the exponential function. Not only is the exponential function equal to its own derivative, it is the only function (apart from a multiplication constant) that has this property. Because of this, the exponential function plays a central role in mathematics and sciences.

1.3.2 The Natural Logarithm

If $e^y = x$, then by definition

$$y = \log_e x.$$

The logarithm to the base e is known as the natural logarithm. It appears with amazing frequency in mathematics and its applications. So we give it a special symbol. It is written as $\ln x$. That is

$$y = \log_e x = \ln x.$$

Thus

$$e^{\ln x} = x.$$

Furthermore,

$$\ln e^x = x \ln e = x.$$

In this sense, the exponential function and the natural logarithm are inverses of each other.

Example 1.3.1 Find the value of $\ln 10$.

Solution 1.3.1 Since

$$10^{\frac{1}{2.3026}} = e, \quad \Rightarrow \quad 10 = e^{2.3026},$$

it follows

$$\ln 10 = \ln e^{2.3026} = 2.3026.$$

The derivative of $\ln x$ is of special interests.

$$\frac{d(\ln x)}{dx} = \lim_{\Delta x \to 0} \frac{\ln(x + \Delta x) - \ln x}{\Delta x},$$

$$\ln(x + \Delta x) - \ln x = \ln \frac{x + \Delta x}{x} = \ln(1 + \frac{\Delta x}{x}).$$

Now (1.10) can be written as

$$t = \ln(1 + t)$$

for a very small t. Since Δx approaches zero as a limit, for any fixed x, $\frac{\Delta x}{x}$ can certainly be made as small as we wish. Therefore, we can set $\frac{\Delta x}{x} = t$, and conclude

$$\frac{\Delta x}{x} = \ln(1 + \frac{\Delta x}{x}).$$

Thus,

$$\frac{d(\ln x)}{dx} = \lim_{\Delta x \to 0} \frac{1}{\Delta x} \ln(1 + \frac{\Delta x}{x}) = \lim_{\Delta x \to 0} \frac{1}{\Delta x} \frac{\Delta x}{x} = \frac{1}{x}.$$

This in turn means

$$d(\ln x) = \frac{dx}{x},$$

or

$$\int \frac{dx}{x} = \ln x + c, \tag{1.12}$$

where c is the constant of integration. It is well known that because of

$$\frac{dx^{n+1}}{dx} = (n + 1)x^n,$$

we have

$$\int x^n dx = \frac{x^{n+1}}{(n + 1)} + c.$$

This formula holds for all values of n except for $n = -1$, since then the denominator $n + 1$ is zero. This had been a difficult problem, but now we see that (1.12) provides the "missing case".

In numerous phenomena, ranging from population growth to the decay of radioactive material, in which the rate of change of some quantity is proportional to the quantity itself. Such phenomenon is governed by the differential equation

$$\frac{dy}{dt} = ky$$

where k is a constant that is positive if y is increasing and negative if y is decreasing. To solve this equation, we write it as

$$\frac{dy}{y} = kdt$$

and then integrate both sides to get

$$\ln y = kt + c, \quad or \quad y = e^{kt+c} = e^{kt}e^{c}.$$

If y_0 denotes the value of y when $t = 0$, then $y_0 = e^c$ and

$$y = y_0e^{kt}.$$

This equation is called the law of exponential change.

1.3.3 Approximate Value of e

The number e is found of such great importance, but what is the numerical value of e, which we have, so far, defined as $10^{(1/2.3025)}$? We can use our table of successive square root of 10 to calculate this number. The powers of 10 are given in the first column of Table 1.1. If we can find a series of numbers n_1, n_2, $n_3 \cdots$ in this column, such that

$$\frac{1}{2.3026} = n_1 + n_2 + n_3 + \cdots ,$$

then

$$10^{\frac{1}{2.3026}} = 10^{n_1+n_2+n_3+\cdots} = 10^{n_1}10^{n_2}10^{n_3}\cdots .$$

We can read from the second column of the table 10^{n_1}, and 10^{n_2}, and 10^{n_3} and so on, and multiply them together. Let us do just that.

$$\frac{1}{2.3026} = 0.43429 = 0.25 + 0.125 + 0.03125 + 0.015625$$

$$+0.0078125 + 0.00390625 + 0.00048828 + 0.00012207$$

$$+0.000061035 + 0.000026535.$$

From the table we find $10^{0.25} = 1.77828$, $10^{0.125} = 1.33352$, $\cdots$ etc., except for the last term for which we use (1.7). Thus

$$e = 10^{\frac{1}{2.3026}} = 1.77828 \times 1.33352 \times 1.074607 \times 1.036633 \times 1.018152$$
$$\times\ 1.009035 \times 1.0011249 \times 1.000281117 \times 1.000140548$$
$$\times\ (1 + 2.3026 \times 0.000026535) = 2.71826.$$

Since $\frac{1}{2.3026}$ is only accurate to 5 significant digits, we cannot expect our result to be accurate more than that. (The accurate result is $2.71828\cdots$) Thus what we get is only an approximation. Is there a more precise definition of e? The answer is yes. We will discuss this question in the next section.

1.4   The Exponential Function as an Infinite Series

1.4.1   Compound Interest

The origins of the number e are not very clear. The existence of this peculiar number could be extracted from the logarithmic table as we did. In fact there is an indirect reference to e in the second edition of Napier's table. But most probably the peculier property of the number e was noticed even earlier in connection with compound interest.

A sum of money invested at x percent annual interest rate (x expressed as a decimal, for example, $x = 0.06$ for 6%) means that at the end of the year the sum grows by a factor $(1 + x)$. Some banks compute the accrued interest not once a year but several times a year. For example, if an annual interest rate of x percent is compounded semiannually, the bank will use one-half of the annual rate as the rate per period. Hence, if P is the original sum, at the end of the half-year, the sum grows to $P\left(1 + \frac{x}{2}\right)$, and at the end of the year the sum becomes

$$\left[P\left(1 + \frac{x}{2}\right)\right]\left(1 + \frac{x}{2}\right) = P\left(1 + \frac{x}{2}\right)^2.$$

In the banking industry one finds all kinds of compounding schemes – annually, semiannually, quarterly, monthly, weekly, and even daily. Suppose the compounding is done n times a year, at the end of the year, the principal P will yield the amount

$$S = P(1 + \frac{x}{n})^n.$$

It is interesting to compare the amounts of money a given principal will yield after one year for different conversion periods. Table 1.2 shows that the amounts of money one will get for \$100 invested for one year at 6% annual interest rate at different

Table 1.2 The yields of 100 dollars invested for one year at 6% annual interest rate at different conversion periods

	n	x/n	$100(1 + x/n)^n$
Annually	1	0.06	106.00
Semiannually	2	0.03	106.09
Quarterly	4	0.015	106.136
Monthly	12	0.005	106.168
Weekly	52	0.0011538	106.180
Daily	365	0.0001644	106.183

conversion periods. The result is quite surprising. As we see, a principal of $100 compounded daily or weekly yield practically the same. But will this pattern go on? Is it possible that no matter how large n is, the values of $(1 + \frac{x}{n})^n$ will settle on the same number? To answer this question, we must use methods other than merely computing individual values. Fortunately, such a method is available. With the binomial formula,

$$(a + b)^n = a^n + na^{n-1}b + \frac{n(n-1)}{2!}a^{n-2}b^2$$
$$+ \frac{n(n-1)(n-2)}{3!}a^{n-3}b^3 + \cdots + b^n,$$

we have

$$\left(1 + \frac{x}{n}\right)^n = 1 + n\left(\frac{x}{n}\right) + \frac{n(n-1)}{2!}\left(\frac{x}{n}\right)^2$$
$$+ \frac{n(n-1)(n-2)}{3!}\left(\frac{x}{n}\right)^3 + \cdots + \left(\frac{x}{n}\right)^n$$
$$= 1 + x + \frac{(1-1/n)}{2!}x^2 + \frac{(1-1/n)(1-2/n)}{3!}x^3 + \cdots \left(\frac{x}{n}\right)^n.$$

Now as $n \to \infty$, $\frac{k}{n} \to 0$. Therefore

$$\lim_{n\to\infty}\left(1 + \frac{x}{n}\right)^n = 1 + x + \frac{x^2}{2!} + \frac{x^3}{3!} + \frac{x^4}{4!} + \cdots. \tag{1.13}$$

becomes an infinite series. Standard tests for convergence show that this is a convergent series for all real values of x. In other words, the value of $(1 + \frac{x}{n})^n$ does settle on a specific limit as n increase without bound.

1.4.2 The Limiting Process Representing e

In early eighteenth century, Euler used the letter e to represent the series (1.13) for the case of $x = 1$,

$$e = \lim_{n \to \infty} \left(1 + \frac{1}{n}\right)^n = 1 + 1 + \frac{1}{2!} + \frac{1}{3!} + \frac{1}{4!} + \cdots . \qquad (1.14)$$

This choice, like many other symbols of his, such as i, π, $f(x)$, became universally accepted.

It is important to note that when we say that the limit of $\frac{1}{n}$ as $n \to \infty$ is 0, it does not mean that $\frac{1}{n}$ itself will ever be equal to 0, in fact, it will not. Thus, if we let $t = \frac{1}{n}$, then as $n \to \infty$, $t \to 0$. So (1.14) can be written as

$$e = \lim_{t \to 0} (1 + t)^{1/t} .$$

In words, it says that if t is very small, then

$$e^t = \left[(1 + t)^{1/t}\right]^t = 1 + t, \quad t \to 0.$$

This is exactly the same equation as shown in (1.10). Therefore, e is the same number previously written as $10^{1/2.3026}$. Now the formal definition of e is given by the limiting process

$$e = \lim_{n \to \infty} \left(1 + \frac{1}{n}\right)^n ,$$

which can be written as an infinite series as shown in (1.14). The series converges rather fast. With seven terms, it gives us 2.71825. This approximation can be improved by adding more terms until the desired accuracy is reached. Since it is monotonously convergent, each additional term brings it closer to the limit: $2.71828 \cdots$.

1.4.3 The Exponential Function e^x

Raising e to x power, we have

$$e^x = \lim_{n \to \infty} \left(1 + \frac{1}{n}\right)^{nx} .$$

Let $nx = m$, then $\frac{x}{m} = \frac{1}{n}$. As n goes to ∞, so does m. Thus the above equation becomes

$$e^x = \lim_{m \to \infty} \left(1 + \frac{x}{m}\right)^m.$$

Now m may not be an integer, but the binomial formula is equally valid for non-integer power (one of the early discoveries of Isaac Newton). Therefore by the same reason as in (1.13), we can express the exponential function as an infinite series,

$$e^x = 1 + x + \frac{x^2}{2!} + \frac{x^3}{3!} + \frac{x^4}{4!} + \cdots . \tag{1.15}$$

It is from this series that the numerical values of e^x are most easily obtained, the first few terms usually suffice to obtain the desired accuracy.

We have shown in (1.11) that the derivative of e^x must be equal to itself. This is clearly the case as we take derivative of (1.15) term by term,

$$\frac{d}{dx}e^x = 0 + 1 + x + \frac{x^2}{2!} + \frac{x^3}{3!} + \cdots = e^x.$$

1.5 Unification of Algebra and Geometry

1.5.1 The Remarkable Euler Formula

Leonhard Euler (1707 – 1783) was born in Basel, a border town between Switzerland, France, and Germany. He is one of the great mathematicians and certainly the most prolific scientist of all time. His immense output fills at least seventy volumes. In 1771, after he became blind, he published three volumes of a profound treatise of optics. For almost 40 years after his death, the Academy at St. Petersburg continued to publish his manuscripts. Euler played with formulas like a child playing with toys, making all kinds of substitutions until he got something interesting. Often the results were sensational.

He took the infinite series of e^x, and boldly replaced the real variable x in (1.15) with the imaginary expression $i\theta$ and got

$$e^{i\theta} = 1 + i\theta + \frac{(i\theta)^2}{2!} + \frac{(i\theta)^3}{3!} + \frac{(i\theta)^4}{4!} + \cdots .$$

Since $i^2 = -1$, $i^3 = -i$, $i^4 = 1$, and so on, this equation became

$$e^{i\theta} = 1 + i\theta - \frac{\theta^2}{2!} - \frac{i\theta^3}{3!} + \frac{\theta^4}{4!} + \cdots .$$

He then changed the order of terms, collecting all the real terms separately from the imaginary terms, and arrived at the series

$$e^{i\theta} = \left(1 - \frac{\theta^2}{2!} + \frac{\theta^4}{4!} + \cdots\right) + i\left(\theta - \frac{\theta^3}{3!} + \frac{\theta^5}{5!} + \cdots\right).$$

Now it was already known in Euler's time that the two series appearing in the parentheses are the power series of the trigonometric functions $\cos\theta$ and $\sin\theta$, respectively. Thus Euler arrived at the remarkable formula (1.2)

$$e^{i\theta} = \cos\theta + i\sin\theta,$$

which at once links the exponential function to ordinary trigonometry.

Strictly speaking, Euler played the infinite series rather loosely. Collecting all the real terms separately from the imaginary terms, he changed the order of terms. To do so with an infinite series can be dangerous. It may affect its sum, or even change a convergent series into a divergent series. But this result has withstood the test of rigor.

Euler derived hundreds of formulas, but this one is often called the most famous formula of all formulas. Feynman called it the amazing jewel.

1.5.2 The Complex Plane

The acceptance of complex number as a bona fide members of our number system was greatly helped by the realization that a complex number could be given a simple, concrete geometric interpretation.

In a two-dimensional rectangular coordinate system, a point is specified by its x and y components. If we interpret the x- and y-axes as the real and imaginary axes, respectively, then the complex number $z = x + iy$ is represented by the point (x, y). The horizontal position of the point is x, the vertical position of the point is y, as shown in Fig. 1.1. We can then add and subtract complex numbers by separately adding or subtracting the real and imaginary components. When thought of in this way, the plane is called the complex plane, or the Argand plane.

Fig. 1.1 Complex plane, also known as Argand diagram. The real part of a complex number is along the x-axis, and the imaginary part, along the y-axis

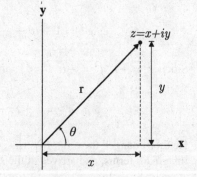

This graphic representation was independently suggested around 1800 by C. Wessel of Norway, J.R. Argand of France, and C.F. Gauss. The publications by Wessel and Argand went all but unnoticed, but the reputation of Gauss ensured wide dissemination and acceptance of the complex numbers as points in the complex plane.

At the time when this interpretation was suggested, the Euler formula (1.2) had already been known for at least 50 years. It might have played the role of guiding principle for this suggestion. The geometric interpretation of the complex number is certainly consistent with the Euler formula. We can derive the Euler formula by expressing $e^{i\theta}$ as a point in the complex plane.

Since the most general number is a complex number in the form of a real part plus an imaginary part, so let us express $e^{i\theta}$ as

$$e^{i\theta} = a\,(\theta) + ib\,(\theta). \tag{1.16}$$

Note that both the real part a and the imaginary part b must be functions of θ. Here θ is any real number. Changing i to $-i$, in both sides of this equation, we get the complex conjugate

$$e^{-i\theta} = a\,(\theta) - ib\,(\theta).$$

Since

$$e^{i\theta}e^{-i\theta} = e^{i\theta - i\theta} = e^0 = 1,$$

it follows that

$$e^{i\theta}e^{-i\theta} = (a + ib)(a - ib) = a^2 + b^2 = 1.$$

Furthermore

$$\frac{d}{d\theta}e^{i\theta} = ie^{i\theta} = i(a + ib) = ia - b,$$

but

$$\frac{d}{d\theta}e^{i\theta} = \frac{d}{d\theta}a + i\frac{d}{d\theta}b = a' + ib',$$

equating the real part to real part and imaginary part to imaginary part of the last two equations, we have

$$a' = \frac{d}{d\theta}a = -b, \quad b' = \frac{d}{d\theta}b = a.$$

Thus

$$a'b = -b^2, \quad b'a = a^2,$$

and

$$b'a - a'b = a^2 + b^2 = 1.$$

Now let $a\,(\theta)$ represent the abscissa (x-coordinate) and $b\,(\theta)$ represent the ordinate (y-coordinate) of a point in the complex plane as shown in Fig. 1.2. Let α be the angle

Fig. 1.2 The Argand
diagram of the complex
number $z = e^{i\theta} = a + ib$.
The distance between the
origin and the point (a, b)
must be 1

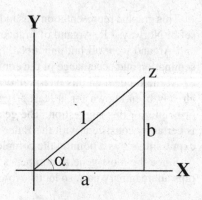

between the x-axis and the vector from the origin to the point. Since the length of
this vector is given by the Pythagorean theorem

$$r^2 = a^2 + b^2 = 1,$$

clearly

$$\cos \alpha = \frac{a(\theta)}{1} = a(\theta), \quad \sin \alpha = \frac{b(\theta)}{1} = b(\theta), \quad \tan \alpha = \frac{b(\theta)}{a(\theta)}. \qquad (1.17)$$

Now

$$\frac{d \tan \alpha}{d\theta} = \frac{d \tan \alpha}{d\alpha} \frac{d\alpha}{d\theta} = \frac{1}{\cos^2 \alpha} \frac{d\alpha}{d\theta} = \frac{1}{a^2} \frac{d\alpha}{d\theta},$$

but

$$\frac{d \tan \alpha}{d\theta} = \frac{d}{d\theta} \left(\frac{b}{a} \right) = \frac{b'a - a'b}{a^2} = \frac{1}{a^2}.$$

It is clear from the last two equations that

$$\frac{d\alpha}{d\theta} = 1.$$

In other words,

$$\alpha = \theta + c.$$

To determine the constant c, let us look at the case $\theta = 0$. Since $e^{i0} = 1 = a + ib$
means $a = 1$ and $b = 0$, in this case it is clear from the diagram that $\alpha = 0$. Therefore
c must be equal to zero, so

$$\alpha = \theta.$$

It follows from (1.17) that

$$a(\theta) = \cos \alpha = \cos \theta, \quad b(\theta) = \sin \alpha = \sin \theta.$$

Putting them back to (1.16), we obtain again

$$e^{i\theta} = \cos \theta + i \sin \theta.$$

Note that we have derived the Euler formula without the series expansion. Previously we have derived this formula in a purely algebraic manner. Now we see that $\cos \theta$ and $\sin \theta$ are the cosine and sine functions naturally defined in geometry. This is the unification of algebra and geometry.

It took 250 years for mathematicians to get comfortable with complex numbers. Once fully accepted, the advance in the theory of complex variables was rather rapid. In a short span of forty years, Augustin Louis Cauchy (1789–1857) of France and Georg Friedrich Bernhard Riemann (1826–1866) of Germany developed a beautiful and powerful theory of complex functions, which we will describe in the next chapter.

In this introductory chapter, we have presented some pieces of historic notes for showing that the logical structure of mathematics is as interesting as any other human endeavor. Now we must leave history behind because of our limited space. For more detailed information, we recommend the following references, from which much of our accounts are taken.

Richard Feynman, Robert B. Leighton, and Mathew Sands, *The Feynman Lectures on Physics,* Vol. 1, Chap. 22, (1963) Addison Wesley

Eli Maor, *e: the Story of a Number,* (1994) Princeton University Press

Tristan Needham, *Visual Complex Analysis,* Chap. 1, (1997) Oxford University Press

1.6 Polar Form of Complex Numbers

In terms of polar coordinates (r, θ), the variable x and y are

$$x = r \cos \theta, \quad y = r \sin \theta.$$

The complex variable z is then written as

$$z = x + iy = r(\cos \theta + i \sin \theta) = re^{i\theta}. \tag{1.18}$$

The quantity r, known as the modulus, is the absolute value of z and is given by

$$r = |z| = (zz^*)^{1/2} = (x^2 + y^2)^{1/2}.$$

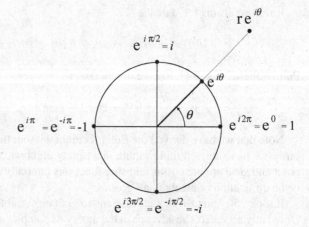

Fig. 1.3 Polar form of complex numbers. The unit circle in the complex plane is described by $e^{i\theta}$. A general complex number is given by $re^{i\theta}$

The angle θ is known as the argument, or phase, of z. Measured in radians, it is given by

$$\theta = \tan^{-1}\frac{y}{x}.$$

If z is in the second or third quadrants, one has to use this equation with care. In the second quadrant, $\tan\theta$ is negative, but in a hand-held calculator, or a computer code, a negative arctangent is interpreted as an angle in the fourth quadrant. In the third quadrant, $\tan\theta$ is positive, but a calculator will interpret a positive arctangent as an angle in the first quadrant. Since an angle is fixed by its sine and cosine, θ is uniquely determined by the pair of equations

$$\cos\theta = \frac{x}{|z|}, \quad \sin\theta = \frac{y}{|z|}.$$

But in practice we usually compute $\tan^{-1}(y/x)$ and adjust for the quadrant problem by adding and subtracting π. Because of its identification as an angle, θ is determined only up to an integer multiple of 2π. We shall make the usual choice of limiting θ to the interval of $0 \le \theta < 2\pi$ as its principal value. However, in computer codes the principal value is usually chosen in the open interval of $-\pi \le \theta < \pi$.

Equation (1.18) is called the polar form of z. It is immediately clear that the complex conjugate of z in the polar form is

$$z^*(r, \theta) = z(r, -\theta) = re^{-i\theta}.$$

In the complex plane, z^* is the reflection of z across the x-axis.

It is helpful to always keep the complex plane in mind. As θ increases, $e^{i\theta}$ describes an unit circle in the complex plane as shown in Fig. 1.3. To reach a general complex number z, we must take the unit vector $e^{i\theta}$ that points at z and stretch it by the length $|z| = r$.

It is very convenient to multiply or divide two complex numbers in polar forms. Let

$$z_1 = r_1 e^{i\theta_1}, \quad z_2 = r_2 e^{i\theta_2},$$

then

$$z_1 z_2 = r_1 e^{i\theta_1} r_2 e^{i\theta_2} = r_1 r_2 e^{i(\theta_1 + \theta_2)} = r_1 r_2 [\cos(\theta_1 + \theta_2) + i \sin(\theta_1 + \theta_2)],$$

$$\frac{z_1}{z_2} = \frac{r_1 e^{i\theta_1}}{r_2 e^{i\theta_2}} = \frac{r_1}{r_2} e^{i(\theta_1 - \theta_2)} = \frac{r_1}{r_2} [\cos(\theta_1 - \theta_2) + i \sin(\theta_1 - \theta_2)].$$

1.6.1 Powers and Roots of Complex Numbers

To obtain the nth power of a complex number, we take the nth power of the modulus and multiply the phase angle by n,

$$z^n = \left(r e^{i\theta} \right)^n = r^n e^{in\theta} = r^n (\cos n\theta + i \sin n\theta).$$

This is a correct formula for both positive and negative integer n. But if n is a fraction number, we must use this formula with care. For example, we can interpret $z^{1/4}$ as the fourth root of z. In other words, we want to find a number whose $4th$ power is equal to z. It is instructive to work out the details for the case of $z = 1$. Clearly

$$1^4 = \left(e^{i0} \right)^4 = e^{i0} = 1,$$
$$i^4 = \left(e^{i\pi/2} \right)^4 = e^{i2\pi} = 1,$$
$$(-1)^4 = \left(e^{i\pi} \right)^4 = e^{i4\pi} = 1,$$
$$(-i)^4 = \left(e^{i3\pi/2} \right)^4 = e^{i6\pi} = 1.$$

Therefore there are four distinct answers,

$$1^{1/4} = \begin{cases} 1 \\ i \\ -1 \\ -i \end{cases}.$$

The multiplicity of roots is tied to the multiple ways of representing 1 in the polar form: e^{i0}, $e^{i2\pi}$, $e^{i4\pi}$, etc. Thus to compute all the nth roots of z, we must express z as

$$z = r e^{i\theta + ik2\pi}, \quad (k = 0, 1, 2 \ldots n - 1)$$

and

$$z^{\frac{1}{n}} = \sqrt[n]{r}e^{i\theta/n+ik2\pi/n}, \quad (k = 0, 1, 2 \cdots n - 1).$$

The reason that k stops at $n - 1$ is because once k reaches n, $e^{ik2\pi/n} = e^{i2\pi} = 1$ and the root repeats itself. Therefore there are n distinct roots.

In general, if n and m are positive integers that have no common factor, then

$$z^{m/n} = \sqrt[n]{|z|^m}e^{i\frac{m}{n}(\theta+2k\pi)} = \sqrt[n]{|z|^m}[\cos\frac{m}{n}(\theta + 2k\pi) + i\sin\frac{m}{n}(\theta + 2k\pi)]$$

where $z = |z|e^{i\theta}$ and $k = 0, 1, 2, \ldots n - 1$.

Example 1.6.1 Express $(1 + i)^8$ in the form of $a + bi$.

Solution 1.6.1 Let $z = (1 + i) = re^{i\theta}$, where

$$r = (zz^*)^{1/2} = \sqrt{2}, \quad \theta = \tan^{-1}\frac{1}{1} = \frac{\pi}{4}.$$

It follows that

$$(1 + i)^8 = z^8 = r^8e^{i8\theta} = 16e^{i2\pi} = 16.$$

Example 1.6.2 Express the following in the form of $a + bi$

$$\frac{\left(\frac{3}{2}\sqrt{3} + \frac{3}{2}i\right)^6}{\left(\sqrt{\frac{5}{2}} + i\sqrt{\frac{5}{2}}\right)^3}.$$

Solution 1.6.2 Let us denote

$$z_1 = \left(\frac{3}{2}\sqrt{3} + \frac{3}{2}i\right) = r_1e^{i\theta_1},$$

$$z_2 = \left(\sqrt{\frac{5}{2}} + i\sqrt{\frac{5}{2}}\right) = r_2e^{i\theta_2},$$

where

$$r_1 = (z_1z_1^*)^{1/2} = 3, \quad \theta_1 = \tan^{-1}\left(\frac{1}{\sqrt{3}}\right) = \frac{\pi}{6},$$

$$r_2 = (z_2z_2^*)^{1/2} = \sqrt{5}, \quad \theta_2 = \tan^{-1}(1) = \frac{\pi}{4}.$$

Thus

$$\frac{\left(\frac{3}{2}\sqrt{3}+\frac{3}{2}i\right)^6}{\left(\sqrt{\frac{5}{2}}+i\sqrt{\frac{5}{2}}\right)^3} = \frac{z_1^6}{z_2^3} = \frac{\left(3e^{i\pi/6}\right)^6}{\left(\sqrt{5}e^{i\pi/4}\right)^3} = \frac{3^6 e^{i\pi}}{\left(\sqrt{5}\right)^3 e^{i3\pi/4}}$$

$$= \frac{729}{5\sqrt{5}}e^{i(\pi-3\pi/4)} = \frac{729}{5\sqrt{5}}e^{i\pi/4}$$

$$= \frac{729}{5\sqrt{5}}\left(\cos\frac{\pi}{4}+i\sin\frac{\pi}{4}\right) = \frac{729}{5\sqrt{10}}\left(1+i\right).$$

Example 1.6.3 Find all the cube roots of 8.

Solution 1.6.3 Express 8 as a complex number z in the complex plane,

$$z = 8e^{ik2\pi}, \quad k = 0, 1, 2, \ldots.$$

Therefore

$$z^{1/3} = (8)^{1/3} e^{ik2\pi/3} = 2e^{ik2\pi/3}, \quad k = 0, 1, 2.$$

$$z^{1/3} = \begin{cases} 2e^{i0} = 2, & k = 0 \\ 2e^{i2\pi/3} = 2\left(\cos\frac{2\pi}{3}+i\sin\frac{2\pi}{3}\right) = -1+i\sqrt{3}, & k = 1 \\ 2e^{i4\pi/3} = 2\left(\cos\frac{4\pi}{3}+i\sin\frac{4\pi}{3}\right) = -1-i\sqrt{3}, & k = 2. \end{cases}$$

Note that the three roots are on a circle of radius 2 centered at the origin. They are 120° apart.

Example 1.6.4 Find all the cube roots of $\sqrt{2}+i\sqrt{2}$.

Solution 1.6.4 The polar form of $\sqrt{2}+i\sqrt{2}$ is

$$z = \sqrt{2}+i\sqrt{2} = 2e^{i\pi/4+ik2\pi}.$$

The cube roots of $\sqrt{2}+i\sqrt{2}$ are given by

$$z^{1/3} = \begin{cases} (2)^{1/3} e^{i\pi/12} = (2)^{1/3}\left(\cos\frac{\pi}{12}+i\sin\frac{\pi}{12}\right), & k = 0 \\ (2)^{1/3} e^{i(\pi/12+2\pi/3)} = (2)^{1/3}\left(\cos\frac{3\pi}{4}+i\sin\frac{3\pi}{4}\right), & k = 1 \\ (2)^{1/3} e^{i(\pi/12+4\pi/3)} = (2)^{1/3}\left(\cos\frac{17\pi}{12}+i\sin\frac{17\pi}{12}\right), & k = 2. \end{cases}$$

Again the three roots are on a circle 120° apart.

Example 1.6.5 Find all the values of z that satisfy the equation $z^4 = -64$.

Solution 1.6.5 Express -64 as a point in the complex plane

$$-64 = 64e^{i\pi + ik2\pi}, \quad k = 0, 1, 2, \ldots.$$

It follows that

$$z = (-64)^{1/4} = (64)^{1/4} \, e^{i(\pi + 2k\pi)/4}, \quad k = 0, 1, 2, 3.$$

$$z = \begin{cases} 2\sqrt{2}(\cos \frac{\pi}{4} + i \sin \frac{\pi}{4}) = 2 + 2i, & k = 0 \\ 2\sqrt{2}(\cos \frac{3\pi}{4} + i \sin \frac{3\pi}{4}) = -2 + 2i, & k = 1 \\ 2\sqrt{2}(\cos \frac{5\pi}{4} + i \sin \frac{5\pi}{4}) = -2 - 2i, & k = 2 \\ 2\sqrt{2}(\cos \frac{7\pi}{4} + i \sin \frac{7\pi}{4}) = 2 - 2i, & k = 3. \end{cases}$$

Note that the four roots are on a circle of radius $\sqrt{8}$ centered at the origin. They are $90°$ apart.

Example 1.6.6 Find all the values of $(1 - i)^{3/2}$.

Solution 1.6.6

$$(1 - i) = \sqrt{2}e^{i\theta}, \quad \theta = \tan^{-1}(-1) = -\frac{\pi}{4}.$$

$$(1 - i)^3 = 2\sqrt{2}e^{i3\theta + ik2\pi}, \quad k = 0, 1, 2, \ldots.$$

$$(1 - i)^{3/2} = \sqrt[4]{8}e^{i(3\theta/2 + k\pi)}, \quad k = 0, 1.$$

$$(1 - i)^{3/2} = \begin{cases} \sqrt[4]{8}\left[\cos\left(-\frac{3\pi}{8}\right) + i \sin\left(-\frac{3\pi}{8}\right)\right], & k = 0 \\ \sqrt[4]{8}\left[\cos\left(\frac{5\pi}{8}\right) + i \sin\left(\frac{5\pi}{8}\right)\right], & k = 1. \end{cases}$$

1.6.2   Trigonometry and Complex Numbers

Many trigonometric identities can be most elegantly proved with complex numbers. For example, taking the complex conjugate of the Euler formula

$$(e^{i\theta})^* = (\cos \theta + i \sin \theta)^*,$$

we have

$$e^{-i\theta} = \cos\theta - i\sin\theta.$$

It is interesting to write this equation as

$$e^{-i\theta} = e^{i(-\theta)} = \cos(-\theta) + i\sin(-\theta).$$

Comparing the last two equations, we find that

$$\cos(-\theta) = \cos\theta$$
$$\sin(-\theta) = -\sin\theta$$

which is consistent with what we know about the cosine and sine functions of trigonometry.

Adding and subtracting $e^{i\theta}$ and $e^{-i\theta}$, we have

$$e^{i\theta} + e^{-i\theta} = (\cos\theta + i\sin\theta) + (\cos\theta - i\sin\theta) = 2\cos\theta,$$
$$e^{i\theta} - e^{-i\theta} = (\cos\theta + i\sin\theta) - (\cos\theta - i\sin\theta) = 2i\sin\theta.$$

Using them one can easily express the powers of cosine and sine in terms of $\cos n\theta$ and $\sin n\theta$. For example, with $n = 2$

$$\cos^2\theta = \left[\frac{1}{2}\left(e^{i\theta} + e^{-i\theta}\right)\right]^2 = \frac{1}{4}\left(e^{i2\theta} + 2e^{i\theta}e^{-i\theta} + e^{-i2\theta}\right)$$
$$= \frac{1}{2}\left[\frac{1}{2}\left(e^{i2\theta} + e^{-i2\theta}\right) + e^{i0}\right] = \frac{1}{2}\left(\cos 2\theta + 1\right),$$

$$\sin^2\theta = \left[\frac{1}{2i}\left(e^{i\theta} - e^{-i\theta}\right)\right]^2 = -\frac{1}{4}\left(e^{i2\theta} - 2e^{i\theta}e^{-i\theta} + e^{-i2\theta}\right)$$
$$= \frac{1}{2}\left[-\frac{1}{2}\left(e^{i2\theta} + e^{-i2\theta}\right) + e^{i0}\right] = \frac{1}{2}\left(-\cos 2\theta + 1\right).$$

To find an identity for $\cos(\theta_1 + \theta_2)$ and $\sin(\theta_1 + \theta_2)$, we can view them as components of $\exp[i(\theta_1 + \theta_2)]$. Since

$$e^{i\theta_1}e^{i\theta_2} = e^{i(\theta_1+\theta_2)} = \cos(\theta_1 + \theta_2) + i\sin(\theta_1 + \theta_2),$$

and

$$e^{i\theta_1}e^{i\theta_2} = [\cos\theta_1 + i\sin\theta_1][\cos\theta_2 + i\sin\theta_2]$$
$$= (\cos\theta_1\cos\theta_2 - \sin\theta_1\sin\theta_2) + i(\sin\theta_1\cos\theta_2 + \cos\theta_1\sin\theta_2),$$

equating the real and imaginary parts of these equivalent expressions, we get the familiar formulas

$$\cos(\theta_1 + \theta_2) = \cos\theta_1 \cos\theta_2 - \sin\theta_1 \sin\theta_2,$$
$$\sin(\theta_1 + \theta_2) = \sin\theta_1 \cos\theta_2 + \cos\theta_1 \sin\theta_2.$$

From these two equations, it follows that

$$\tan(\theta_1 + \theta_2) = \frac{\sin(\theta_1 + \theta_2)}{\cos(\theta_1 + \theta_2)} = \frac{\sin\theta_1 \cos\theta_2 + \cos\theta_1 \sin\theta_2}{\cos\theta_1 \cos\theta_2 - \sin\theta_1 \sin\theta_2}.$$

Dividing top and bottom by $\cos\theta_1 \cos\theta_2$, we obtain

$$\tan(\theta_1 + \theta_2) = \frac{\tan\theta_1 + \tan\theta_2}{1 - \tan\theta_1 \tan\theta_2}.$$

This formula can be derived directly with complex numbers. Let z_1 and z_2 be two points in the complex plane whose x-components are both equal to 1.

$$z_1 = 1 + iy_1 = r_1 e^{i\theta_1}, \quad \tan\theta_1 = \frac{y_1}{1} = y_1,$$
$$z_2 = 1 + iy_2 = r_2 e^{i\theta_2}, \quad \tan\theta_2 = \frac{y_2}{1} = y_2.$$

The product of the two is given by

$$z_1 z_2 = r_1 r_2 e^{i(\theta_1 + \theta_2)}, \quad \tan(\theta_1 + \theta_2) = \frac{\mathrm{Im}(z_1 z_2)}{\mathrm{Re}(z_1 z_2)}.$$

But

$$z_1 z_2 = (1 + iy_1)(1 + iy_2) = (1 - y_1 y_2) + i(y_1 + y_2),$$

therefore

$$\tan(\theta_1 + \theta_2) = \frac{\mathrm{Im}(z_1 z_2)}{\mathrm{Re}(z_1 z_2)} = \frac{y_1 + y_2}{1 - y_1 y_2} = \frac{\tan\theta_1 + \tan\theta_2}{1 - \tan\theta_1 \tan\theta_2}.$$

These identities can, of course, be demonstrated geometrically. However, it is much easier to prove them algebraically with complex numbers.

Example 1.6.7 Prove De Moivre formula

$$(\cos\theta + i\sin\theta)^n = \cos n\theta + i\sin n\theta.$$

Solution 1.6.7 Since $(\cos\theta + i\sin\theta) = e^{i\theta}$, it follows that

$$(\cos\theta + i\sin\theta)^n = \left(e^{i\theta}\right)^n = e^{in\theta}$$
$$= \cos n\theta + i\sin n\theta.$$

This theorem was published in 1707 by Abraham De Moivre, a French mathematician working in London.

Example 1.6.8 Use De Moivre's theorem and binomial expansion to express $\cos 4\theta$ and $\sin 4\theta$ in terms of powers of $\cos\theta$ and $\sin\theta$.

Solution 1.6.8

$$\cos 4\theta + i\sin 4\theta = e^{i4\theta} = \left(e^{i\theta}\right)^4 = (\cos\theta + i\sin\theta)^4$$
$$= \cos^4\theta + 4\cos^3\theta(i\sin\theta) + 6\cos^2\theta(i\sin\theta)^2 + 4\cos\theta(i\sin\theta)^3 + (i\sin\theta)^4$$
$$= \left(\cos^4\theta - 6\cos^2\theta\sin^2\theta + \sin^4\theta\right) + i\left(4\cos^3\theta\sin\theta - 4\cos\theta\sin^3\theta\right)$$

Equating the real and imaginary parts of these complex expressions, we obtain

$$\cos 4\theta = \cos^4\theta - 6\cos^2\theta\sin^2\theta + \sin^4\theta,$$
$$\sin 4\theta = 4\cos^3\theta\sin\theta - 4\cos\theta\sin^3\theta.$$

Example 1.6.9 Express $\cos^4\theta$ and $\sin^4\theta$ in terms of multiples of θ.

Solution 1.6.9

$$\cos^4\theta = \left[\frac{1}{2}\left(e^{i\theta} + e^{-i\theta}\right)\right]^4$$
$$= \frac{1}{16}\left[\left(e^{i4\theta} + e^{-i4\theta}\right) + 4\left(e^{i2\theta} + e^{-i2\theta}\right) + 6\right]$$
$$= \frac{1}{8}\cos 4\theta + \frac{1}{2}\cos 2\theta + \frac{3}{8}.$$

$$\sin^4\theta = \left[\frac{1}{2i}\left(e^{i\theta} - e^{-i\theta}\right)\right]^4$$
$$= \frac{1}{16}\left[\left(e^{i4\theta} + e^{-i4\theta}\right) - 4\left(e^{i2\theta} + e^{-i2\theta}\right) + 6\right]$$
$$= \frac{1}{8}\cos 4\theta - \frac{1}{2}\cos 2\theta + \frac{3}{8}.$$

Example 1.6.10   Show that

$$\cos(\theta_1 + \theta_2 + \theta_3) = \cos\theta_1 \cos\theta_2 \cos\theta_3 - \sin\theta_1 \sin\theta_2 \cos\theta_3$$
$$- \sin\theta_1 \sin\theta_3 \cos\theta_2 - \sin\theta_2 \sin\theta_3 \cos\theta_1,$$
$$\sin(\theta_1 + \theta_2 + \theta_3) = \sin\theta_1 \cos\theta_2 \cos\theta_3 + \sin\theta_2 \cos\theta_1 \cos\theta_3$$
$$+ \sin\theta_3 \cos\theta_1 \cos\theta_2 - \sin\theta_1 \sin\theta_2 \sin\theta_3.$$

Solution 1.6.10

$$\cos(\theta_1 + \theta_2 + \theta_3) + i\sin(\theta_1 + \theta_2 + \theta_3) = e^{i(\theta_1 + \theta_2 + \theta_3)} = e^{i\theta_1} e^{i\theta_2} e^{i\theta_3},$$

$$e^{i\theta_1} e^{i\theta_2} e^{i\theta_3} = (\cos\theta_1 + i\sin\theta_1)(\cos\theta_2 + i\sin\theta_2)(\cos\theta_3 + i\sin\theta_3)$$
$$= \cos\theta_1 (1 + i\tan\theta_1)\cos\theta_2 (1 + i\tan\theta_2)\cos\theta_3 (1 + i\tan\theta_3).$$

Since

$$(1 + a)(1 + b)(1 + c) = 1 + (a + b + b) + (ab + bc + ca) + abc,$$

$$(1 + i\tan\theta_1)(1 + i\tan\theta_2)(1 + i\tan\theta_3) = 1 + i(\tan\theta_1 + \tan\theta_2 + \tan\theta_3)$$
$$+ i^2 (\tan\theta_1 \tan\theta_2 + \tan\theta_2 \tan\theta_3 + \tan\theta_3 \tan\theta_1) + i^3 \tan\theta_1 \tan\theta_2 \tan\theta_3$$
$$= [1 - (\tan\theta_1 \tan\theta_2 + \tan\theta_2 \tan\theta_3 + \tan\theta_3 \tan\theta_1)]$$
$$+ i[(\tan\theta_1 + \tan\theta_2 + \tan\theta_3) - \tan\theta_1 \tan\theta_2 \tan\theta_3].$$

Therefore

$$e^{i\theta_1} e^{i\theta_2} e^{i\theta_3} = \cos\theta_1 \cos\theta_2 \cos\theta_3$$
$$\times \left\{ \begin{array}{l} [1 - (\tan\theta_1 \tan\theta_2 + \tan\theta_2 \tan\theta_3 + \tan\theta_3 \tan\theta_1)] \\ + i[(\tan\theta_1 + \tan\theta_2 + \tan\theta_3) - \tan\theta_1 \tan\theta_2 \tan\theta_3] \end{array} \right\}.$$

Equating the real and imaginary parts,

$$\cos(\theta_1 + \theta_2 + \theta_3) = \cos\theta_1 \cos\theta_2 \cos\theta_3$$
$$\times [1 - (\tan\theta_1 \tan\theta_2 + \tan\theta_2 \tan\theta_3 + \tan\theta_3 \tan\theta_1)]$$
$$= \cos\theta_1 \cos\theta_2 \cos\theta_3 - \sin\theta_1 \sin\theta_2 \cos\theta_3$$
$$- \sin\theta_1 \sin\theta_3 \cos\theta_2 - \sin\theta_2 \sin\theta_3 \cos\theta_1,$$

$$\sin(\theta_1 + \theta_2 + \theta_3) = \cos\theta_1 \cos\theta_2 \cos\theta_3$$
$$\times [(\tan\theta_1 + \tan\theta_2 + \tan\theta_3) - \tan\theta_1 \tan\theta_2 \tan\theta_3]$$
$$= \sin\theta_1 \cos\theta_2 \cos\theta_3 + \sin\theta_2 \cos\theta_1 \cos\theta_3$$
$$+ \sin\theta_3 \cos\theta_1 \cos\theta_2 - \sin\theta_1 \sin\theta_2 \sin\theta_3.$$

Example 1.6.11 If $\theta_1, \theta_2, \theta_3$ are the three interior angles of a triangle, show that

$$\tan\theta_1 + \tan\theta_2 + \tan\theta_3 = \tan\theta_1 \tan\theta_2 \tan\theta_3.$$

Solution 1.6.11 Since

$$\tan(\theta_1 + \theta_2 + \theta_3) = \frac{\sin(\theta_1 + \theta_2 + \theta_3)}{\cos(\theta_1 + \theta_2 + \theta_3)},$$

using the results of the previous problem and dividing the top and bottom by $\cos\theta_1 \cos\theta_2 \cos\theta_3$, we have

$$\tan(\theta_1 + \theta_2 + \theta_3) = \frac{\tan\theta_1 + \tan\theta_2 + \tan\theta_3 - \tan\theta_1 \tan\theta_2 \tan\theta_3}{1 - \tan\theta_1 \tan\theta_2 - \tan\theta_2 \tan\theta_3 - \tan\theta_3 \tan\theta_1}.$$

Now $\theta_1, \theta_2, \theta_3$ are the three interior angles of a triangle, so $\theta_1 + \theta_2 + \theta_3 = \pi$ and $\tan(\theta_1 + \theta_2 + \theta_3) = \tan\pi = 0$. Therefore

$$\tan\theta_1 + \tan\theta_2 + \tan\theta_3 = \tan\theta_1 \tan\theta_2 \tan\theta_3.$$

Example 1.6.12 Show that

$$\cos\theta + \cos 3\theta + \cos 5\theta + \cdots + \cos(2n-1)\theta = \frac{\sin n\theta \cos n\theta}{\sin\theta},$$
$$\sin\theta + \sin 3\theta + \sin 5\theta + \cdots + \sin(2n-1)\theta = \frac{\sin^2 n\theta}{\sin\theta}.$$

Solution 1.6.12 Let

$$C = \cos\theta + \cos 3\theta + \cos 5\theta + \cdots + \cos(2n-1)\theta,$$
$$S = \sin\theta + \sin 3\theta + \sin 5\theta + \cdots + \sin(2n-1)\theta.$$

$$Z = C + iS = (\cos\theta + i\sin\theta) + (\cos 3\theta + i\sin 3\theta)$$
$$+ (\cos 5\theta + i\sin 5\theta) + \cdots + (\cos(n-1)\theta + i\sin(2n-1)\theta)$$
$$= e^{i\theta} + e^{i3\theta} + e^{i5\theta} + \cdots + e^{i(2n-1)\theta}.$$

$$e^{i2\theta} Z = e^{i3\theta} + e^{i5\theta} + \cdots + e^{i(2n-1)\theta} + e^{i(2n+1)\theta}$$

$$Z - e^{i2\theta} Z = e^{i\theta} - e^{i(2n+1)\theta}$$

$$Z = \frac{e^{i\theta} - e^{i(2n+1)\theta}}{1 - e^{i2\theta}} = \frac{e^{i\theta}(1 - e^{i2n\theta})}{e^{i\theta}\left(e^{-i\theta} - e^{i\theta}\right)} = \frac{e^{in\theta}\left(e^{-in\theta} - e^{in\theta}\right)}{\left(e^{-i\theta} - e^{i\theta}\right)}$$

$$= \frac{e^{in\theta}\sin n\theta}{\sin\theta} = \frac{\cos n\theta \sin n\theta}{\sin\theta} + i\frac{\sin n\theta \sin n\theta}{\sin\theta}.$$

Therefore

$$C = \frac{\cos n\theta \sin n\theta}{\sin\theta}, \quad S = \frac{\sin^2 n\theta}{\sin\theta}.$$

Example 1.6.13 For $r < 1$, show that

$$\left(\sum_{n=0}^{\infty} r^{2n}\cos n\theta\right)^2 + \left(\sum_{n=0}^{\infty} r^{2n}\sin n\theta\right)^2 = \frac{1}{1 - 2r^2\cos\theta + r^4}.$$

Solution 1.6.13 Let

$$Z = \sum_{n=0}^{\infty} r^{2n}\cos n\theta + i\sum_{n=0}^{\infty} r^{2n}\sin n\theta$$

$$= \sum_{n=0}^{\infty} r^{2n}(\cos n\theta + i\sin n\theta) = \sum_{n=0}^{\infty} r^{2n}e^{in\theta}$$

$$= 1 + r^2 e^{i\theta} + r^4 e^{i2\theta} + r^6 e^{i3\theta} + \cdots.$$

Since $r < 1$, so this is a convergent series.

$$r^2 e^{i\theta} Z = r^2 e^{i\theta} + r^4 e^{i2\theta} + r^6 e^{i3\theta} + \cdots.$$

$$Z - r^2 e^{i\theta} Z = 1,$$

$$Z = \frac{1}{1 - r^2 e^{i\theta}}.$$

$$|Z|^2 = ZZ^* = \frac{1}{1 - r^2 e^{i\theta}} \times \frac{1}{1 - r^2 e^{-i\theta}}$$

$$= \frac{1}{1 - r^2(e^{i\theta} + e^{-i\theta}) + r^4} = \frac{1}{1 - 2r^2 \cos\theta + r^4},$$

But

$$|Z|^2 = \left(\sum_{n=0}^{\infty} r^{2n} \cos n\theta + i \sum_{n=0}^{\infty} r^{2n} \sin n\theta \right) \left(\sum_{n=0}^{\infty} r^{2n} \cos n\theta - i \sum_{n=0}^{\infty} r^{2n} \sin n\theta \right)$$

$$= \left(\sum_{n=0}^{\infty} r^{2n} \cos n\theta \right)^2 + \left(\sum_{n=0}^{\infty} r^{2n} \sin n\theta \right)^2.$$

Therefore

$$\left(\sum_{n=0}^{\infty} r^{2n} \cos n\theta \right)^2 + \left(\sum_{n=0}^{\infty} r^{2n} \sin n\theta \right)^2 = \frac{1}{1 - 2r^2 \cos\theta + r^4}.$$

This is the intensity of the light, transmitted through a film after multiple reflections at the surfaces of the film and r is the fraction of light reflected each time.

1.6.3 Geometry and Complex Numbers

There are three geometric representations of the complex number $z = x + iy$:

(a) as the point $P\ (x, y)$ in the xy-plane,

(b) as the vector OP from the origin to the point P,

(c) as any vector that is of the same length and same direction as OP.

For example, $z_A = 3 + i$ can be represented by the point A in Fig. 1.4. It can also be represented by the vector z_A. Similarly $z_B = -2 + 3i$ can be represented by the point B and the vector z_B. Now let us define z_C as $z_A + z_B$,

$$z_C = z_A + z_B = (3 + i) + (-2 + 3i) = 1 + 4i.$$

So z_C is represented by the point C and the vector z_C. Clearly the two shaded triangles in Fig. 1.4 are identical. The vector AC (from A to C) is not only parallel to z_B, it is also of the same length as z_B. In this sense, we say that the vector AC can also represent z_B. Thus z_A, z_B, and $z_A + z_B$ are three sides of the triangle OAC. Since the sum of two sides of a triangle must be greater or equal to the third side, it follows that

$$|z_A| + |z_B| \geq |z_A + z_B|.$$

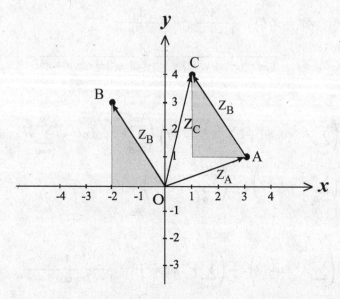

Fig. 1.4 Addition and subtraction of complex numbers in the complex plane. A complex number can be represented by a point in the complex plane, or by the vector from the origin to that point. The vector can be moved parallel to itself

Since $z_B = z_C - z_A$ and z_B is the same as the AC, we can interpret $z_C - z_A$ as the vector from the tip of z_A to the tip of z_C. The distance between C and A is simply $|z_C - z_A|$.

If z is a variable and z_A is fixed, then a circle of radius r centered at z_A is described by the equation

$$|z - z_A| = r.$$

If the two segments AB and CD are parallel, then

$$z_B - z_A = k(z_D - z_C),$$

where k is a real number. If $k = 1$, then A, B, C, D must be the vertices of a parallelogram.

If the two segments AB and CD are perpendicular to each other, then the ratio $\frac{z_D - z_C}{z_B - z_A}$ must be a pure imaginary number. This can be seen as follows.

The segment AB in Fig. 1.5 can be expressed as

$$z_B - z_A = |z_B - z_A| e^{i\beta},$$

and segment CD as

$$z_D - z_C = |z_D - z_C| e^{i\alpha}.$$

Fig. 1.5 Perpendicular segments. If AB and CD are perpendicular, then the ratio of $z_B - z_A$ and $z_D - z_C$ must be purely imaginary

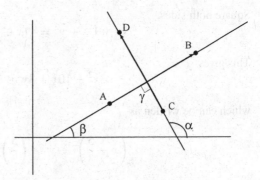

So

$$\frac{z_D - z_C}{z_B - z_A} = \frac{|z_D - z_C| e^{i\alpha}}{|z_B - z_A| e^{i\beta}} = \frac{|z_D - z_C|}{|z_B - z_A|} e^{i(\alpha - \beta)}.$$

It is well known that the exterior angle is equal to the sum of the two interior angles, that is, in Fig. 1.5 $\alpha = \beta + \gamma$, or $\gamma = \alpha - \beta$. If AB is perpendicular to CD, then $\gamma = \frac{\pi}{2}$, and

$$e^{i(\alpha - \beta)} = e^{i\gamma} = e^{i\pi/2} = i.$$

Thus

$$\frac{z_D - z_C}{z_B - z_A} = \frac{|z_D - z_C|}{|z_B - z_A|} i.$$

Since $\dfrac{|z_D - z_C|}{|z_B - z_A|}$ is real, so $\dfrac{z_D - z_C}{z_B - z_A}$ must be imaginary.

The following examples will illustrate how to use these principles to solve problems in geometry.

Example 1.6.14 Determine the curve in the complex plane that is described by

$$\left| \frac{z+1}{z-1} \right| = 2.$$

Solution 1.6.14 $\left| \dfrac{z+1}{z-1} \right| = 2$ can be written as $|z + 1| = 2|z - 1|$. With $z = x + iy$, this equation becomes

$$|(x + 1) + iy| = 2|(x - 1) + iy|,$$

$$\{[(x + 1) + iy][(x + 1) - iy]\}^{1/2} = 2\{[(x - 1) + iy][(x - 1) - iy]\}^{1/2}.$$

Square both sides,

$$(x + 1)^2 + y^2 = 4 (x - 1)^2 + 4y^2.$$

This gives

$$3x^2 - 10x + 3y^2 + 3 = 0,$$

which can be written as

$$\left(x - \frac{5}{3}\right)^2 + y^2 - \left(\frac{5}{3}\right)^2 + 1 = 0,$$

or

$$\left(x - \frac{5}{3}\right)^2 + y^2 = \left(\frac{4}{3}\right)^2.$$

This represents a circle of radius $\frac{4}{3}$ with a center at $\left(\frac{5}{3}, 0\right)$.

Example 1.6.15 In the parallelogram shown in Fig. 1.6, the base is fixed along the x-axis and is of length a. The length of the other side is b. As the angle θ between the two sides changes determine the locus of the center of the parallelogram.

Solution 1.6.15 Let the origin of the coordinates be at the left bottom corner of the parallelogram. So

$$z_A = a, \qquad z_B = be^{i\theta}.$$

Let the center of the parallelogram be z which is at the midpoint of the diagonal OC. Thus

$$z = \frac{1}{2}(z_A + z_B) = \frac{1}{2}a + \frac{1}{2}be^{i\theta}.$$

Or

$$z - \frac{1}{2}a = \frac{1}{2}be^{i\theta}$$

Fig. 1.6 The curve is described by the center of a parallelogram. If the base is fixed, the locus of the center is a circle

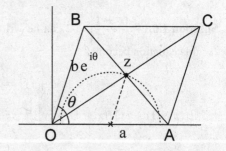

It follows that

$$\left| z - \frac{1}{2}a \right| = \left| \frac{1}{2}be^{i\theta} \right| = \frac{1}{2}b.$$

Therefore the locus of the center is a circle of radius $\frac{1}{2}b$ centered at $\frac{1}{2}a$. Half of the circle is shown in Fig. 1.6.

Example 1.6.16 If E, F, G, H are midpoints of the quadrilateral ABCD. Prove that EFGH is a parallelogram.

Solution 1.6.16 Let the vector from origin to any point P be z_P, then from Fig. 1.7 we see that

$$z_E = z_A + \frac{1}{2}(z_B - z_A),$$

$$z_F = z_B + \frac{1}{2}(z_C - z_B).$$

$$z_F - z_E = z_B + \frac{1}{2}(z_C - z_B) - z_A - \frac{1}{2}(z_B - z_A) = \frac{1}{2}(z_C - z_A).$$

$$z_G = z_D + \frac{1}{2}(z_C - z_D),$$

$$z_H = z_A + \frac{1}{2}(z_D - z_A).$$

$$z_G - z_H = z_D + \frac{1}{2}(z_C - z_D) - z_A - \frac{1}{2}(z_D - z_A) = \frac{1}{2}(z_C - z_A).$$

Fig. 1.7 Parallelogram formed by the midpoints of a quadrilateral

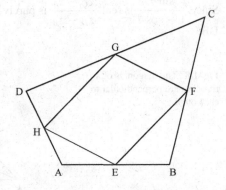

Thus

$$z_F - z_E = z_G - z_H.$$

Therefore EFGH is a parallelogram.

Example 1.6.17 Use the complex number to show that the diagonals of a rhombus (a parallelogram with equal sides) are perpendicular to each other (Fig. 1.8).

Solution 1.6.17 The diagonal AC is given by $z_C - z_A$, and the diagonal DB is given by $z_B - z_D$. Let the length of each side of the rhombus be a, and the origin of the coordinates coincide with A. Furthermore let the x-axis be along the line AB. Thus

$$z_A = 0, \quad z_B = a, \quad z_D = ae^{i\theta}.$$

Furthermore,

$$z_C = z_B + z_D = a + ae^{i\theta}.$$

Therefore

$$z_C - z_A = a + ae^{i\theta} = a\left(1 + e^{i\theta}\right),$$
$$z_B - z_D = a - ae^{i\theta} = a\left(1 - e^{i\theta}\right).$$

Thus

$$\frac{z_C - z_A}{z_B - z_D} = \frac{a\left(1 + e^{i\theta}\right)}{a\left(1 - e^{i\theta}\right)} = \frac{\left(1 + e^{i\theta}\right)\left(1 - e^{-i\theta}\right)}{\left(1 - e^{i\theta}\right)\left(1 - e^{-i\theta}\right)}$$
$$= \frac{e^{i\theta} - e^{-i\theta}}{2 - \left(e^{i\theta} + e^{-i\theta}\right)} = i\frac{\sin\theta}{1 - \cos\theta}.$$

Since $\dfrac{\sin\theta}{1 - \cos\theta}$ is real, $\dfrac{z_C - z_A}{z_D - z_B}$ is purely imaginary. Hence AC is perpendicular to DB.

Fig. 1.8 The diagonals of a rhombus are perpendicular to each other

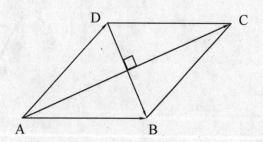

Example 1.6.18 In the triangle AOB, the angle between AO and OB is 90° and the length of AO is the same as the length of OB. The point D trisects the line AB such that $AD = 2DB$, and C is the midpoint of OB. Show that AC is perpendicular to OD (Fig. 1.9).

Solution 1.6.18 Let the real axis be along OA and the imaginary axis along OB. Let the length of OA and OB be a. Thus

$$z_O = 0, \quad z_A = a, \quad z_B = ai, \quad z_C = \frac{1}{2}ai,$$

$$z_D = z_A + \frac{2}{3}(z_B - z_A) = a + \frac{2}{3}(ai - a) = \frac{1}{3}a(1 + 2i),$$

$$z_D - z_O = \frac{1}{3}a(1 + 2i) - 0 = \frac{1}{3}a(1 + 2i).$$

The vector AC is given by $z_C - z_A$,

$$z_C - z_A = \frac{1}{2}ai - a = i\frac{1}{2}a(1 + 2i).$$

Thus

$$\frac{z_C - z_A}{z_D - z_O} = i\frac{3}{2}.$$

Since this is purely imaginary, therefore AC is perpendicular to OD.

Fig. 1.9 A problem in geometry. If OA is perpendicular to OB and $OA = OB$, then the line CA is perpendicular to the line OD where C is the midpoint of OB and $DA = 2BD$

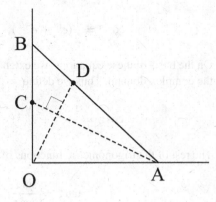

1.7   Elementary Functions of Complex Variable

1.7.1   Exponential and Trigonometric Functions of z

The exponential function e^z is of fundamental importance, not only for its own sake, but also as a basis for defining all the other elementary functions. The exponential function of real variable is well known. Now we wish to give meaning to e^z when $z = x + iy$. In the spirit of Euler, we can work our way in a purely manipulative manner. Assuming that e^z obeys all the familiar rules of the exponential function of a real number, we have

$$e^z = e^{x+iy} = e^x e^{iy} = e^x (\cos y + i \sin y). \tag{1.19}$$

Thus we can define e^z as $e^x (\cos y + i \sin y)$. It reduces to e^x when the imaginary part of z vanishes. It is also easy to show that

$$e^{z_1} e^{z_2} = e^{z_1+z_2}.$$

Furthermore, in the next chapter we shall consider in detail the meaning of derivatives with respect to a complex z. Now it suffices to know that

$$\frac{d}{dz}e^z = e^z.$$

Therefore the definition of (1.19) preserves all the familiar properties of the exponential function.

We have already seen that

$$\cos \theta = \frac{1}{2}(e^{i\theta} + e^{-i\theta}), \quad \sin \theta = \frac{1}{2i}(e^{i\theta} - e^{-i\theta}).$$

On the basis of these equations, we extend the definitions of the cosine and sine into the complex domain. Thus we define

$$\cos z = \frac{1}{2}(e^{iz} + e^{-iz}), \quad \sin z = \frac{1}{2i}(e^{iz} - e^{-iz}).$$

The rest of the trigonometric functions of z are defined in a usual way. For example,

$$\tan z = \frac{\sin z}{\cos z}, \quad \cot z = \frac{\cos z}{\sin z},$$
$$\sec z = \frac{1}{\cos z}, \quad \csc z = \frac{1}{\sin z}.$$

With these definitions we can show that all the familiar formulas of trigonometry remain valid when real variable x is replaced by complex variable z :

$$\cos(-z) = \cos z, \quad \sin(-z) = -\sin z,$$
$$\cos^2 z + \sin^2 z = 1,$$

$$\cos(z_1 \pm z_2) = \cos z_1 \cos z_2 \mp \sin z_1 \sin z_2,$$
$$\sin(z_1 \pm z_2) = \sin z_1 \cos z_2 \pm \cos z_1 \sin z_2,$$
$$\frac{d}{dz} \cos z = -\sin z, \quad \frac{d}{dz} \sin z = \cos z.$$

To prove them, we must start with their definitions. For example,

$$\cos^2 z + \sin^2 z = \left[\frac{1}{2}(e^{iz} + e^{-iz}) \right]^2 + \left[\frac{1}{2i}(e^{iz} - e^{-iz}) \right]^2$$
$$= \frac{1}{4}(e^{i2z} + 2 + e^{-i2z}) - \frac{1}{4}(e^{i2z} - 2 + e^{-i2z}) = 1.$$

Example 1.7.1 Express e^{1-i} in the form of $a + bi$, accurate to three decimal places.

Solution 1.7.1
$$e^{1-i} = e^1 e^{-i} = e(\cos 1 - i \sin 1).$$

Using a hand-held calculator, we find

$$e^{1-i} \simeq 2.718(0.5403 - 0.8415i)$$
$$= 1.469 - 2.287i.$$

Example 1.7.2 Show that
$$\sin 2z = 2 \sin z \cos z.$$

Solution 1.7.2

$$2 \sin z \cos z = 2\frac{1}{2i}(e^{iz} - e^{-iz})\frac{1}{2}(e^{iz} + e^{-iz})$$
$$= \frac{1}{2i}(e^{i2z} - e^{-i2z}) = \sin 2z.$$

Example 1.7.3 Compute $\sin(1 - i)$.

Solution 1.7.3 By definition

$$
\begin{aligned}
\sin(1 - i) &= \frac{1}{2i}\left(e^{i(1-i)} - e^{-i(1-i)}\right) = \frac{1}{2i}\left(e^{1+i} - e^{-1-i}\right) \\
&= \frac{1}{2i}\left\{e\left[\cos(1) + i\sin(1)\right]\right\} - \frac{1}{2i}\left\{e^{-1}\left[\cos(1) - i\sin(1)\right]\right\} \\
&= \frac{1}{2i}\left(e - e^{-1}\right)\cos(1) + \frac{1}{2}\left(e + e^{-1}\right)\sin(1).
\end{aligned}
$$

We can get the same result by using the trigonometric addition formula.

1.7.2 Hyperbolic Functions of z

The following particular combinations of exponentials arise frequently,

$$
\cosh z = \frac{1}{2}\left(e^z + e^{-z}\right), \qquad \sinh z = \frac{1}{2}\left(e^z - e^{-z}\right).
$$

They are called hyperbolic cosine (*abbreviated* cosh) and hyperbolic sine (*abbreviated* sinh). Clearly

$$
\cosh(-z) = \cosh z, \qquad \sinh(-z) = -\sinh z.
$$

The other hyperbolic functions are defined in a similar way to parallel the trigonometric functions:

$$
\tanh z = \frac{\sinh z}{\cosh z}, \qquad \coth z = \frac{1}{\tanh z},
$$

$$
\sec hz = \frac{1}{\cosh z}, \qquad \csc hz = \frac{1}{\sinh z}.
$$

With these definitions, all identities involving hyperbolic functions of real variable are preserved when the variable is complex. For example,

$$
\cosh^2 z - \sinh^2 z = \frac{1}{4}\left(e^z + e^{-z}\right)^2 - \frac{1}{4}\left(e^z - e^{-z}\right)^2 = 1,
$$

$$
\sinh 2z = \frac{1}{2}\left(e^{2z} - e^{-2z}\right) = \frac{1}{2}\left(e^z - e^{-z}\right)\left(e^z + e^{-z}\right)
$$

$$
= 2\sinh z \cosh z.
$$

There is a close relationship between the trigonometric and hyperbolic functions when the variable is complex. For example,

$$\sin iz = \frac{1}{2i}\left(e^{i(iz)} - e^{-i(iz)}\right) = \frac{1}{2i}\left(e^{-z} - e^{z}\right)$$
$$= \frac{i}{2}\left(e^{z} - e^{-z}\right) = i \sinh z.$$

Similarly we can show

$$\cos iz = \cosh z,$$
$$\sinh iz = i \sin z, \quad \cosh iz = \cos z.$$

Furthermore,

$$\sin z = \sin(x + iy) = \sin x \cos iy + \cos x \sin iy$$
$$= \sin x \cosh y + i \cos x \sinh y,$$
$$\cos z = \cos x \cosh y - i \sin x \sinh y.$$

Example 1.7.4 Show that

$$\frac{d}{dz}\cosh z = \sinh z, \quad \frac{d}{dz}\sinh z = \cosh z.$$

Solution 1.7.4

$$\frac{d}{dz}\cosh z = \frac{d}{dz}\frac{1}{2}\left(e^{z} + e^{-z}\right) = \frac{1}{2}\left(e^{z} - e^{-z}\right) = \sinh z,$$
$$\frac{d}{dz}\sinh z = \frac{d}{dz}\frac{1}{2}\left(e^{z} - e^{-z}\right) = \frac{1}{2}\left(e^{z} + e^{-z}\right) = \cosh z.$$

Example 1.7.5 Evaluate $\cos(1 + 2i)$.

Solution 1.7.5

$$\cos(1 + 2i) = \cos 1 \cosh 2 - i \sin 1 \sinh 2$$
$$= (0.5403)(3.7622) - i(0.8415)(3.6269) = 2.033 - 3.052i.$$

Example 1.7.6 Evaluate $\cos{(\pi - i)}$.

Solution 1.7.6 By definition,

$$\cos(\pi - i) = \frac{1}{2}\left(e^{i(\pi-i)} + e^{-i(\pi-i)}\right) = \frac{1}{2}\left(e^{i\pi+1} + e^{-i\pi-1}\right)$$
$$= \frac{1}{2}(-e - e^{-1}) = -\cosh{(1)} = -1.543.$$

We get the same result by the expansion,

$$\cos(\pi - i) = \cos\pi\cosh{(1)} + i\sin\pi\sinh{(1)}$$
$$= -\cosh{(1)} = -1.543.$$

1.7.3 Logarithm and General Power of z

The natural logarithm of $z = x + iy$ is denoted $\ln z$ and is defined in a similar way as in the real variable, namely, as the inverse of the exponential function. However, there is an important difference. A real valued exponential $y = e^x$ is a one to one function, since two different x always produce two different values of y. Strictly speaking, only one to one function has an inverse, because only then will each value of y can be the image of exactly one x value. But the complex exponential e^z is a multi-valued function, since

$$e^z = e^{x+iy} = e^x\left(\cos y + i\sin y\right).$$

When y is increased by an integer multiple of 2π, the exponential returns to its original value. Therefore to define a complex logarithm we have to relax the one to one restriction. Thus,

$$w = \ln z$$

is defined for $z \neq 0$ by the relation

$$e^w = z.$$

If we set

$$w = u + iv, \qquad z = re^{i\theta},$$

this becomes

$$e^w = e^{u+iv} = e^u e^{iv}_* = re^{i\theta}.$$

Since

$$|e^w| = \left[(e^w)(e^w)^*\right]^{1/2} = \left(e^{u+iv}e^{u-iv}\right)^{1/2} = e^u,$$

$$|e^w| = \left[\left(re^{i\theta}\right)\left(re^{i\theta}\right)^*\right]^{1/2} = \left[\left(re^{i\theta}\right)\left(re^{-i\theta}\right)\right]^{1/2} = r,$$

Therefore

$$e^u = r.$$

By definition,

$$u = \ln r.$$

Since $e^w = z$,

$$e^u e^{iv} = re^{iv} = re^{i\theta}$$

it follows that

$$v = \theta.$$

Thus

$$w = u + iv = \ln r + i\theta.$$

Therefore the rule of logarithm is preserved,

$$\ln z = \ln re^{i\theta} = \ln r + i\theta. \tag{1.20}$$

Since θ is the polar angle, after it is increased by 2π in the z complex plane, it comes back to the same point and z will have the same value. However, the logarithm of z will not return to its original value. Its imaginary part will increase by $2\pi i$. If the argument of z in a particular interval of 2π is denoted as θ_0, then (1.20) can be written as

$$\ln z = \ln r + i(\theta_0 + 2\pi n), \quad n = 0, \pm 1, \pm 2, \ldots.$$

By specifying such an interval, we say that we have selected a particular branch of θ as the principal branch. The value corresponding to $n = 0$ is known as the principal value and is commonly denoted as $Ln\ z$, that is

$$Ln\ z = \ln r + i\theta_0.$$

The choice of the principal branch is somewhat arbitrary.

Figure 1.10 illustrates two possible branch selections. Fig. 1.10a depicts the branch that selects the value of the argument of z from the interval $-\pi < \theta \leq \pi$. The values in this branch are most commonly used in complex algebra computer codes. The argument θ is inherently discontinuous, jumping by 2π as z crosses the

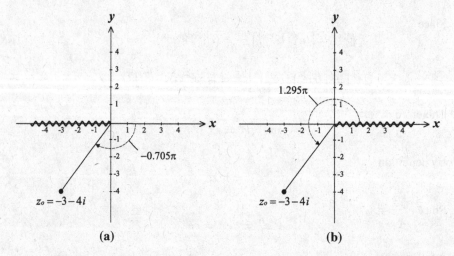

(a)   **(b)**

Fig. 1.10 Two possible branch selections. **a** Branch cut on the negative x-axis. The point $-3 - 4i$ has argument -0.705π. **b** Branch cut on the positive x-axis. The point $-3 - 4i$ has argument 1.295π

negative x-axis. This line of discontinuities is known as the branch cut. The cut ends at the origin, which is known as the branch point.

With the branch cut along the negative real axis, the principal value of the logarithm of $z_0 = -3 - 4i$ is given by $\ln(|z_0| e^{i\theta})$ where $\theta = \tan^{-1} \frac{4}{3} - \pi$, thus the principal value is

$$\ln(-3 - 4i) = \ln 5 e^{i\theta} = \ln 5 + i(\tan^{-1} \frac{4}{3} - \pi) = 1.609 - 0.705\pi i.$$

However, if we select the interval $0 \leq \theta < 2\pi$ as the principal branch, then the branch cut is along the positive x-axis, as shown in Fig. 1.10b. In this case the principal value of the logarithm of z_0 is

$$\ln(-3 - 4i) = \ln 5 + i(\tan^1 \frac{4}{3} + \pi) = 1.609 + 1.295\pi i.$$

Unless otherwise specified, we shall use the interval $0 \leq \theta < 2\pi$ as the principal branch.

It can be easily checked that the familiar laws of logarithm which hold for real variables can be established for complex variables as well. For example,

$$\begin{aligned} \ln z_1 z_2 &= \ln r_1 e^{i\theta_1} r_2 e^{i\theta_2} = \ln r_1 r_2 + i(\theta_1 + \theta_2) \\ &= \ln r_1 + \ln r_2 + i\theta_1 + i\theta_2 \\ &= (\ln r_1 + i\theta_1) + (\ln r_2 + i\theta_2) = \ln z_1 + \ln z_2. \end{aligned}$$

This relation is always true as long as infinitely many values of logarithms are taken into consideration. However, if only the principal values are taken, then the sum of the two principal values $\ln z_1 + \ln z_2$ may fall outside of the principal branch of $\ln(z_1 z_2)$.

Example 1.7.7 Find all values of $\ln 2$.

Solution 1.7.7 The real number 2 is also the complex number $2 + i0$, and

$$2 + i0 = 2e^{in2\pi}, \quad n = 0, \pm 1, \pm 2, \ldots .$$

Thus

$$\ln 2 = Ln\ 2 + n2\pi i$$
$$= 0.693 + n2\pi i, \quad n = 0, \pm 1, \pm 2, \ldots .$$

Even positive real numbers now have infinitely many logarithms. Only one of them is real, corresponding to $n = 0$ principal value.

Example 1.7.8 Find all values of $\ln(-1)$.

Solution 1.7.8

$$\ln(-1) = \ln e^{i(\pi \pm 2\pi n)} = i(\pi + 2\pi n), \quad n = 0, \pm 1, \pm 2, \ldots .$$

The principal value is $i\pi$ for $n = 0$.

Since $\ln a = x$ means $e^x = a$, so long as the variable x is real, a is always positive. Thus, in the domain of real numbers, the logarithm of a negative number does not exist. Therefore the answer must come from the complex domain. The situation was still sufficiently confused in the eighteenth century that it was possible for so great a mathematician as D'Alembert (1717–1783) to think $\ln(-x) = \ln(x)$, so $\ln(-1) = \ln(1) = 0$. His reason was the following. Since $(-x)(-x) = x^2$, therefore $\ln[(-x)(-x)] = \ln x^2 = 2\ln x$. But $\ln[(-x)(-x)] = \ln(-x) + \ln(-x) = 2\ln(-x)$, so we get $\ln(-x) = \ln x$. This is incorrect, because it applies the rule of ordinary algebra to the domain of complex numbers. It was Euler who pointed out that $\ln(-1)$ must be equal to the complex number $i\pi$, which is in accordance with his equation $e^{i\pi} = -1$.

Example 1.7.9 Find the principal value of $\ln(1 + i)$.

Solution 1.7.9 Since

$$1 + i = \sqrt{2}e^{i\pi/4},$$

$$\ln(1 + i) = \ln \sqrt{2} + \frac{\pi}{4}i = 0.3466 + 0.7854i.$$

We are now in a position to consider the general power of a complex number. First let us see how to find i^i. Since

$$i = e^{i(\frac{\pi}{2} + 2\pi n)},$$

$$i^i = \left[e^{i(\frac{\pi}{2} + 2\pi n)}\right]^i = e^{-(\frac{\pi}{2} + 2\pi n)}, \quad n = 0, \pm 1, \pm 2, \ldots.$$

We get infinitely many values—all of them real. In a literal sense, Euler showed that the imaginary power of an imaginary number can be real.

In general, since $a = e^{\ln a}$, so

$$a^b = \left(e^{\ln a}\right)^b = e^{b \ln a}.$$

In this formula, both a and b can be complex numbers. For example, to find $(1 + i)^{1-i}$, first we write

$$(1 + i)^{1-i} = \left[e^{\ln(1+i)}\right]^{1-i} = e^{(1-i)\ln(1+i)}.$$

Since

$$\ln(1 + i) = \ln \sqrt{2}e^{i(\frac{\pi}{4} + 2\pi n)} = \ln \sqrt{2} + i(\frac{\pi}{4} + 2\pi n), \quad n = 0, \pm 1, \pm 2, \ldots,$$

now

$$(1 + i)^{1-i} = e^{\left(\ln \sqrt{2} + i(\frac{\pi}{4} + 2\pi n) - i \ln \sqrt{2} + (\frac{\pi}{4} + 2\pi n)\right)}$$

$$= e^{\ln \sqrt{2} + \frac{\pi}{4} + 2\pi n} e^{i(\frac{\pi}{4} + 2\pi n - \ln \sqrt{2})}.$$

Using

$$e^{i2\pi n} = 1, \quad e^{\ln \sqrt{2}} = \sqrt{2},$$

we have

$$(1+i)^{1-i} = \sqrt{2}e^{\frac{\pi}{4}+2\pi n}\left[\cos\left(\frac{\pi}{4}-\ln\sqrt{2}\right)+i\sin\left(\frac{\pi}{4}-\ln\sqrt{2}\right)\right].$$

Using a calculator, this expression is found to be

$$(1+i)^{1-i} = e^{2\pi n}\left(2.808 + 1.318i\right), \qquad n = 0, \pm 1, \pm 2, \ldots.$$

Example 1.7.10 Find all values of $i^{1/2}$.

Solution 1.7.10

$$i^{1/2} = \left[e^{i\left(\frac{\pi}{2}+2\pi n\right)}\right]^{1/2} = e^{i\pi/4}e^{in\pi}, \qquad n = 0, \pm 1, \pm 2, \ldots.$$

Since $e^{in\pi} = 1$ for n even, and $e^{in\pi} = -1$ for n odd, thus

$$i^{1/2} = \pm e^{i\pi/4} = \pm\left(\cos\frac{\pi}{4} + i\sin\frac{\pi}{4}\right) = \pm\frac{\sqrt{2}}{2}\left(1 + i\right).$$

Notice that although n can be any of the infinitely many integers, we find only two values for $i^{1/2}$ as we should, for it is the square root of i.

Example 1.7.11 Find the principal value of 2^i.

Solution 1.7.11

$$2^i = \left[e^{\ln 2}\right]^i = e^{i\ln 2} = \cos\left(\ln 2\right) + i\sin\left(\ln 2\right)$$
$$= 0.769 + 0.639i.$$

Example 1.7.12 Find the principal value of $(1+i)^{2-i}$.

Solution 1.7.12
$$(1+i)^{2-i} = \exp\left[(2-i)\ln\left(1+i\right)\right].$$

The principal value of $\ln(1+i)$ is

$$\ln\left(1+i\right) = \ln\sqrt{2}e^{i\pi/4} = \ln\sqrt{2} + i\frac{\pi}{4}.$$

Therefore

$$(1+i)^{2-i} = \exp\left[(2-i)\left(\ln\sqrt{2}+i\frac{\pi}{4}\right)\right]$$

$$= \exp\left(2\ln\sqrt{2}+\frac{\pi}{4}\right)\exp\left[i\left(\frac{\pi}{2}-\ln\sqrt{2}\right)\right]$$

$$= 2e^{\pi/4}\left[\cos\left(\frac{\pi}{2}-\ln\sqrt{2}\right)+i\sin\left(\frac{\pi}{2}-\ln\sqrt{2}\right)\right]$$

$$= 4.3866\,(\sin 0.3466 + i\cos 0.3466) = 1.490 + 4.126i.$$

1.7.4 Inverse Trigonomeric and Hyperbolic Functions

Starting from their definitions, we can work out sensible expressions for the inverse of trigonometric and inverse hyperbolic functions. For example, to find

$$w = \sin^{-1} z,$$

we write this as

$$z = \sin w = \frac{1}{2i}\left(e^{iw} - e^{-iw}\right).$$

Multiplying e^{iw}, we have

$$ze^{iw} = \frac{1}{2i}\left(e^{i2w} - 1\right).$$

Rearranging, we get a quadratic equation in e^{iw},

$$\left(e^{iw}\right)^2 - 2ize^{iw} - 1 = 0.$$

The solution of this equation is

$$e^{iw} = \frac{1}{2}\left(2iz \pm \sqrt{-4z^2 + 4}\right) = iz \pm \left(1 - z^2\right)^{1/2}.$$

Taking logarithm of both sides,

$$iw = \ln\left[iz \pm \left(1 - z^2\right)^{1/2}\right].$$

Therefore

$$w = \sin^{-1} z = -i\ln\left[iz \pm \left(1 - z^2\right)^{1/2}\right].$$

Because of the logarithm, this expression is multi-valued. Even in the principal branch, $\sin^{-1} z$ has two values for $z \neq 1$ because of the square roots.

Similarly, we can show

$$\cos^{-1} z = -i \ln \left[z \pm \left(z^2 - 1 \right)^{1/2} \right],$$

$$\tan^{-1} z = \frac{i}{2} \ln \frac{i+z}{i-z},$$

$$\sinh^{-1} z = \ln \left[z \pm \left(1 + z^2 \right)^{1/2} \right],$$

$$\cosh^{-1} z = \ln \left[z \pm \left(z^2 - 1 \right)^{1/2} \right],$$

$$\tanh^{-1} z = \frac{1}{2} \ln \frac{1+z}{1-z}.$$

Example 1.7.13 Evaluate $\cos^{-1} 2$.

Solution 1.7.13 Let $w = \cos^{-1} 2$, so $\cos w = 2$. It follows

$$\frac{1}{2} \left(e^{iw} + e^{-iw} \right) = 2$$

Multiplying e^{iw}, we have a quadratic equation in e^{iw}

$$\left(e^{iw} \right)^2 + 1 = 4 e^{iw}.$$

Solving for e^{iw},

$$e^{iw} = \frac{1}{2} (4 \pm \sqrt{16 - 4}) = 2 \pm \sqrt{3}.$$

Thus

$$iw = \ln \left(2 \pm \sqrt{3} \right).$$

Therefore

$$\cos^{-1} 2 = w = -i \ln \left(2 \pm \sqrt{3} \right).$$

Now

$$\ln(2 + \sqrt{3}) = 1.317, \quad \ln \left(2 - \sqrt{3} \right) = -1.317.$$

Note only in this particular case, $-\ln(2 + \sqrt{3}) = \ln \left(2 - \sqrt{3} \right)$, since

$$-\ln(2 + \sqrt{3}) = \ln(2 + \sqrt{3})^{-1} = \ln \frac{1}{2 + \sqrt{3}} = \ln \frac{2 - \sqrt{3}}{2^2 - (\sqrt{3})^2} = \ln(2 - \sqrt{3}).$$

Thus the principal values of $\ln\left(2 \pm \sqrt{3}\right) = \pm 1.317$. Therefore

$$\cos^{-1} 2 = \mp 1.317i + 2\pi n, \quad n = 0, \pm 1, \pm 2, \ldots.$$

In real variable domain, the maximum value of cosine is one. Therefore we expect $\cos^{-1} 2$ to be complex numbers. Also note that $\pm$ solutions may be expected since $\cos(-z) = \cos(z)$.

Example 1.7.14 Show that

$$\tan^{-1} z = \frac{i}{2}[\ln(i + z) - \ln(i - z)].$$

Solution 1.7.14 Let $w = \tan^{-1} z$, so

$$z = \tan w = \frac{\sin w}{\cos w} = \frac{e^{iw} - e^{-iw}}{i\left(e^{iw} + e^{-iw}\right)},$$

$$iz\left(e^{iw} + e^{-iw}\right) = e^{iw} - e^{-iw},$$

$$(iz - 1)\,e^{iw} + (iz + 1)\,e^{-iw} = 0.$$

Multiplying e^{iw} and rearranging, we have

$$e^{i2w} = \frac{1 + iz}{1 - iz}.$$

Taking logarithm on both sides,

$$i2w = \ln\frac{1 + iz}{1 - iz} = \ln\frac{i - z}{i + z}.$$

Thus

$$w = \frac{1}{2i}\ln\frac{i - z}{i + z} = -\frac{i}{2}\ln\frac{i - z}{i + z} = \frac{i}{2}\ln\frac{i + z}{i - z},$$

$$\tan^{-1} z = w = \frac{i}{2}[\ln(i + z) - \ln(i - z)].$$

Exercises

1. Approximate $\sqrt{2}$ as 1.414 and use the table of successive square root of 10 to compute $10^{\sqrt{2}}$.

   Ans. 25.94

2. Use the table of successive square root of 10 to compute $\log 2$.

   Ans. 0.3010

3. How long will it take for a sum of money to double if invested at 20% interest rate compounded annually? (This question was posted in a clay tablet dated 1700 B.C. now at Louvre.)

   Hint: Solve $(1.2)^x = 2$.

   Ans. 3.8 years, or 3 years 9 months and 18 d.

4. Suppose the annual interest rate is fixed at 5%. Banks are competing by offering compound interests with increasing number of conversions, monthly, daily, hourly, and so on. With a principal of $100, what is the maximum amount of money one can get after one year?

   Ans. $100e^{0.05} = 105.13$

5. Simplify (express it in the form of $a + ib$)

$$\frac{\cos 2\alpha + i \sin 2\alpha}{\cos \alpha + i \sin \alpha}.$$

   Ans. $\cos \alpha + i \sin \alpha$.

6. Simplify (express it in the form of $a + ib$)

$$\frac{(\cos \theta - i \sin \theta)^2}{(\cos \theta + i \sin \theta)^3}.$$

   Ans. $\cos 5\theta - i \sin 5\theta$.

7. Find the roots of

$$x^4 + 1 = 0.$$

   Ans. $\frac{\sqrt{2}}{2} + i\frac{\sqrt{2}}{2}$, $-\frac{\sqrt{2}}{2} + i\frac{\sqrt{2}}{2}$, $\frac{\sqrt{2}}{2} - i\frac{\sqrt{2}}{2}$, $-\frac{\sqrt{2}}{2} - i\frac{\sqrt{2}}{2}$.

8. Find all the distinct fourth roots of $8 - i8\sqrt{3}$.

   Ans. $2\left(\cos \frac{5\pi}{12} + i \sin \frac{5\pi}{12}\right)$, $2\left(\cos \frac{11\pi}{12} + i \sin \frac{11\pi}{12}\right)$, $2\left(\cos \frac{17\pi}{12} + i \sin \frac{17\pi}{12}\right)$, $2\left(\cos \frac{23\pi}{12} + i \sin \frac{23\pi}{12}\right)$.

9. Find all the values of the following in the form of $a + ib$.

$$(a)\ i^{2/3}, \quad (b)\ (-1)^{1/3}, \quad (c)\ (3 + 4i)^4.$$

   Ans. (a) -1, $(1 \pm i\sqrt{3})/2$, (b) -1, $(1 \pm i\sqrt{3})/2$, (c) $-527 - 336i$.

10. Use complex numbers to show

$$\cos 3\theta = 4\cos^3\theta - 3\cos\theta,$$
$$\sin 3\theta = 3\sin\theta - 4\sin^3\theta.$$

11. Use complex numbers to show

$$\cos^2\theta = \frac{1}{2}\left(\cos 2\theta + 1\right),$$
$$\sin^2\theta = \frac{1}{2}\left(1 - \cos 2\theta\right).$$

12. Show that

$$\sum_{k=0}^{n}\cos k\theta = \frac{1}{2} + \frac{\sin\left[\left(n+\frac{1}{2}\right)\theta\right]}{2\sin\frac{1}{2}\theta},$$

$$\sum_{k=0}^{n}\sin k\theta = \frac{1}{2}\cot\frac{1}{2}\theta - \frac{\cos\left[\left(n+\frac{1}{2}\right)\theta\right]}{2\sin\frac{1}{2}\theta}.$$

13. Find the location of the center and the radius of the following circle:

$$\left|\frac{z-1}{z+1}\right| = 3.$$

Ans. $(-\frac{5}{4},\ 0)$ $r = \frac{3}{4}$.

14. Use complex numbers to show that the diagonals of a parallelogram bisect each other.

15. Use complex numbers to show that the line segment joining the two midpoints of two sides of any triangle is parallel to the third side and half its length.

16. Use complex numbers to prove that medians of a triangle intersect at a point two-thirds of the way from any vertex to the midpoint of the opposite side.

17. Let ABC be an isosceles triangle such that $AB = AC$. Use complex numbers to show that the line from A to the midpoint of BC is perpendicular to BC.

18. Express the principal value of the following in the form of $a + ib$:

$$(a)\ \exp(\frac{i\pi}{4} + \frac{\ln 2}{2}),\quad (b)\ \cos(\pi - 2i\ln 3),\quad (c)\ \ln(-i).$$

Ans. $(a)\ 1 + i$, $(b)\ -\frac{41}{9}$, $(c)\ -i\frac{\pi}{2}$ or $i\frac{3\pi}{2}$.

19. Express the principal value of the following in the form of $a + ib$:

$$(a)\ i^{3+i}, \quad (b)\ (2i)^{1+i}, \quad (c)\ \left(\frac{1+i\sqrt{3}}{2}\right)^i.$$

Ans. $(a)\ -0.20788i,\ (b)\ -0.2657+0.3189i,\ (c)\ 0.35092.$

20. Find all the values of the following expressions:

$$(a)\ \sin\left(i\ln\frac{1-i}{1+i}\right), \quad (b)\ \tan^{-1}(2i), \quad (c)\ \cosh^{-1}\left(\frac{1}{2}\right)$$

Ans. $(a)\ 1,\ (b)\ \frac{1+2n}{2}\pi+i\frac{1}{2}\ln 3,\ (c)\ i\left(\pm\frac{\pi}{3}+2n\pi\right).$

21. With $z = x + iy$, verify the following

$$\sin z = \sin x \cosh y + i\cos x \sinh y,$$
$$\cos z = \cos x \cosh y - i\sin x \sinh y,$$
$$\sinh z = \sinh x \cos y + i\cosh x \sin y,$$
$$\cosh z = \cosh x \cos y + i\sinh x \sin y.$$

22. Show that

$$\sin 2z = 2\sin z \cos z,$$
$$\cos 2z = \cos^2 z - \sin^2 z,$$
$$\cosh^2 z - \sinh^2 z = 1.$$

23. Show that

$$\cos^{-1} z = -i\ln\left[z \pm \left(z^2 - 1\right)^{1/2}\right],$$
$$\sinh^{-1} z = \ln\left[z \pm \left(1 + z^2\right)^{1/2}\right].$$

Chapter 2
Complex Functions

Complex numbers were first used to simplify calculations. In the course of time, it became clear that the theory of complex functions is a very effective tool in engineering and sciences. Often the most elegant solutions of important problems in heat conduction, elasticity, electrostatics, and hydrodynamics are produced by complex function methods. In modern physics, complex variables have even become an intrinsic part of the physical theory. For example, it is a fundamental postulate in quantum mechanics that wave functions reside in a complex vector space.

In engineering and sciences the ultimate test is in the laboratory. When you make a measurement, the result you get is, of course, a real number. But the theoretical formulation of the problem often leads us into the realm of complex numbers. It is almost a miracle that, if the theory is correct, further mathematical analysis with complex functions will always lead us to an answer that is real. Therefore the theory of complex functions is an essential tool in modern sciences.

Complex functions to which the concepts and structure of calculus can be applied are called analytic functions. It is the analytic functions that dominate complex analysis. Many interesting properties and applications of analytic functions are studied in this chapter.

2.1 Analytic Functions

The theory of analytic functions is an extension of the differential and integral calculus to realms of complex variables. However, the notion of a derivative of a complex function is far more subtle than that of a real function. This is because of the intrinsically two-dimensional nature of the complex numbers. The success made in analyzing this question by Cauchy and Riemann left a deep imprint on the whole of mathematics. It also had a far reaching consequences in several branches of mathematical physics.

© The Author(s), under exclusive license to Springer Nature Switzerland AG 2022   65
K.-T. Tang, *Mathematical Methods for Engineers and Scientists 1*,
https://doi.org/10.1007/978-3-031-05678-9_2

2.1.1 Complex Function as Mapping Operation

From the complex variable $z = x + iy$, one can construct complex functions $f(z)$. Formally we can define functions of complex variables in exactly the same way as functions of real variables are defined, except allowing the constants and variables to assume complex values.

Let $w = f(z)$ denote some functional relationship connecting w and z. These functions may then be resolved into real and imaginary parts

$$w = f(x + iy) = u(x, y) + iv(x, y)$$

in which both $u(x, y)$ and $v(x, y)$ are real functions. For example, if

$$w = f(z) = z^2,$$

then

$$w = (x + iy)^2 = (x^2 - y^2) + i2xy.$$

So the real and imaginary parts of w (u, v) are respectively

$$u(x, y) = (x^2 - y^2), \tag{2.1}$$
$$v(x, y) = 2xy. \tag{2.2}$$

Since two dimensions are needed to specify the independent variable $z(x, y)$ and another two dimensions to specify the dependent variable w (u, v), a complex function cannot be represented by a single two or three-dimensional plot. The functional relationship $w = f(z)$ is perhaps best pictured as a mapping, or a transformation, operation. A set of points (x, y) in the z-plane $(z = x + iy)$ are mapped into another set of points (u, v) in the w-plane $(w = u + iv)$. If we allow the variable x and y to trace some curve in the z-plane, this will force the variable u and v to trace an image curve in the w-plane.

In the above example, if the point (x, y) in the z-plane moves along the hyperbola $x^2 - y^2 = c$ (where c is a constant), the image point given by (2.1) will move along the curve $u = c$, that is a vertical line in the w-plane. Similarly, if the point moves along the hyperbola $2xy = k$, the image point given by (2.2) will trace the horizontal line $v = k$ in the w-plane. The hyperbolas $x^2 - y^2 = c$ and $2xy = k$ form two families of curves in the z-plane, each curve corresponding to a given value of the constant c or k. Their image curves form a rectangular grid of horizontal and vertical lines in the w-plane, as shown in Fig. 2.1.

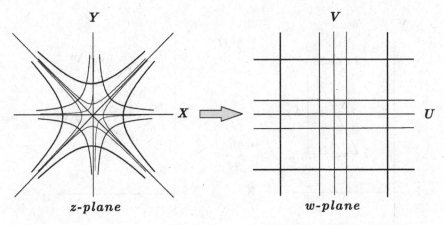

Fig. 2.1 The function $w = z^2$ maps hyperbolas in the z-plane onto horizontal and vertical lines in the w-plane

2.1.2 Differentiation of a Complex Function

To discuss the differentiation of a complex function $f(z)$ at certain point z_0, the function must be defined in some neighborhood of the point z_0. By the neighborhood we mean the set of all points in a sufficiently small circular region with center at z_0. If $z_0 = x_0 + iy_0$ and $z = z_0 + \Delta z$ are two nearby points in the z-plane with $\Delta z = \Delta x + i\Delta y$, the corresponding image points in the w-plane are $w_0 = u_0 + iv_0$ and $w = w_0 + \Delta w$, where $w_0 = f(z_0)$ and $w = f(z) = f(z_0 + \Delta z)$. The change Δw caused by the increment Δz in z_0 is

$$\Delta w = f(z_0 + \Delta z) - f(z_0).$$

These functional relationships are shown in Fig. 2.2.

Now we define the derivative $f'(z) = \dfrac{dw}{dz}$ by the usual formula

$$f'(z_0) = \lim_{\Delta z \to 0} \frac{\Delta w}{\Delta z} = \lim_{\Delta z \to 0} \frac{f(z_0 + \Delta z) - f(z_0)}{\Delta z}. \tag{2.3}$$

It is most important to note that in this formula $z = z_0 + \Delta z$ can assume any position in the neighborhood of z_0 and Δz can approach zero along any of the infinitely many paths joining z with z_0. Hence if the derivative is to have a unique value, we must demand that the limit be independent of the way in which Δz is made to approach zero. This restriction greatly narrows down the class of complex functions that possess derivatives.

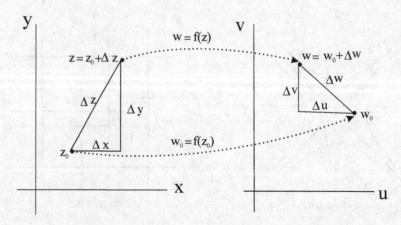

Fig. 2.2 The neighborhood of z_0 in the z-plane is mapped onto the neighborhood of w_0 in the w-plane by the function $w = f(z)$

For example, if $f(z) = |z|^2$, then $w = zz^*$, and

$$\frac{\Delta w}{\Delta z} = \frac{|z + \Delta z|^2 - |z|^2}{\Delta z} = \frac{(z + \Delta z)(z^* + \Delta z^*) - zz^*}{\Delta z}$$

$$= z^* + \Delta z^* + \frac{z \Delta z^*}{\Delta z} = x - iy + \Delta x - i \Delta y + (x + iy)\frac{\Delta x - i \Delta y}{\Delta x + i \Delta y}.$$

For the derivative $f'(z)$ to exist, the limit of this quotient must be the same no matter how Δz approaches zero. Since $\Delta z = \Delta x + i \Delta y$, $\Delta z \to 0$ means, of course, both $\Delta x \to 0$ and $\Delta y \to 0$. However the way they go to zero may make a difference. If we let Δz approach zero along path I in Fig. 2.3, so that first $\Delta y \to 0$ and then $\Delta x \to 0$, we get

Fig. 2.3 To be differentiable at z, the same limit must be obtained no matter which path Δz is taken to approach zero

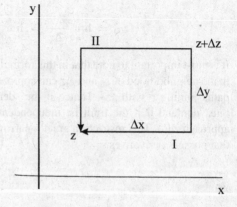

$$\lim_{\Delta z \to 0} \frac{\Delta w}{\Delta z} = \lim_{\Delta x \to 0} \left\{ \lim_{\Delta y \to 0} \left[x - iy + \Delta x - i\Delta y + (x + iy) \frac{\Delta x - i\Delta y}{\Delta x + i\Delta y} \right] \right\}$$
$$= 2x.$$

But if we take path II and first allow $\Delta x \to 0$ and then $\Delta y \to 0$, we obtain

$$\lim_{\Delta z \to 0} \frac{\Delta w}{\Delta z} = \lim_{\Delta y \to 0} \left\{ \lim_{\Delta x \to 0} \left[x - iy + \Delta x - i\Delta y + (x + iy) \frac{\Delta x - i\Delta y}{\Delta x + i\Delta y} \right] \right\}$$
$$= -2iy.$$

These limits are different, and hence $w = |z|^2$ has no derivative except possibly at $z = 0$.

On the other hand, if we consider $w = z^2$, then

$$w + \Delta w = (z + \Delta z)^2 = z^2 + 2z\Delta z + (\Delta z)^2,$$

so that

$$\frac{\Delta w}{\Delta z} = \frac{2z\Delta z + (\Delta z)^2}{\Delta z} = 2z + \Delta z.$$

The limit of this quotient as $\Delta z \to 0$ is invariably $2z$, whatever may be the path along which Δz approaches zero. Therefore the derivative exists everywhere and

$$\frac{dw}{dz} = \lim_{\Delta z \to 0} \frac{\Delta w}{\Delta z} = \lim_{\Delta z \to 0} (2z + \Delta z) = 2z.$$

It is clear that not every combination of $u(x, y) + iv(x, y)$ can be differentiated with respect to z. If a complex function $f(z)$ whose derivative $f'(z)$ exists at z_0 and at every point in the neighborhood of z_0, then the function is said to be analytic at z_0. An analytic function is a function that is analytic in some region (domain) of the complex plane. A function that is analytic in the whole complex plane is called an entire function. A point at which an analytic function ceases to have a derivative is called a singular point.

2.1.3   Cauchy-Riemann Conditions

We will now investigate the conditions that a complex function must satisfy in order to be differentiable.

It follows from the definition

$$f(z) = f(x + iy) = u(x, y) + iv(x, y),$$

that

$$f(z + \Delta z) = f((x + \Delta x) + i(y + \Delta y))$$
$$= u(x + \Delta x, y + \Delta y) + i v(x + \Delta x, y + \Delta y).$$

Since $w = f(z)$ and $w + \Delta w = f(z + \Delta z)$, so

$$\Delta w = f(z + \Delta z) - f(z) = \Delta u + i \Delta v$$

where

$$\Delta u = u(x + \Delta x, y + \Delta y) - u(x, y),$$
$$\Delta v = v(x + \Delta x, y + \Delta y) - v(x, y).$$

We can add $0 = -u(x, y + \Delta y) + u(x, y + \Delta y)$ to Δu without changing its value

$$\Delta u = u(x + \Delta x, y + \Delta y) - u(x, y)$$
$$= u(x + \Delta x, y + \Delta y) - u(x, y + \Delta y) + u(x, y + \Delta y) - u(x, y).$$

Recall the definition of partial derivative

$$\lim_{\Delta x \to 0} \frac{1}{\Delta x} [u(x + \Delta x, y + \Delta y) - u(x, y + \Delta y)] = \frac{\partial u}{\partial x}.$$

In this expression only x variable is increased by Δx and y variable remains the same. If it is implicitly understood that the symbol Δx carries the meaning that it is approaching zero as a limit, then we can move it to the right-hand side,

$$u(x + \Delta x, y + \Delta y) - u(x, y + \Delta y) = \frac{\partial u}{\partial x} \Delta x.$$

Similarly, in the following expression only y variable is increased by Δy, so

$$u(x, y + \Delta y) - u(x, y) = \frac{\partial u}{\partial y} \Delta y.$$

Therefore

$$\Delta u = \frac{\partial u}{\partial x} \Delta x + \frac{\partial u}{\partial y} \Delta y. \tag{2.4}$$

Likewise,

$$\Delta v = \frac{\partial v}{\partial x} \Delta x + \frac{\partial v}{\partial y} \Delta y.$$

Hence the derivative given by (2.3) can be written as

Fig. 2.4 Infinitely many paths Δz can approach zero, each characterized by its slope

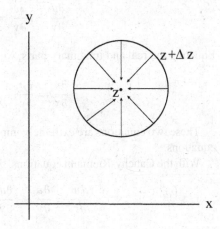

$$\frac{dw}{dz} = \lim_{\Delta z \to 0} \frac{\Delta u + i \Delta v}{\Delta x + i \Delta y} = \lim_{\Delta z \to 0} \frac{(\frac{\partial u}{\partial x} + i \frac{\partial v}{\partial x})\Delta x + (\frac{\partial u}{\partial y} + i \frac{\partial v}{\partial y})\Delta y}{\Delta x + i \Delta y}.$$

Dividing both the numerator and denominator by Δx, we have

$$\frac{dw}{dz} = \lim_{\Delta z \to 0} \frac{(\frac{\partial u}{\partial x} + i \frac{\partial v}{\partial x}) + (\frac{\partial u}{\partial y} + i \frac{\partial v}{\partial y})\frac{\Delta y}{\Delta x}}{1 + i \frac{\Delta y}{\Delta x}}$$

$$= \lim_{\Delta z \to 0} \frac{(\frac{\partial u}{\partial x} + i \frac{\partial v}{\partial x})}{1 + i \frac{\Delta y}{\Delta x}} \left[1 + \frac{(\frac{\partial u}{\partial y} + i \frac{\partial v}{\partial y})}{(\frac{\partial u}{\partial x} + i \frac{\partial v}{\partial x})} \frac{\Delta y}{\Delta x} \right].$$

There are infinitely many paths that Δz can approach zero, each path is characterized by its slope $\frac{\Delta y}{\Delta x}$ as shown in Fig. 2.4. For all these paths to give the same limit, $\frac{\Delta y}{\Delta x}$ must be eliminated from this expression. This will be the case if and only if

$$\frac{(\frac{\partial u}{\partial y} + i \frac{\partial v}{\partial y})}{(\frac{\partial u}{\partial x} + i \frac{\partial v}{\partial x})} = i, \tag{2.5}$$

since then the expression becomes

$$\frac{dw}{dz} = \lim_{\Delta z \to 0} \frac{(\frac{\partial u}{\partial x} + i \frac{\partial v}{\partial x})}{1 + i \frac{\Delta y}{\Delta x}} \left[1 + i \frac{\Delta y}{\Delta x} \right] = \frac{\partial u}{\partial x} + i \frac{\partial v}{\partial x}. \tag{2.6}$$

which is independent of $\frac{\Delta y}{\Delta x}$.

From (2.5), we have

$$\frac{\partial u}{\partial y} + i\frac{\partial v}{\partial y} = i\frac{\partial u}{\partial x} - \frac{\partial v}{\partial x}.$$

Equating the real and imaginary parts, we arrive at the following pair of equations:

$$\frac{\partial u}{\partial x} = \frac{\partial v}{\partial y}, \quad \frac{\partial u}{\partial y} = -\frac{\partial v}{\partial x}.$$

These two equations are extremely important and are known as Cauchy-Riemann equations.

With the Cauchy-Riemann equations, the derivative shown in (2.6) can be written as

$$\frac{dw}{dz} = \frac{\partial v}{\partial y} - i\frac{\partial u}{\partial y} = \frac{\partial u}{i\partial y} + i\frac{\partial v}{i\partial y}. \tag{2.7}$$

The expression in (2.6) is the derivative with Δz approaching zero along the real x-axis and the expression in (2.7) is the derivative with Δz approaching zero along the imaginary y-axis. For an analytic function, they must be the same.

Thus if $u(x, y)$, $v(x, y)$ are continuous and satisfy the Cauchy-Riemann equations in some region of the complex plane, then $f(z) = u(x, y) + iv(x, y)$ is an analytic function in that region. In other words, Cauchy-Riemann equations are necessary and sufficient conditions for the function to be differentiable.

2.1.4 Cauchy-Riemann Equations in Polar Coordinates

Often the function $f(z)$ is expressed in polar coordinates, so it is convenient to express the Cauchy-Riemann equations in polar form.

Since $x = r\cos\theta$ and $y = r\sin\theta$, so by chain rule

$$\frac{\partial u}{\partial r} = \frac{\partial u}{\partial x}\frac{\partial x}{\partial r} + \frac{\partial u}{\partial y}\frac{\partial y}{\partial r} = \frac{\partial u}{\partial x}\cos\theta + \frac{\partial u}{\partial y}\sin\theta,$$

$$\frac{\partial u}{\partial \theta} = \frac{\partial u}{\partial x}\frac{\partial x}{\partial \theta} + \frac{\partial u}{\partial y}\frac{\partial y}{\partial \theta} = -\frac{\partial u}{\partial x}r\sin\theta + \frac{\partial u}{\partial y}r\cos\theta.$$

Similarly,

$$\frac{\partial v}{\partial r} = \frac{\partial v}{\partial x}\frac{\partial x}{\partial r} + \frac{\partial v}{\partial y}\frac{\partial y}{\partial r} = \frac{\partial v}{\partial x}\cos\theta + \frac{\partial v}{\partial y}\sin\theta,$$

$$\frac{\partial v}{\partial \theta} = \frac{\partial v}{\partial x}\frac{\partial x}{\partial \theta} + \frac{\partial v}{\partial y}\frac{\partial y}{\partial \theta} = -\frac{\partial v}{\partial x}r\sin\theta + \frac{\partial v}{\partial y}r\cos\theta.$$

With the Cauchy-Riemann conditions

$$\frac{\partial u}{\partial x} = \frac{\partial v}{\partial y}, \quad \frac{\partial u}{\partial y} = -\frac{\partial v}{\partial x},$$

we have

$$\frac{\partial v}{\partial r} = -\frac{\partial u}{\partial y}\cos\theta + \frac{\partial u}{\partial x}\sin\theta,$$

$$\frac{\partial v}{\partial \theta} = \frac{\partial u}{\partial y}r\sin\theta + \frac{\partial u}{\partial x}r\cos\theta.$$

Thus,

$$\frac{\partial u}{\partial r} = \frac{1}{r}\frac{\partial v}{\partial \theta}, \quad \frac{1}{r}\frac{\partial u}{\partial \theta} = -\frac{\partial v}{\partial r}$$

are the Cauchy-Riemann equations in the polar form.

It is instructive to derive these equations from the definition of the derivative

$$f'(z) = \lim_{\Delta z \to 0} \frac{\Delta w}{\Delta z}$$

In the polar coordinates, $z = re^{i\theta}$,

$$\Delta w = \Delta u\,(r, \theta) + i\,\Delta v\,(r, \theta),$$

$$\Delta z = (r + \Delta r)\,e^{i(\theta + \Delta\theta)} - re^{i\theta}.$$

For $\Delta z \to 0$, we can first let $\Delta\theta \to 0$ and obtain

$$\Delta z = (r + \Delta r)\,e^{i\theta} - re^{i\theta} = \Delta re^{i\theta},$$

and then let $\Delta r \to 0$, so

$$f'(z) = \lim_{\Delta r \to 0} \frac{\Delta u\,(r, \theta) + i\,\Delta v\,(r, \theta)}{\Delta re^{i\theta}} = \frac{1}{e^{i\theta}}\left(\frac{\partial u}{\partial r} + i\frac{\partial v}{\partial r}\right).$$

But if we let $\Delta r \to 0$ first, we get

$$\Delta z = re^{i(\theta + \Delta\theta)} - re^{i\theta}.$$

Since

$$e^{i(\theta + \Delta\theta)} - e^{i\theta} = \frac{de^{i\theta}}{d\theta}\Delta\theta = ie^{i\theta}\,\Delta\theta,$$

so Δz can be written as

$$\Delta z = rie^{i\theta}\,\Delta\theta,$$

and when we take the limit $\Delta\theta \to 0$, the derivative becomes

$$f'(z) = \lim_{\Delta\theta\to 0} \frac{\Delta u\,(r,\theta) + i\,\Delta v\,(r,\theta)}{r i e^{i\theta}\,\Delta\theta} = \frac{1}{i r e^{i\theta}}\left(\frac{\partial u}{\partial\theta} + i\frac{\partial v}{\partial\theta}\right).$$

For an analytic function, the two expressions of derivative must be the same,

$$\frac{1}{e^{i\theta}}\left(\frac{\partial u}{\partial r} + i\frac{\partial v}{\partial r}\right) = \frac{1}{i r e^{i\theta}}\left(\frac{\partial u}{\partial\theta} + i\frac{\partial v}{\partial\theta}\right).$$

Therefore

$$\frac{\partial u}{\partial r} = \frac{1}{r}\frac{\partial v}{\partial\theta}, \qquad \frac{\partial v}{\partial r} = -\frac{1}{r}\frac{\partial u}{\partial\theta},$$

which is what we obtained by direct transformation.

Furthermore, the derivative is given by either of the equivalent expressions

$$f'(z) = e^{-i\theta}\left(\frac{\partial u}{\partial r} + i\frac{\partial v}{\partial r}\right)$$

$$= \frac{1}{ir}e^{-i\theta}\left(\frac{\partial u}{\partial\theta} + i\frac{\partial v}{\partial\theta}\right).$$

2.1.5 Analytic Function as a Function of z Alone

In any analytic function $w = u(x, y) + iv(x, y)$, the variables $x,\ y$ can be replaced by their equivalents in terms of $z,\ z^*$:

$$x = \frac{1}{2}\left(z + z^*\right) \quad and \quad y = \frac{1}{2i}\left(z - z^*\right),$$

since the complex variable $z = x + iy$ and $z^* = x - iy$. Thus an analytic function can be regarded formally as a function of z and z^*. To show that w depends only on z and does not involve z^*, it is sufficient to compute $\frac{\partial w}{\partial z^*}$ and verify that it is identically zero. Now by chain rule,

$$\frac{\partial w}{\partial z^*} = \frac{\partial(u + iv)}{\partial z^*} = \frac{\partial u}{\partial z^*} + i\frac{\partial v}{\partial z^*}$$

$$= \left(\frac{\partial u}{\partial x}\frac{\partial x}{\partial z^*} + \frac{\partial u}{\partial y}\frac{\partial y}{\partial z^*}\right) + i\left(\frac{\partial v}{\partial x}\frac{\partial x}{\partial z^*} + \frac{\partial v}{\partial y}\frac{\partial y}{\partial z^*}\right).$$

Since, from the equations expressing x and y in terms of z and z^*, we have

$$\frac{\partial x}{\partial z^*} = \frac{1}{2} \quad and \quad \frac{\partial y}{\partial z^*} = \frac{i}{2},$$

we can write

$$
\begin{aligned}
\frac{\partial w}{\partial z^*} &= \left(\frac{1}{2} \frac{\partial u}{\partial x} + \frac{i}{2} \frac{\partial u}{\partial y} \right) + i \left(\frac{1}{2} \frac{\partial v}{\partial x} + \frac{i}{2} \frac{\partial v}{\partial y} \right) \\
&= \frac{1}{2} \left(\frac{\partial u}{\partial x} - \frac{\partial v}{\partial y} \right) + \frac{i}{2} \left(\frac{\partial u}{\partial y} + \frac{\partial v}{\partial x} \right).
\end{aligned}
$$

Since w, by hypothesis, is an analytic function, u and v satisfy the Cauchy-Riemann conditions, therefore each of the quantities in parentheses in the last expression vanishes. Thus

$$
\frac{\partial w}{\partial z^*} = 0. \tag{2.8}
$$

Hence, w is independent of z^*, that is, it depends on x and y only through the combination $x + iy$.

Therefore if w is an analytic function, then it can be written as

$$
w = f(z),
$$

and its derivative is defined as

$$
\frac{dw}{dz} = \lim_{\Delta z \to 0} \frac{f(z + \Delta z) - f(z)}{\Delta z}.
$$

This definition is formally identical with that for the derivatives of a function of a real variable. Since the general theory of limits is phrased in terms of absolute values, so if it is valid for real variables, it will also be valid for complex variables. Hence formulas in real variables will have counterparts in complex variables. For example, formulas such as

$$
\frac{d(w_1 \pm w_2)}{dz} = \frac{dw_1}{dz} \pm \frac{dw_2}{dz}
$$

$$
\frac{d(w_1 w_2)}{dz} = w_1 \frac{dw_2}{dz} + w_2 \frac{dw_1}{dz}
$$

$$
\frac{d(w_1/w_2)}{dz} = \frac{w_2(dw_1/dz) - w_1(dw_2/dz)}{w_2^2}, \qquad w_2 \neq 0
$$

$$
\frac{d(w^n)}{dz} = nw^{n-1} \frac{dw}{dz}
$$

are all valid as long as w_1, w_2, and w are analytic functions.

Specifically, any polynomial in z

$$
w(z) = a_n z^n + a_{n-1} z^{n-1} + \cdots + a_1 z + a_0
$$

is analytic in the whole complex plane and therefore is an entire function. Its derivative is

$$w'(z) = na_n z^{n-1} + (n-1)a_{n-1}z^{n-2} + \cdots + a_1.$$

Consequently any rational function of z (a polynomial over another polynomial) is analytic at every point for which its denominator is not zero. At the zeros of the denominator, the function blows up and is not differentiable. Therefore the zeros of the denominator are the singular points of the function.

In fact we can take (2.8) as an alternative statement of the differentiability condition. Thus, the elementary functions defined in the previous chapter are all analytic functions, (some with singular points), since they are functions of z alone. It can be easily shown that they satisfy the Cauchy-Riemann conditions.

Example 2.1.1 Show that the real part u and the imaginary part v of $w = z^2$ satisfy the Cauchy-Riemann equations. Find the derivative of w through the partial derivatives of u and v.

Solution 2.1.1 Since

$$w = z^2 = (x + iy)^2 = \left(x^2 - y^2\right) + i2xy,$$

so the real and imaginary parts are

$$u(x, y) = x^2 - y^2, \qquad v(x, y) = 2xy.$$

Therefore

$$\frac{\partial u}{\partial x} = 2x = \frac{\partial v}{\partial y}, \quad \frac{\partial u}{\partial y} = -2y = -\frac{\partial v}{\partial x}.$$

Thus the Cauch-Riemann equations are satisfied. It is differentiable and

$$\frac{dw}{dz} = \frac{\partial u}{\partial x} + i\frac{\partial v}{\partial x} = 2x + i2y = 2z$$

which is what we found before regarding z as a single variable.

Example 2.1.2 Show that the real part u and the imaginary part v of $f(z) = e^z$ satisfy the Cauchy-Riemann equations. Find the derivative of $f(z)$ through the partial derivatives of u and v.

Solution 2.1.2 Since

$$e^z = e^{x+iy} = e^x (\cos y + i \sin y),$$

the real and imaginary parts are, respectively,

$$u = e^x \cos y \quad and \quad v = e^x \sin y.$$

It follows that

$$\frac{\partial u}{\partial x} = e^x \cos y = \frac{\partial v}{\partial y}, \quad \frac{\partial u}{\partial y} = -e^x \sin y = -\frac{\partial v}{\partial x}.$$

So the Cauchy-Riemann equations are satisfied, and

$$f'(z) = \frac{\partial u}{\partial x} + i \frac{\partial v}{\partial x} = e^x \cos y + i e^x \sin y = e^z,$$

which is what we expect by regarding z as a single variable.

Example 2.1.3 Show that the real part u and the imaginary part v of $\ln z$ satisfy the Cauchy-Riemann equations, and find $\frac{d}{dz} \ln z$ through the partial derivatives of u and v. (a) use rectangular coordinates, (b) use polar coordinates.

Solution 2.1.3 (a) With rectangular coordinates, $z = x + iy$,

$$\ln z = u(x, y) + iv(x, y) = \ln \left(x^2 + y^2\right)^{1/2} + i(\tan^{-1} \frac{y}{x} + 2n\pi).$$

So

$$u = \ln \left(x^2 + y^2\right)^{1/2}, \quad v = (\tan^{-1} \frac{y}{x} + 2n\pi),$$

$$\frac{\partial u}{\partial x} = \frac{1}{2} \frac{2x}{\left(x^2 + y^2\right)} = \frac{x}{\left(x^2 + y^2\right)},$$

$$\frac{\partial u}{\partial y} = \frac{1}{2} \frac{2y}{\left(x^2 + y^2\right)} = \frac{y}{\left(x^2 + y^2\right)},$$

$$\frac{\partial v}{\partial x} = \frac{-y/x^2}{1 + (y/x)^2} = -\frac{y}{\left(x^2 + y^2\right)},$$

$$\frac{\partial v}{\partial y} = \frac{1/x}{1 + (y/x)^2} = \frac{x}{\left(x^2 + y^2\right)}.$$

Therefore

$$\frac{\partial u}{\partial x} = \frac{\partial v}{\partial y}, \qquad \frac{\partial u}{\partial y} = -\frac{\partial v}{\partial x}.$$

The Cauchy-Riemann equations are satisfied, and

$$\frac{d}{dz} \ln z = \frac{\partial u}{\partial x} + i \frac{\partial v}{\partial x} = \frac{x}{(x^2 + y^2)} - i \frac{y}{(x^2 + y^2)}$$

$$= \frac{x - iy}{(x^2 + y^2)} = \frac{x - iy}{(x + iy)(x - iy)} = \frac{1}{(x + iy)} = \frac{1}{z}.$$

(b) With polar coordinates, $z = re^{i\theta}$,

$$\ln z = u(r, \theta) + iv(r, \theta) = \ln r + i(\theta + 2n\pi).$$

$$u = \ln r, \qquad v = \theta + 2n\pi,$$

$$\frac{\partial u}{\partial r} = \frac{1}{r}, \qquad \frac{\partial v}{\partial r} = 0,$$

$$\frac{\partial u}{\partial \theta} = 0, \qquad \frac{\partial v}{\partial \theta} = 1.$$

Therefore the Cauchy-Riemann conditions in polar coordinates

$$\frac{\partial u}{\partial r} = \frac{1}{r} = \frac{1}{r} \frac{\partial v}{\partial \theta}, \qquad \frac{1}{r} \frac{\partial u}{\partial \theta} = 0 = -\frac{\partial v}{\partial r},$$

are satisfied. The derivative is given by

$$\frac{d}{dz} \ln z = e^{-i\theta} \left(\frac{\partial u}{\partial r} + i \frac{\partial v}{\partial r} \right) = e^{-i\theta} \frac{1}{r} = \frac{1}{re^{i\theta}} = \frac{1}{z},$$

as expected.

Example 2.1.4 Show that the real part u and the imaginary part v of z^n satisfy the Cauchy-Riemann equations, and find $\frac{d}{dz} z^n$ through the partial derivatives of u and v.

Solution 2.1.4 For this problem, it is much easier to work with polar coordinates with $z = re^{i\theta}$,

$$z^n = u(r, \theta) + iv(r, \theta) = r^n e^{in\theta} = r^n (\cos n\theta + i \sin n\theta).$$

$$u = r^n \cos n\theta, \qquad v = r^n \sin n\theta,$$

$$\frac{\partial u}{\partial r} = nr^{n-1}\cos n\theta, \qquad \frac{\partial u}{\partial \theta} = -nr^n \sin n\theta,$$

$$\frac{\partial v}{\partial r} = nr^{n-1}\sin n\theta, \qquad \frac{\partial v}{\partial \theta} = nr^n \cos n\theta.$$

Therefore the Cauchy-Riemann conditions in polar coordinates

$$\frac{\partial u}{\partial r} = nr^{n-1}\cos n\theta = \frac{1}{r}\frac{\partial v}{\partial \theta},$$

$$\frac{1}{r}\frac{\partial u}{\partial \theta} = -nr^{n-1}\sin n\theta = -\frac{\partial v}{\partial r},$$

are satisfied and the derivative of z^n is given by

$$\frac{d}{dz}z^n = e^{-i\theta}\left(\frac{\partial u}{\partial r} + i\frac{\partial v}{\partial r}\right) = e^{-i\theta}\left(nr^{n-1}\cos n\theta + inr^{n-1}\sin n\theta\right)$$

$$= e^{-i\theta}nr^{n-1}e^{in\theta} = nr^{n-1}e^{i(n-1)\theta} = n\left(re^{i\theta}\right)^{n-1} = nz^{n-1},$$

as one would get regarding z as a single variable.

2.1.6 Analytic Function and Laplace's Equation

Analytic functions have many interesting, important properties and applications. One of them is that both the real part and imaginary part of an analytic function satisfy the two-dimensional Laplace equation

$$\frac{\partial^2 \phi}{\partial x^2} + \frac{\partial^2 \phi}{\partial y^2} = 0.$$

A great many physical problems lead to Laplace's equation, naturally we are very much interested in its solution.

If $f(z) = u(x, y) + iv(x, y)$ is analytic, then u and v satisfy the Cauchy-Riemann conditions

$$\frac{\partial u}{\partial x} = \frac{\partial v}{\partial y}, \qquad \frac{\partial u}{\partial y} = -\frac{\partial v}{\partial x}.$$

Differentiate the first equation with respect to x and the second equation with respect to y, we have

$$\frac{\partial^2 u}{\partial x^2} = \frac{\partial^2 v}{\partial x \partial y}$$

$$\frac{\partial^2 u}{\partial y^2} = -\frac{\partial^2 v}{\partial y \partial x}.$$

Adding the two equations, we get

$$\frac{\partial^2 u}{\partial x^2} + \frac{\partial^2 u}{\partial y^2} = \frac{\partial^2 v}{\partial x \partial y} - \frac{\partial^2 v}{\partial y \partial x}.$$

As long as they are continuous, the order of differentiation can be interchanged

$$\frac{\partial^2 v}{\partial x \partial y} = \frac{\partial^2 v}{\partial y \partial x},$$

therefore it follows that

$$\frac{\partial^2 u}{\partial x^2} + \frac{\partial^2 u}{\partial y^2} = 0.$$

This is the Laplace equation for u. Similarly, if we differentiate the first Cauchy-Riemann equation with respect to y, and the second one with respect to x, we can show that v also satisfies the Laplace equation

$$\frac{\partial^2 v}{\partial x^2} + \frac{\partial^2 v}{\partial y^2} = 0.$$

Functions satisfying the Laplace equation are called harmonic functions. Two functions that satisfy both the Laplace equation and the Cauchy-Riemann equations are known as conjugate harmonic functions. We have shown that real and imaginary parts of an analytic function are conjugate harmonic functions.

A family of two-dimensional curves can be represented by the equation

$$u(x, y) = k.$$

For example if $u(x, y) = x^2 + y^2$ and $k = 4$, then this equation represents a circle centered at the origin with radius 2. By changing the constant k, we change the radius of the circle. Thus the equation $x^2 + y^2 = k$ represents a family of circles all centered at the origin with various radii.

Each of the conjugate harmonic functions forming the real and imaginary parts of an analytic function $f(z)$ generates a family of curves in the x-y-plane. That is, if $f(z) = u(x, y) + iv(x, y)$, then $u(x, y) = k$ and $v(x, y) = c$, where k and c are constants, are two families of curves.

If Δu is the difference of u at two nearby points, then by (2.4)

$$\Delta u = \frac{\partial u}{\partial x} \Delta x + \frac{\partial u}{\partial y} \Delta y.$$

Now if the two points are on the same curve, that is

$$u(x + \Delta x, y + \Delta y) = k, \quad u(x, y) = k,$$

then

$$\Delta u = u(x + \Delta x, y + \Delta y) - u(x, y) = 0.$$

In this case,

$$0 = \frac{\partial u}{\partial x} \Delta x + \frac{\partial u}{\partial y} \Delta y.$$

To find the slope of this curve, we divide both sides of this equation by Δx

$$0 = \frac{\partial u}{\partial x} + \frac{\partial u}{\partial y} \frac{\Delta y}{\Delta x},$$

therefore the slope of the curve $u(x, y) = k$ is given by

$$\frac{\Delta y}{\Delta x} \Big|_u = -\frac{\partial u / \partial x}{\partial u / \partial y}.$$

Similarly, the slope of the curve $v(x, y) = c$ is given by

$$\frac{\Delta y}{\Delta x} \Big|_v = -\frac{\partial v / \partial x}{\partial v / \partial y}.$$

Since u and v satisfy the Cauchy-Riemann equations

$$\frac{\partial u}{\partial x} = \frac{\partial v}{\partial y}, \quad \frac{\partial u}{\partial y} = -\frac{\partial v}{\partial x},$$

the slope of the curve $v(x, y) = c$ can be written as

$$\frac{\Delta y}{\Delta x} \Big|_v = \frac{\partial u / \partial y}{\partial u / \partial x},$$

which, at any common point, is just the negative reciprocal of the slope of the curve $u(x, y) = k$. From the analytic geometry, we know that the two families of curves are orthogonal (perpendicular) to each other. For example, the real part of the analytic function z^2 is $u(x, y) = x^2 - y^2$, the family of curves of $u = k$ is the hyperbolas asymptotic to the line $y = \pm x$ as shown in the z-plane of Fig. 2.1. The imaginary part of z^2 is $v(x, y) = 2xy$, the family of curves of $v = c$ is the hyperbolas asymptotic to the x- and y-axes, also shown in the z-plane of Fig. 2.1. It is seen that they are indeed orthogonal to each other at the points of intersections.

These remarkable properties of analytic functions serve as basis for many important methods used in fluid dynamics, electrostatics, and other branches of physics. We will go into more details of solving specific problems in a later chapter.

Example 2.1.5 Let $f(z) = u(x, y) + iv(x, y)$ be an analytic function. If $u(x, y) = xy$, find $v(x, y)$ and $f(z)$

Solution 2.1.5
$$\frac{\partial u}{\partial x} = y = \frac{\partial v}{\partial y}, \qquad \frac{\partial u}{\partial y} = x = -\frac{\partial v}{\partial x}.$$

Method 1: Find $f(z)$ from its derivatives:

$$\frac{df}{dz} = \frac{\partial u}{\partial x} + i\frac{\partial v}{\partial x} = y - ix = -i(x + iy) = -iz$$

$$f(z) = -i\frac{1}{2}z^2 + C.$$

$$f(z) = -\frac{i}{2}(x + iy)^2 + C = xy - \frac{i}{2}(x^2 - y^2) + C,$$

$$v(x, y) = -\frac{1}{2}(x^2 - y^2) + C'.$$

Method 2: Find $v(x, y)$ first:

$$\frac{\partial v}{\partial y} = y \quad\Longrightarrow\quad v(x, y) = \int y\, dy = \frac{1}{2}y^2 + k(x),$$

$$\frac{\partial v}{\partial x} = -x \quad\Longrightarrow\quad \frac{\partial v}{\partial x} = \frac{dk(x)}{dx} = -x, \quad\Longrightarrow k(x) = -\frac{1}{2}x^2 + C;$$

$$v(x, y) = \frac{1}{2}y^2 - \frac{1}{2}x^2 + C.$$

$$f(z) = xy + i\frac{1}{2}\left(y^2 - x^2 + 2C\right)$$

$$x = \frac{1}{2}(z + z^*), \quad y = \frac{1}{2i}(z - z^*) \quad\Longrightarrow\quad f(z) = -\frac{1}{2}z^2 i + C'.$$

Example 2.1.6 Let $f(z) = u(x, y) + iv(x, y)$ be an analytic function. If $u(x, y) = \ln(x^2 + y^2)$, find $v(x, y)$ and $f(z)$

Solution 2.1.6

$$\frac{\partial u}{\partial x} = \frac{2x}{(x^2 + y^2)} = \frac{\partial v}{\partial y}, \quad \frac{\partial u}{\partial y} = \frac{2y}{(x^2 + y^2)} = -\frac{\partial v}{\partial x}.$$

Method 1: Find $f(z)$ first from its derivatives:

$$\frac{df}{dz} = \frac{\partial u}{\partial x} + i\frac{\partial v}{\partial x} = \frac{2x}{(x^2 + y^2)} - i\frac{2y}{(x^2 + y^2)}$$

$$= 2\frac{x - iy}{(x^2 + y^2)} = 2\frac{x - iy}{(x - iy)(x + iy)} = 2\frac{1}{x + iy} = \frac{2}{z}.$$

$$f(z) = 2\ln z + C = \ln z^2 + C.$$

$$z = re^{i\theta}; \quad r = (x^2 + y^2)^{1/2} \quad \theta = \tan^{-1}\frac{y}{x}$$

$$\ln z^2 = \ln(x^2 + y^2)e^{i2\theta} = \ln(x^2 + y^2) + i2\tan^{-1}\frac{y}{x}$$

$$v(x, y) = 2\tan^{-1}\frac{y}{x} + C.$$

Method 2: Find $v(x, y)$ first:

$$\frac{\partial v}{\partial y} = \frac{2x}{(x^2 + y^2)} \implies v(x, y) = \int \frac{2x}{(x^2 + y^2)} dy = 2\tan^{-1}\frac{y}{x} + k(x).$$

$$\frac{\partial v(x, y)}{\partial x} = 2\left(\frac{-y}{x^2}\right)\frac{1}{(1 + y^2/x^2)} + \frac{dk(x)}{dx} = \frac{-2y}{(x^2 + y^2)} + \frac{dk(x)}{dx}.$$

$$\frac{\partial v}{\partial x} = \frac{-2y}{(x^2 + y^2)} \implies \frac{dk(x)}{dx} = 0, \quad k(x) = C.$$

$$v(x, y) = 2\tan^{-1}\frac{y}{x} + C$$

$$f(z) = \ln(x^2 + y^2) + i2\tan^{-1}\frac{y}{x} + iC$$

$$= \ln(x^2 + y^2)e^{i2\theta} + iC$$

$$f(z) = \ln z^2 + C'.$$

Example 2.1.7 Let $f(z) = \frac{1}{z} = u(x, y) + iv(x, y)$, (a) show explicitly that the Cauchy-Riemann equations are satisfied; (b) show explicitly that both real part and imaginary part satisfy the Laplace equation; (c) Describe the family of curves $u(x, y) = k$ and $v(x, y) = c$ and sketch them; (d) show explicitly that the curves

$u(x, y) = k$ and $v(x, y) = c$ are perpendicular to each other at the points they intersect.

Solution 2.1.7

$$f(z) = \frac{1}{z} = \frac{1}{x + iy} = \frac{1}{x + iy} \cdot \frac{x - iy}{x - iy} = \frac{x - iy}{x^2 + y^2} = \frac{x}{x^2 + y^2} - i\frac{y}{x^2 + y^2}.$$

Therefore

$$u(x, y) = \frac{x}{x^2 + y^2}, \quad v(x, y) = -\frac{y}{x^2 + y^2}.$$

(a)

$$\frac{\partial u}{\partial x} = \frac{1}{x^2 + y^2} - \frac{2x^2}{(x^2 + y^2)^2} = \frac{y^2 - x^2}{(x^2 + y^2)^2},$$

$$\frac{\partial u}{\partial y} = \frac{-2xy}{(x^2 + y^2)^2},$$

$$\frac{\partial v}{\partial x} = \frac{2xy}{(x^2 + y^2)^2},$$

$$\frac{\partial v}{\partial y} = \frac{-1}{x^2 + y^2} + \frac{2y^2}{(x^2 + y^2)^2} = \frac{y^2 - x^2}{(x^2 + y^2)^2}.$$

Clearly the Cauchy-Riemann equations are satisfied

$$\frac{\partial u}{\partial x} = \frac{\partial v}{\partial y}, \quad \frac{\partial u}{\partial y} = -\frac{\partial v}{\partial x}.$$

(b)

$$\frac{\partial^2 u}{\partial x^2} = \frac{-2x}{(x^2 + y^2)^2} + \frac{(y^2 - x^2)(-2)(2x)}{(x^2 + y^2)^3} = \frac{2x^3 - 6xy^2}{(x^2 + y^2)^3},$$

$$\frac{\partial^2 u}{\partial y^2} = \frac{-2x}{(x^2 + y^2)^2} + \frac{-2xy(-2)(2y)}{(x^2 + y^2)^3} = \frac{-2x^3 + 6xy^2}{(x^2 + y^2)^3}.$$

Thus the real part satisfies the Laplace equation

$$\frac{\partial^2 u}{\partial x^2} + \frac{\partial^2 u}{\partial y^2} = 0.$$

Furthermore,

$$\frac{\partial^2 v}{\partial x^2} = \frac{2y}{(x^2 + y^2)^2} + \frac{(2xy)(-2)(2x)}{(x^2 + y^2)^3} = \frac{2y^3 - 6x^2 y}{(x^2 + y^2)^3},$$

$$\frac{\partial^2 v}{\partial y^2} = \frac{2y}{(x^2 + y^2)^2} + \frac{(-x^2 + y^2)(-2)(2y)}{(x^2 + y^2)^3} = \frac{-2y^3 + 6x^2 y}{(x^2 + y^2)^3}.$$

The imaginary part also satisfies the Laplace equation

$$\frac{\partial^2 v}{\partial x^2} + \frac{\partial^2 v}{\partial y^2} = 0.$$

(c) The equation

$$u(x, y) = \frac{x}{x^2 + y^2} = k$$

can be written as

$$x^2 + y^2 = \frac{x}{k},$$

or

$$\left(x - \frac{1}{2k}\right)^2 + y^2 = \frac{1}{4k^2},$$

which is a circle for any given constant k. Therefore $u(x, y) = k$ is a family of circles centered at $\left(\frac{1}{2k}, 0\right)$ with radius $\frac{1}{2k}$. This family of circles is shown in Fig. 2.5 as solid circles. Similarly

$$v(x, y) = -\frac{y}{x^2 + y^2} = c$$

can be written as

$$x^2 + y^2 = -\frac{y}{c}, \quad \text{or} \quad x^2 + \left(y + \frac{1}{2c}\right)^2 = \frac{1}{4c^2}.$$

Therefore $v(x, y) = c$ is a family of circles of radius $\frac{1}{2c}$, centered at $(0, -\frac{1}{2c})$. They are shown as the dotted circles in Fig. 2.5.

(d) On the curve represented by

$$u(x, y) = \frac{x}{x^2 + y^2} = k,$$

$$du = \frac{\partial u}{\partial x} dx + \frac{\partial u}{\partial y} dy = 0$$

which is given by

Fig. 2.5 The families of curves described by the real part and imaginary part of the function $f(z) = \dfrac{1}{z}$

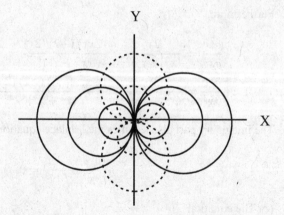

$$du = \left[\frac{1}{x^2 + y^2} - \frac{2x^2}{(x^2 + y^2)^2} \right] dx - \frac{2xy}{(x^2 + y^2)^2} dy = 0.$$

It follows that

$$\frac{dy}{dx} \Big|_u = \frac{y^2 - x^2}{2xy}.$$

Similarly, with

$$v(x, y) = -\frac{y}{x^2 + y^2} = c,$$

$$dv = \frac{2xy}{(x^2 + y^2)^2} dx - \left[\frac{1}{x^2 + y^2} - \frac{2y^2}{(x^2 + y^2)^2} \right] dy = 0,$$

and

$$\frac{dy}{dx} \Big|_v = \frac{2xy}{x^2 - y^2}.$$

Since the two slopes are negative reciprocals of each other, the two curves are perpendicular.

Those who are familiar with electrostatics will recognize that the curves in Fig. 2.5 are electric field lines and equipotential lines of a line dipole.

2.2 Complex Integration

There are some elegant and powerful theorems regarding integrating analytic functions around a loop. It is these theorems that make complex integrations interesting and useful. But before we discuss these theorems, we must define complex integration.

2.2.1 Line Integral of a Complex Function

When a complex variable z moves in the two-dimensional complex plane, it traces out a curve. Therefore to define a integral of complex function $f(z)$ between two points A and B, we must also specify the path (called contour) along which z moves. The value of the integral will be dependent, in general, upon the contour. However, we will find, that under certain conditions, the integral does not depend upon which of the contours is chosen.

We denote the integral of a complex function $f(z) = u(x, y) + iv(x, y)$ along a contour Γ from point A to point B as

$$I = \int_{A,\Gamma}^{B} f(z)\, dz.$$

The integral can be defined in terms of a Riemann sum as in the real variable integration. The contour is subdivided into n segments as shown in Fig. 2.6.

We form the summation

Fig. 2.6 The Riemann sum along the contour Γ which is subdivided into n segments

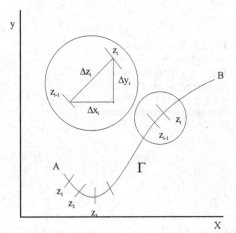

$$I_n = \sum_{i=1}^{n} f(\zeta_i)(z_i - z_{i-1}) = \sum_{i=1}^{n} f(\zeta_i)\,\Delta z_i,$$

where $z_0 = A$, $z_n = B$, and $f(\zeta_i)$ is the function evaluated at a point on Γ between z_{i-1} and z_i. If I_n approaches a limit as $n \to \infty$ and $|\Delta z_i| \to 0$, then we can define the integral as

$$\int_{A,\Gamma}^{B} f(z)\,dz = \lim_{|\Delta z_i| \to 0,\ n \to \infty} \sum_{i=1}^{n} f(z_i)\,\Delta z_i.$$

Since $\Delta z_i = \Delta x_i + i\,\Delta y_i$ as shown in Fig. 2.6, the integral can be written as

$$\int_{A,\Gamma}^{B} f(z)\,dz = \int_{A,\Gamma}^{B} (u+iv)(dx+idy) = \int_{A,\Gamma}^{B} [(udx - vdy) + i(vdx + udy)]$$

$$= \int_{A,\Gamma}^{B} (udx - vdy) + i \int_{A,\Gamma}^{B} (vdx + udy). \tag{2.9}$$

Thus the complex contour integral is expressed in terms of two line integrals.

Example 2.2.1 Evaluate the integral $I = \int_{A}^{B} z^2 dz$ from $z_A = 0$ to $z_B = 1+i$, (a) along the contour $\Gamma_1 : y = x^2$, (b) along y-axis from 0 to i, then along the horizontal line from i to $1+i$, as Γ_2 shown in Fig. 2.7.

Solution 2.2.1

$$f(z) = z^2 = (x+iy)^2 = (x^2 - y^2) + i2xy = u + iv.$$

Fig. 2.7 Two contours Γ_1 and Γ_2 from A ($z_A = 0$) to B ($z_B = 1+i$), Γ_1 : along the curve $y = x^2$, Γ_2 : first along y-axis to C ($z_C = i$), then along a horizontal line to B

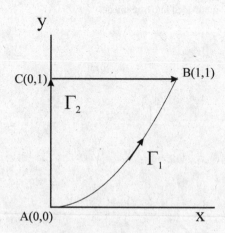

$$\int_{A,\Gamma}^{B} f(z)\,dz = \int_{A,\Gamma}^{B} [(x^2 - y^2)dx - 2xy\,dy] + i \int_{A,\Gamma}^{B} [2xy\,dx + (x^2 - y^2)dy].$$

(a) along Γ_1, $y = x^2$, $dy = 2x\,dx$,

$$\int_{A,\Gamma_1}^{B} f(z)\,dz = \int_0^1 [(x^2 - x^4)dx - 2xx^2 2x\,dx] + i \int_0^1 [2xx^2 dx + (x^2 - x^4)2x\,dx]$$

$$= \int_0^1 (x^2 - 5x^4)dx + i \int_0^1 (4x^3 - 2x^5)dx = -\frac{2}{3} + \frac{2}{3}i.$$

(b) Let $z_C = i$ as shown in Fig. 2.7. So

$$\int_{A,\Gamma_2}^{B} f(z)\,dz = \int_{A,\Gamma_2}^{C} f(z)\,dz + \int_{C,\Gamma_2}^{B} f(z)\,dz$$

From A to $C : x = 0$, $dx = 0$,

$$\int_{A,\Gamma_2}^{C} f(z)\,dz = i \int_0^1 (-y^2)dy = -\frac{1}{3}i.$$

From C to $B : y = 1$, $dy = 0$,

$$\int_{C,\Gamma_2}^{B} f(z)\,dz = \int_0^1 (x^2 - 1)dx + i \int_0^1 2x\,dx = -\frac{2}{3} + i.$$

$$\int_{A,\Gamma_2}^{B} f(z)\,dz = -\frac{1}{3}i - \frac{2}{3} + i = -\frac{2}{3} + \frac{2}{3}i.$$

The integrals along Γ_1 and Γ_2 are observed to be equal.

2.2.2 Parametric Form of Complex Line Integral

If along the contour Γ, z is expressed parametrically, these line integrals can be transformed into ordinary integrals in which there is only one independent variable. For if $z = z(t)$, where t is a parameter, and $A = z(t_A)$, $B = z(t_B)$, then

$$\int_A^B f(z)\,dz = \int_{t_A}^{t_B} f(z(t)) \frac{dz}{dt} dt. \qquad (2.10)$$

For instance, on Γ_1 of the previous example, $y = x^2$, we can set $z(t) = x(t) + iy(t)$ with $x(t) = t$ and $y(t) = t^2$. If follows that $\dfrac{dz}{dt} = 1 + i2t$, and

$$\int_{A,\Gamma_1}^{B} z^2 dz = \int_0^1 \left(t + it^2\right)^2 (1 + i2t)\, dt$$

$$= \int_0^1 [(t^2 - 5t^4) + i(4t^3 - 2t^5)] dt = -\frac{2}{3} + \frac{2}{3}i.$$

Similarly, on Γ_2 of the previous example, from A to C, we can set $z(t) = it$ with $0 \le t \le 1$, and $\dfrac{dz}{dt} = i$. From C to B, we can set $z(t) = (t-1) + i$ with $1 \le t \le 2$, and $\dfrac{dz}{dt} = 1$. Thus,

$$\int_{A,\Gamma_2}^{B} z^2 dz = \int_0^1 (it)^2 i\, dt + \int_1^2 (t - 1 + i)^2 \, dt$$

$$= -\frac{1}{3}i - \frac{2}{3} + i = -\frac{2}{3} + \frac{2}{3}i.$$

Parametrization of a Circular Contour A circular contour can be easily parameterized with the angular variable of the polar coordinates. This is of considerable importance because through the principle of deformation of contours, which we will soon see, other contour integrations can also be carried out by changing the contour into a circle.

Consider the integral $I = \oint_C f(z)\, dz$, where C is a circle of radius r centered at the origin. Clearly we can express z as

$$z(\theta) = r\cos\theta + ir\sin\theta = re^{i\theta},$$
$$\frac{dz}{d\theta} = -r\sin\theta + ir\cos\theta = ire^{i\theta}.$$

This means $dz = ire^{i\theta} d\theta$, so the integral becomes

$$I = \int_0^{2\pi} f\left(re^{i\theta}\right) ire^{i\theta} d\theta.$$

The following example will illustrate how this is done.

Example 2.2.2 Evaluate the integral $\oint_C z^n dz$, where n is an integer and C is a circle of radius r around the origin.

Solution 2.2.2

$$\oint_C z^n dz = \int_0^{2\pi} \left(re^{i\theta}\right)^n ire^{i\theta} d\theta = ir^{n+1} \int_0^{2\pi} e^{i(n+1)\theta} d\theta$$

For $n \neq -1$,

$$\int_0^{2\pi} e^{i(n+1)\theta} d\theta = \frac{1}{i(n+1)} \left[e^{i(n+1)\theta} \right]_0^{2\pi} = \frac{1}{i(n+1)} [1-1] = 0.$$

For $n = -1$,

$$\int_0^{2\pi} \left(re^{i\theta}\right)^n ire^{i\theta} d\theta = i \int_0^{2\pi} d\theta = 2\pi i.$$

This means

$$\oint_C z^n dz = 0, \quad for \ n \neq -1,$$

$$\oint_C \frac{dz}{z} = 2\pi i.$$

Note that these results are independent of the radius r.

Some Properties of Complex Line Integral The parametric form of the complex line integral enables us to see immediately that many formulas of ordinary integration of real variables can be directly applied to the complex integration. For example, the complex integral from B to A along the same path Γ is given by the right hand side of (2.10) with t_A and t_B interchanged, introducing a negative sign to the equation. Therefore

$$\int_{A,\Gamma}^B f(z)\, dz = - \int_{B,\Gamma}^A f(z)\, dz.$$

Similarly, if C is on Γ, then

$$\int_{A,\Gamma}^B f(z)\, dz = \int_{A,\Gamma}^C f(z)\, dz + \int_{C,\Gamma}^B f(z)\, dz.$$

If the integral from A to B is along Γ_1 and from B back to A is along a different contour Γ_2, we can write the sum of the two integrals as

$$\int_{A,\Gamma_1}^B f(z)\, dz + \int_{B,\Gamma_2}^A f(z)\, dz = \oint_\Gamma f(z)\, dz$$

where $\Gamma = \Gamma_1 + \Gamma_2$ and the symbol $\oint_\Gamma$ is to signify that the integration is taken counter clock wise along the closed contour Γ. Thus

$$\oint_{c.c.w.} f(z)\,dz = \int_{A,\Gamma_1}^{B} f(z)\,dz + \int_{B,\Gamma_2}^{A} f(z)\,dz$$

$$= -\int_{B,\Gamma_1}^{A} f(z)\,dz - \int_{A,\Gamma_2}^{B} f(z)\,dz = -\oint_{c.w.} f(z)\,dz,$$

where $c.c.w.$ means counter clock wise and $c.w.$ means clock wise.

Furthermore, we can show that

$$\left| \int_{A,\Gamma}^{B} f(z)\,dz \right| \le ML \tag{2.11}$$

where M is the maximum value of $|f(z)|$ on Γ and L is the length of Γ. This is because

$$\left| \int_{t_A}^{t_B} f(z)\frac{dz}{dt}\,dt \right| \le \int_{t_A}^{t_B} \left| f(z)\frac{dz}{dt} \right| dt$$

which is a generalization of $|z_1 + z_2| \le |z_1| + |z_2|$. By the definition of M, we have

$$\int_{t_A}^{t_B} \left| f(z)\frac{dz}{dt} \right| dt \le M \int_{t_A}^{t_B} \left| \frac{dz}{dt} \right| dt = M \int_{A}^{B} |dz| = ML.$$

Thus, starting with (2.10), we have

$$\left| \int_{A}^{B} f(z)\,dz \right| = \left| \int_{t_A}^{t_B} f(z)\frac{dz}{dt}\,dt \right| \le ML.$$

2.3 Cauchy's Integral Theorem

As we have seen, the results of integrations of z^2 along Γ_1 and Γ_2 of Fig. 2.7 are exactly the same. Therefore a closed-loop integration from A to B along Γ_1 and returning from B to A along Γ_2 is equal to zero. In 1825, Cauchy proved a theorem which enables us to see that this must be the case without carrying out the integration. Before we discuss this theorem, let us first review Green's lemma of real variables.

2.3.1 Green's Lemma

There is an important relation that allows us to transform a line integral into an area integral for lines and areas in the xy-plane. It is often referred to as Green's lemma, which states that

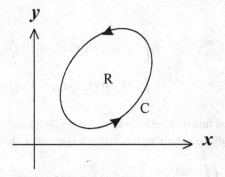

Fig. 2.8 The closed contour C of the line integral in Green's lemma. C is counter clock wise and is defined as the positive direction with respect to the interior of R

$$\oint_C [P(x,y)dx + Q(x,y)dy] = \iint_R \left[\frac{\partial Q(x,y)}{\partial x} - \frac{\partial P(x,y)}{\partial y} \right] dxdy, \quad (2.12)$$

where C is a closed curve surrounding the region R. The curve C is traversed counter clock wise, that is with the region R always to the left as shown in Fig. 2.8.

To prove Green's lemma, let us use Fig. 2.9, part (a) to carry out the first part of the area double integral

$$\iint_R \frac{\partial Q(x,y)}{\partial x} dxdy = \int_c^d \left[\int_{x=g_1(y)}^{x=g_2(y)} \frac{\partial Q(x,y)}{\partial x} dx \right] dy$$

$$= \int_c^d [Q(x,y)]_{x=g_1(y)}^{x=g_2(y)} dy.$$

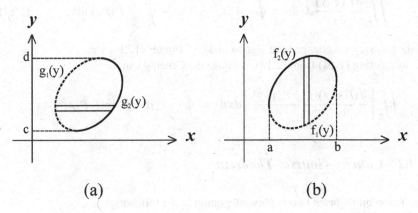

(a)   (b)

Fig. 2.9 Same contour but with two different ways to carry out the area double integral in Green's lemma

Now

$$\int_c^d [Q(x, y)]_{x=g_1(y)}^{x=g_2(y)} \, dy = \int_c^d Q(g_2(y), y) dy - \int_c^d Q(g_1(y), y) dy$$

$$= \int_c^d Q(g_2(y), y) dy + \int_d^c Q(g_1(y), y) dy$$

The contour of the last line integral is from $y = c$ going through $g_2(y)$ to $y = d$ and then returning through $g_1(y)$ to $y = c$. Clearly it is counter clock wise closed-loop integral

$$\iint_R \frac{\partial Q(x, y)}{\partial x} dx dy = \oint_{c.c.w} Q(x, y) \, dy. \tag{2.13}$$

Next we will use Fig. 2.9, part (b) to carry out the second part of the area double integral

$$\iint_R \frac{\partial P(x, y)}{\partial y} dx dy = \int_a^b \left[\int_{y=f_1(x)}^{y=f_2(x)} \frac{\partial P(x, y)}{\partial y} dy \right] dx = \int_a^b [P(x, y)]_{y=f_1(x)}^{y=f_2(x)} \, dx.$$

$$\int_a^b [P(x, y)]_{y=f_1(x)}^{y=f_2(x)} \, dx = \int_a^b P(x, f_2(x)) dx - \int_a^b P(x, f_1(x)) dx$$

$$= \int_a^b P(x, f_2(x)) dx + \int_b^a P(x, f_1(x)) dx.$$

In this case the contour is from $x = a$ going through $f_2(x)$ to $x = b$ and then returning to $x = a$ through $f_1(x)$. Therefore it is clockwise.

$$\iint_R \frac{\partial P(x, y)}{\partial y} dx dy = \oint_{c.w.} P(x, y) dx = - \oint_{c.c.w.} P(x, y) dx. \tag{2.14}$$

In the last step we changed the sign to make it counter clock wise.

Subtracting (2.14) from (2.13), we have the Green's lemma

$$\iint_R \left[\frac{\partial Q(x, y)}{\partial x} - \frac{\partial P(x, y)}{\partial y} \right] dx dy = \oint_{c.c.w.} [Q(x, y) dy + P(x, y) dx].$$

2.3.2 Cauchy-Goursat Theorem

An important theorem in complex integration is the following:

If C is a closed contour and $f(z)$ is analytic on and inside C, then

$$\oint_C f(z)dz = 0. \tag{2.15}$$

This is known as Cauchy's theorem. The proof goes as follows. Starting with

$$\oint_C f(z)dz = \oint_C (udx - vdy) + i\oint_C (vdx + udy), \tag{2.16}$$

making use of the Green's lemma of (2.12) and identifying P as u and Q as $-v$, we have

$$\oint_C (udx - vdy) = \int\int_R \left[-\frac{\partial v}{\partial x} - \frac{\partial u}{\partial y} \right] dxdy.$$

Since $f(z)$ is analytic, so u and v satisfy Cauchy-Riemann conditions. In particular

$$-\frac{\partial v}{\partial x} = \frac{\partial u}{\partial y},$$

therefore the area double integral is equal to zero, thus

$$\oint_C (udx - vdy) = 0.$$

Similarly, identifying u as Q and v as P, from Green's lemma we have

$$\oint_C (vdx + udy) = \int\int_R [\frac{\partial u}{\partial x} - \frac{\partial v}{\partial y}]dxdy.$$

Because of the other Cauch-Riemann condition

$$\frac{\partial u}{\partial x} = \frac{\partial v}{\partial y},$$

the integral on the left-hand side is also equal to zero

$$\oint_C (vdx + udy) = 0.$$

Thus both line integrals on the right-hand side of (2.16) are zero, therefore

$$\oint_C f(z)dz = 0,$$

which is known as Cauchy's integral theorem.

In this proof, we have used Green's lemma which requires the first partial derivatives of u and v to be continuous. Therefore we have implicitly assumed that the derivative of $f(z)$ is continuous. In 1903, Goursat proved this theorem without assuming the continuity of $f'(z)$. Therefore this theorem is also called Cauchy-Goursat theorem. Mathematically Goursat's removal of the continuity assumption from the proof of the theorem is very important because it enables us to rigorously establish that derivatives of analytic functions are analytic, and they are automatically continuous. A version of Goursat's proof can be found in *Complex Variables and Applications,* by J.W. Brown and R.V. Churchill, 6th Edition, McGraw-Hill, 1996.

2.3.3   *Fundamental Theorem of Calculus*

If the closed contour Γ is divided into two parts Γ_1 and Γ_2, as shown in Fig. 2.7, and $f(z)$ is analytic on and between Γ_1 and Γ_2, then Cauchy's integral theorem can be written as

$$\oint_\Gamma f(z)dz = \int_{A\,\Gamma_1}^B f(z)dz + \int_{B\,\Gamma_2}^A f(z)dz$$

$$= \int_{A\,\Gamma_1}^B f(z)dz - \int_{A\,\Gamma_2}^B f(z)dz = 0,$$

where the negative sign appears since we have exchanged the limit on the last integral. Thus we have

$$\int_{A\,\Gamma_1}^B f(z)dz = \int_{A\,\Gamma_2}^B f(z)dz, \tag{2.17}$$

showing that the value of a line integral between two points is independent of the path provided that the integrand is an analytic function in the domain on and between the contours.

With this in mind, we can show that, as long as $f(z)$ is analytic in a region containing A and B,

$$\int_A^B f(z)\,dz = F(B) - F(A),$$

where

$$\frac{dF(z)}{dz} = \lim_{\Delta z \to 0} \frac{F(z+\Delta z) - F(z)}{\Delta z} = f(z).$$

The integral

$$F(z) = \int_{z_0}^z f(z')\,dz' \tag{2.18}$$

uniquely define the function $F(z)$ if z_0 is a fixed point and $f(z')$ is analytic throughout the region containing the path between z_0 and z. Similarly, we can define

$$F\left(z+\Delta z\right) = \int_{z_0}^{z+\Delta z} f\left(z'\right) dz' = \int_{z_0}^{z} f\left(z'\right) dz' + \int_{z}^{z+\Delta z} f\left(z'\right) dz'.$$

Clearly,

$$F\left(z+\Delta z\right) - F\left(z\right) = \int_{z}^{z+\Delta z} f\left(z'\right) dz'.$$

For a small Δz, the right-hand side reduces to

$$\int_{z}^{z+\Delta z} f\left(z'\right) dz' \rightarrow f\left(z\right) \Delta z$$

which implies that

$$\frac{F\left(z+\Delta z\right) - F\left(z\right)}{\Delta z} = f\left(z\right).$$

Thus

$$\frac{dF\left(z\right)}{dz} = f\left(z\right),$$

and the fundamental theorem of calculus follows:

$$\int_{A}^{B} f\left(z\right) dz = \int_{A}^{B} dF\left(z\right) = F\left(B\right) - F\left(A\right).$$

Example 2.3.1 Find the value of the integral $\int_0^{1+i} z^2 dz$.

Solution 2.3.1

$$\int_0^{1+i} z^2 dz = \left[\frac{1}{3}z^3\right]_0^{1+i} = \frac{1}{3}\left(1+i\right)^3 = -\frac{2}{3} + \frac{2}{3}i.$$

Note that the result is the same as in example 2.2.1

Example 2.3.2 Find the values of the following integrals:

$$I_1 = \int_{-\pi i}^{\pi i} \cos z\, dz, \qquad I_2 = \int_{4+\pi i}^{4-3\pi i} e^{z/2} dz.$$

Solution 2.3.2

$$I_1 = \int_{-\pi i}^{\pi i} \cos z\, dz = [\sin z]_{-\pi i}^{\pi i} = \sin\left(\pi i\right) - \sin\left(-\pi i\right)$$

$$= 2\sin\left(\pi i\right) = 2\frac{1}{2i}\left(e^{i(i\pi)} - e^{-i(i\pi)}\right) = \left(e^{\pi} - e^{-\pi}\right) i \simeq 23.097i.$$

$$I_2 = \int_{4+\pi i}^{4-3\pi i} e^{z/2}dz = \left[2e^{z/2}\right]_{4+\pi i}^{4-3\pi i} = 2\left(e^{2-i3\pi/2} - e^{2+i\pi/2}\right)$$
$$= 2e^2\left(e^{-i3\pi/2} - e^{i\pi/2}\right) = 2e^2(i - i) = 0.$$

Example 2.3.3 Find the values of the following integral:

$$\int_{-i}^{i} \frac{dz}{z}.$$

Solution 2.3.3 Since $z = 0$ is a singular point, the path of integration must not pass through the origin. Furthermore

$$\int \frac{dz}{z} = \ln z + C,$$

where $\ln z$ is a multi-valued function, therefore there is a branch cut. To evaluate this definite integral we must specify the path of z going from $-i$ to i. There are two possibilities as shown in (a) and (b) of the following figure:

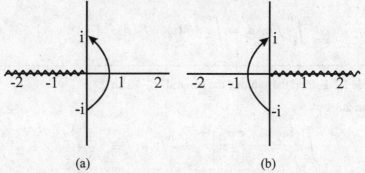

(a)   (b)

(a) To go from $-i$ to i in the right half of the complex plane, we must take the negative real axis as the branch cut. In the principal branch, $-\pi < \theta < \pi$. Thus

$$\int_{-i}^{i} \frac{dz}{z} = [\ln z]_{-i}^{i} = \left[\ln e^{i\theta}\right]_{\theta=-\frac{1}{2}\pi}^{\theta=\frac{1}{2}\pi} = [i\theta]_{\theta=-\frac{1}{2}\pi}^{\theta=\frac{1}{2}\pi} = i\frac{1}{2}\pi + i\frac{1}{2}\pi = i\pi.$$

(b) To go from $-i$ to i in the left half of the complex plane, we must take the positive real axis as the branch cut. Therefore $0 < \theta < 2\pi$. Thus

$$\int_{-i}^{i} \frac{dz}{z} = [\ln z]_{-i}^{i} = \left[\ln e^{i\theta}\right]_{\theta=\frac{3}{2}\pi}^{\theta=\frac{1}{2}\pi} = [i\theta]_{\theta=\frac{3}{2}\pi}^{\theta=\frac{1}{2}\pi} = i\frac{1}{2}\pi - i\frac{3}{2}\pi = -i\pi.$$

2.4 Consequences of Cauchy's Theorem

2.4.1 Principle of Deformation of Contours

There is an immediate, practical consequence of the Cauchy Integral Theorem. The contour of a complex integral can be arbitrarily deformed through an analytic region without changing the value of the integral.

Consider the integration along the two contours shown on the left side of Fig. 2.10. If $f(z)$ is analytic, then

$$\oint_{abcda} f(z)dz = 0$$

$$\oint_{efghe} f(z)dz = 0.$$

Naturally the sum of them is also equal to zero

$$\oint_{abcda} f(z)dz + \oint_{efghe} f(z)dz = 0.$$

Notice that the integrals along ab and along he are in the opposite direction. If ab coincides with he, their contributions will cancel each other. Thus if the gaps between ab and he, and between cd and fg are shrinking to zero, the sum of these two integrals becomes the sum of the integral along the outer contour C_1 and the integral along the inner contour C_2 but in the opposite direction. If we change the direction of C_2, we must change the sign of the integral. Therefore

$$\oint_{abcda} f(z)dz + \oint_{efghe} f(z)dz = \oint_{C_1} f(z)dz - \oint_{C_2} f(z)dz = 0.$$

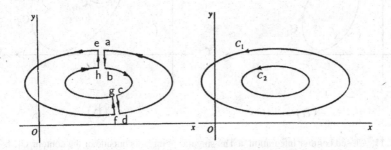

Fig. 2.10 Contour deformation

It follows

$$\oint_{C_1} f(z)dz = \oint_{C_2} f(z)dz.$$

Thus we have shown that the line integral of an analytic function around any closed curve C_1 is equal to the line integral of the same function around any other closed curve C_2 into which C_1 can be continuously deformed as long as $f(z)$ is analytic between C_1 and C_2 and is single-valued on C_1 and C_2.

2.4.2 The Cauchy Integral Formula

The Cauchy integral formula is a natural extension of the Cauchy integral theorem. Consider the integral

$$I_1 = \oint_{C_1} \frac{f(z)}{z - z_0} dz,$$

where $f(z)$ is analytic everywhere in the z-plane, and C_1 is a closed contour that does not include the point z_0 as shown in Fig. 2.11a.

Since $(z - z_0)^{-1}$ is analytic everywhere except at $z = z_0$, and z_0 is outside of C_1, therefore $f(z)/(z - z_0)$ is analytic inside C_1. It follows from Cauchy's integral theorem that

$$I_1 = \oint_{C_1} \frac{f(z)}{z - z_0} dz = 0.$$

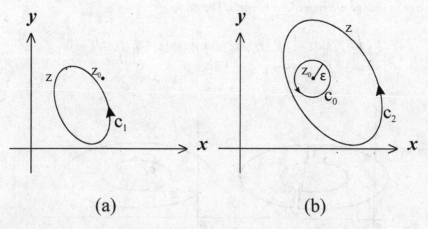

(a)   (b)

Fig. 2.11 Closed contour integration. **a** The singular point z_0 is outside of the contour C_1. **b** The contour C_2 encloses z_0, C_2 can be deformed into the circle C_0 without changing the value of the integral

Now consider a second integral

$$I_2 = \oint_{C_2} \frac{f(z)}{z - z_0} dz,$$

similar to the first, except now the contour C_2 encloses z_0, as shown in Fig. 2.11b. The integrand in this integral is not analytic at $z = z_0$ which is inside C_2, so we cannot invoke the Cauchy integral theorem to argue that $I_2 = 0$. However, the integrand is analytic everywhere, except at the point $z = z_0$, so we can deform the contour into an infinitesimal circle of radius ε centered at z_0, without changing its value

$$I_2 = \lim_{\varepsilon \to 0} \oint_{C_0} \frac{f(z)}{z - z_0} dz.$$

This deformation is also shown in Fig. 2.11b.

This last integral can be evaluated. In order to see more clearly, we enlarge the contour in Fig. 2.12.

Since z is on the circle C_0, with the notation shown in Fig. 2.12, it is clear that

$$z = x + iy$$
$$x = x_0 + \varepsilon \cos\theta$$
$$y = y_0 + \varepsilon \sin\theta.$$

Therefore

$$z = (x_0 + iy_0) + \varepsilon(\cos\theta + i \sin\theta). \tag{2.19}$$

Fig. 2.12 Circular contour for the Cauchy integral formula

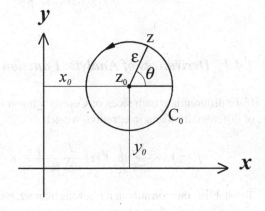

Since

$$z_0 = x_0 + iy_0,$$
$$e^{i\theta} = \cos\theta + i\sin\theta,$$

we can write

$$z = z_0 + \varepsilon e^{i\theta}. \tag{2.20}$$

On C_0, ε is a constant, and θ goes from 0 to 2π. Therefore

$$dz = i\varepsilon e^{i\theta}d\theta. \tag{2.21}$$

and

$$\oint_{C_0} \frac{f(z)}{z - z_0}dz = \int_0^{2\pi} \frac{f\left(z_0 + \varepsilon e^{i\theta}\right)}{\varepsilon e^{i\theta}} i\varepsilon e^{i\theta}d\theta = i\int_0^{2\pi} f\left(z_0 + \varepsilon e^{i\theta}\right)d\theta. \tag{2.22}$$

As $\varepsilon \to 0$, $f\left(z_0 + \varepsilon e^{i\theta}\right) \to f(z_0)$ and can be taken outside the integral:

$$I_2 = \oint_{C_2} \frac{f(z)}{z - z_0}dz = \lim_{\varepsilon \to 0} \oint_{C_0} \frac{f(z)}{z - z_0}dz$$
$$= \lim_{\varepsilon \to 0} i\int_0^{2\pi} f\left(z_0 + \varepsilon e^{i\theta}\right)d\theta = if(z_0)\int_0^{2\pi} d\theta = 2\pi if(z_0). \tag{2.23}$$

where C is any closed, counterclockwise path that encloses z_0, and $f(z)$ is analytic inside C. This result is known as Cauchy's integral formula, usually written as

$$f(z_0) = \frac{1}{2\pi i} \oint_C \frac{f(z)}{z - z_0}dz. \tag{2.24}$$

2.4.3 Derivatives of Analytic Function

If we differentiate both sides of Cauchy's integral formula, interchanging the order of differentiation and integration, we get

$$f'(z_0) = \frac{1}{2\pi i} \oint_C f(z) \frac{d}{dz_0} \frac{1}{(z - z_0)}dz = \frac{1}{2\pi i} \oint_C \frac{f(z)}{(z - z_0)^2}dz.$$

To establish this formula in a rigorous manner, we may start with the formal expression of the derivative

$$f'(z_0) = \lim_{\Delta z_0 \to 0} \frac{f(z_0 + \Delta z_0) - f(z_0)}{\Delta z_0} = \lim_{\Delta z_0 \to 0} \frac{1}{\Delta z_0} [f(z_0 + \Delta z_0) - f(z_0)]$$

$$= \lim_{\Delta z_0 \to 0} \frac{1}{\Delta z_0} \left[\frac{1}{2\pi i} \oint_C \frac{f(z)}{z - z_0 - \Delta z_0} dz - \frac{1}{2\pi i} \oint_C \frac{f(z)}{z - z_0} dz \right].$$

Now

$$\oint_C \frac{f(z)}{z - z_0 - \Delta z_0} dz - \oint_C \frac{f(z)}{z - z_0} dz = \oint_C f(z) \left(\frac{1}{z - z_0 - \Delta z_0} - \frac{1}{z - z_0} \right) dz$$

$$= \oint_C f(z) \frac{\Delta z_0}{(z - z_0 - \Delta z_0)(z - z_0)} dz = \Delta z_0 \oint_C \frac{f(z) dz}{(z - z_0 - \Delta z_0)(z - z_0)}.$$

Therefore

$$f'(z_0) = \lim_{\Delta z_0 \to 0} \frac{1}{\Delta z_0} \left[\frac{\Delta z_0}{2\pi i} \oint_C \frac{f(z) dz}{(z - z_0 - \Delta z_0)(z - z_0)} \right]$$

$$= \lim_{\Delta z_0 \to 0} \frac{1}{2\pi i} \oint_C \frac{f(z) dz}{(z - z_0 - \Delta z_0)(z - z_0)} = \frac{1}{2\pi i} \oint_C \frac{f(z)}{(z - z_0)^2} dz.$$

In a like manner we can show that

$$f''(z_0) = \frac{2}{2\pi i} \oint_C \frac{f(z)}{(z - z_0)^3} dz, \tag{2.25}$$

and in general

$$f^{(n)}(z_0) = \frac{n!}{2\pi i} \oint_C \frac{f(z)}{(z - z_0)^{n+1}} dz. \tag{2.26}$$

Thus we have established the fact that analytic functions possess derivatives of all orders. Also, all derivatives of analytic functions are analytic. This is quite different from our experience with real variables, where we have encountered functions that possess first and second derivatives at a particular point, but yet the third derivative is not defined.

Cauchy's integral formula allows us to determine the value of an analytic function at any point z interior to a simply connected region by integrating around a curve C surrounding the region. Only values of the function on the boundary are used. Thus, we note that if an analytic function is prescribed on the entire boundary of a simply connected region, the function and all its derivatives can be determined at all interior points. Cauchy's integral formula can be written in the form of

$$f(z) = \frac{1}{2\pi i} \oint_C \frac{f(\varsigma)}{\varsigma - z} d\varsigma \tag{2.27}$$

where z is any interior point inside C. The complex variable ς is on C and is simply a dummy variable of integration that disappears in the integration process. Cauchy's integral formula is often used in this form.

Example 2.4.1 Evaluate the integrals

$$(a) \oint \frac{z^2 \sin \pi z}{z - \frac{1}{2}} dz, \quad (b) \oint \frac{\cos z}{z^3} dz$$

around the circle $|z| = 1$.

Solution 2.4.1 (a) The singular point is at $z = \frac{1}{2}$ which is inside the circle $|z| = 1$. Therefore

$$\oint \frac{z^2 \sin \pi z}{z - \frac{1}{2}} dz = 2\pi i \left[z^2 \sin \pi z \right]_{z=1/2} = 2\pi i \left(\frac{1}{2} \right)^2 \sin \left(\pi \frac{1}{2} \right) = \frac{1}{2}\pi i.$$

(b) The singular point is at $z = 0$ which is inside the circle $|z| = 1$. Therefore

$$\oint \frac{\cos z}{z^3} dz = \frac{2\pi i}{2!} \left[\frac{d^2}{dz^2} \cos z \right]_{z=0} = \pi i [-\cos(0)] = -\pi i.$$

Example 2.4.2 Evaluate the integral

$$\oint \frac{z^2 - 1}{(z - 2)^2} dz$$

around (a) the circle $|z| = 1$, (b) the circle $|z| = 3$.

Solution 2.4.2 (a) The singular point is at $z = 2$. It is outside the circle as shown in Fig. 2.13a. Inside the circle $|z| = 1$, the function $\frac{z^2-1}{(z-2)^2}$ is analytic, therefore

$$\oint \frac{z^2 - 1}{(z - 2)^2} dz = 0.$$

(b) Since $z = 2$ is inside the circle $|z| = 3$, as shown in Fig. 2.13b we can write the integral as

$$\oint \frac{z^2 - 1}{(z - 2)^2} dz = \oint \frac{f(z)}{(z - 2)^2} dz = 2\pi i f'(2),$$

where

$$f(z) = z^2 - 1, \quad f'(z) = 2z, \quad and \quad f'(2) = 4.$$

Thus

$$\oint \frac{z^2 - 1}{(z - 2)^2} dz = 2\pi i 4 = 8\pi i.$$

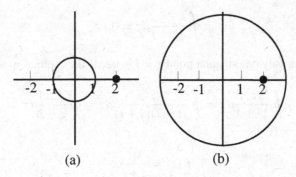

Fig. 2.13 a $|z| = 1$, b $|z| = 3$

Example 2.4.3 Evaluate the integral

$$\oint \frac{z^2}{z^2 + 1} dz$$

(*a*) around the circle $|z - 1| = 1$, (*b*) around the circle $|z - i| = 1$, (*c*) around the circle $|z - 1| = 2$ (Fig. 2.14).

Solution 2.4.3 Unless the relationship between the singular points and the contour is clear as in previous examples, to solve problems of closed contour integration, it is best to first find the singular points (known as poles) and display them on the complex plane, then draw the contour. In this particular problem, the singular points are at $z = \pm i$, which are the solutions of $z^2 + 1 = 0$. The three contours are shown in the following figure (Fig. 2.14).

(*a*) It is seen that both singular points are outside of the contour $|z - 1| = 1$, therefore

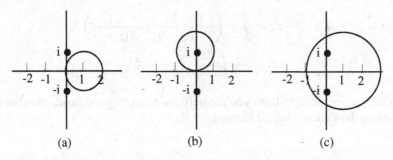

Fig. 2.14 a $|z - 1| = 1$, b $|z - i| = 1$, c $|z - 1| = 2$

$$\oint \frac{z^2}{z^2+1} dz = 0.$$

(b) In this case, only one singular point $z = i$ is inside the contour, so we can write

$$\oint \frac{z^2}{z^2+1} dz = \oint \frac{z^2}{(z-i)(z+i)} dz = \oint \frac{f(z)}{z-i} dz$$

where

$$f(z) = \frac{z^2}{z+i}.$$

Thus, it follows

$$\oint \frac{z^2}{z^2+1} dz = \oint \frac{f(z)}{z-i} dz = 2\pi i f(i) = 2\pi i \frac{(i)^2}{i+i} = -\pi.$$

(c) In this case, both singular points are inside the contour. To make use of the Cauchy integral formula, we first take the partial fraction of $\frac{1}{z^2+1}$,

$$\frac{1}{z^2+1} = \frac{1}{(z-i)(z+i)} = \frac{A}{(z-i)} + \frac{B}{(z+i)}$$
$$= \frac{A(z+i) + B(z-i)}{(z-i)(z+i)} = \frac{(A+B)z + (A-B)i}{(z-i)(z+i)}.$$

So

$$A + B = 0, \quad (A-B)i = 1,$$
$$B = -A, \quad 2Ai = 1,$$
$$A = \frac{1}{2i} = -\frac{i}{2}, \quad B = \frac{i}{2}.$$

It follows that

$$\oint \frac{z^2}{z^2+1} dz = \oint z^2 \left(-\frac{i}{2} \frac{1}{(z-i)} + \frac{i}{2} \frac{1}{(z+i)} \right) dz$$
$$= -\frac{i}{2} \oint \frac{z^2}{z-i} dz + \frac{i}{2} \oint \frac{z^2}{z+i} dz.$$

Each integral on the right-hand side has only one singular point inside the contour. According the Cauchy integral formula,

$$\oint \frac{z^2}{z-i} dz = 2\pi i \, (i)^2 = -2\pi i,$$

$$\oint \frac{z^2}{z+i} dz = 2\pi i \, (-i)^2 = -2\pi i.$$

Therefore

$$\oint \frac{z^2}{z^2+1} dz = -\frac{i}{2}(-2\pi i) + \frac{i}{2}(-2\pi i) = 0.$$

Example 2.4.4 Evaluate the integral

$$\oint \frac{z-1}{2z^2+3z-2} dz$$

around the square whose vertices are $(1, 1)$, $(-1, 1)$, $(-1, -1)$, $(1, -1)$.

Solution 2.4.4 To find the singular points, we set the denominator to zero

$$2z^2 + 3z - 2 = 0,$$

which gives the singular points at

$$z = \frac{1}{4}\left(-3 \pm \sqrt{9+16}\right) = \left\{ \begin{matrix} \frac{1}{2} \\ -2 \end{matrix} \right. .$$

The denominator can be written as

$$2z^2 + 3z - 2 = 2(z - \frac{1}{2})(z+2).$$

The singular points and the contour are shown in the following figure:

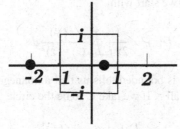

Since only the singular point at $z = \frac{1}{2}$ is inside the contour, we can write the integral as

$$\oint \frac{z-1}{2z^2+3z-2} dz = \oint \frac{z-1}{2(z-\frac{1}{2})(z+2)} dz = \oint \frac{f(z)}{(z-\frac{1}{2})} dz = 2\pi i f\left(\frac{1}{2}\right)$$

where

$$f(z) = \frac{z - 1}{2(z + 2)}, \quad f\left(\frac{1}{2}\right) = -\frac{1}{10}.$$

Therefore

$$\oint \frac{z - 1}{2z^2 + 3z - 2} dz = -\frac{1}{5}\pi i.$$

Several important theorems can be easily proved by Cauchy's integral formula and its derivatives.

Gauss' Mean Value Theorem If $f(z)$ is analytic inside and on a circle C with center at z_0, then the mean value of $f(z)$ on C is $f(z_0)$.

This theorem follows directly from the Cauchy's integral formula

$$f(z_0) = \frac{1}{2\pi i} \oint_C \frac{f(z)}{z - z_0} dz.$$

Let the circle C be $|z - z_0| = r$, thus

$$z = z_0 + re^{i\theta}, \quad and \quad dz = ire^{i\theta} d\theta.$$

Therefore

$$f(z_0) = \frac{1}{2\pi i} \oint_C \frac{f(z)}{z - z_0} dz = \frac{1}{2\pi i} \oint_C \frac{f(z_0 + re^{i\theta})}{re^{i\theta}} ire^{i\theta} d\theta$$

$$= \frac{1}{2\pi} \oint_C f(z_0 + re^{i\theta}) d\theta,$$

which is the mean value of $f(z)$ on C.

Liouville's Theorem If $f(z)$ is analytic in the entire complex plane and $|f(z)|$ is bounded for all values of z, then $f(z)$ is a constant.

To prove this theorem, we start with

$$f'(z) = \frac{1}{2\pi i} \oint_C \frac{f(z')}{(z' - z)^2} dz'.$$

The condition that $|f(z)|$ is bounded tells us that a nonnegative constant M exists such that $|f(z)| \leq M$ for all z. If we take C to be the circle $|z' - z| = R$, then

$$|f'(z)| \leq \left|\frac{1}{2\pi i}\right| \oint_C \frac{|f(z')|}{|(z' - z)^2|} |dz'|$$

$$\leq \frac{1}{2\pi} \frac{1}{R^2} M 2\pi R = \frac{M}{R},$$

Since $f(z')$ is analytic everywhere, we may take R as large as we like. It is clear that $\dfrac{M}{R} \to 0$, as $R \to \infty$. Therefore $|f'(z)| = 0$, which implies that $f'(z) = 0$ for all z, so $f(z)$ is a constant.

Fundamental Theorem of Algebra The following theorem is now known as the fundamental theorem of algebra. In the last chapter we mentioned that this theorem is of critical importance in our number system.

Every polynomial equation

$$P_n(z) = a_0 + a_1 z + \cdots a_n z^n = 0$$

of degree one or greater has at least one root.

To prove this theorm, let us first assume the contrary, namely that $P_n(z) \neq 0$ for any z. Then the function

$$f(z) = \frac{1}{P_n(z)}$$

is analytic everywhere. Since nowhere will $f(z)$ go to infinity and $f(z) \to 0$ as $z \to \infty$, so $|f(z)|$ is bounded for all z. By Liouville's theorem we conclude that $f(z)$ must be a constant, and hence $P_n(z)$ must be a constant. This is a contradiction, since $P_n(z)$ is given as a polynomial of z. Therefore, $P_n(z) = 0$ must have at least one root.

It follows from this theorem that $P_n(z) = 0$ has exactly n roots. Since $P_n(z) = 0$ has at least one root, let us denote that root z_1. Thus

$$P_n(z) = (z - z_1)Q_{n-1}(z)$$

where $Q_{n-1}(z)$ is a polynomial of degree $n - 1$. By the same argument, we conclude that $Q_{n-1}(z)$ must have at least one root, which we denote it as z_2. Repeating this procedure n times we find

$$P_n(z) = (z - z_1)(z - z_2) \cdots (z - z_n) = 0.$$

Hence $P_n(z) = 0$ has exactly n roots.

Exercises

1. Show that the real and the imaginary parts of the following functions $f(z)$ satisfy the Cauchy-Reimann conditions

(a) z^2, (b) e^z, (c) $\dfrac{1}{z+2}$.

2. Show that both the real part $u(x, y)$ and the imaginary part $v(x, y)$ of the analytic function $e^z = u(x, y) + iv(x, y)$ satisfy the Laplace equation

$$\frac{\partial^2 \phi}{\partial x^2} + \frac{\partial^2 \phi}{\partial y^2} = 0.$$

3. Show that the derivative of $\dfrac{1}{z+2}$ calculated in the following three different ways gives the same result.

(a) Let $\Delta y = 0$, so that $\Delta z \to 0$ parallel to the x-axis. In this case

$$f'(z) = \frac{\partial u}{\partial x} + i \frac{\partial v}{\partial x}.$$

(b) Let $\Delta x = 0$, so that $\Delta z \to 0$ parallel to the y-axis. In this case

$$f'(z) = \frac{\partial u}{i \partial y} + \frac{\partial v}{\partial y}.$$

(c) Use the same rule as if z were a real variable. That is

$$f'(z) = \frac{df}{dz}.$$

4. Let $z^2 = u(x, y) + iv(x, y)$, find the point of intersection of $u(x, y) = 1$ and $v(x, y) = 2$. Show that at the point of intersection the curve $u(x, y) = 1$ is perpendicular to $v(x, y) = 2$.

5. Let $f(z) = u(x, y) + iv(x, y)$ be an analytic function. If $u(x, y)$ is given by the following function

(a) $x^2 - y^2$;   (b) $e^y \sin x$,

show that they satisfy the Laplace equation. Find the corresponding conjugate harmonic function $v(x, y)$. Express $f(z)$ as a function of z only.

Ans. (a) $v(x, y) = 2xy + c$, $f(z) = z^2 + c$. (b) $v(x, y) = e^y \cos x + c$, $f(z)$ $= ie^{-iz} + c$.

6. In which quadrants of the complex plane is $f(z) = |x| - i\,|y|$ an analytic function?

Hint: In first quadrant, $x > 0$, so $\frac{\partial u}{\partial x} = \frac{\partial |x|}{\partial x} = \frac{\partial x}{\partial x} = 1$, in the second quadrant, $x < 0$, so $\frac{\partial u}{\partial x} = \frac{\partial |x|}{\partial x} = \frac{\partial(-x)}{\partial x} = -1$, and so on.

Ans. $f(z)$ is analytic only in the second and fourth quadrants.

7. Express the real part and the imaginary part of $(z+1)^2$ in terms of polar coordinates, that is, find $u(r, \theta)$ and $v(r, \theta)$ in the expression

$$(z+1)^2 = u(r, \theta) + iv(r, \theta).$$

Show that they satisfy the Cauchy-Riemann equations in the polar form:

$$\frac{\partial u\,(r,\theta)}{\partial r} = \frac{1}{r}\frac{\partial v\,(r,\theta)}{\partial \theta}, \qquad \frac{1}{r}\frac{\partial u\,(r,\theta)}{\partial \theta} = -\frac{\partial v\,(r,\theta)}{\partial r}.$$

8. Show that when an analytic function is expressed in terms of polar coordinates, both its real part and its imaginary part satisfy Laplace's equation in polar coordinates

$$\frac{\partial^2 \phi}{\partial r^2} + \frac{1}{r}\frac{\partial \phi}{\partial r} + \frac{1}{r^2}\frac{\partial^2 \phi}{\partial \theta^2} = 0.$$

9. To show that line integral are, in general, dependent on the path of integration, evaluate

$$\int_{-1}^{i} |z|^2\, dz$$

(a) along the straight line from the initial point -1 to the final point i, (b) along the arc of the unit circle $|z| = 1$ traversed in the clockwise direction from the initial point -1 to the final point i.

Hint: (a) Parameterize the line segment by $z = -1 + (1+i)t,\ 0 \le t \le 1$. (b) Parameterize the arc by $z = e^{i\theta},\ \pi \ge \theta \ge \pi/2$.

Ans. (a) $2(1+i)/3$, (b) $1+i$.

10. To verify that the line integral of an analytic function is independent of the path, evaluate

$$\int_0^{3+i} z^2\, dz$$

(a) along the line $y = x/3$, (b) along the real axis to 3 and then vertically to $3+i$, (c) along the imaginary axis to i and then horizontally to $3+i$.

Ans. (a) $6 + \frac{26}{3}i$, (b) $6 + \frac{26}{3}i$, (c) $6 + \frac{26}{3}i$.

11. Verify Green's lemma

$$\oint [A(x,y)dx + B(x,y)dy] = \iint_R \left[\frac{\partial B(x,y)}{\partial x} - \frac{\partial A(x,y)}{\partial y}\right] dxdy,$$

for the integral

$$\oint [(x^2 + y)dx - xy^2 dy]$$

taken around the boundary of the square with vertices at $(0,0)$, $(1,0)$, $(0,1)$, $(1,1)$.

12. Verify Green's lemma for the integral

$$\oint [(x - y)dx + (x + y)dy]$$

taken around the boundary of the area in the first quadrant between the curve $y = x^2$ and $y^2 = x$.

13. Evaluate

$$\int_0^{3+i} z^2 dz$$

with fundamental theorem of calculus. That is,

$$if \quad \frac{dF(z)}{dz} = f(z), \quad then \quad \int_A^B f(z)\,dz = F(B) - F(A),$$

provided $f(z)$ is analytic in a region between A and B.

Ans. $6 + \frac{26}{3}i$.

14. What is the value of

$$\oint_C \frac{3z^2 + 7z + 1}{z + 1}\,dz$$

(a) if C is the circle $|z + 1| = 1$? (b) if C is the circle $|z + i| = 1$? (c) if C is the ellipse $x^2 + 2y^2 = 8$?

Ans. (a) $-6\pi i$, (b) 0, (c) $-6\pi i$.

15. What is the value of

$$\oint_C \frac{z + 4}{z^2 + 2z + 5}\,dz$$

(a) if C is the circle $|z| = 1$? (b) if C is the circle $|z + 1 - i| = 2$? (c) if C is the circle $|z + 1 + i| = 2$?

Ans. (a) 0,   (b) $\frac{1}{2}(3 + 2i)\pi$,   (c) $\frac{1}{2}(-3 + 2i)\pi$.

16. What is the value of

$$\oint_C \frac{e^z}{(z + 1)^2}\,dz$$

around the circle $|z - 1| = 3$?

Ans. $2\pi i e^{-1}$.

17. What is the value of

$$\oint_C \frac{z + 1}{z^3 - 2z^2}\,dz$$

(a) If C is the circle $|z| = 1$? (b) If C is the circle $|z - 2 - i| = 2$? (c) If C is the circle $|z - 1 - 2i| = 2$?

Ans. (a) $-\frac{3}{2}\pi i$,   (b) $\frac{3}{2}\pi i$,   (c) 0.

18. Find the value of the closed-loop integral

$$\oint \frac{z^3 + \sin z}{(z - i)^3}\,dz$$

taken around the boundary of the triangle with vertices at ± 2, $2i$.

Ans. $\pi\left(e - e^{-1}\right)/2 - 6\pi$.

19. What is the value of

$$\oint_C \frac{\tan z}{z^2} dz$$

if C is the circle $|z| = 1$?
   Ans. $2\pi i$.

20. What is the value of

$$\oint_C \frac{\ln z}{(z-2)^2} dz$$

if C is the circle $|z - 3| = 2$?
   Ans. πi.

Chapter 3
Complex Series and Theory of Residues

Series expansions are ubiquitous in science and engineering. In the theory of complex functions, series expansions play a crucial role because they are the basis for deriving and using the theory of residues, which provide a powerful method for calculating both complex contour integrals and some difficult integrals of real variable. Before the formal development, we will first review a basic geometric series.

3.1 A Basic Geometric Series

Let

$$S = 1 + z + z^2 + z^3 + \cdots\cdots + z^n. \tag{3.1}$$

Multiplying by z,

$$zS = z + z^2 + z^3 + \cdots\cdots + z^n + z^{n+1},$$

and subtracting the two series

$$(1 - z)S = 1 - z^{n+1},$$

we get

$$S = \frac{1 - z^{n+1}}{1 - z}.$$

Now if $|z| < 1$, $z^{n+1} \to 0$ as $n \to \infty$. Thus, if n goes to infinity,

$$S = \frac{1}{1 - z},$$

© The Author(s), under exclusive license to Springer Nature Switzerland AG 2022
K.-T. Tang, *Mathematical Methods for Engineers and Scientists 1*,
https://doi.org/10.1007/978-3-031-05678-9_3

and it follows from (3.1) that

$$\frac{1}{1-z} = 1 + z + z^2 + z^3 + \cdots = \sum_{k=0}^{\infty} z^k. \tag{3.2}$$

Clearly, it will diverge for $|z| \geq 1$. It is important to remember that this series converges only for $|z| < 1$. Under this condition, the following alternative series is also convergent,

$$\frac{1}{1+z} = \frac{1}{1-(-z)} = \sum_{k=0}^{\infty} (-z)^k = 1 - z + z^2 - z^3 + \cdots. \tag{3.3}$$

3.2   Taylor Series

Taylor series is perhaps the most familiar series in real variables. Taylor series in complex variables is even more interesting.

3.2.1   The Complex Taylor Series

In many applications of complex variables, we wish to expand an analytic function $f(z)$ into a series around a particular point $z = z_0$. We will show that if $f(z)$ is analytic in the neighborhood of z_0 including the point at $z = z_0$, then $f(z)$ can be represented as a series of positive powers of $(z - z_0)$.

First, let us recall

$$f(z) = \frac{1}{2\pi i} \oint_C \frac{f(t)}{t-z} dt, \tag{3.4}$$

where t is the integration variable and it is on the enclosed contour C, inside which $f(z)$ is analytic. The quantity $(z - z_0)$ can be introduced into the integral through the identity

$$\frac{1}{t-z} = \frac{1}{(t-z_0)+(z_0-z)} = \frac{1}{(t-z_0)\left(1 - \dfrac{z-z_0}{t-z_0}\right)}.$$

If

$$\left| \frac{z-z_0}{t-z_0} \right| < 1, \tag{3.5}$$

then by the basic geometric series (3.2)

$$\frac{1}{1 - \dfrac{z - z_0}{t - z_0}} = 1 + \left(\frac{z - z_0}{t - z_0}\right) + \left(\frac{z - z_0}{t - z_0}\right)^2 + \left(\frac{z - z_0}{t - z_0}\right)^3 + \cdots . \qquad (3.6)$$

Therefore, (3.4) can be written as

$$
\begin{aligned}
f(z) &= \frac{1}{2\pi i} \oint_C \frac{f(t)}{t - z} dt \\
&= \frac{1}{2\pi i} \oint_C \frac{f(t)}{t - z_0} \left[1 + \left(\frac{z - z_0}{t - z_0}\right) + \left(\frac{z - z_0}{t - z_0}\right)^2 + \left(\frac{z - z_0}{t - z_0}\right)^3 + \cdots \right] dt \\
&= \frac{1}{2\pi i} \oint_C \frac{f(t)}{t - z_0} dt + \left[\frac{1}{2\pi i} \oint_C \frac{f(t)}{(t - z_0)^2} dt \right] (z - z_0) \\
&\quad + \left[\frac{1}{2\pi i} \oint_C \frac{f(t)}{(t - z_0)^3} dt \right] (z - z_0)^2 + \left[\frac{1}{2\pi i} \oint_C \frac{f(t)}{(t - z_0)^4} dt \right] (z - z_0)^3 \cdots .
\end{aligned}
$$
$$(3.7)$$

According to Cauchy's integral formula and its derivatives

$$f^{(n)}(z_0) = \frac{n!}{2\pi i} \oint_C \frac{f(t)}{(t - z_0)^{n+1}} dt,$$

the above Eq. (3.7) becomes

$$
\begin{aligned}
f(z) &= \sum_{n=0} \frac{f^{(n)}(z_0)}{n!} (z - z_0)^n \\
&= f(z_0) + f'(z_0)(z - z_0) + \frac{f''(z_0)}{2}(z - z_0)^2 + \cdots .
\end{aligned}
$$
$$(3.8)$$

This is the well-known Taylor series.

3.2.2 Convergence of Taylor Series

To discuss the convergence of the Taylor series, let us first recall the definition of singular points.

Singularity. If $f(z)$ is analytic at all points in the neighborhood of z_s but is not differentiable at z_s, then z_s is called a singular point. We also say that $f(z)$ has a singularity at $z = z_s$. For example:

$\dfrac{1}{z^2 + 1}$ has singularities at $z = i, -i$.

$\tan z = \dfrac{\sin z}{\cos z}$ has singularities at $z = \pm\dfrac{\pi}{2}, \pm\dfrac{3\pi}{2}, \pm\dfrac{5\pi}{2}, \ldots$

Fig. 3.1 Radius of
convergence of the Taylor
series. The expansion center
is at z_0. The singular point at
s limits the region of
convergence within the
interior of the circle of radius
R

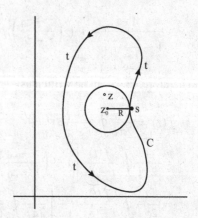

$$\frac{1+2z}{z^2 - 5z + 6} \text{ has singularities at } z = 2, \ 3.$$

$$\frac{1}{e^z + 1} \text{ has singularities at } z = \pm i\pi, \ \pm i3\pi, \ \pm i5\pi, \ldots.$$

Radius of Convergence. The Cauchy integral formula of (3.4) is, of course, valid for
all z inside the Contour C, if $f(t)$ is analytic in and on C. However, in developing the
Taylor series around $z = z_0$, we have used (3.6), which is true only if the condition
of (3.5) is satisfied. This means $|z - z_0|$ must be less than $|t - z_0|$. Since t is on the
contour C as shown in Fig. 3.1, the distance $|t - z_0|$ is changing as t is moving around
C. With the contour shown in the figure, the smallest $|t - z_0|$ is $|s - z_0|$, where s is
the point on C closest to z_0. For $|z - z_0|$ to be less than all possible $|t - z_0|$, $|z - z_0|$
must be less than $|s - z_0|$. This means the Taylor series of (3.8) is valid only for those
points of z which are inside the circle centered at z_0, with a radius $R = |s - z_0|$.

If $f(z)$ is analytic everywhere, we can draw the contour C as large as we want.
Therefore, the Taylor series is convergent in the entire complex plane. However, if
$f(z)$ has a singular point at $z = s$, then the contour must be so drawn in such a way
that the point $z = s$ is outside of C. In Fig. 3.1, the contour C can be infinitesimally
close to s, but s must not be on or inside C. For such a case, the largest possible
radius of convergence is $|s - z_0|$. Therefore, the radius of convergence of a Taylor
series is equal to the distance between its expansion center and the nearest singular
point.

The discussion above applies equally well to a circular region about the origin,
$z_0 = 0$. The Taylor series about the origin

$$f(z) = f(0) + f'(0)z + \frac{f''(0)}{2!}z^2 + \cdots$$

is called the Maclaurin series.

Even in the expansion of a function of a real variable, the radius of convergence
is equally important. To illustrate, consider

$$f(z) = \frac{1}{1+z^2} = 1 - z^2 + z^4 - z^6 \cdots .$$

This series converges throughout the interior of the largest circle around the origin in which $f(z)$ is analytic. Now, $f(z)$ has two singular points at $z = \pm i$, and even though one may be concerned solely with real values of z, for which $1/(1+x^2)$ is everywhere infinitely differentiable with respect to x, these singularities in the complex plane set an inescapable limit to the interval of convergence on the x-axis. Since the distance between the expansion center at $z = 0$ and the nearest singular point, i or $-i$ is $|i - 0| = 1$, the radius of convergence is equal to one. The series is convergent only inside the circle of radius 1, centered at origin. Thus, the interval of convergence on the x-axis is between $x = \pm 1$. In other words, the Maclaurin series

$$\frac{1}{1+x^2} = 1 - x^2 + x^4 - x^6 \cdots$$

is valid only for $-1 < x < 1$, although $\dfrac{1}{1+x^2}$ and its derivatives of all orders are well defined along the real axis x. Now if we expand the real function $1/(1+x^2)$ into Taylor series around $x = x_0$, then the radius of convergence is equal to $|i - x_0| = \sqrt{1 + x_0^2}$. This means that this series will converge only in the interval between $x = x_0 - \sqrt{1 + x_0^2}$ and $x = x_0 + \sqrt{1 + x_0^2}$.

3.2.3 Analytic Continuation

If we know the values of an analytic function in some small region around z_0, we can use the Taylor expansion about z_0 to find the values of the function in a larger region. Although the Taylor expansion is valid only inside the circle of radius of convergence which is determined by the location of the nearest singular point, a chain of Taylor expansions can be used to determine the function throughout the entire complex plane except at the singular points of the function. This process is illustrated in Fig. 3.2.

Suppose we know the values around z_0 and the singular point nearest to z_0 is s_0. The Taylor expansion about z_0 holds within a circular region of radius $|z_0 - s_0|$. Since the Taylor expansion gives the values of the function and all its derivatives at every point in this circle, we can use any point in this circle as the new expansion center. For example, we may expand another Taylor series about z_1 as shown in Fig. 3.2. We can do this because $f^n(z_1)$ is known for all n from the first Taylor expansion about z_0. The radius of convergence of this second Taylor series is determined by the distance from z_1 to the nearest singular point s_1. Continue this way, as indicated in Fig. 3.2, we can cover the whole complex plane except at the singular points $s_0, s_1, s_2 \cdots$. In other

Fig. 3.2 Analytic
Continuation. A series of
Taylor expansions which
analytically continue a
function originally known in
the region around z_0. The
first expansion about z_0 is
valid only inside the circle of
radius $|z_0 - s_0|$, where s_0 is
the singular point nearest to
z_0. The next Taylor
expansion is around z_1
which is inside the first
circle. The second Taylor
expansion is limited by the
singular point s_1, and so on

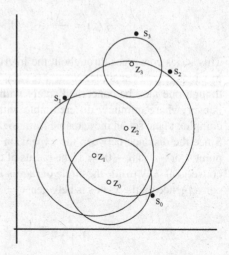

words, the analytic function everywhere can be constructed from the knowledge of
the function in a small region. This process is called analytic continuation.

An immediate consequence of analytic continuation is the so-called identity theorem. It states that if $f(z)$ and $g(z)$ are analytic and $f(z) = g(z)$ along a curve L in
a region D, then $f(z) = g(z)$ throughout D. We can show this by considering the
analytic function $h(z) = f(z) - g(z)$. If we can show that $h(z)$ is identically zero
throughout the region, then the theorem is proved.

If we choose a point $z = z_0$ on L, then we can expand $h(z)$ in a Taylor series
about z_0,

$$h(z) = h(z_0) + h'(z_0)(z - z_0) + \frac{1}{2!}h''(z_0)(z - z_0)^2 + \cdots,$$

which will converge inside the some circle that extends as far as the nearest point
of the boundary of D. But since z_0 is on L, $h(z_0) = 0$. Furthermore, the derivatives
of h must also be zero if z is approaching z_0 along L. Since $h(z)$ is analytic, its
derivatives are independent on the way how z is approaching z_0, this means

$$h'(z_0) = h''(z_0) = \cdots = 0.$$

Therefore, $h(z) = 0$ inside the circle. We may now expand on a new point, which
can lie anywhere inside the circle. Thus, by analytic continuation, we may show that
$h(z) = 0$ throughout the region D.

3.2.4 Uniqueness of Taylor Series

If there are constants a_n $(n = 0, 1, 2, \ldots)$ such that

$$f(z) = \sum_{n=0}^{\infty} a_n (z - z_0)^n$$

is convergent for all points z interior to some circle centered at z_0, then this power series must be the Taylor series, regardless of how those constants are obtained. This is quite easy to show, since

$$f(z) = a_0 + a_1 (z - z_0) + a_2 (z - z_0)^2 + a_3 (z - z_0)^3 + \cdots,$$
$$f'(z) = a_1 + 2a_2 (z - z_0) + 3a_3 (z - z_0)^2 + 4a_4 (z - z_0)^3 \cdots,$$
$$f''(z) = 2a_2 + 3 \cdot 2a_3 (z - z_0) + 4 \cdot 3 (z - z_0)^2 + \cdots,$$

clearly

$$f(z_0) = a_0, \quad f'(z_0) = a_1, \quad f''(z_0) = 2a_2, \quad f'''(z_0) = 3 \cdot 2a_3, \quad \cdots.$$

It follows that

$$a_n = \frac{1}{n!} f^{(n)}(z_0),$$

which are the Taylor coefficients. Thus, Taylor series is unique. Thus, no matter how the power series is obtained, if it is convergent in some circular region, it is the Taylor series. The following examples illustrate some of the techniques of expanding a function into its Taylor series.

Example 3.2.1 Find the Taylor series about the origin and its radius of convergence for

$$(a) \sin z, \quad (b) \cos z, \quad (c) \ e^z.$$

Solution 3.2.1 (a) Since $f(z) = \sin z$,

$$f'(z) = \cos z, \quad f''(z) = -\sin z, \quad f'''(z) = -\cos z, \quad f^4(z) = \sin z, \quad \ldots.$$

Hence,

$$f(0) = 0, \quad f'(0) = 1, \quad f''(0) = 0, \quad f'''(0) = -1, \quad f^4(0) = 0, \quad \ldots.$$

Thus,

$$\sin z = \sum_{n=0}^{\infty} \frac{1}{n!} f^{(n)}(0) z^n$$

$$= z - \frac{1}{3!} z^3 + \frac{1}{5!} z^5 + \cdots.$$

This series is valid for all z, since $\sin z$ is an entire function (analytic for the entire complex plane).

(b) If $f(z) = \cos z$, then

$$f'(z) = -\sin z, \quad f''(z) = -\cos z, \quad f'''(z) = \sin z, \quad f^4(z) = \cos z, \quad \ldots$$

$$f(0) = 1, \quad f'(0) = 0, \quad f''(0) = -1, \quad f'''(0) = 0, \quad f^4(0) = 1, \quad \ldots.$$

Therefore,

$$\cos z = 1 - \frac{1}{2!} z^2 + \frac{1}{4!} z^4 - \cdots.$$

This series is also valid for all z.

(c) For $f(z) = e^z$, then

$$f^{(n)}(z) = \frac{d^n}{dz^n} e^z = e^z, \quad \text{and} \quad f^{(n)}(0) = 1.$$

It follows

$$e^z = 1 + z + \frac{1}{2!} z^2 + \frac{1}{3!} z^3 + \cdots.$$

This series converges for all z, since e^z is an entire function.

Example 3.2.2 Find the Taylor series about the origin and its radius of convergence for

$$f(z) = \frac{e^z}{\cos z}.$$

Solution 3.2.2 The singular points of the function are at the zeros of the denominator. Since $\cos \frac{\pi}{2} = 0$, the singular point nearest to the origin is at $z = \pm \frac{\pi}{2}$. Therefore, the Taylor series about the origin is valid for $|z| < \frac{\pi}{2}$. We can find the constants a_n of

$$\frac{e^z}{\cos z} = a_0 + a_1 z + a_2 z^2 + \cdots$$

from $f^{(n)}(0)$, but the repeated differentiations become increasingly tedious. So, we take the advantage of the fact that the Taylor series for e^z and $\cos z$ are already known. Replacing e^z and $\cos z$ with their respective Taylor series in the equation

$$e^z = \left(a_0 + a_1 z + a_2 z^2 + a_3 z^3 + \cdots\right) \cos z,$$

we obtain

$$1 + z + \frac{1}{2!}z^2 + \frac{1}{3!}z^3 + \cdots = \left(a_0 + a_1 z + a_2 z^2 + a_3 z^3 + \cdots\right)\left(1 - \frac{1}{2!}z^2 + \frac{1}{4!}z^4 - \cdots\right).$$

Multiplying out and collecting terms, we have

$$1 + z + \frac{1}{2!}z^2 + \frac{1}{3!}z^3 + \cdots = a_0 + a_1 z + \left(a_2 - \frac{1}{2}a_0\right)z^2 + \left(a_3 - \frac{1}{2}a_1\right)z^3 + \cdots.$$

Therefore,

$$a_0 = 1, \quad a_1 = 1, \quad a_2 - \frac{1}{2}a_0 = \frac{1}{2}, \quad a_3 - \frac{1}{2}a_1 = \frac{1}{3!}, \quad \ldots.$$

It follows that

$$a_2 = \frac{1}{2} + \frac{1}{2}a_0 = 1, \quad a_3 = \frac{1}{3!} + \frac{1}{2}a_1 = \frac{2}{3}, \quad \ldots$$

and

$$\frac{e^z}{\cos z} = 1 + z + z^2 + \frac{2}{3}z^3 + \cdots, \qquad |z| < \frac{\pi}{2}.$$

Example 3.2.3 Find the Taylor series about $z = 2$ for

$$(a)\ \frac{1}{z}, \quad (b)\ \frac{1}{z^2}.$$

Solution 3.2.3 (a) The function $\frac{1}{z}$ has a singular point at $z = 0$, the distance between this point and the expansion center is 2. Therefore, the Taylor series about $z = 2$ is convergent for $|z - 2| < 2$ and has the form

$$\frac{1}{z} = a_0 + a_1(z - 2) + a_2(z - 2)^2 + \cdots.$$

We can write the function as

$$\frac{1}{z} = \frac{1}{2 + (z - 2)} = \frac{1}{2}\frac{1}{1 + \left(\dfrac{z - 2}{2}\right)}.$$

For $|z - 2| < 2$, $\left|\dfrac{z-2}{2}\right| < 1$. Therefore, we can use the geometric series (3.3) to expand

$$\frac{1}{1 + \left(\dfrac{z-2}{2}\right)} = 1 - \left(\frac{z-2}{2}\right) + \left(\frac{z-2}{2}\right)^2 - \left(\frac{z-2}{2}\right)^3 + \cdots.$$

It follows that for $|z - 2| < 2$

$$\frac{1}{z} = \frac{1}{2}\left[1 - \left(\frac{z-2}{2}\right) + \left(\frac{z-2}{2}\right)^2 - \left(\frac{z-2}{2}\right)^3 + \left(\frac{z-2}{2}\right)^4 - \cdots\right]$$

$$= \frac{1}{2} - \frac{1}{4}(z - 2) + \frac{1}{8}(z - 2)^2 - \frac{1}{16}(z - 2)^3 + \frac{1}{32}(z - 2)^4 - \cdots.$$

(b) Since

$$\frac{1}{z^2} = -\frac{d}{dz}\frac{1}{z},$$

therefore

$$\frac{1}{z^2} = -\frac{d}{dz}\left(\frac{1}{2} - \frac{1}{4}(z - 2) + \frac{1}{8}(z - 2)^2 - \frac{1}{16}(z - 2)^3 + \frac{1}{32}(z - 2)^4 - \cdots\right)$$

$$= \frac{1}{4} - \frac{1}{4}(z - 2) + \frac{3}{16}(z - 2)^2 - \frac{1}{8}(z - 2)^3 + \cdots.$$

Example 3.2.4 Find the Taylor series about the origin for

$$f(z) = \frac{1}{1 + z - 2z^2}.$$

Solution 3.2.4 Since $1 + z - 2z^2 = (1 - z)(1 + 2z)$, the function $f(z)$ has two singular points at $z = 1$ and $z = -\frac{1}{2}$. The Taylor series expansion about $z = 0$ will be convergent for $|z| < \frac{1}{2}$. Furthermore,

$$\frac{1}{1 + z - 2z^2} = \frac{1/3}{1 - z} + \frac{2/3}{1 + 2z}$$

For $|z| < \frac{1}{2}$ and $|2z| < 1$,

$$\frac{1}{1 - z} = 1 + z + z^2 + z^3 + \cdots,$$

$$\frac{1}{1 + 2z} = 1 - 2z + 4z^2 - 8z^3 + \cdots.$$

Thus,

$$f(z) = \frac{1}{3}\left(1 + z + z^2 + z^3 + \cdots\right) + \frac{2}{3}\left(1 - 2z + 4z^2 - 8z^3 + \cdots\right)$$
$$= 1 - z + 3z^2 - 5z^3 + \cdots .$$

Example 3.2.5 Find the Taylor series about the origin for

$$f(z) = \ln(1 + z).$$

Solution 3.2.5 First note that

$$\frac{d}{dz}\ln(1 + z) = \frac{1}{1 + z},$$

and

$$\frac{1}{1 + z} = 1 - z + z^2 - z^3 + \cdots ,$$

so

$$d\ln(1 + z) = \left(1 - z + z^2 - z^3 + \cdots\right) dz.$$

Integrating both sides, we have

$$\ln(1 + z) = z - \frac{1}{2}z^2 + \frac{1}{3}z^3 + \cdots + k.$$

The integration constant $k = 0$, since at $z = 0$, $\ln(1) = 0$. Therefore,

$$\ln(1 + z) = z - \frac{1}{2}z^2 + \frac{1}{3}z^3 + \cdots .$$

This series converges for $|z| < 1$, since at $z = -1$, $f(z)$ is singular.

3.3 Laurent Series

In many applications, it is necessary to expand functions around points at which, or in the neighborhood of which, the functions are not analytic. The method of Taylor series is obviously inapplicable in such cases. A new type of series known as Laurent expansion is required. This series furnishes us with a representation which is valid in the annular ring bounded by two concentric circles, provided that the function being expanded is analytic everywhere on and between the two circles.

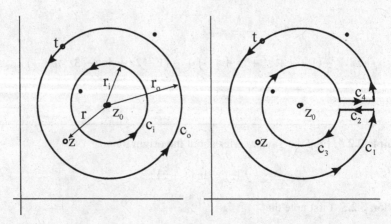

Fig. 3.3 Annular region between two circles where the function is analytic and the Laurent series is valid. Inside the inner circle and outside the outer circle, the function may have singular points

Consider the annulus bounded by two circles of C_o and C_i with a common center z_0 as shown in Fig. 3.3a. The function $f(z)$ is analytic inside the annular region; however, there may be singular points inside the smaller circle or outside the larger circle. We can apply the Cauchy's integral formula to the region which is cut up as shown in Fig. 3.3b. The region is now simply connected and is bounded by the curve $C' = C_1 + C_2 + C_3 + C_4$. Cauchy's integral formula is then

$$f(z) = \frac{1}{2\pi i} \oint_{C'} \frac{f(t)}{t - z} dt$$
$$= \frac{1}{2\pi i} \left(\int_{C_1} \frac{f(t)}{t - z} dt + \int_{C_2} \frac{f(t)}{t - z} dt + \int_{C_3} \frac{f(t)}{t - z} dt + \int_{C_4} \frac{f(t)}{t - z} dt \right),$$

where t is on C' and z is a point inside C'. Now let the gap between C_2 and C_4 shrink to zero, then the integrals along C_2 and C_4 will cancel each other, since they are oriented in the opposite directions, if $f(z)$ is single valued. Furthermore, the contour C_1 becomes C_o and the contour C_3 is identical to C_i turning the opposite direction. Therefore,

$$f(z) = \frac{1}{2\pi i} \oint_{C_o} \frac{f(t)}{t - z} dt - \frac{1}{2\pi i} \oint_{C_i} \frac{f(t)}{t - z} dt, \qquad (3.9)$$

where C_o and C_i are both traversed in the counterclockwise direction. The negative sign results because the direction of integration was reversed on C_i.

We can introduce z_0, the common center of C_o and C_i, as the expansion center. For the first integral in (3.9) with t on C_o, we can write

$$\frac{1}{t-z} = \frac{1}{t-z_0+z_0-z} = \frac{1}{(t-z_0)-(z-z_0)}$$

$$= \frac{1}{(t-z_0)\left(1-\dfrac{z-z_0}{t-z_0}\right)}. \tag{3.10}$$

Since t is on C_o and z is inside C_o, as shown in Fig. 3.3a,

$$\left|\frac{z-z_0}{t-z_0}\right| = \frac{r}{r_o} < 1,$$

so we can expand $\left(1-\dfrac{z-z_0}{t-z_0}\right)^{-1}$ with the geometric series of (3.2), and (3.10) becomes

$$\frac{1}{t-z} = \frac{1}{t-z_0}[1+\frac{z-z_0}{t-z_0}+\left(\frac{z-z_0}{t-z_0}\right)+\left(\frac{z-z_0}{t-z_0}\right)^2$$

$$+\left(\frac{z-z_0}{t-z_0}\right)^3+\cdots\cdots], \quad for\ t\ on\ C_o. \tag{3.11}$$

For the second integral with t on C_i and z is between C_o and C_i, we can write

$$\frac{1}{t-z} = -\frac{1}{z-t} = -\frac{1}{z-z_0+z_0-t}$$

$$= -\frac{1}{(z-z_0)-(t-z_0)} = -\frac{1}{(z-z_0)\left[1-\dfrac{t-z_0}{z-z_0}\right]}.$$

Since

$$\left|\frac{t-z_0}{z-z_0}\right| = \frac{r_i}{r} < 1$$

as shown in Fig. 3.3a, we can again expand $\left(1-\dfrac{t-z_0}{z-z_0}\right)^{-1}$ with the geometric series, and write

$$\frac{1}{t-z} = -\frac{1}{z-z_0}\left[1+\frac{t-z_0}{z-z_0}+\left(\frac{t-z_0}{z-z_0}\right)^2+\left(\frac{t-z_0}{z-z_0}\right)^3+\cdots\right] \tag{3.12}$$

for t on C_i.

Putting (3.11) and (3.12) into (3.9), we have

$$f(z) = I_{C_0} + I_{C_i},$$

where

$$I_{C_0} = \frac{1}{2\pi i} \oint_{C_o} \frac{f(t)}{t - z_0} \left[1 + \frac{z - z_0}{t - z_0} + \left(\frac{z - z_0}{t - z_0} \right)^2 + \left(\frac{z - z_0}{t - z_0} \right)^3 + \cdots \right] dt$$

$$= \frac{1}{2\pi i} \oint_{C_o} \frac{f(t)}{t - z_0} dt + \left(\frac{1}{2\pi i} \oint_{C_o} \frac{f(t)}{(t - z_0)^2} dt \right) (z - z_0)$$

$$+ \left(\frac{1}{2\pi i} \oint_{C_o} \frac{f(t)}{(t - z_0)^3} dt \right) (z - z_0)^2 + \left(\frac{1}{2\pi i} \oint_{C_o} \frac{f(t)}{(t - z_0)^4} dt \right) (z - z_0)^3 + \cdots,$$

and

$$I_{C_i} = \frac{1}{2\pi i} \oint_{C_i} \frac{f(t)}{z - z_0} \left[1 + \frac{t - z_0}{z - z_0} + \left(\frac{t - z_0}{z - z_0} \right)^2 + \left(\frac{t - z_0}{z - z_0} \right)^3 + \cdots \right] dt$$

$$= \left(\frac{1}{2\pi i} \oint_{C_i} f(t) dt \right) \frac{1}{z - z_0} + \left(\frac{1}{2\pi i} \oint_{C_i} f(t)(t - z_0) dt \right) \frac{1}{(z - z_0)^2}$$

$$+ \left(\frac{1}{2\pi i} \oint_{C_i} f(t)(t - z_0)^2 dt \right) \frac{1}{(z - z_0)^3} + \cdots\cdots\cdots.$$

Therefore, in the region between C_i and C_0, $f(z)$ can be expressed as

$$f(z) = \sum_{n=0}^{\infty} a_n (z - z_0)^n + \sum_{k=1}^{\infty} b_n \frac{1}{(z - z_0)^k},$$

where

$$a_n = \frac{1}{2\pi i} \oint_{C_o} \frac{f(t)}{(t - z_0)^{n+1}} dt, \qquad b_k = \frac{1}{2\pi i} \oint_{C_i} f(t)(t - z_0)^{k-1} dt.$$

Because of the principle of deformation of contours, we can replace both C_i and C_0 by a closed contour C between C_i and C_o without changing the values of the integrals. Thus, we can write this series as

$$f(z) = \sum_{n=-\infty}^{\infty} a_n (z - z_0)^n,$$

with

$$a_n = \frac{1}{2\pi i} \oint_C \frac{f(t)}{(t - z_0)^{n+1}} dt. \tag{3.13}$$

This expansion is known as the Laurent series which contains both negative and positive powers of $(z - z_0)$.

It should be noted that the coefficients of positive powers of $(z - z_0)$ cannot be replaced by the derivative expressions, since $f(z)$ is not analytic inside C. However, if there is no singular point inside C_i, then these coefficients can indeed be replaced by $f^{(n)}(z_0)/n!$, at the same time the coefficients of the negative powers of $(z - z_0)$ are identically equal to zero by the Cauchy theorem, since $f(t)(t - z_0)^{-n-1}$ for $n \leq -1$ are analytic inside C. In such a case, the Laurent expansion reduces to the Taylor expansion.

3.3.1 Uniqueness of Laurent Series

Just as Taylor series, Laurent series is unique. If a series

$$\sum_{n=-\infty}^{\infty} a_n (z - z_0)^n = \sum_{n=1}^{\infty} \frac{a_{-n}}{(z - z_0)^n} + \sum_{n=0}^{\infty} a_n (z - z_0)^n$$

converges to $f(z)$ at all points in some annular domain about z_0, then regardless how the constants a_n are obtained, the series is the Laurent expansion for $f(z)$ in powers of $(z - z_0)$ for that domain. This statement is proved if we can show that

$$a_n = \frac{1}{2\pi i} \oint_C \frac{f(t)}{(t - z_0)^{n+1}} dt. \tag{3.14}$$

We now show that this is indeed the case. Let $g_k(t)$ be defined as

$$g_k(t) = \frac{1}{2\pi i} \frac{1}{(t - z_0)^{k+1}},$$

where k is an integer, either positive or negative, or zero. Furthermore, let C be a circle inside the annulus centered at z_0 and taken in the counterclockwise direction, so

$$\oint_C g_k(t) f(t) dt = \frac{1}{2\pi i} \oint_C \frac{f(t)}{(t - z_0)^{k+1}} dt. \tag{3.15}$$

Now if $f(t)$ is expressible as

$$f(t) = \sum_{n=-\infty}^{\infty} a_n (t - z_0)^n,$$

then

$$\oint_C g_k(t) f(t) dt = \frac{1}{2\pi i} \oint_C \frac{1}{(t-z_0)^{k+1}} \left(\sum_{n=-\infty}^{\infty} a_n (t-z_0)^n \right) dt$$

$$= \sum_{n=-\infty}^{\infty} a_n \frac{1}{2\pi i} \oint_C \frac{1}{(t-z_0)^{k-n+1}} dt.$$

The last integral can be easily evaluated by setting $t - z_0 = re^{i\theta}$, so $dt = ire^{i\theta} d\theta$ and

$$\oint_C \frac{1}{(t-z_0)^{k-n+1}} dt = \int_0^{2\pi} \frac{ire^{i\theta}}{r^{k-n+1} e^{i(k-n+1)\theta}} d\theta$$

$$= \frac{i}{r^{k-n}} \int_0^{2\pi} e^{i(n-k)} d\theta = 2\pi i \delta_{nk} = \begin{cases} 0 & n \neq k \\ 2\pi i & n = k \end{cases} \quad (3.16)$$

Thus,

$$\oint_C g_k(t) f(t) dt = \sum_{n=-\infty}^{\infty} a_n \frac{1}{2\pi i} 2\pi i \delta_{nk} = a_k. \quad (3.17)$$

It follows from (3.15) that

$$a_k = \frac{1}{2\pi i} \oint_C \frac{f(t)}{(t-t_0)^{k+1}} dt. \quad (3.18)$$

Since k is an arbitrary integer, (3.14) must hold.

Thus, no matter how the expansion is obtained, as long as it is valid in the specified annular domain, it is the Laurent series. This enables us to determine the Laurent coefficients with elementary techniques, as illustrated in the following examples. The integral representations of the Laurent coefficients (3.18) are important, not as means of finding the coefficients, but as means of using the coefficients to evaluate these integrals. We will elaborate this aspect of the Laurent series in the following sections on the theory of residues.

Example 3.3.1 Find the Laurent series about $z = 0$ for the function

$$f(z) = e^{1/z}.$$

Solution 3.3.1 Since $f(z)$ is analytic for all z, except for $z = 0$, the expansion of $f(z)$ about $z = 0$ will be a Laurent series valid in the annulus $0 < |z| < \infty$. To obtain the expansion, let $\frac{1}{z} = t$, and note

$$e^t = 1 + t + \frac{1}{2!}t^2 + \frac{1}{3!}t^3 + \cdots.$$

Therefore,

$$e^{1/z} = 1 + \frac{1}{z} + \frac{1}{2!}\left(\frac{1}{z}\right)^2 + \frac{1}{3!}\left(\frac{1}{z}\right)^3 + \cdots.$$

Example 3.3.2 Find all possible Laurent expansions about $z = 0$ of

$$f(z) = \frac{1 + 2z^2}{z^3 + z^5},$$

and specify the regions in which they are valid.

Solution 3.3.2 By setting the denominator to zero $z^3 + z^5 = z^3(1 + z^2) = 0$, We get three singular points, $z = 0$, and $z = \pm i$. They are shown in Fig. 3.4. Therefore, we can expand the function about $z = 0$ in two different Laurent series, one is valid for the region $0 < |z| < 1$ as shown in (a), the other is valid in the region $|z| > 1$ as shown in (b).

The function can be written as

$$f(z) = \frac{1 + 2z^2}{z^3 + z^5} = \frac{1 + 2z^2}{z^3(1 + z^2)} = \frac{2(1 + z^2) - 1}{z^3(1 + z^2)} = \frac{1}{z^3}\left(2 - \frac{1}{1 + z^2}\right).$$

In the case of (a), $|z| < 1$, so is $|z^2| < 1$. We can use the geometric series to express

$$\frac{1}{1 + z^2} = 1 - z^2 + z^4 - z^6 + \cdots.$$

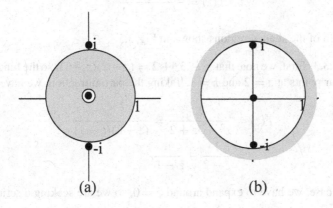

(a)   (b)

Fig. 3.4 If the function has three singular points at $z = 0$ and $z = \pm i$, then the function can be expanded into two different Laurent series about $z = 0$. **a** One series is valid in the region $0 < |z| < 1$, **b** the other series is valid in the the region $1 < |z|$

Therefore,

$$f(z) = \frac{1}{z^3} \left(2 - \left[1 - z^2 + z^4 - z^6 + \cdots \right] \right)$$

$$= \frac{1}{z^3} + \frac{1}{z} - z + z^3 + \cdots, \quad for \;\; 0 < |z| < 1.$$

In the case of (b), $|z^2| > 1$, we first write

$$\frac{1}{1 + z^2} = \frac{1}{z^2 \left(1 + \frac{1}{z^2} \right)}.$$

Since $\left| \frac{1}{z^2} \right| < 1$, again we can use the geometric series

$$\frac{1}{\left(1 + \frac{1}{z^2} \right)} = 1 - \frac{1}{z^2} + \frac{1}{z^4} - \frac{1}{z^6} + \cdots.$$

Thus,

$$f(z) = \frac{1}{z^3} \left(2 - \frac{1}{z^2} \left[1 - \frac{1}{z^2} + \frac{1}{z^4} - \frac{1}{z^6} + \cdots \right] \right)$$

$$= \frac{2}{z^3} - \frac{1}{z^5} + \frac{1}{z^7} - \frac{1}{z^9} + \cdots, \quad for \;\; |z| > 1.$$

Example 3.3.3 Find the Laurent series expansion of

$$f(z) = \frac{1}{z^2 - 3z + 2}$$

valid in each of the shaded regions shown in Fig. 3.5.

Solution 3.3.3 First, we note that $z^2 - 3z + 2 = (z - 2)(z - 1)$, so the function has two singular points at $z = 2$ and $z = 1$. Taking the partial fraction, we have

$$f(z) = \frac{1}{z^2 - 3z + 2} = \frac{1}{(z - 2)(z - 1)}$$

$$= \frac{1}{z - 2} - \frac{1}{z - 1}.$$

(a) In this case, we have to expand around $z = 0$, so we are seeking a series in the form of

$$f(z) = \sum_{n=-\infty}^{\infty} a_n z^n.$$

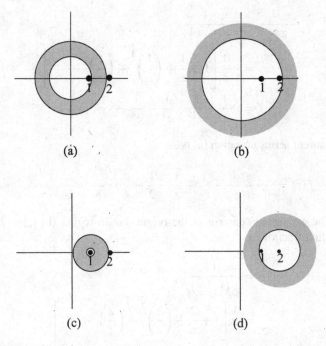

Fig. 3.5 The function with two singular points at $z = 1$ and $z = 2$ can be expanded into different Laurent series in different regions: **a** expanded about $z = 0$ valid in the region $1 < |z| < 2$, **b** expanded about $z = 0$ valid in the region $2 < |z|$, **c** expanded about $z = 1$ valid in the region $0 < |z - 1| < 1$, **d** expanded $z = 2$ valid in the region $1 < |z - 2|$

The values of $|z|$ between the two circles are such that $1 < |z| < 2$. To make use of the basic geometric series, we write

$$\frac{1}{z-2} = -\frac{1}{2(1-\frac{z}{2})}.$$

Since $|z/2| < 1$, so

$$\frac{1}{z-2} = -\frac{1}{2}\left[1 + \frac{z}{2} + \left(\frac{z}{2}\right)^2 + \left(\frac{z}{2}\right)^3 + \cdots\right]$$

$$= -\frac{1}{2} - \frac{z}{4} - \frac{z^2}{8} - \frac{z^3}{16} - \cdots.$$

As for the second fraction, we note that $|z| > 1$, so $|1/z| < 1$. Therefore, we write

$$-\frac{1}{z-1} = -\frac{1}{z(1-\frac{1}{z})}$$

$$= -\frac{1}{z}\left[1 + \frac{1}{z} + \left(\frac{1}{z}\right)^2 + \left(\frac{1}{z}\right)^3 + \cdots\right]$$

$$= -\frac{1}{z} - \frac{1}{z^2} - \frac{1}{z^3} - \frac{1}{z^4} - \cdots.$$

Thus, the Laurent series in region (a) is

$$f(z) = \cdots - \frac{1}{z^3} - \frac{1}{z^2} - \frac{1}{z} - \frac{1}{2} - \frac{z}{4} - \frac{z^2}{8} - \frac{z^3}{16} - \cdots.$$

(b) Again the expansion center is at the origin, but in region (b) $|z| > 2$. So, we expand the first fraction as

$$\frac{1}{z-2} = \frac{1}{z\left(1 - \frac{2}{z}\right)}$$

$$= \frac{1}{z}\left[1 + \frac{2}{z} + \left(\frac{2}{z}\right)^2 + \left(\frac{2}{z}\right)^3 + \cdots\right]$$

$$= \frac{1}{z} + \frac{2}{z^2} + \frac{4}{z^3} + \frac{8}{z^4} + \cdots.$$

Note that the expansion of the second fraction we worked out in part (a) is still valid in this case

$$-\frac{1}{z-1} = -\frac{1}{z} - \frac{1}{z^2} - \frac{1}{z^3} - \frac{1}{z^4} - \cdots.$$

Thus, the Laurent series in region (b) is the sum of these two expressions

$$f(z) = \frac{1}{z-2} - \frac{1}{z-1}$$

$$= \frac{1}{z^2} + \frac{3}{z^3} + \frac{7}{z^4} + \cdots.$$

(c) In this region, we are expanding around $z = 1$, so we are seeking a series of the form

$$f(z) = \sum_{n=-\infty}^{\infty} a_n (z-1)^n.$$

Since in this region $0 < |z - 1| < 1$, so we choose to write the function as

$$f(z) = \frac{1}{(z-1)(z-2)} = \frac{1}{(z-1)\left[(z-1)-1\right]} = -\frac{1}{(z-1)\left[1-(z-1)\right]},$$

and use the geometric series for

$$\frac{1}{1-(z-1)} = 1 + (z-1) + (z-1)^2 + (z-1)^3 + \cdots$$

Therefore, the desired Laurent series valid in region (c) is

$$f(z) = -\frac{1}{(z-1)}\left(1 + (z-1) + (z-1)^2 + (z-1)^3 + \cdots\right)$$

$$= -\frac{1}{(z-1)} - 1 - (z-1) - (z-1)^2 - \cdots .$$

(d) In this region, we are seeking an expansion about $z = 2$ in form of

$$f(z) = \sum_{n=-\infty}^{\infty} a_n (z-2)^n .$$

that is valid for $|z-2| > 1$. So, we choose to write the function as

$$f(z) = \frac{1}{(z-1)(z-2)} = \frac{1}{[(z-2)+1](z-2)} = \frac{1}{(z-2)^2 \left[1 + \dfrac{1}{z-2}\right]} .$$

Since $\left|\dfrac{1}{z-2}\right| < 1$, we can use the geometric series for

$$\left[1 + \frac{1}{z-2}\right]^{-1} = 1 - \frac{1}{(z-2)} + \frac{1}{(z-2)^2} - \frac{1}{(z-2)^3} + \cdots .$$

Therefore, the desired Laurent series valid in region (d) is

$$f(z) = \frac{1}{(z-2)^2}\left(1 - \frac{1}{(z-2)} + \frac{1}{(z-2)^2} - \frac{1}{(z-2)^3} + \cdots\right)$$

$$= \frac{1}{(z-2)^2} - \frac{1}{(z-2)^3} + \frac{1}{(z-2)^4} - \frac{1}{(z-2)^5} + \cdots .$$

3.4 Theory of Residues

3.4.1 Zeros and Poles

Zeros. If $f(z_0) = 0$, then the point z_0 is said to be a zero of the function $f(z)$. If $f(z)$ is analytic at z_0, then we can expand it in a Taylor series

$$f(z) = \sum_{n=0}^{\infty} a_n(z - z_0)^n = a_0 + a_1(z - z_0) + a_2(z - z_0)^2 + \cdots.$$

Since z_0 is a zero of the function, clearly $a_0 = 0$. If $a_1 \neq 0$, then z_0 is said to be a simple zero. If both a_0 and a_1 are zero and $a_2 \neq 0$, then z_0 is a zero of order two, and so on.

If $f(z)$ has a zero of order m at z_0, that is, $a_0, a_1, \ldots a_{m-1}$ are all zero and $a_m \neq 0$, then $f(z)$ can be written as

$$f(z) = (z - z_0)^m g(z),$$

where

$$g(z) = a_m + a_{m+1}(z - z_0) + a_{m+2}(z - z_0)^2 + \cdots.$$

It is clear that $g(z)$ is analytic (therefore continuous) at z_0, and $g(z_0) = a_m \neq 0$. It follows that in the immediate neighborhood of z_0, there is no other zero, because $g(z)$ cannot suddenly drop to zero, since it is continuous. Therefore, there exists a disk of finite radius δ surrounds z_0, within which $g(z) \neq 0$. In other words,

$$f(z) \neq 0 \quad for \quad 0 < |z - z_0| < \delta.$$

In this sense, z_0 is said to be an isolated zero of $f(z)$.

Isolated Singularities. As we recall, a singularity of a function $f(z)$ is a point at which $f(z)$ is not analytic. A point at which $f(z)$ is analytic is called a regular point. A point z_0 is said to be an isolated singularity of $f(z)$ if there exists a neighborhood of z_0 in which z_0 is the only singular point of $f(z)$. For example, a rational function $P(z)/Q(z)$, (the ratio of two polynomials), is analytic everywhere except at zeros of $Q(z)$. If all the zeros of $Q(z)$ are isolated, then all the singularities of $P(z)/Q(z)$ are isolated.

Poles. If $f(z)$ has an isolated singular point at z_0, then in the immediate neighborhood of z_0, $f(z)$ can be expanded in a Laurent series

$$f(z) = \sum_{n=0}^{\infty} a_n(z - z_0)^n + \frac{a_{-1}}{(z - z_0)} + \frac{a_{-2}}{(z - z_0)^2} + \frac{a_{-3}}{(z - z_0)^3} \cdots.$$

The portion of the series involving negative powers of $(z - z_0)$ is called the principal part of $f(z)$ at z_0. If the principal part contains at least one nonzero term but the number of such terms are finite, then there exists an integer m such that

$$a_{-m} \neq 0 \quad \text{and} \quad a_{-(m+1)} = a_{-(m+2)} = \cdots = 0.$$

That is, the expansion takes the form

$$f(z) = \sum_{n=0}^{\infty} a_n (z - z_0)^n + \frac{a_{-1}}{(z - z_0)} + \frac{a_{-2}}{(z - z_0)^2} + \cdots \frac{a_{-m}}{(z - z_0)^m},$$

where $a_{-m} \neq 0$. In this case, the isolated singular point z_0 is called a pole of order m. A pole of order one is usually referred to as a simple pole.

If an infinite number of coefficients of negative powers are nonzero, then z_0 is called an essential singular point.

3.4.2 Definition of the Residue

If z_0 is an isolated singular point of $f(z)$, then the function $f(z)$ is analytic in the neighborhood of $z = z_0$ with the exception of the point $z = z_0$ itself. In the immediate neighborhood of z_0, $f(z)$ can be expanded in a Laurent series

$$f(z) = \sum_{n=-\infty}^{\infty} a_n (z - z_0)^n. \tag{3.19}$$

The coefficients a_n are expressed in terms of contour integrals of (3.13). Among the coefficients, a_{-1} is of particular interest,

$$a_{-1} = \frac{1}{2\pi i} \oint_C f(z) dz, \tag{3.20}$$

where C is a closed contour in the counterclockwise direction around z_0. It is the coefficient of $(z - z_0)^{-1}$ term in the expansion, and is called the residue of $f(z)$ at the isolated singular point z_0. We emphasize once again, for a_{-1} of (3.20) to be called the residue at z_0, the closed contour C must not contain any singularity other than z_0. We shall denote this residue as

$$a_{-1} = Res_{z=z_0} [f(z)].$$

The reason for the name "residue" is that if we integrate the Laurent series term by term over a circular contour, the only term which survives the integration process is the a_{-1} term. This follows from (3.19) that

$$\oint f(z)\,dz = \sum_{n=-\infty}^{\infty} a_n \oint (z - z_0)^n dz.$$

These integrals can be easily evaluated by setting $z - z_0 = re^{i\theta}$ and $dz = ire^{i\theta}d\theta$,

$$\oint (z - z_0)^n dz = \int_0^{2\pi} ir^{n+1} e^{i(n+1)\theta} d\theta = \begin{cases} 0 & n \neq -1 \\ 2\pi i & n = -1. \end{cases}$$

Thus, only the term with $n = -1$ is left. The coefficient of this term is called the residue,

$$\frac{1}{2\pi i} \oint f(z)\,dz = a_{-1}.$$

3.4.3 Methods of Finding Residues

Residues are defined in (3.20). In some cases, we can carry out this integral directly. However, in general, residues can be found by much easier methods. It is because of these methods, residues are so useful.

Laurent Series. If it is easy to write down the Laurent series for $f(z)$ about $z = z_0$ that is valid in the immediate neighborhood of z_0, then the residue is just the coefficient a_{-1} of the term $1/(z - z_0)$. For example,

$$f(z) = \frac{3}{z - 2}$$

is already in the form of a Laurent series about $z = 2$ with $a_{-1} = 3$ and $a_n = 0$ for $n \neq -1$. Therefore, the residue at 2 is simply 3.

It is also easy to find the residue of $\exp\left(\frac{1}{z^2}\right)$ at $z = 0$, since

$$e^{1/z^2} = 1 + \frac{1}{z^2} + \frac{1}{2}\frac{1}{z^4} + \frac{1}{3!}\frac{1}{z^6} + \cdots.$$

There is no $1/z$ term, therefore the residue is equal to zero.

Simple Pole. Suppose that $f(z)$ has a simple, or first-order, pole at $z = z_0$, so we can write

$$f(z) = \frac{a_{-1}}{z - z_0} + a_0 + a_1(z - z_0) + \cdots.$$

If we multiply this identity by $(z - z_0)$, we get

$$(z - z_0)f(z) = a_{-1} + a_0(z - z_0) + a_1(z - z_0)^2 + \cdots.$$

Now if we let z approach z_0, we obtain for the residue

$$a_{-1} = \lim_{z \to a} (z - z_0) f(z).$$

For example, if

$$f(z) = \frac{4 - 3z}{z(z-1)(z-2)},$$

the residue at $z = 0$ is

$$Res_{z=0} [f(z)] = \lim_{z \to 0} z \frac{4 - 3z}{z(z-1)(z-2)} = \frac{4}{(-1)(-2)} = 2,$$

the residue at $z = 1$ is

$$Res_{z=1} [f(z)] = \lim_{z \to 1} (z-1) \frac{4 - 3z}{z(z-1)(z-2)} = \frac{4-3}{1(-1)} = -1,$$

and the residue at $z = 2$ is

$$Res_{z=2} [f(z)] = \lim_{z \to 2} (z-2) \frac{4 - 3z}{z(z-1)(z-2)} = \frac{4-6}{2(1)} = -1.$$

These results can also be understood in terms of partial fractions. It can be readily verified that

$$f(z) = \frac{4 - 3z}{z(z-1)(z-2)} = \frac{2}{z} + \frac{-1}{z-1} + \frac{-1}{z-2}.$$

In the region $|z| < 1$, both $\frac{-1}{z-1}$ and $\frac{-1}{z-2}$ are analytic. Therefore, they can be expressed in terms of Taylor series, which has no negative power terms. Thus, the Laurent series of $f(z)$ about $z = 0$ in the region $0 < |z| < 1$ is of the form

$$f(z) = \frac{2}{z} + a_0 + a_1 z + a_2 z^2 + \cdots.$$

It is seen that a_{-1} comes solely from the first term. Therefore, the residue at $z = 0$ must equal to 2.

Similarly, the Laurent series of $f(z)$ about $z = 1$ in the region $0 < |z - 1| < 1$ is of the form

$$f(z) = \frac{-1}{z-1} + a_0 + a_1(z-1) + a_2(z-1)^2 + \cdots.$$

Hence, a_{-1} is equal to -1. For the same reason, the residue at $z = 2$ comes from the term $\frac{-1}{z-2}$, and is clearly equal to -1.

Multiple Order Pole. If $f(z)$ has a third-order pole at $z = z_0$, then

$$f(z) = \frac{a_{-3}}{(z - z_0)^3} + \frac{a_{-2}}{(z - z_0)^2} + \frac{a_{-1}}{z - z_0} + a_0 + a_1(z - z_0) + \cdots.$$

To obtain the residue a_{-1}, we must multiply this identity by $(z - a)^3$

$$(z - z_0)^3 f(z) = a_{-3} + a_{-2}(z - z_0) + a_{-1}(z - z_0)^2 + a_0(z - z_0)^3 + \cdots,$$

and differentiate twice with respect to z

$$\frac{d}{dz}[(z - z_0)^3 f(z)] = a_{-2} + 2a_{-1}(z - z_0) + 3a_0(z - z_0)^2 + \cdots,$$

$$\frac{d^2}{dz^2}[(z - z_0)^3 f(z)] = 2a_{-1} + 3 \cdot 2a_0 (z - z_0) + \cdots.$$

Next, we let z approach z_0

$$\lim_{z \to z_0} \frac{d^2}{dz^2}[(z - z_0)^3 f(z)] = 2a_{-1},$$

and finally, divide it by 2,

$$\frac{1}{2} \lim_{z \to z_0} \frac{d^2}{dz^2}[(z - z_0)^3 f(z)] = a_{-1}.$$

Thus, if $f(z)$ has a pole of order m at $z = z_0$, then the residue of $f(z)$ at $z = z_0$ is

$$Res_{z=z_0}[f(z)] = \frac{1}{(m - 1)!} \lim_{z \to z_0} \frac{d^{m-1}}{dz^{m-1}}[(z - a)^m f(z)].$$

For example,

$$f(z) = \frac{1}{z(z - 2)^4}$$

clearly has a fourth-order pole at $z = 2$. Thus,

$$Res_{z=2}[f(z)] = \frac{1}{3!} \lim_{z \to 2} \frac{d^3}{dz^3}[(z - 2)^4 \frac{1}{z(z - 2)^4}]$$

$$= \frac{1}{6} \lim_{z \to 2} \frac{d^3}{dz^3} \frac{1}{z} = \frac{1}{6} \lim_{z \to 2} \frac{(-1)(-2)(-3)}{z^4} = -\frac{1}{16}.$$

To check this result, we can expand $f(z)$ in a Laurent series about $z = 2$ in the region $0 < |z - 2| < 2$. For this purpose, let us write $f(z)$ as

$$f(z) = \frac{1}{z(z-2)^4} = \frac{1}{(z-2)^4[2+(z-2)]} = \frac{1}{2(z-2)^4} \frac{1}{\left(1+\frac{z-2}{2}\right)}.$$

Since $\left|\frac{z-2}{2}\right| < 1$, so we have

$$f(z) = \frac{1}{2(z-2)^4}\left[1 - \frac{z-2}{2} + \left(\frac{z-2}{2}\right)^2 - \left(\frac{z-2}{2}\right)^3 + \cdots\right]$$

$$= \frac{1}{2}\frac{1}{(z-2)^4} - \frac{1}{4}\frac{1}{(z-2)^3} + \frac{1}{8}\frac{1}{(z-2)^2} - \frac{1}{16}\frac{1}{(z-2)} + \frac{1}{32} - \cdots.$$

It is seen that the coefficient of the $(z-2)^{-1}$ is indeed $-\frac{1}{16}$.

Derivative of the Denominator. If $p(z)$ and $q(z)$ are analytic functions, and $q(z)$ has a simple zero at z_0 and $p(z_0) \neq 0$, then

$$f(z) = \frac{p(z)}{q(z)}$$

has a simple pole at z_0. As $q(z)$ is analytic, so it can be expressed as a Taylor series about z_0

$$q(z) = q(z_0) + q'(z_0)(z - z_0) + \frac{q''(z_0)}{2!}(z - z_0)^2 \cdots.$$

But it has a zero at z_0, so $q(z_0) = 0$, and

$$q(z) = q'(z_0)(z - z_0) + \frac{q''(z_0)}{2!}(z - z_0)^2 \cdots.$$

Since $f(z)$ has a simple pole at z_0, its residue at z_0 is

$$Res_{z=z_0}[f(z)] = \lim_{z \to z_0}(z - z_0)\frac{p(z)}{q(z)}$$

$$= \lim_{z \to z_0}(z - z_0)\frac{p(z)}{q'(z_0)(z - z_0) + \frac{q''(z_0)}{2!}(z - z_0)^2 \cdots}$$

$$= \frac{p(z_0)}{q'(z_0)}.$$

This formula is very often the most efficient way of finding the residue. For example, the function

$$f(z) = \frac{z}{z^4 + 4}$$

has four simple poles, located at the zeros of the denominator

$$z^4 + 4 = 0.$$

The four roots of this equation are

$$z_1 = \sqrt{2}e^{i\pi/4} = 1 + i, \qquad z_2 = \sqrt{2}e^{i(\pi/4+\pi/2)} = -1 + i,$$
$$z_3 = \sqrt{2}e^{i(\pi/4+\pi)} = -1 - i, \qquad z_4 = \sqrt{2}e^{i(\pi/4+3\pi/2)} = 1 - i.$$

The residues at z_1, z_2, z_3 and z_4 are

$$Res_{z=z_1}[f(z)] = \lim_{z \to z_1} \frac{z}{(z^4 + 4)'} = \lim_{z \to z_1} \frac{z}{4z^3} = \lim_{z \to z_1} \frac{1}{4z^2}$$

$$= \lim_{z \to (1+i)} \frac{1}{4z^2} = \frac{1}{4(1+i)^2} = -\frac{1}{8}i,$$

$$Res_{z=z_2}[f(z)] = \lim_{z \to (-1+i)} \frac{1}{4z^2} = \frac{1}{4(-1+i)^2} = \frac{1}{8}i,$$

$$Res_{z=z_3}[f(z)] = \lim_{z \to (-1-i)} \frac{1}{4z^2} = \frac{1}{4(-1-i)^2} = -\frac{1}{8}i,$$

$$Res_{z=z_4}[f(z)] = \lim_{z \to (1-i)} \frac{1}{4z^2} = \frac{1}{4(1-i)^2} = \frac{1}{8}i.$$

It can be readily verified that

$$\frac{z}{z^4 + 4} = \frac{-i/8}{z - (1+i)} + \frac{i/8}{z - (-1+i)} + \frac{-i/8}{z - (-1-i)} + \frac{i/8}{z - (1-i)}.$$

Therefore, the calculated residues must be correct.

3.4.4 Cauchy's Residue Theorem

Consider a simple closed curve C containing in its interior a number of isolated singular points, z_1, z_2, ..., of a function $f(z)$. If around each singular point we draw a circle so small that it encloses no other singular points as shown in Fig. 3.6, so $f(z)$ is analytic in the region between C and these small circles. Then introducing cuts as in the proof of Laurent series, we find by the Cauchy theorem that the integral around C counterclockwise plus the integral around the small circles clockwise is zero, since the integrals along the cuts cancel. Thus,

$$\frac{1}{2\pi i} \oint_C f(z)dz - \frac{1}{2\pi i} \oint_{C_1} f(z)dz + \cdots - \frac{1}{2\pi i} \oint_{C_n} f(z)dz = 0.$$

Fig. 3.6 The circles
$C_1, C_2, \ldots, C_N$ enclosing,
respectively, the singular
points $z_1, z_2, \ldots, z_N$ within
a simple closed curve

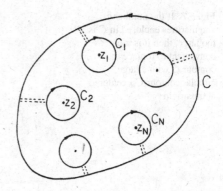

where all integrals are counterclockwise, the minus sign is to account for the clock-wise direction of the small circles. It follows that

$$\frac{1}{2\pi i} \oint_C f(z)dz = \frac{1}{2\pi i} \oint_{C_1} f(z)dz + \cdots + \frac{1}{2\pi i} \oint_{C_n} f(z)dz.$$

The integrals on the right are, by definition, just the residues of $f(z)$ at the various isolated singularities within C. Hence, we have established the important residue theorem:

   If there are n number of singular points of $f(z)$ inside the contour C, then

$$\oint_C f(z)dz = 2\pi i \left\{ Res_{z=z_1}[f(z)] + Res_{z=z_2}[f(z)] + \cdots Res_{z=z_n}[f(z)] \right\}.$$

$$(3.21)$$

This theorem is known as Cauchy's residue theorem or just the residue theorem.

3.4.5 Second Residue Theorem

If the number of singular points inside the enclosed contour C is too large, or there are nonisolated singular points interior in C, it will be difficult to carry out the contour integration with the Cauchy's residue theorem. For such cases, there is another residue theorem that is more efficient.

   Suppose $f(z)$ has many singular points in C and no singular point outside of C, as shown in Fig. 3.7.

   If we want to evaluate the integral $\oint_C f(z)\,dz$, we can first construct a circular contour C_R outside of C, centered at the origin with a radius R. Then by the principle of deformation of contours,

$$\oint_C f(z)\,dz = \oint_{C_R} f(z)\,dz.$$

Fig. 3.7 If the number of
singularities enclosed in C is
too large, then it is more
convenient to replace the
contour C with a large
circular contour C_R centered
at the origin

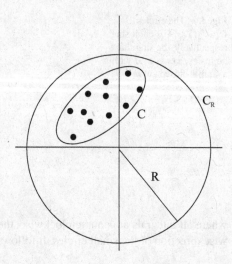

Now if we expand $f(z)$ in terms of Laurent series about $z = 0$ in the region $|z| > R$,

$$f(z) = \cdots \frac{a_{-3}}{z^3} + \frac{a_{-2}}{z^2} + \frac{a_{-1}}{z} + a_0 + a_1 z + a_2 z^2 + \cdots,$$

the coefficient a_{-1} is given by the integral

$$a_{-1} = \frac{1}{2\pi i} \oint_{C_R} f(z)\, dz.$$

Note that a_{-1} in this equation is not the residue of $f(z)$ about $z = 0$, because the
series is not valid in the immediate neighborhood of $z = 0$. However, if we change
z to $1/z$, then

$$f\left(\frac{1}{z}\right) = \cdots a_{-3} z^3 + a_{-2} z^2 + a_{-1} z + a_0 + \frac{a_1}{z} + \frac{a_2}{z^2} + \cdots$$

is convergent for $|z| < 1/R$. It is seen that a_{-1} is the residue at $z = 0$ of the function
$\frac{1}{z^2} f\left(\frac{1}{z}\right)$, since

$$\frac{1}{z^2} f\left(\frac{1}{z}\right) = \cdots a_{-3} z + a_{-2} + \frac{a_{-1}}{z} + \frac{a_0}{z^2} + \frac{a_1}{z^3} + \cdots$$

is a Laurent series valid in the region $0 < |z| < \frac{1}{R}$. Hence, we arrived at the following
theorem.

If $f(z)$ is analytic everywhere except for a number of singular points interior to a positive oriented closed contour C, then

$$\oint_C f(z)\, dz = 2\pi i\, Res_{z=0}\left[\frac{1}{z^2} f\left(\frac{1}{z}\right)\right].$$

Example 3.4.1 Evaluate the integal $\oint_C f(z)\, dz$ for

$$f(z) = \frac{5z - 2}{z(z - 1)},$$

where C is along the circle $|z| = 2$ in the counterclockwise direction. (*a*) Use the Cauchy residue theorem. (*b*) Use the second residue theorem.

Solution 3.4.1 (*a*) The function has two simple poles at $z = 0$, $z = 1$. Both lie within the circle $|z| = 2$. So

$$\oint_C f(z)\, dz \stackrel{.}{=} 2\pi i\, \{Res_{z=0}[f(z)] + Res_{z=1}[f(z)]\}.$$

Since

$$Res_{z=0}[f(z)] = \lim_{z \to 0} z\frac{5z - 2}{z(z - 1)} = \frac{-2}{-1} = 2,$$

$$Res_{z=1}[f(z)] = \lim_{z \to 1}(z - 1)\frac{5z - 2}{z(z - 1)} = \frac{3}{1} = 3,$$

thus

$$\oint_C f(z)\, dz = 2\pi i\, (2 + 3) = 10\pi i.$$

If C is in the clockwise direction, the answer would be $-10\pi i$.

(*b*) According to the second residue theorem,

$$\oint_C f(z)\, dz = 2\pi i\, Res_{z=0}\left[\frac{1}{z^2} f\left(\frac{1}{z}\right)\right].$$

Now

$$f\left(\frac{1}{z}\right) = \frac{5/z - 2}{1/z(1/z - 1)} = \frac{(5 - 2z)z}{1 - z},$$

$$\frac{1}{z^2} f\left(\frac{1}{z}\right) = \frac{5 - 2z}{z(1 - z)},$$

which has a simple pole at $z = 0$. Thus,

$$Res_{z=0}\left[\frac{1}{z^2}f\left(\frac{1}{z}\right)\right] = \lim_{z\to 0} z\frac{5 - 2z}{z(1 - z)} = \frac{5}{1} = 5.$$

Therefore,

$$\oint_C f(z)\, dz = 2\pi i \cdot 5 = 10\pi i.$$

Not surprisingly, this is the same result obtained in (a).

Example 3.4.2 Find the value of the integral

$$\oint_C \frac{dz}{z^3(z + 4)}$$

taken counterclockwise around the circle (a) $|z| = 2$, (b) $|z + 2| = 3$.

Solution 3.4.2 (a) The function has a third order pole at $z = 0$ and a simple pole at $z = -4$. Only $z = 0$ is inside the circle $|z| = 2$. Therefore,

$$\oint_C \frac{dz}{z^3(z + 4)} = 2\pi i\, Res_{z=0}[f(z)].$$

For the third-order pole,

$$Res_{z=0}[f(z)] = \frac{1}{2!}\lim_{z\to 0}\frac{d^2}{dz^2}z^3\frac{1}{z^3(z + 4)} = \frac{1}{2}\lim_{z\to 0}\frac{2}{(z + 4)^3} = \frac{1}{64}.$$

Therefore,

$$\oint_C \frac{dz}{z^3(z + 4)} = 2\pi i\frac{1}{64} = \frac{\pi}{32}i.$$

(b) For the circle $|z + 2| = 3$, the center is at $z = -2$ and the radius is 3. Both singular points are inside the circle. Thus,

$$\oint_C \frac{dz}{z^3(z + 4)} = 2\pi i\left\{Res_{z=0}[f(z)] + Res_{z=-4}[f(z)]\right\}.$$

Since

$$Res_{z=-4}[f(z)] = \lim_{z\to -4}(z + 4)\frac{1}{z^3(z + 4)} = \frac{1}{(-4)^3} = -\frac{1}{64},$$

so

$$\oint_C \frac{dz}{z^3 (z+4)} = 2\pi i \left\{ \frac{1}{64} - \frac{1}{64} \right\} = 0.$$

Example 3.4.3 Find the value of the integral

$$\oint_C \tan \pi z \, dz$$

taken counterclockwise around the unit circle $|z| = 1$.

Solution 3.4.3 Since

$$f(z) = \tan \pi z = \frac{\sin \pi z}{\cos \pi z},$$

and

$$\cos \frac{2n+1}{2} \pi = 0, \quad for \quad n = \cdots - 2, -1, 0, 1, 2 \cdots,$$

therefore $z = (2n+1)/2$ are zeros of $\cos \pi z$. Expanding $\cos \pi z$ about any of these zeros in Taylor series, one can readily see that $f(z)$ has a simple pole at each of these singular points. Among them, $z = \frac{1}{2}$ and $z = -\frac{1}{2}$ are inside $|z| = 1$. Hence

$$\oint_C \tan \pi z \, dz = 2\pi i \left\{ Res_{z=1/2}[f(z)] + Res_{z=-1/2}[f(z)] \right\}.$$

The simplest way to find these residues is by the "derivative of the denominator" method,

$$Res_{z=1/2}[f(z)] = \left[\frac{\sin \pi z}{(\cos \pi z)'} \right]_{z=\frac{1}{2}} = \left[\frac{\sin \pi z}{-\pi \sin \pi z} \right]_{z=\frac{1}{2}} = -\frac{1}{\pi},$$

$$Res_{z=-1/2}[f(z)] = \left[\frac{\sin \pi z}{(\cos \pi z)'} \right]_{z=-\frac{1}{2}} = \left[\frac{\sin \pi z}{-\pi \sin \pi z} \right]_{z=-\frac{1}{2}} = -\frac{1}{\pi}.$$

Therefore,

$$\oint_C \tan \pi z \, dz = 2\pi i \left\{ -\frac{1}{\pi} - \frac{1}{\pi} \right\} = -4i.$$

Example 3.4.4 Evaluate the integal $\oint_C f(z) \, dz$ for

$$f(z) = z^2 \exp \left(\frac{1}{z} \right),$$

where C is counterclockwise around the unit circle $|z| = 1$.

Solution 3.4.4 The function $f(z)$ has an essential singularity at $z = 0$. Thus,

$$\oint_C z^2 \exp\left(\frac{1}{z}\right) dz = 2\pi i \, Res_{z=0}[f(z)].$$

The residue is simply the coefficient of the z^{-1} term in the Laurent series about $z = 0$,

$$z^2 \exp\left(\frac{1}{z}\right) = z^2 \left(1 + \frac{1}{z} + \frac{1}{2!}\frac{1}{z^2} + \frac{1}{3!}\frac{1}{z^3} + \frac{1}{4!}\frac{1}{z^4} \cdots\right)$$

$$= z^2 + z + \frac{1}{2} + \frac{1}{3!}\frac{1}{z} + \frac{1}{4!}\frac{1}{z^2} + \cdots.$$

Therefore,

$$Res_{z=0}[f(z)] = \frac{1}{3!} = \frac{1}{6}.$$

Hence,

$$\oint_C z^2 \exp\left(\frac{1}{z}\right) dz = 2\pi i \frac{1}{6} = \frac{\pi}{3}i.$$

Example 3.4.5 Evaluate the integal $\oint_C f(z)\, dz$ for

$$f(z) = \frac{z^{99} \exp\left(\frac{1}{z}\right)}{z^{100} + 1},$$

where C is counterclockwise around the circle $|z| = 2$.

Solution 3.4.5 There are 100 singular points located on the circumference of the unit circle $|z| = 1$ and an essential singular point at $z = 0$. Obviously, the second residue theorem is more convenient. That is,

$$\oint_C f(z)\, dz = 2\pi i \, Res_{z=0}\left[\frac{1}{z^2} f\left(\frac{1}{z}\right)\right].$$

Now

$$f\left(\frac{1}{z}\right) = \frac{(1/z)^{99} \exp(z)}{(1/z)^{100} + 1} = \frac{z \exp(z)}{1 + z^{100}},$$

$$\frac{1}{z^2} f\left(\frac{1}{z}\right) = \frac{\exp(z)}{z(1 + z^{100})}.$$

So

$$Res_{z=0}\left[\frac{1}{z^2} f\left(\frac{1}{z}\right)\right] = \lim_{z \to 0} z \frac{\exp(z)}{z(1 + z^{100})} = 1.$$

Therefore,

$$\oint_C \frac{z^{99} \exp\left(\frac{1}{z}\right)}{z^{100} + 1} dz = 2\pi i.$$

Example 3.4.6 (*a*) Show that if $z = 1$ and $z = 2$ are inside the closed contour C, then

$$\oint_C \frac{1}{(z - 1)(z - 2)} dz = 0.$$

(*b*) Show that if all the singular points $s_1, s_2, \ldots, s_n$ of the following function:

$$f(z) = \frac{1}{(z - s_1)(z - s_2) \cdots (z - s_n)}$$

are inside the closed contour C, then

$$I = \oint_C f(z)\, dz = 0.$$

Solution 3.4.6 (*a*) Taking partial fraction, we have

$$\frac{1}{(z - 1)(z - 2)} = \frac{A}{(z - 1)} + \frac{B}{(z - 2)}.$$

So

$$\oint_C \frac{1}{(z - 1)(z - 2)} dz = \oint_C \frac{A}{(z - 1)} dz + \oint_C \frac{B}{(z - 2)} dz$$
$$= 2\pi i (A + B).$$

Since

$$\frac{A}{(z - 1)} + \frac{B}{(z - 2)} = \frac{A(z - 2) + B(z - 1)}{(z - 1)(z - 2)} = \frac{(A + B)z - (2A + B)}{(z - 1)(z - 2)}$$

and

$$\frac{(A + B)z - (2A + B)}{(z - 1)(z - 2)} = \frac{1}{(z - 1)(z - 2)},$$

it follows that

$$A + B = 0.$$

Therefore,

$$\oint_C \frac{1}{(z - 1)(z - 2)} dz = 0.$$

(b) The partial fraction of $f(z)$ is of the form

$$\frac{1}{(z-s_1)(z-s_2)\cdots(z-s_n)} = \frac{r_1}{(z-s_1)} + \frac{r_2}{(z-s_2)} + \cdots + \frac{r_n}{(z-s_n)}.$$

Therefore,

$$\oint_C f(z)\,dz = \oint_C \frac{r_1}{(z-s_1)}dz + \oint_C \frac{r_2}{(z-s_2)}dz + \cdots + \oint_C \frac{r_n}{(z-s_n)}dz$$
$$= 2\pi i\,(r_1 + r_2 + \cdots + r_n).$$

Now

$$\frac{r_1}{(z-s_1)} + \frac{r_2}{(z-s_2)} + \cdots + \frac{r_n}{(z-s_n)} = \frac{(r_1 + r_2 + \cdots + r_n)\,z^{n-1} + \cdots}{(z-s_1)(z-s_2)\cdots(z-s_n)},$$

and

$$\frac{(r_1 + r_2 + \cdots + r_n)\,z^{n-1} + \cdots}{(z-s_1)(z-s_2)\cdots(z-s_n)} = \frac{1}{(z-s_1)(z-s_2)\cdots(z-s_n)}.$$

Since the numerator of the right-hand side has no z^{n-1} term, therefore

$$(r_1 + r_2 + \cdots + r_n) = 0,$$

and

$$\oint_C \frac{1}{(z-s_1)(z-s_2)\cdots(z-s_n)}dz = 0.$$

3.5   Evaluation of Real Integrals with Residues

A very surprising fact is that we can use the residue theorem to evaluate integrals of real variable. For certain classes of complicated real integrals, residue theorem offers a simple and elegant way of carrying out the integration.

3.5.1   Integrals of Trigonometric Functions

Let us consider the integral of the type

$$I = \int_0^{2\pi} F(\cos\theta, \sin\theta)\,d\theta.$$

If we make the substitution

$$z = e^{i\theta}, \quad \frac{dz}{d\theta} = ie^{i\theta} = iz,$$

then

$$\cos\theta = \frac{1}{2}(e^{i\theta} + e^{-i\theta}) = \frac{1}{2}\left(z + \frac{1}{z}\right)$$

$$\sin\theta = \frac{1}{2i}(e^{i\theta} - e^{-i\theta}) = \frac{1}{2i}\left(z - \frac{1}{z}\right),$$

and

$$d\theta = \frac{1}{iz}dz.$$

The given integral takes the form

$$I = \oint_C f(z)\frac{dz}{iz},$$

the integration being taken counterclockwise around the unit circle centered at $z = 0$. We illustrate this method with the following examples.

Example 3.5.1 Show that

$$I = \int_0^{2\pi} \frac{d\theta}{\sqrt{2} - \cos\theta} = 2\pi.$$

Solution 3.5.1 With the transformation just discussed, we can write the integral as

$$I = \oint_C \frac{dz}{\left[\sqrt{2} - \frac{1}{2}\left(z + \frac{1}{z}\right)\right]iz} = \oint_C \frac{-2dz}{i(z^2 - 2\sqrt{2}z + 1)}$$

$$= -\frac{2}{i}\oint_C \frac{dz}{(z - \sqrt{2} - 1)(z - \sqrt{2} + 1)}.$$

The integrand has two simple poles. The one at $\sqrt{2} + 1$ lies outside the unit circle and is thus of no interest. The one at $\sqrt{2} - 1$ is inside the unit circle, and the residue at that point is

$$Res_{z=\sqrt{2}-1}[f(z)] = \lim_{z \to \sqrt{2}-1}(z - \sqrt{2} + 1)\frac{1}{(z - \sqrt{2} - 1)(z - \sqrt{2} + 1)} = -\frac{1}{2}.$$

Thus,

$$I = -\frac{2}{i} 2\pi i \left(-\frac{1}{2}\right) = 2\pi.$$

Example 3.5.2 Evaluate the integral

$$I = \int_0^\pi \frac{d\theta}{a - b \cos \theta}, \quad a > b > 0.$$

Solution 3.5.2 Since the integrand is symmetric about $\theta = \pi$, so we can extend the integration interval to $[0, 2\pi]$,

$$I = \frac{1}{2} \int_0^{2\pi} \frac{d\theta}{a - b \cos \theta},$$

which can be written as an integral around an unit circle in the complex plane

$$I = \frac{1}{2} \oint \frac{1}{a - b\frac{1}{2}(z + \frac{1}{z})} \frac{dz}{iz} = \oint \frac{1}{2az - bz^2 - b} \frac{dz}{i}.$$

Now

$$\oint \frac{1}{2az - bz^2 - b} \frac{dz}{i} = -\frac{1}{bi} \oint \frac{1}{z^2 - \frac{2a}{b}z + 1} dz,$$

taking this seemingly trivial step of making the coefficient of z^2 to be 1 can actually avoid many pitfalls of what follows. The singular points of the integrand are at the zeros of the denominator,

$$z^2 - \frac{2a}{b}z + 1 = 0.$$

Let z_1 and z_2 be the roots of this equation. They are easily found to be

$$z_1 = \frac{1}{b}\left(a - \sqrt{a^2 - b^2}\right), \quad z_2 = \frac{1}{b}\left(a + \sqrt{a^2 - b^2}\right).$$

Since

$$(z - z_1)(z - z_2) = z^2 - (z_1 + z_2) + z_1 z_2 = z^2 - \frac{2a}{b}z + 1,$$

it follows that

$$z_1 z_2 = 1.$$

This means that one root must be greater than 1, and the other less than 1. Furthermore, $z_1 < z_2$, z_1 must be less than 1 and z_2 greater than 1. Therefore, only z_1 is inside the unit circle. Thus,

$$\oint \frac{dz}{(z - z_1)(z - z_2)} = 2\pi i \, Res_{z=z_1}[f(z)]$$

$$= 2\pi i \lim_{z \to z_1} (z - z_1) \frac{1}{(z - z_1)(z - z_2)} = 2\pi i \frac{1}{(z_1 - z_2)},$$

and

$$\frac{1}{(z_1 - z_2)} = -\frac{b}{2\sqrt{a^2 - b^2}}.$$

Therefore,

$$I = -\frac{1}{bi} 2\pi i \left(-\frac{b}{2\sqrt{a^2 - b^2}} \right) = \frac{\pi}{\sqrt{a^2 - b^2}}.$$

Example 3.5.3 Show that

$$\int_0^{2\pi} \cos^{2n} \theta \, d\theta = \frac{2\pi \, (2n)!}{2^{2n} \, (n!)^2}.$$

Solution 3.5.3 The integral can be written as

$$I = \int_0^{2\pi} \cos^{2n} \theta \, d\theta = \oint_C \left[\frac{1}{2} \left(z + \frac{1}{z} \right) \right]^{2n} \frac{dz}{iz} = \frac{1}{2^{2n} i} \oint_C \left[\sum_{k=0}^{2n} C_k^{2n} z^{2n-k} \frac{1}{z^k} \right] \frac{dz}{z},$$

where C_k^{2n} are the binomial coefficients

$$C_k^{2n} = \frac{(2n)!}{k!(2n - k)!}.$$

Carrying out the integration term by term, the only nonvanishing term is the term of z^{-1}. Since

$$\left[\sum_{k=0}^{2n} C_k^{2n} z^{2n-k} \frac{1}{z^k} \right] \frac{1}{z} = \left[\sum_{k=0}^{2n} C_k^{2n} z^{2n-2k} \right] \frac{1}{z},$$

it is clear that the coefficient of z^{-1} is given by term with $k = n$. Thus,

$$I = \frac{1}{2^{2n} i} \oint_C \left[z^{2n-1} + 2n z^{2n-3} + \cdots \frac{C_n^{2n}}{z} + \cdots \frac{1}{z^{2n+1}} \right] dz$$

$$= \frac{1}{2^{2n} i} 2\pi i C_n^{2n} = \frac{2\pi \, (2n)!}{2^{2n} (n!)^2}.$$

3.5.2   Improper Integrals I: Closing the Contour with a Semicircle at Infinity

We consider the real integrals of the type

$$I = \int_{-\infty}^{\infty} f(x)dx.$$

Such an integral, for which the interval of integration is not finite, is called improper integral, and it has the meaning

$$\int_{-\infty}^{\infty} f(x)dx = \lim_{R \to \infty} \int_{-R}^{R} f(x)dx.$$

Under certain conditions, this type of integral can be evaluated with the residue theorem. The idea is to close the contour by adding additional pieces along which the integral is either zero or some multiple of the original integral along the real axis.

If $f(x)$ is a rational function (i.e. ratio of two polynomials) with no singularity on the real axis and

$$\lim_{z \to \infty} z f(z) = 0,$$

then it can be shown that the integral along the real axis from $-\infty$ to ∞ is equal to the integral around a closed contour which consists of (a) the straight line along the real axis and (b) the semicircle C_R at infinity as shown in Fig. 3.8.

This is so, because with

$$z = Re^{i\theta}, \quad dz = iRe^{i\theta}d\theta = izd\theta,$$

$$\left| \int_{C_R} f(z)\,dz \right| = \left| \int_0^{\pi} f(z)\,izd\theta \right| \leq Max\,|f(z)z|\,\pi,$$

which goes to zero as $R \to \infty$, since $\lim_{z \to \infty} z f(z) = 0$. Therefore,

$$\lim_{R \to \infty} \int_{C_R} f(z)\,dz = 0.$$

It follows that

$$\int_{-\infty}^{\infty} f(x)dx = \lim_{R \to \infty} \left[\int_{-R}^{R} f(x)dx + \int_{C_R} f(z)\,dz \right] = \oint_{u.h.p} f(z)\,dz,$$

where $u.h.p$ means the entire upper half-plane. As $R \to \infty$, all the poles of $f(z)$ in the upper half-plane will be inside the contour. Hence,

Fig. 3.8 As $R \to \infty$, the semicircle C_R is at infinity. The contour consists of the real axis and C_R encloses the entire upper half-plane

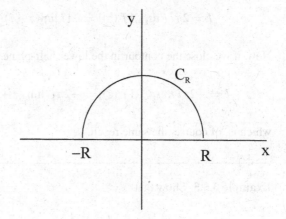

$$\int_{-\infty}^{\infty} f(x)dx = 2\pi i \text{ (sum of residues of } f(z) \text{ in the upper half-plane)}.$$

By the same token, we can, of course, close the contour in the lower half-plane. However, in that case, the direction of integration will be clockwise. Therefore,

$$\int_{-\infty}^{\infty} f(x)dx = \oint_{l.h.p} f(z)\,dz$$
$$= -2\pi i \text{ (sum of residues of } f(z) \text{ in the lower half-plane)}.$$

Example 3.5.4 Evaluate the integral

$$I = \int_{-\infty}^{\infty} \frac{1}{1+x^2}dx.$$

Solution 3.5.4 Since

$$\lim_{z \to \infty} z \frac{1}{1+z^2} = 0,$$

we can evaluate this integral with contour integration. That is

$$\int_{-\infty}^{\infty} \frac{1}{1+x^2}dx = \oint_{u.h.p} \frac{1}{1+z^2}dz.$$

The singular points of

$$f(z) = \frac{1}{1+z^2}$$

are at $z = i$ and $z = -i$. Only $z = i$ is in the upper half-plan. Therefore,

$$I = 2\pi i\, Res_{z=i}[f\,(z)] = 2\pi i \lim_{z\to i}(z-i)\frac{1}{(z-i)(z+i)} = \pi.$$

Now, if we close the contour in the lower half-plane, then

$$I = -2\pi i\, Res_{z=-i}[f\,(z)] = -2\pi i \lim_{z\to -i}(z+i)\frac{1}{(z-i)(z+i)} = \pi,$$

which is, of course, the same result.

Example 3.5.5 Show that

$$\int_{-\infty}^{\infty} \frac{1}{x^4+1}dx = \frac{\pi}{\sqrt{2}}.$$

Solution 3.5.5 The four singular points of the function

$$f\,(z) = \frac{1}{z^4+1}$$

are $e^{i\pi/4}$, $e^{i3\pi/4}$, $e^{i5\pi/4}$, $e^{i7\pi/4}$. Only two, $e^{i\pi/4}$, $e^{i3\pi/4}$ are in the upper half-plane. Therefore,

$$\oint_{u.h.p.} \frac{1}{z^4+1}dz = 2\pi i\{Res_{z=\exp(i\pi/4)}[f\,(z)] + Res_{z=\exp(i3\pi/4)}[f\,(z)]\}.$$

For problems of this type, it is much easier to find the residue by the method of $p(a)/q'(a)$. If we use the method of $\lim_{z\to a}(z-a)f(z)$, the calculation will be much more cumbersome. Since

$$Res_{z=\exp(i\pi/4)}[f\,(z)] = \left[\frac{1}{(z^4+1)'}\right]_{z=e^{i\pi/4}} = \left[\frac{1}{4z^3}\right]_{z=\exp(i\pi/4)} = \frac{1}{4}e^{-i3\pi/4},$$

$$Res_{z=\exp(i3\pi/4)}[f\,(z)] = \left[\frac{1}{4z^3}\right]_{z=\exp(i3\pi/4)} = \frac{1}{4}e^{-i9\pi/4} = \frac{1}{4}e^{-i\pi/4},$$

so

$$\int_{-\infty}^{\infty} \frac{1}{x^4+1}dx = 2\pi i\left[\frac{1}{4}e^{-i3\pi/4} + \frac{1}{4}e^{-i\pi/4}\right]$$

$$= \frac{\pi}{2}\left[e^{-i\pi/4} + e^{i\pi/4}\right] = \pi\cos\left(\frac{\pi}{4}\right) = \frac{\pi}{\sqrt{2}}.$$

3.5.3 Fourier Integral and Jordan's Lemma

Another very important class of integrals of the form

$$I = \int_{-\infty}^{\infty} e^{ikx} f(x)\,dx$$

can also be evaluated with the residue theorem. This class is known as the Fourier integral of $f(x)$. We will show that as long as $f(x)$ has no singularity along the real axis and

$$\lim_{z \to \infty} f(z) = 0, \tag{3.22}$$

the contour of this integral can be closed with an infinitely large semicircle in the upper half-plane if k is positive, and in the lower half-plane if k is negative. This statement is based on the Jordan's lemma, which states that, under the condition (3.22),

$$\lim_{R \to \infty} \int_{C_R} e^{ikz} f(z)\,dz = 0$$

where k is a positive real number and C_R is the semicircle in the upper half-plane with infinitely large radius R.

To prove this lemma, we first make the following observation. In Fig. 3.9, $y = \sin\theta$ and $y = \frac{2}{\pi}\theta$ are shown together. It is seen that in the interval $[0, \frac{\pi}{2}]$, the curve $y = \sin\theta$ is concave and always lies on or above the straight line $y = \frac{2}{\pi}\theta$. Therefore,

$$\sin\theta \geq \frac{2}{\pi}\theta \quad for \quad 0 \leq \theta \leq \frac{\pi}{2}.$$

With

$$z = Re^{i\theta} = R\cos\theta + iR\sin\theta, \quad dz = iRe^{i\theta}d\theta,$$

Fig. 3.9 Visualization of the inequality $\sin\theta \geq 2\theta/\pi$ for $0 \leq \theta \leq \pi/2$

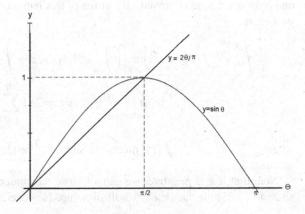

we have

$$\left| \int_{C_R} e^{ikz} f(z)\, dz \right| = \left| \int_0^\pi e^{ikz} f(z)\, i Re^{i\theta} d\theta \right| \leq \int_0^\pi \left| e^{ikz} \right| \left| f(z) \right| R \left| e^{i\theta} \right| d\theta.$$

Since

$$\left| e^{ikz} \right| = \left| e^{ik(R\cos\theta + iR\sin\theta)} \right| = \left| e^{ikR\cos\theta} \right| \left| e^{-kR\sin\theta} \right| = e^{-kR\sin\theta},$$

so

$$\left| \int_{C_R} e^{ikz} f(z)\, dz \right| \leq Max\, |f(z)|\, R \int_0^\pi e^{-kR\sin\theta} d\theta.$$

Using $\sin(\pi - \theta) = \sin\theta$, we can write the last integral as

$$\int_0^\pi e^{-kR\sin\theta} d\theta = 2 \int_0^{\pi/2} e^{-kR\sin\theta} d\theta.$$

Now, $\sin\theta \geq \frac{2}{\pi}\theta$ in the interval $[0, \pi/2]$, therefore

$$2 \int_0^{\pi/2} e^{-kR\sin\theta} d\theta \leq 2 \int_0^{\pi/2} e^{-kR2\theta/\pi} d\theta = \frac{\pi}{kR} \left(1 - e^{-kR} \right). \tag{3.23}$$

Thus,

$$\left| \int_{C_R} e^{ikz} f(z)\, dz \right| \leq Max\, |f(z)|\, \frac{\pi}{k} \left(1 - e^{-kR} \right).$$

As $z \to \infty$, $R \to \infty$ and the right-hand side of the last equation goes to zero, since $\lim_{z \to \infty} f(z) = 0$. It follows that

$$\lim_{z \to \infty} \int_{C_R} e^{ikz} f(z)\, dz = 0,$$

and Jordan's lemma is proved. By virtue of this lemma, the Fourier integral can be written as

$$\int_{-\infty}^{\infty} e^{ikx} f(x) dx = \lim_{R \to \infty} \left(\int_{-R}^{R} e^{ikx} f(x) dx + \int_{C_R} e^{ikz} f(z)\, dz \right)$$

$$= \oint_{u.h.p} e^{ikz} f(z)\, dz = 2\pi i \sum_{i=1}^{all} R_{u.h.p} \left[e^{ikz} f(z) \right],$$

where $\sum_{i=1}^{all} R_{u.h.p} \left[e^{ikz} f(z) \right]$ means the sum of all residues of $e^{ikz} f(z)$ in the upper half-plane.

Note that if k is negative, we cannot close the contour in the upper half-plane, since in (3.23) the factor e^{-kR} will blow up. However, in this case we can close the contour in the lower half-plane, because integrating from $\theta = 0$ to $\theta = -\pi$ will introduce another minus sign to make the large semicircular integral in the lower half-plane vanish. Therefore,

$$\int_{-\infty}^{\infty} e^{-i|k|x} f(x)dx = -2\pi i \sum_{i=1}^{all} R_{l.h.p} \left[e^{-i|k|z} f(z) \right],$$

where $\sum_{i=1}^{all} R_{l.h.p} \left[e^{-i|k|z} f(z) \right]$ means the sum of all residues of $e^{-i|k|z} f(z)$ in the lower half-plane. The minus sign is due to the fact that in this case the closed contour integration is clockwise.

Since $\sin kx$ and $\cos kx$ are linear combinations of e^{ikx} and e^{-ikx}, the real integrals of the form

$$\int_{-\infty}^{\infty} \cos kx f(x)dx \quad and \quad \int_{-\infty}^{\infty} \sin kx f(x)dx$$

can be obtained easily from this class of integrals,

$$\int_{-\infty}^{\infty} \cos kx f(x)dx = \frac{1}{2} \left[\int_{-\infty}^{\infty} e^{ikx} f(x)dx + \int_{-\infty}^{\infty} e^{-ikx} f(x)dx \right], \quad (3.24)$$

$$\int_{-\infty}^{\infty} \sin kx f(x)dx = \frac{1}{2i} \left[\int_{-\infty}^{\infty} e^{ikx} f(x)dx - \int_{-\infty}^{\infty} e^{-ikx} f(x)dx \right]. \quad (3.25)$$

If it is certain that the result of the integration is a finite real value, then we may write

$$\int_{-\infty}^{\infty} \cos kx f(x)dx = \mathrm{Re} \int_{-\infty}^{\infty} e^{ikx} f(x)dx, \quad (3.26)$$

$$\int_{-\infty}^{\infty} \sin kx f(x)dx = \mathrm{Im} \int_{-\infty}^{\infty} e^{ikx} f(x)dx. \quad (3.27)$$

These formulae must be used with caution. While (3.24) and (3.25) are always valid, (3.26) and (3.27) are valid only if there is no imaginary number in $f(x)$.

Example 3.5.6 Evaluate the integral

$$I = \int_{-\infty}^{\infty} \frac{\sin x}{x+i} dx.$$

Solution 3.5.6 There is a simple pole located in the lower half-plane, and

$$\int_{-\infty}^{\infty} \frac{\sin x}{x+i} dx = \frac{1}{2i} \int_{-\infty}^{\infty} \frac{e^{ix}}{x+i} dx - \frac{1}{2i} \int_{-\infty}^{\infty} \frac{e^{-ix}}{x+i} dx$$

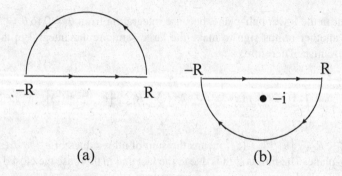

(a)   (b)

Fig. 3.10 Closing the contour with an infinitely large semicircle. **a** Contour closed in the upper half-plane, **b** contour closed in the lower half-plane

To evaluate the first integral in the right-hand side, we must close the contour in the upper half-plane as shown in Fig. 3.10a.

Since the function is analytic everywhere in the upper half-plane, therefore

$$\int_{-\infty}^{\infty} \frac{e^{ix}}{x+i} dx = \oint_{u.h.p} \frac{e^{iz}}{z+i} dz = 0.$$

To evaluate the second integral, we must close the contour in the lower half-plane as shown in Fig. 3.10b. Since there is a simple pole located at $z = -i$, we have

$$\int_{-\infty}^{\infty} \frac{e^{-ix}}{x+i} dx = \oint_{l.h.p} \frac{e^{-iz}}{z+i} dz = -2\pi i \, Res_{z=-i} \left[\frac{e^{-iz}}{z+i} \right]$$

$$= -2\pi i \lim_{z \to -i} (z+i) \frac{e^{-iz}}{z+i} = -2\pi i e^{-1}.$$

Thus,

$$I = \int_{-\infty}^{\infty} \frac{\sin x}{x+i} dx = \frac{1}{2i} \left[0 - (-2\pi i e^{-1}) \right] = \frac{\pi}{e}.$$

Clearly,

$$I \neq Im \int_{-\infty}^{\infty} \frac{e^{ix}}{x+i} dx,$$

this is because there is the imaginary number i in the function.

Example 3.5.7 Evaluate the integral

$$I = \int_{-\infty}^{\infty} \frac{1}{x^2+4} e^{-i\omega x} dx, \quad \omega > 0.$$

Solution 3.5.7

$$I = \oint_{l.h.p} \frac{1}{z^2+4} e^{-i\omega z} dz.$$

The only singular point in the lower half-plane is at $z = -2i$, therefore

$$I = -2\pi i \, Res_{z=-2i} \left[\frac{e^{-i\omega z}}{z^2+4} \right]$$

$$= -2\pi i \lim_{z \to -2i} (z+2i) \frac{e^{-i\omega z}}{(z+2i)(z-2i)} = -2\pi i \frac{e^{-2\omega}}{-4i} = \frac{\pi}{2} e^{-2\omega}.$$

This integral happens to be the Fourier transform of $\frac{1}{x^2+4}$, as we shall see in a later chapter.

Example 3.5.8 Evaluate the integral

$$I(t) = \frac{A}{2\pi} \int_{-\infty}^{\infty} \frac{e^{i\omega t}}{R+i\omega L} d\omega,$$

for both $t > 0$ and $t < 0$.

Solution 3.5.8

$$I = \frac{A}{2\pi} \frac{1}{iL} \int_{-\infty}^{\infty} \frac{e^{i\omega t}}{\left(\frac{R}{iL}\right)+\omega} d\omega = \frac{A}{2\pi} \frac{1}{iL} \int_{-\infty}^{\infty} \frac{e^{i\omega t}}{\omega - i\frac{R}{L}} d\omega$$

For $t > 0$, we can close the contour in the upper half-plane.

$$I = \frac{A}{2\pi} \frac{1}{iL} \oint_{u.h.p} \frac{e^{itz}}{z - i\frac{R}{L}} dz$$

The only singular point is located in the upper half-plane at $z = i\frac{R}{L}$. Therefore

$$I = \frac{A}{2\pi} \frac{1}{iL} 2\pi i \lim_{z \to i\frac{R}{L}} \left(z - i\frac{R}{L} \right) \frac{e^{itz}}{z - i\frac{R}{L}} = \frac{A}{L} e^{it(i\frac{R}{L})} = \frac{A}{L} e^{-\frac{R}{L}t}.$$

For $t < 0$, we must close the contour in the lower half-plane. Since there is no singular point in the lower half-plane, the integral is zero. Thus,

$$I(t) = \begin{cases} \frac{A}{L} e^{-\frac{R}{L}t} & t > 0 \\ 0 & t < 0 \end{cases}.$$

For those who are familiar with AC circuits, this integral $I(t)$ is the current in a circuit with resistance R and inductance L connected in series under a voltage impulse V. A high pulse in a short duration can be expressed as

$$V(t) = \frac{A}{2\pi} \int_{-\infty}^{\infty} e^{i\omega t} d\omega,$$

and the impedance of the circuit is $Z = R + i\omega L$ for the ω component, and the corresponding current is given by $\frac{V}{Z}$. Thus, the total current is the integral we have evaluated.

Example 3.5.9 Evaluate the integral

$$I = \int_{-\infty}^{\infty} \frac{x \sin x}{(x^2 + 1)} dx.$$

Solution 3.5.9

$$I = \int_{-\infty}^{\infty} \frac{x \sin x}{(x^2 + 1)} dx = \operatorname{Im} \int_{-\infty}^{\infty} \frac{x e^{ix}}{(x^2 + 1)} dx,$$

$$\int_{-\infty}^{\infty} \frac{x e^{ix}}{(x^2 + 1)} dx = \oint_{u.h.p} \frac{z e^{iz}}{(z^2 + 1)} dz.$$

There is only one singular point in the upper half-plane located at $z = i$. So,

$$\oint_{u.h.p} \frac{z e^{iz}}{(z^2 + 1)} dz = 2\pi i \, Res_{z=i} \left[\frac{z e^{iz}}{(z^2 + 1)} \right]$$

$$= 2\pi i \lim_{z \to i} (z - i) \frac{z e^{iz}}{(z^2 + 1)} = 2\pi i \frac{i e^{-1}}{2i} = \frac{\pi}{e} i,$$

and

$$I = \operatorname{Im} \left(\frac{\pi}{e} i \right) = \frac{\pi}{e}.$$

Example 3.5.10 (*a*) Show that

$$I = \int_0^{\infty} \frac{\cos bx}{x^2 + a^2} dx = \frac{\pi}{2a} e^{-ba}, \qquad a > 0, \quad b > 0.$$

(*b*) Use the result of (*a*) to find the value of

$$\int_0^{\infty} \frac{\cos bx}{(x^2 + a^2)^2} dx.$$

Solution 3.5.10 (a) Since the integrand is an even function, so

$$\int_0^\infty \frac{\cos bx}{x^2 + a^2} dx = \frac{1}{2} \int_{-\infty}^\infty \frac{\cos bx}{x^2 + a^2} dx.$$

$$\int_{-\infty}^\infty \frac{\cos bx}{x^2 + a^2} dx = \text{Re} \int_{-\infty}^\infty \frac{e^{ibx}}{x^2 + a^2} dx = \text{Re} \oint_{u.h.p} \frac{e^{ibz}}{z^2 + a^2} dz$$

The singular point in the upper half-plane is at $z = ia$, so

$$\oint_{u.h.p} \frac{e^{ibz}}{z^2 + a^2} dz = 2\pi i \, \text{Res}_{z=ia} \left[\frac{e^{ibz}}{z^2 + a^2} \right]$$

$$= 2\pi i \lim_{z \to ia} (z - ia) \frac{e^{ibz}}{(z - ia)(z + ia)} = 2\pi i \frac{e^{ib(ia)}}{2ai} = \frac{\pi}{a} e^{-ba}.$$

Thus,

$$\int_0^\infty \frac{\cos bx}{x^2 + a^2} dx = \frac{1}{2} \text{Re} \left(\frac{\pi}{a} e^{-ba} \right) = \frac{\pi}{2a} e^{-ba}.$$

(b) Taking derivative of both sides with respect to a,

$$\frac{d}{da} \int_0^\infty \frac{\cos bx}{x^2 + a^2} dx = \frac{d}{da} \left(\frac{\pi}{2a} e^{-ba} \right),$$

we have

$$\int_0^\infty \frac{-2a \cos bx}{(x^2 + a^2)^2} dx = \frac{-\pi}{2a^2} e^{-ba} + \frac{\pi(-b)}{2a} e^{-ba}.$$

Therefore,

$$\int_0^\infty \frac{\cos bx}{(x^2 + a^2)^2} dx = \frac{\pi}{4a^3} (1 + ab) e^{-ba}.$$

3.5.4 Improper Integrals II: Closing the Contour with Rectangular and Pie-Shaped Contour

If the integrand does not go to zero fast enough on the infinitely large contour C_R, then the contour cannot be closed with a large semicircle, up or down. For such a case, there may be other types of closed contours that will enable us to eliminate all parts of the integral but the desired portion. However, selecting an appropriate contour requires considerable ingenuity. Here, we present two additional kinds of contours that are known to be useful.

Rectangular Contour. If the height of the rectangle can be so chosen that the integral along the top side of the rectangle is equal to a constant multiple of the integral along the real axis, then such a contour may be useful for evaluating integrals whose integrand vanishes as the absolute value of the real variable goes to infinity. Generally, integrands containing exponential function or hyperbolic functions are good candidates for this method. Again the method is best illustrated by an example.

Example 3.5.11 Show that

$$I = \int_{-\infty}^{\infty} \frac{e^{ax}}{1 + e^x} dx = \frac{\pi}{\sin a\pi}, \qquad for \ \ 0 < a < 1.$$

Solution 3.5.11 First, we analytically continue the integrand to the complex plane,

$$f(z) = \frac{e^{az}}{1 + e^z}.$$

The denominator of $f(z)$ is unchanged if z is increased by $2\pi i$, whereas the numerator changes by a factor of $e^{a2\pi i}$. Thus, a rectangular contour shown in Fig. 3.11 may be appropriate.

Integrating around the rectangular loop, we have

$$\oint f(z)\,dz = J_1 + J_2 + J_3 + J_4,$$

where

$$J_1 = \int_{L_1} \frac{e^{az}}{1+e^z} dz = \int_{-R}^{R} \frac{e^{ax}}{1+e^x} dx,$$

$$J_2 = \int_{L_2} \frac{e^{az}}{1+e^z} dz = \int_{0}^{2\pi} \frac{e^{a(R+iy)}}{1+e^{R+iy}} i\,dy,$$

$$J_3 = \int_{L_3} \frac{e^{az}}{1+e^z} dz = \int_{R}^{-R} \frac{e^{a(x+i2\pi)}}{1+e^{(x+i2\pi)}} dx = e^{i2\pi a} \int_{R}^{-R} \frac{e^{ax}}{1+e^x} dx,$$

$$J_4 = \int_{L_3} \frac{e^{az}}{1+e^z} dz = \int_{2\pi}^{0} \frac{e^{a(-R+iy)}}{1+e^{-R+iy}} i\,dy.$$

As $R \to \infty$,

$$\lim_{R\to\infty} J_1 = \lim_{R\to\infty} \int_{-R}^{R} \frac{e^{ax}}{1+e^x} dx = \int_{-\infty}^{\infty} \frac{e^{ax}}{1+e^x} dx = I,$$

$$\lim_{R\to\infty} J_3 = \lim_{R\to\infty} e^{i2\pi a} \int_{R}^{-R} \frac{e^{ax}}{1+e^x} dx = -e^{i2\pi a} \int_{-\infty}^{\infty} \frac{e^{ax}}{1+e^x} dx = -e^{i2\pi a} I.$$

Fig. 3.11 A closed rectangular contour

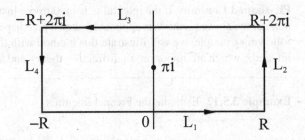

Furthermore, since $\left|e^{a(R+iy)}\right| = e^{aR}$ and the minimum value of $\left|1 + e^{R+iy}\right|$ is $\left|1 - e^R\right|$, hence

$$\lim_{R \to \infty} |J_2| \leq \lim_{R \to \infty} \left|\frac{e^{aR}}{1 - e^R}\right| 2\pi = \lim_{R \to \infty} \frac{2\pi}{e^{(1-a)R}} \to 0, \quad \text{since } a < 1.$$

Similarly,

$$\lim_{R \to \infty} |J_4| \leq \lim_{R \to \infty} \left|\frac{e^{-aR}}{1 - e^{-R}}\right| 2\pi = \lim_{R \to \infty} 2\pi e^{-aR} \to 0, \quad \text{since } a > 0.$$

Therefore,

$$\lim_{R \to \infty} \oint \frac{e^{az}}{1 + e^z} dz = (1 - e^{i2\pi a})I.$$

Now inside the loop, there is a simple pole at $z = i\pi$, since

$$1 + e^z = 1 + e^{i\pi} = 1 - 1 = 0.$$

By the residue theorem, we have

$$\oint \frac{e^{az}}{1 + e^z} dz = 2\pi i \, \text{Res}_{z=i\pi} \left[\frac{e^{az}}{1 + e^z}\right] = 2\pi i \left[\frac{e^{az}}{(1 + e^z)'}\right]_{z=i\pi}$$

$$= 2\pi i \frac{e^{i\pi a}}{e^{i\pi}} = -2\pi i e^{i\pi a}.$$

Thus,

$$(1 - e^{i2\pi a})I = -2\pi i e^{i\pi a},$$

so

$$I = \frac{-2\pi i e^{i\pi a}}{1 - e^{i2\pi a}} = \frac{-2\pi i}{e^{-i\pi a} - e^{i\pi a}} = \frac{\pi}{\sin \pi a}.$$

Pie-shaped Contour. If the integral is from 0 to ∞, instead of from $-\infty$ to ∞ and none of the above methods is applicable, then a pie-shaped contour may work. In the following example, we will illustrate this method with the evaluation of the Fresnel integrals, which are important in diffraction theory and signal propagation.

Example 3.5.12 Evaluate the Fresnel integrals

$$I_c = \int_0^\infty \cos\left(x^2\right) dx, \quad I_s = \int_0^\infty \sin\left(x^2\right) dx.$$

Solution 3.5.12 The two Fresnel integrals are the real and imaginary parts of the exponential integral,

$$\int_0^\infty e^{ix^2} dx = I_c + i I_s.$$

We integrate the complex function e^{iz^2} around the pie-shaped contour shown in Fig. 3.12. Since the function is analytic within the closed contour, the loop integral must be zero,

$$\oint e^{iz^2} dz = 0.$$

This loop integral naturally divides into three parts. First, from 0 to R along the real x-axis, then along the path of an arc C_R from R to R'. Finally, returning to 0 along the straight radial line with $\theta = \pi/4$.

$$\int_0^R e^{ix^2} dx + \int_{C_R} e^{iz^2} dz + \int_{R'}^0 e^{iz^2} dz = 0.$$

In the limit of $R \to \infty$, the first integral is what we want to find,

Fig. 3.12 A pie-shaped contour. In the complex plane, R' is at $z\left(R'\right) = Re^{i\pi/4}$

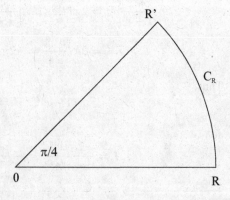

$$\lim_{R\to\infty}\int_0^R e^{ix^2}dx = \int_0^\infty e^{ix^2}dx.$$

On the path of the third integral, with $z = re^{i\theta}$ and $\theta = \pi/4$,

$$z^2 = r^2e^{i2\theta} = r^2e^{i\pi/2} = ir^2,$$
$$dz = e^{i\theta}dr = e^{i\pi/4}dr,$$

so the third integral becomes

$$\int_{R'}^0 e^{iz^2}dz = \int_R^0 e^{-r^2}e^{i\pi/4}dr = -e^{i\pi/4}\int_0^R e^{-r^2}dr.$$

We will now show that the second integral along C_R is equal to zero in the limit of $R \to \infty$. On C_R

$$z = Re^{i\theta}, \qquad dz = iRe^{i\theta}d\theta,$$
$$z^2 = R^2e^{i2\theta} = R^2(\cos 2\theta + i \sin 2\theta),$$

so the second integral can be written as

$$\int_{C_R} e^{iz^2}dz = iR\int_0^{\pi/4} e^{iR^2(\cos 2\theta + i\sin 2\theta)}e^{i\theta}d\theta$$
$$= iR\int_0^{\pi/4} e^{i(R^2\cos 2\theta + \theta)}e^{-R^2\sin 2\theta}d\theta.$$

Thus,

$$\left|\int_R^{R'} e^{iz^2}dz\right| \le R\int_0^{\pi/4} e^{-R^2\sin 2\theta}d\theta = \frac{R}{2}\int_0^{\pi/2} e^{-R^2\sin\phi}d\phi,$$

where $\phi = 2\theta$. According to (3.23) of Jordan's lemma,

$$\int_0^{\pi/2} e^{-R^2\sin\phi}d\phi \le \frac{\pi}{2R^2}(1 - e^{-R^2}).$$

Therefore, it goes to zero as $1/R^2$ for $R \to \infty$. With the second integral equal to zero, we are left with

$$\int_0^\infty e^{ix^2}dx - e^{i\pi/4}\int_0^\infty e^{-r^2}dr = 0. \qquad (3.28)$$

Now it is well known that

$$\int_0^\infty e^{-x^2}dx = \frac{\sqrt{\pi}}{2}.$$

To verify this expression, define

$$I = \int_0^\infty e^{-x^2}dx = \int_0^\infty e^{-y^2}dy,$$

so

$$I^2 = \int_0^\infty e^{-x^2}dx \int_0^\infty e^{-y^2}dy = \int_0^\infty \int_0^\infty e^{-(x^2+y^2)}dxdy.$$

In polar coordinates

$$I^2 = \int_0^\infty \int_0^{\pi/2} e^{-\rho^2}\rho d\varphi d\rho = \frac{\pi}{2}\int_0^\infty e^{-r^2}\rho d\rho = \frac{\pi}{4},$$

so $I = \sqrt{\pi}/2$.

It follows from (3.28) that

$$\int_0^\infty e^{ix^2}dx = e^{i\pi/4}\int_0^\infty e^{-r^2}dr = e^{i\pi/4}\frac{\sqrt{\pi}}{2} = \left(\cos\frac{\pi}{4} + i\sin\frac{\pi}{4}\right)\frac{\sqrt{\pi}}{2}$$

$$= \left(\frac{1}{\sqrt{2}} + i\frac{1}{\sqrt{2}}\right)\frac{\sqrt{\pi}}{2} = \sqrt{\frac{\pi}{8}} + i\sqrt{\frac{\pi}{8}}.$$

Therefore,

$$\int_0^\infty \cos\left(x^2\right)dx = \int_0^\infty \sin\left(x^2\right)dx = \sqrt{\frac{\pi}{8}}.$$

3.5.5   Integration Along a Branch Cut

Some integrals of multivalued functions can also be evaluated by Cauchy's residue theorem. For example, the integrand of the integral

$$I = \int_0^\infty x^{-\alpha}f(x)dx$$

is multivalued if α is not an integer. In the complex plane, $z^{-\alpha}$ is multivalued because with z expressed as

$$z = re^{i(\theta+n2\pi)}$$

where n is an integer, $z^{-\alpha}$ becomes

$$z^{-\alpha} = e^{-\alpha \ln z} = e^{-\alpha(\ln r + i\theta + in2\pi)}.$$

Fig. 3.13 A contour that excludes the branch cut along the positive x-axis

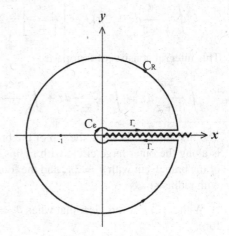

It is seen that $z^{-\alpha}$ is a multivalued function. For instance, with $\alpha = 1/3$,

$$z^{-\frac{1}{3}} = \begin{cases} e^{-\frac{1}{3}(\ln r + i\theta)} & n = 0 \\ e^{-\frac{1}{3}(\ln r + i\theta)}e^{i2\pi/3} = (-\frac{1}{2} + i\frac{\sqrt{3}}{2})e^{-\frac{1}{3}(\ln r + i\theta)} & n = 1 \\ e^{-\frac{1}{3}(\ln r + i\theta)}e^{i4\pi/3} = (-\frac{1}{2} - i\frac{\sqrt{3}}{2})e^{-\frac{1}{3}(\ln r + i\theta)} & n = 2. \end{cases}$$

To define $z^{-\alpha}$ as a single-valued function, the angle θ must be restricted in an interval of 2π by a branch cut. If we choose the branch cut along the positive x-axis, then our real integral is an integral along the top of the branch cut. Very often the problem can be solved with a closed contour as shown in Fig. 3.13, in which the entire branch cut is excluded from the interior of the contour. Again let us illustrate the method with an example.

Example 3.5.13 Evaluate the integral

$$I = \int_0^\infty \frac{x^{-\alpha}}{1+x} dx, \qquad 0 < \alpha < 1.$$

Solution 3.5.13 Consider the contour integral

$$\oint \frac{z^{-\alpha}}{1+z} dz$$

around the closed contour shown in Fig. 3.13. Since the branch point at $z = 0$ and the entire branch cut are excluded, the only singular point inside this contour is at $z = -1$. Therefore,

$$\oint \frac{z^{-\alpha}}{1+z}dz = 2\pi i \, Res\left[\frac{z^{-\alpha}}{1+z}\right] = 2\pi i \, (-1)^{-\alpha} = 2\pi i \, e^{i\pi(-\alpha)} = \frac{2\pi i}{e^{i\pi\alpha}}.$$

This integral consists of four parts,

$$\oint \frac{z^{-\alpha}}{1+z}dz = \int_{\Gamma_+} \frac{z^{-\alpha}}{1+z}dz + \int_{C_R} \frac{z^{-\alpha}}{1+z}dz + \int_{\Gamma_-} \frac{z^{-\alpha}}{1+z}dz + \int_{C_\epsilon} \frac{z^{-\alpha}}{1+z}dz.$$

The first integral is along the top of the branch cut with $\theta = 0$, the second integral is along the outer large circle with radius R, the third integral is along the bottom of the branch cut with $\theta = 2\pi$, and the fourth integral is along the inner small circle with radius ϵ.

With $z = re^{i\theta}$, it is clear that when $\theta = 0$ and $\theta = 2\pi$, r is the same as x. Therefore,

$$\int_{\Gamma_+} \frac{z^{-\alpha}}{1+z}dz = \int_\epsilon^R \frac{x^{-\alpha}}{1+x}dx,$$

$$\int_{\Gamma_-} \frac{z^{-\alpha}}{1+z}dz = \int_R^\epsilon \frac{x^{-\alpha}e^{i2\pi(-\alpha)}}{1+xe^{i2\pi}}e^{i2\pi}dx = -e^{-i2\pi\alpha}\int_\epsilon^R \frac{x^{-\alpha}}{1+x}dx.$$

On C_R, $z = Re^{i\theta}$,

$$\left|\int_{C_R} \frac{z^{-\alpha}}{1+z}dz\right| \le \left|\frac{R^{-\alpha}}{1-R}2\pi R\right|,$$

where $R^{-\alpha}$ is maximum of the numerator, $1 - R$ is the minimum of denominator and $2\pi R$ is the length of C_R. As $R \to \infty$,

$$\left|\frac{R^{-\alpha}}{1-R}2\pi R\right| \sim R^{-\alpha} \to 0, \quad \text{since } \alpha > 0.$$

Similarly, on C_ϵ, $z = \epsilon e^{i\theta}$,

$$\left|\int_{C_\epsilon} \frac{z^{-\alpha}}{1+z}dz\right| \le \frac{\epsilon^{-\alpha}}{1-\epsilon}2\pi\epsilon.$$

As $\epsilon \to 0$,

$$\frac{\epsilon^{-\alpha}}{1-\epsilon}2\pi\epsilon \sim \epsilon^{1-\alpha} \to 0, \quad \text{since } \alpha < 1.$$

On taking the limit $R \to \infty$ and $\epsilon \to 0$, we are left with

$$\oint \frac{z^{-\alpha}}{1+z}dz = \int_0^\infty \frac{x^{-\alpha}}{1+x}dx - e^{-i2\pi\alpha}\int_0^\infty \frac{x^{-\alpha}}{1+x}dx = \frac{2\pi i}{e^{i\pi\alpha}}.$$

Thus,

$$\left(1 - e^{-i2\pi\alpha}\right) \int_0^\infty \frac{x^{-\alpha}}{1+x} dx = \frac{2\pi i}{e^{i\pi\alpha}},$$

and

$$\int_0^\infty \frac{x^{-\alpha}}{1+x} dx = \frac{2\pi i}{e^{i\pi\alpha}\left(1 - e^{-i2\pi\alpha}\right)} = \frac{2\pi i}{e^{i\pi\alpha} - e^{-i\pi\alpha}} = \frac{\pi}{\sin \pi\alpha}.$$

3.5.6 Principal Value and Indented Path Integrals

Sometimes we have to deal with integrals $\int f(x) dx$ whose integrand becomes infinite at a point $x = x_0$ in the range of integration

$$\lim_{x \to x_0} f(x) = \infty.$$

In order to make sense of this kind of integral, we define the principal value integral as

$$P \int_{-R}^R f(x) dx = \lim_{\varepsilon \to 0} \left[\int_{-R}^{x_0 - \varepsilon} f(x) dx + \int_{x_0 + \varepsilon}^R f(x) dx \right].$$

It is a way to avoid singularity. One integrates within a small distance ε of the singularity in question, skips over the singularity, and begins integrating again at a distance ε beyond the singularity.

When evaluating the integral using the residue theorem, we are not allowed to have a singularity on the contour, however, with a principal value integrals we can accommodate simple poles on the contour by deforming the contour so as to avoid the poles.

The principal value integral

$$P \int_{-\infty}^\infty f(x) dx$$

can be evaluated by the theorem of residue for a function $f(z)$ that satisfies the asymptotic conditions that we have discussed. That is, either $zf(z) \to 0$ as $z \to \infty$, or $f(z) = e^{imz} g(z)$ and $g(z) \to 0$ as $z \to \infty$. Let us first assume that $f(z)$ has one simple pole on the real axis at $z = x_0$ and is analytic everywhere else. In this case, it is clear that the closed contour integral around the indented path shown in Fig. 3.14 is equal to zero,

$$\oint f(z) dz = 0,$$

since the only singular point is outside the contour.

Fig. 3.14 The closed contour consists of a large semicircle C_R in the upper half plane of radius R, the line segments from $-R$ to $x_0 - \varepsilon$ and from $x_0 + \varepsilon$ to R along the real axis, and a small semicircle C_ε of radius ϵ above the singular point x_0

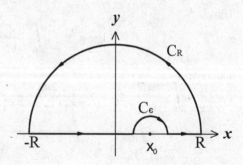

The integral can be written as

$$\oint f(z)\, dz = \int_{-R}^{x_0-\varepsilon} f(x)\, dx + \int_{C_\varepsilon} f(z)\, dz + \int_{x_0+\varepsilon}^{R} f(x)\, dx + \int_{C_R} f(z)\, dz.$$

In the limit of $R \to \infty$, with $f(z)$ satisfying the specified conditions, we have shown

$$\int_{C_R} f(z)\, dz = 0.$$

Furthermore, the two line integrals along the x-axis become the principal value integral as $\varepsilon \to 0$,

$$\lim_{\varepsilon \to 0} \left[\int_{-\infty}^{x_0-\varepsilon} f(x)\, dx + \int_{x_0+\varepsilon}^{\infty} f(x)\, dx \right] = P \int_{-\infty}^{\infty} f(x)\, dx.$$

Since $f(z)$ has a simple pole at $z = x_0$, so in the immediate neighborhood of x_0, the Laurent series of $f(z)$ has the form

$$f(z) = \frac{a_{-1}}{z - x_0} + \sum_{n=0}^{\infty} a_n (z - x_0)^n.$$

On the semicircle C_ε around x_0,

$$z - x_0 = \varepsilon e^{i\theta}, \qquad dz = i\varepsilon e^{i\theta} d\theta,$$

where ε is the radius of the semicircle. The integral around C_ε can thus be written as

$$\int_{C_\varepsilon} f(z)\, dz = \int_{\pi}^{0} \left(\frac{a_{-1}}{\varepsilon e^{i\theta}} + \sum_{n=0}^{\infty} a_n \left(\varepsilon e^{i\theta} \right)^n \right) i\varepsilon e^{i\theta} d\theta$$

On taking the limit $\varepsilon \to 0$, every term vanishes except the first. Therefore

$$\lim_{\varepsilon \to 0} \int_{C_\varepsilon} f(z)\,dz = \int_\pi^0 a_{-1} i\,d\theta = -i\pi a_{-1} = -i\pi\, Res_{z=x_0}\left[f(z)\right].$$

It follows that in the limit $R \to \infty$ and $\varepsilon \to 0$,

$$\oint f(z)\,dz = P \int_{-\infty}^\infty f(x)\,dx - i\pi\, Res\left[f(z)\right] = 0.$$

Therefore,

$$P \int_{-\infty}^\infty f(x)\,dx = i\pi\, Res\left[f(z)\right].$$

Note that to avoid the singular point, we can just as well go below it instead above. For the semicircle below the x-axis, the direction of integration is counterclockwise,

$$\lim_{\varepsilon \to 0} \int_{C_\varepsilon} f(z)\,dz = \int_\pi^{2\pi} a_{-1} i\,d\theta = i\pi a_{-1} = i\pi\, Res_{z=x_0}\left[f(z)\right].$$

However, in that case, the singular point is inside the closed contour, and the loop integral is equal to $2\pi i$ times the residue at $z = x_0$. So, we have

$$\oint f(z)\,dz = P \int_{-\infty}^\infty f(x)\,dx + i\pi\, Res_{z=x_0}\left[f(z)\right] = 2\pi i\, Res_{z=x_0}\left[f(z)\right].$$

Not surprisingly, we get the same result,

$$P \int_{-\infty}^\infty f(x)\,dx = 2\pi i\, Res_{z=x_0}\left[f(z)\right] - i\pi\, Res_{z=x_0}\left[f(z)\right] = i\pi\, Res_{z=x_0}\left[f(z)\right].$$

Now if $f(z)$ has more than one pole on the real axis, (all of them first order), furthermore, it has other singularities in the upper half-plane, (not necessary first order), then by the same argument one can show that

$$P \int_{-\infty}^\infty f(x)\,dx = \pi i \left(\sum residues\ on\ x\ axis\right)$$

$$+ 2\pi i \left(\sum resudues\ in\ upper\ half\ plane\right).$$

Example 3.5.14 Find the principal value of

$$P \int_{-\infty}^\infty \frac{e^{ix}}{x}\,dx,$$

and use the result to show

$$\int_{-\infty}^{\infty} \frac{\sin x}{x} dx = \pi.$$

Solution 3.5.14 The only singular point is at $x = 0$, therefore

$$P \int_{-\infty}^{\infty} \frac{e^{ix}}{x} dx = \pi i \, Res_{z=0} \left[\frac{e^{iz}}{z} \right] = \pi i \left[\frac{e^{iz}}{z'} \right]_{z=0} = \pi i.$$

Since

$$P \int_{-\infty}^{\infty} \frac{e^{ix}}{x} dx = P \left[\int_{-\infty}^{\infty} \frac{\cos x}{x} dx + i \int_{-\infty}^{\infty} \frac{\sin x}{x} dx \right],$$

therefore

$$P \int_{-\infty}^{\infty} \frac{\sin x}{x} dx = \text{Im} \left(P \int_{-\infty}^{\infty} \frac{e^{ix}}{x} dx \right) = \pi.$$

We note that $x = 0$ is actually a removable singularity of $\sin x/x$, since as $x \to 0$, $\sin x/x = 1$. This means that ε, instead of approaching zero, can be set equal to exactly zero. Therefore, the principal value of the integral is the integral itself,

$$\int_{-\infty}^{\infty} \frac{\sin x}{x} dx = \pi.$$

It is instructive to check this result in the following way. Since $\sin x/x$ is continuous at $x = 0$, if we move the path of integration an infinitesimal amount at $x = 0$, the value of the integral will not be changed. So let the path go through an infinitesimally small semicircle C_ε on top of the point at $x = 0$. Let us call the indented path the path from $-\infty$ to ε along x-axis, followed by C_ε and then continue from ε to ∞ along the x-axis. Then

$$\int_{-\infty}^{\infty} \frac{\sin x}{x} dx = \int_{Indented} \frac{\sin z}{z} dz.$$

Now let us make use of the identity

$$\sin z = \frac{1}{2i} \left(e^{iz} - e^{-iz} \right),$$

so

$$\int_{-\infty}^{\infty} \frac{\sin x}{x} dx = \frac{1}{2i} \int_{Indented} \frac{e^{iz}}{z} dz - \frac{1}{2i} \int_{Indented} \frac{e^{-iz}}{z} dz.$$

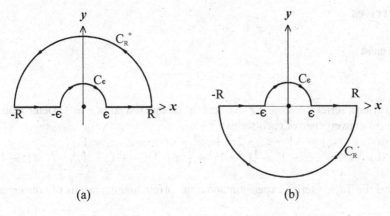

Fig. 3.15 a The indented path from $-R$ to R is closed by a large semicircle C_R^+ in the upper half-plane. **b** The same indented path from $-R$ to R is closed by a large semicircle C_R^- in the lower half-plane

For the first integral in the right-hand side, we can close the contour with an infinitely large semicircle C_R^+ in the upper half-plane, as shown in Fig. 3.15a. Since the singular point is outside the closed contour, so the contour integral vanishes,

$$\int_{Indented} \frac{e^{iz}}{z} dz = \oint_{u.h.p} \frac{e^{iz}}{z} dz = 0.$$

For the second indented path integral, we cannot close the contour in the upper half-plane because of e^{-iz}, so we have to close the contour in the lower half-plane with C_R^-, as shown in Fig. 3.15b. In this case, the singular point at $z = 0$ is inside the contour, therefore

$$\int_{Indented} \frac{e^{-iz}}{z} dz = \oint_{l.h.p} \frac{e^{-iz}}{z} dz = -2\pi i \, Res_{z=0} \left[\frac{e^{-iz}}{z} \right] = -2\pi i.$$

The minus sign accounts for the clockwise direction. It follows that

$$\int_{-\infty}^{\infty} \frac{\sin x}{x} dx = \frac{1}{2i} 0 - \frac{1}{2i} (-2\pi i) = \pi,$$

which is the same as we obtained before.

Exercises

1. Expand

$$f(z) = \frac{z - 1}{z + 1}$$

in a Taylor's series (a) around $z = 0$ and (b) around the point $z = 1$. Determine the radius of convergence of each series.

  Ans: (a) $f(z) = -1 + 2z - 2z^2 + 2z^3 - \cdots \cdots$. $|z| < 1$
  (b) $f(z) = \frac{1}{2}(z - 1) - \frac{1}{4}(z - 1)^2 + \frac{1}{8}(z - 1)^3 - \frac{1}{16}(z - 1)^4 \cdots$. $|z - 1| <$ 2.

2. Find the Taylor series expansion about the origin and the radius of convergence for

$$(a) \quad f(z) = \sin z, \qquad (b) \quad f(z) = \cos z,$$

$$(c) \quad f(z) = e^z, \qquad (d) \quad f(z) = \frac{1}{(1 - z)^m}.$$

  Ans: (a) $\sin z = z - \frac{1}{3!}z^3 + \frac{1}{5!}z^5 - \frac{1}{7!}z^7 + \cdots \cdots$ all z.
  (b) $\cos z = 1 - \frac{1}{2}z^2 + \frac{1}{4!}z^4 - \frac{1}{6!}z^6 + \cdots \cdots$ all z.
  (c) $e^z = 1 + z + \frac{1}{2}z^2 + \frac{1}{3!}z^3 + \cdots \cdots \cdots$ all z.
  (d) $\frac{1}{(1-z)^m} = 1 + mz + \frac{m(m+1)}{2}z^2 + \frac{m(m+1)(m+2)}{3!}z^3 + \cdots$. $|z| < 1$.

3. Find the Taylor series expansion of

$$f(z) = \ln z$$

around $z = 1$ by noting that

$$\frac{d}{dz} \ln z = \frac{1}{z}.$$

  Ans: $\ln z = (z - 1) - \frac{1}{2}(z - 1)^2 + \frac{1}{3}(z - 1)^3 - \frac{1}{4}(z - 1)^4 \cdots$. $|z - 1| < 1$.

4. Expand

$$f(z) = \frac{1}{(z + 1)(z + 2)}$$

in a Taylor's series (a) around $z = 0$ and (b) around the point $z = 2$. Determine the radius of convergence of each series.

  Ans: (a) $f(z) = \frac{1}{2} - \frac{3}{4}z + \frac{7}{8}z^2 - \frac{15}{16}z^3 + \cdots \cdots$. $|z| < 1$.
  (b) $f(z) = (\frac{1}{3} - \frac{1}{4}) - (\frac{1}{3^2} - \frac{1}{4^2})(z - 2) + (\frac{1}{3^3} - \frac{1}{4^3})(z - 2)^2 \cdots \cdots$. $|z - 2| <$ 3.

5. Without obtaining the series, determine the radius of convergence of each of the following expansions:

$$(a) \ \tan^{-1} z \quad around \ z = 1, \quad (b) \ \frac{1}{e^z - 1} \quad around \ z = 4i,$$

$$(c) \ \frac{x}{x^2 + 2x + 10} \quad around \ x = 0.$$

Ans: (a) $\sqrt{2}$, (b) $2\pi - 4$, (c) $\sqrt{10}$.

6. Find the Laurent series for

$$f(z) = \frac{1}{z^2 - 3z + 2},$$

in the region of

$$(a) \ |z| < 1, \quad (b) \ 1 < |z| < 2, \quad (c) \ 0 < |z - 1| < 1,$$
$$(d) \ 2 < |z|, \quad (e) \ |z - 1| > 1, \quad (f) \ 0 < |z - 2| < 1.$$

Ans: (a) $f(z) = \frac{1}{2} + \frac{3}{4}z + \frac{7}{8}z^2 + \frac{15}{16}z^3 + \cdots\cdots$
(b) $f(z) = \cdots\cdots - \frac{1}{z^3} - \frac{1}{z^2} - \frac{1}{z} - \frac{1}{2} - \frac{1}{4}z - \frac{1}{8}z^2 - \frac{1}{16}z^3$
(c) $f(z) = -\frac{1}{(z-1)} - 1 - (z - 1) - (z - 1)^2 - \cdots$
(d) $f(z) = \cdots + \frac{15}{z^5} + \frac{7}{z^4} + \frac{3}{z^3} + \frac{1}{z^2}$
(e) $f(z) = \cdots\cdots + \frac{1}{(z-1)^4} + \frac{1}{(z-1)^3} + \frac{1}{(z-1)^2}$
(f) $f(z) = \frac{1}{z-2} - 1 + (z - 2) - (z - 2)^2 + (z - 2)^3 - \cdots\cdots$.

7. Expand

$$f(z) = \frac{1}{z^2(z - i)}$$

in two different Laurent expansions around $z = i$ and tell where each converges.
Ans: $f(z) = -\frac{1}{z-i} - 2i + 3(z - i) - 4i(z - i)^2 + \cdots\cdots \qquad 0 < |z - i| < 1$
$f(z) = \cdots\cdots \frac{4i}{(z-i)^6} - \frac{3}{(z-i)^5} - \frac{2i}{(z-i)^4} + \frac{1}{(z-i)^3} \qquad |z - i| > 1.$

8. Find the values of $\oint_C f(z)dz$ where C is the circle $|z| = 3$, for the following functions

$$(a) \ f(z) = \frac{1}{z(z + 2)}, \quad (b) \ f(z) = \frac{z + 2}{z(z + 1)}, \quad (c) \ f(z) = \frac{z}{(z + 1)(z + 2)},$$

$$(d) \ f(z) = \frac{1}{z(z + 1)^2}, \quad (e) \ f(z) = \frac{1}{(z + 1)^2}, \quad (f) \ f(z) = \frac{1}{z(z + 1)(z + 4)},$$

by expanding them in an appropriate Laurent series $f(z) = \sum_{n=-\infty}^{n=\infty} a_n z^n$ and using $a_{-1} = \frac{1}{2\pi i} \oint_C f(z)dz$.

Ans: (a) 0, (b) $2\pi i$, (c) $2\pi i$, (d) 0, (e) 0, (f) $-\frac{i\pi}{6}$.

9. Find the residue of

$$f(z) = \frac{z}{z^2 + 1}$$

(a) at $z = i$ and (b) at $z = -i$.
     Ans: (a) 1/2; (b) 1/2.
10. Find the residue of

$$f(z) = \frac{z+1}{z^2(z-2)}$$

(a) at $z = 0$ and (b) at $z = 2$.
     Ans: (a) $-3/4$; (b) 3/4.
11. Find the residue of

$$f(z) = \frac{z}{z^2 + 2z + 5}$$

at each of its poles.
     Ans: $r(-1 + 2i) = (2 + i)/4$; $r(-1 - 2i) = (2 - i)/4$
12. What is the residue of

$$f(z) = \frac{1}{(z+1)^3}$$

at $z = -1$?
     Ans: 0.
13. What is the residue of

$$f(z) = \tan z$$

at $z = \pi/2$?
     Ans: -1
14. What is the residue of

$$f(z) = \frac{1}{z - \sin z}$$

at $z = 0$?
     Ans: 3/10
15. Use the theory of residue to evaluate $\oint_C f(z)dz$ if C is the circle $|z| = 4$ for each of the following functions:

$$(a)\ \frac{z}{z^2-1}, \quad (b)\ \frac{z+1}{z^2(z+2)}, \quad (c)\ \frac{1}{z(z-2)^3},$$

$$(d)\ \frac{1}{z^2+z+1} \quad (e)\ \frac{1}{z(z^2+6z+4)}.$$

Ans. (a) $2\pi i$, (b) 0,   (c) 0,   (d) 0,   (e) $(5 - 3\sqrt{5})i\pi/20$.
16. Show that

$$\oint_C \frac{1}{(z^{100} + 1)(z - 4)} dz = \frac{-2\pi i}{4^{100} + 1},$$

if C is the circle $|z| = 3$.

Hint: First find the value of the integral along $|z| = 5$, then do the integration along $|z| = 5$ and $|z| = 3$ with a cut between them.

17. Use the theory of residue to evaluate the following definite integrals

$$\text{(a)} \int_0^{2\pi} \frac{d\theta}{2 + \cos\theta}, \qquad \text{(b)} \int_0^{2\pi} \frac{\cos 3\theta \, d\theta}{5 - 4\cos\theta},$$

$$\text{(c)} \int_0^{\pi} \frac{\cos 2\theta \, d\theta}{1 - 2a\cos\theta + a^2} \quad (-1 < a < 1), \qquad \text{(d)} \int_0^{\pi} \sin^{2n}\theta \, d\theta \quad n = 1, 2, \dots \ .$$

Ans. (a) $2\pi/\sqrt{3}$, (b) $\pi/12$, (c) $\pi a^2 / \left(1 - a^2\right)$, (d) $\pi (2n)! / \left(2^{2n}(n!)^2\right)$.

18. Show that

$$\text{(a)} \int_0^\infty \frac{dx}{x^2 + 1} = \frac{\pi}{2}, \qquad \text{(b)} \int_{-\infty}^\infty \frac{x^2 + 1}{x^4 + 1} dx = \sqrt{2}\pi$$

$$\text{(c)} \int_0^\infty \frac{dx}{(x^2 + 1)^2} = \frac{\pi}{4}, \qquad \text{(d)} \int_0^\infty \frac{ab}{(x^2 + a^2)(x^2 + b^2)} dx = \frac{\pi}{2(a + b)}.$$

19. Evaluate

$$\text{(a)} \int_{-\infty}^\infty \frac{e^{i3x}}{x - 2i} dx, \qquad \text{(b)} \int_0^\infty \frac{\cos kx}{x^2 + 1} dx,$$

$$\text{(c)} \int_{-\infty}^\infty \frac{\cos mx}{(x - a)^2 + b^2} dx, \qquad \text{(d)} \int_{-\infty}^\infty \frac{\cos mx}{(x^2 + a^2)(x^2 + b^2)} dx .$$

Ans. (a) $2\pi i/e^6$, (b) $\frac{\pi}{2}e^{-|k|}$, (c) $\frac{\pi}{b}e^{-mb}\cos ma$, (d) $\frac{\pi}{a^2 - b^2}\left(\frac{e^{-bm}}{b} - \frac{e^{-am}}{a}\right)$.

20. Use a rectangular contour to show that

$$\int_{-\infty}^\infty \frac{\cos mx}{e^{-x} + e^x} dx = \frac{\pi}{e^{m\pi/2} + e^{-m\pi/2}}.$$

21. Use the "integration along the branch cut" method to show that

$$\int_0^\infty \frac{x^{1/3}}{(1 + x)^2} dx = \frac{2\pi}{3\sqrt{3}}.$$

22. Use a pie-shaped contour with $\theta = 2\pi/3$ to show that

$$\int_0^\infty \frac{1}{x^3 + 1} dx = \frac{2\sqrt{3}\pi}{9}.$$

23. Find the principal value of the following

$$P \int_{-\infty}^{\infty} \frac{1}{(x+1)(x^2+2)} dx.$$

Ans. $\sqrt{2}\pi/6$.

24. Show that

$$\frac{1 - e^{2iz}}{z^2}$$

has a simple pole at $z = 0$. Find the principal value of

$$P \int_{-\infty}^{\infty} \frac{1 - e^{2ix}}{x^2} dx.$$

Use the result to show that

$$\int_{0}^{\infty} \frac{\sin^2 x}{x^2} dx = \frac{\pi}{2}.$$

Ans. 2π.

Chapter 4
Conformal Mapping

Conformal mapping is a powerful technique to solve two-dimensional Laplace's equation with complicated boundary conditions. Although with the advent of fast computers, these problems are increasingly solved by numerical methods. Therefore, the importance of conformal mapping methods has diminished somewhat. Nevertheless, it is instructive to see the solutions to some important problems expressed in closed forms, which the conformal transformations can provide.

As we have seen in Chap. 2, a complex function $w = f(z) = u(x, y) + iv(x, y)$ can be regarded as a mapping from its domain in the z-plane ($z = x + iy$) to its range in the w plane. In this chapter, we will analyze the geometrical properties associated with mappings represented by complex functions. For example, the linkage between the analyticity of a complex function and the conformality of mapping will be studied.

All these efforts are centered around the idea that we can transform a problem with an awkward boundary into one with a simple boundary using conformal mapping $w = f(z)$. We then solve the problem with the simple boundary in the variables u and v. Then we transform back to original variables x and y to get the desired solution. That this is possible is underpinned by the invariance of Laplace's equation under a conformal mapping.

We live in a three-dimensional world, any solution to two-dimensional problems can only be an approximation. However, under certain circumstances, they can be very good approximations. There are three categories of problems that can be approximated by a two-dimensional formulation and solved by the conformal transformations; namely, electrostatic potentials, steady-state temperature distributions, and hydrodynamic flows. Among them, the hydrodynamic flow is probably the least familiar and its characteristics are slightly different; therefore, we have a separate section later in the chapter to discuss this problem.

We will explore the geometrical structure of various transformations. Three classes of mappings, the Joukowski transformation, the Möbius transformation (the bilinear

K.-T. Tang, *Mathematical Methods for Engineers and Scientists 1*,
https://doi.org/10.1007/978-3-031-05678-9_4

transformation), and the Schwarz-Christoffel transformation are particularly interesting. Although the full significance of these famous transformations is far beyond the scope of this book, hopefully, we can at least learn what they are and how to use them.

Instead of discussing the procedure in the abstract, we will first use a few specific examples to illustrate the method before going into the detailed study.

4.1 Examples of Problems Solved by Conformal Mappings

The following examples are chosen to illustrate how conformal mapping is used to solve physical problems. In the process, we will encounter some multi-valued functions. In such a case, it is understood that the principal (or prominent) branch of values is taken without further specification to avoid distraction. Of course, restrictions will be specified if there is a chance of confusing.

When the values of the desired function are specified at the boundary, it is generally known as the Dirichlet problem. Neumann boundary condition usually means that the normal derivative of the desired function is zero. Some problems have mixed boundary conditions. As long as we know what they mean, we will avoid using these special names as much as possible.

Example 4.1.1 Use the conformal mapping $w = \frac{1}{z}$ to find the potential between a long metal cylinder of radius $\frac{1}{2}$ and a metal plane, touching, but insulating from, the cylinder. Let the metal plane be held at potential V_0 and the cylinder at potential V_1.

Solution 4.1.1 Usually the symbol used for potential is $V(x, y)$. Here we use $P(x, y)$ for the potential to avoid the confusion with the imaginary part of w. Since

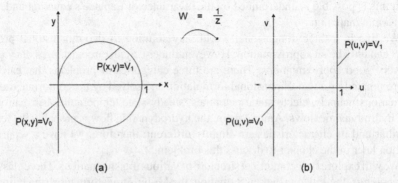

Fig. 4.1 The cross-section of a long metal cylinder, touching but insulated from a planar conducting sheet perpendicular to this page, **a** in the original x-y plane, **b** in the transformed u-v plane

it is known that the potential satisfies Laplace's equation, so we need to solve the equation

$$\frac{\partial^2 P\,(x,\,y)}{\partial x^2} + \frac{\partial^2 P\,(x,\,y)}{\partial y^2} = 0$$

subject to the boundary condition shown in Fig. 4.1a.

With

$$w = f\,(z) = u(x,\,y) + iv(x,\,y)$$

and

$$w = \frac{1}{z} = \frac{1}{x+iy} = \frac{1}{x+iy}\frac{x-iy}{x-iy} = \frac{x}{x^2+y^2} - i\frac{y}{x^2+y^2},$$

therefore

$$u(x,\,y) = \frac{x}{x^2+y^2}, \quad v(x,\,y) = -\frac{y}{x^2+y^2}. \tag{4.1}$$

It is clear that the y-axis ($x = 0$) in the x-y plane is transformed into the v-axis ($u = 0$) in the u-v plane. Furthermore, the cross-section in the x-y plane is a circle, and the center is on the x-axis at $x = \frac{1}{2}$, so

$$\left(x - \frac{1}{2}\right)^2 + y^2 = \left(\frac{1}{2}\right)^2$$

which simplifies to

$$x^2 - x + y^2 = 0.$$

Thus

$$x = x^2 + y^2.$$

Therefore $u(x,\,y) = \frac{x}{x^2+y^2} = 1$.

The circle in Fig. 4.1a is mapped into a straight line $u = 1$ in Fig. 4.1b. Thus the problem in the u-v plane

$$\frac{\partial^2 P(u,\,v)}{\partial u^2} + \frac{\partial^2 P(u,\,v)}{\partial v^2} = 0 \tag{4.2}$$

becomes trivially simple to solve, because obviously $P\,(u,\,v)$ varies with u, but not with v. The Laplace's equation becomes a one-dimensional ordinary differential equation

$$\frac{d^2 P(u,\,v)}{du^2} = 0.$$

The general solution is

$$P(u,\,v) = A + Bu,$$

The boundary conditions are such that at the metal plane the potential is at V_0 and at the metal cylinder the potential is at V_1. Now the metal plane is at $x = 0$, which corresponds to $u = 0$, therefore

$$P(0, v) = V_0 = A.$$

The surface of the metal cylinder is given by the circle $(x - 1/2)^2 + y^2 = (1/2)^2$ which corresponds to $u = 1$, therefore

$$P(1, v) = V_1 = A + B.$$

Hence $A = V_0$ and $B = V_1 - V_0$, thus

$$P(u, v) = V_0 + (V_1 - V_0)u. \tag{4.3}$$

To obtain the solution in the x-y plane, we simply transform u back to x and y coordinates. By Eq. (4.1)

$$P(x, y) = V_0 + (V_1 - V_0)\frac{x}{x^2 + y^2}. \tag{4.4}$$

This is the desired solution.

Now we wish to make a few comments about this problem.

Comment 1: This problem is actually already discussed in Example 2.1.7. There we have explicitly shown that both the real part $u(x, y)$ and the imaginary part $v(x, y)$ satisfy Laplace's equation. Since only $u(x, y)$ satisfies the boundary conditions, so the solution is given by $u(x, y)$. It follows that in Fig. 2.5 the circles of the solid line on the positive x side are the equipotential lines.

Comment 2: We have implicitly assumed that the solution that satisfies the Laplace's equation in the u-v plane will also satisfy Laplace's equation when it is transformed back into x-y plane. Later we will explicitly demonstrate this fact.

Comment 3: If the radius of the cylinder is R instead of $\frac{1}{2}$, what is the solution? This problem can be easily solved by scaling up the transformation to

$$W = \frac{2R}{z}.$$

One can show the solution is given by

$$P(x, y) = V_0 + (V_1 - V_0)\frac{2Rx}{x^2 + y^2}.$$

Comment 4: In Sect. 2.1.6, we have shown that if the transformation function $w = f(z)$ is analytic, then both its real part $u(x, y)$ and imaginary part $v(x, y)$ satisfy the Laplace's equation (That is, they are harmonic). The present example can be regarded as the geometrical interpretation of that statement. Later in this chapter, we will define the term "conformal mapping" more carefully. But before that, let us look at a couple more examples.

Example 4.1.2 Use the conformal mapping of $w = \ln \frac{z-1}{z+1}$ to find the steady temperature $T(x, y)$ in the upper half-plane ($y \geq 0$) of an infinite thin plate whose faces are insulated and whose edge $y = 0$ is kept at temperature zero except for the segment $-1 < x < 1$, where it is kept at temperature unity.

Solution 4.1.2 Since the steady temperature satisfies Laplace's equation, so we need to solve the equation

$$\frac{\partial^2 T(x, y)}{\partial x^2} + \frac{\partial^2 T(x, y)}{\partial y^2} = 0$$

with an awkward boundary condition shown in Fig. 4.2a. To see how can it be transformed into a simple boundary, it is easier to use the polar coordinates

$$w = \ln \frac{z-1}{z+1} = \ln \frac{r_1 e^{i\theta_1}}{r_2 e^{i\theta_2}} = \ln \frac{r_1}{r_2} e^{i(\theta_1 - \theta_2)} = \ln \frac{r_1}{r_2} + i(\theta_1 - \theta_2).$$

The geometry of these coordinates is also shown in Fig. 4.2a.

Simple geometry in Fig. 4.2a shows that $\theta_1 = \theta_2 + \Theta$, so $\Theta = \theta_1 - \theta_2$. With

$$w = u + iv = \ln \frac{r_1}{r_2} + i\Theta, \tag{4.5}$$

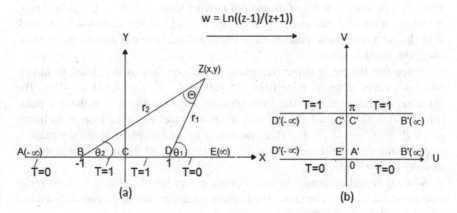

Fig. 4.2 The boundary conditions for the temperature distribution in the upper half-plane. **a** In the original x-y plane. **b** In the u-v plane after transformed by the function $w = \ln \frac{z-1}{z+1}$

we have

$$u = \ln \frac{r_1}{r_2}, \qquad v = \Theta. \tag{4.6}$$

It is clear from Fig. 4.2a that when z falls on x-axis (y = 0),

$$\Theta = 0 \quad for \quad -\infty \le x < -1,$$
$$\Theta = \pi \quad for \quad -1 < x < 1, \tag{4.7}$$
$$\Theta = 0 \quad for \quad \quad 1 < x \le \infty.$$

To see the mapping, let us follow z moving from $A \to B \to C \to D \to E$ along the x-axis in the x-y plane (Fig. 4.2a). Their corresponding positions in the u-v plane are A', B', C', D' and E' (Fig. 4.2b). When A is at $-\infty$, we write it as $A(-\infty)$ and so on. In that case, both r_1 and r_2 are very large, so $r_1 \approx r_2$ and $u = \ln(r_1/r_2) \approx \ln(1) = 0$. This means that in the image w plane, A' is at the origin (0,0). At B where $r_2 = 0$, $r_1 = 2$, so $u = \ln(r_1/r_2) = \ln(2/0) \to \infty$, therefore its image B' is at ∞ in the u-v plane (Fig. 4.2b). In this part of the journey, while u changes from 0 to ∞, v stays at 0, since $v = \Theta = 0$. This means the point moves along the $v = 0$ line as u changes from $u = 0$ to $u = \infty$. At $x = 1$, B moves from $x < -1$ to $x > -1$, therefore by Eq. (4.7) there is a discontinuity as Θ changes from 0 to π. Since $\Theta = v$ by Eq. (4.6), this discontinuity also causes B' jumps from $v = 0$ to $v = \pi$ in the u-v plane. As z moves from B to C to D in the z-plane, in the image plane the point returns from B' to C' to D' along a line of $v = \pi$ parallel to the u-axis $(v = 0)$. The discontinuity at D causes $D'(-\infty)$ jumps down from $v = \pi$ to $v = 0$. Finally from D to E (∞) in the x-y plane corresponds to $D'(-\infty)$ to E' along the $v = 0$ line in the u-v plane.

For a conformal mapping, there is a rule about the transformation of a region. Let A, B, C be on the boundary of the region D, under conformal mapping, they are transformed to the image w plane. While moving from A to B to C on the boundary, if the region D lies on the left (or right), then the image region in the w plane will also lie on the left (or right) of the path of moving from A' to B' to C' on the image boundary. While this rule can be proved [see Polya [35]], the easiest way to check is to choose a convenient point in the z-plane and see where it goes in the w plane under the transformation.

Notice that the entire upper half-plane in the x-y plane is at our left-hand side when we travel along its edge from $A \to B \to C \to D \to E$ (Fig. 4.2a). The entire region inside the infinite strip between $v = 0$ and $v = \pi$ in the u-v plane (Fig. 4.2b) is also at our left-hand side when we travel along the loop of its image $A' \to B' \to C' \to D' \to E'$. Thus, the whole upper half-plane in the x-y plane is mapped unto the infinite strip between u-axis $(v = 0)$ and the line $v = \pi$ in the u-v plane as shown in Fig. 4.2b.

Now the boundary condition in u-v plane is very simple, clearly $T(u, v)$ varies with v but not with u. Therefore, the Laplace equation becomes a one-dimensional ordinary differential equation

$$\frac{d^2 T(u, v)}{dv^2} = 0.$$

A general solution is

$$T(u, v) = A + Bv.$$

The condition $T(u, 0) = 0$ means $A = 0$. The condition $T(u, \pi) = 1$ means $B = 1/\pi$. Thus

$$T(u, v) = \frac{1}{\pi} v$$

This expression satisfies the Laplace's equation and the boundary conditions. To find $T(x.y)$, we have to transform v back to x, y. Since

$$v = \Theta = \theta_1 - \theta_2$$

and

$$\Theta = \tan^{-1} \tan \Theta = \tan^{-1} \tan (\theta_1 - \theta_2).$$

Recall

$$\tan(\theta_1 - \theta_2) = \frac{\tan \theta_1 - \tan \theta_2}{1 + \tan \theta_1 \tan \theta_2},$$

from Fig. 4.3, we see that

$$\tan \theta_1 = \frac{y}{x - 1}, \quad \tan \theta_2 = \frac{y}{x + 1},$$

so

$$\tan (\theta_1 - \theta_2) = \frac{\frac{y}{x-1} - \frac{y}{x+1}}{1 + \frac{y}{x-1}\frac{y}{x+1}} = \frac{2y}{x^2 + y^2 - 1}.$$

Thus

$$\Theta = \tan^{-1} \frac{2y}{x^2 + y^2 - 1}.$$

Therefore

$$T(x, y) = \frac{1}{\pi} v = \frac{1}{\pi} \Theta = \frac{1}{\pi} \tan^{-1} \frac{2y}{x^2 + y^2 - 1}.$$

This is the desired result. We can arrive at this result in another way. Let us write

$$\frac{z - 1}{z + 1} = A + iB = \sqrt{A^2 + B^2} e^{i \tan^{-1} \frac{B}{A}}.$$

so the transformation function is given by

$$w = \ln \frac{z - 1}{z + 1} = \ln(A + iB) = \ln \left(A^2 + B^2 \right)^{1/2} + i \tan^{-1} \frac{B}{A}.$$

Fig. 4.3 Relationship
between θ_1 and θ_2 with x, y

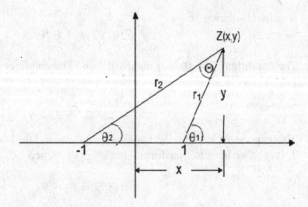

The imaginary part of $w = u + iv$ is clearly

$$v = \tan^{-1} \frac{B}{A}.$$

Since

$$\frac{z-1}{z+1} = \frac{z-1}{z+1} \frac{z^*+1}{z^*+1} = \frac{(x^2+y^2-1)+i2y}{x^2+y^2+2x+1},$$

so

$$\frac{B}{A} = \frac{2y}{x^2+y^2-1}.$$

Therefore

$$T(x, y) = \frac{1}{\pi} v = \frac{1}{\pi} \tan^{-1} \frac{2y}{x^2+y^2-1}.$$

We make a few comments about this problem:

Comment 1: Since the transformation function $\ln \frac{z-1}{z+1}$ is analytic on the upper half-plane except at the singular points on the edge. Section 2.1.6 shows both its real part and its imaginary part are harmonic (satisfying Laplace's equation). In case it is not clear that $T(u, v)$ is independent of u, as claimed, observe a posteriori that the solution does satisfy Laplace's equation and the boundary conditions.

Comment 2: Since the product of a harmonic function and a constant is also harmonic, the function

$$T(x, y) = \frac{T_0}{\pi} \tan^{-1} \frac{2y}{x^2+y^2-1}$$

represents the steady temperatures in the upper half-plane of Fig. 4.2a when the temperature $T = 1$ along the segment $-1 < x < 1$ of the x-axis is replaced by any fixed temperature T_0.

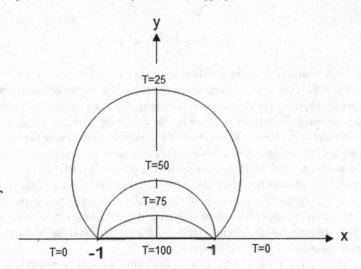

Fig. 4.4 Equal temperature lines in the case the segment $-1 < x < 1$, the temperature is kept at 100 degrees

Comment 3: The best way to develop a general idea of the temperature distribution is to see the map of isotherms, namely a picture of equal temperature lines. To see the line of $T(x, y) = T_1$, let $c = T_1/T_0$ $(0 < c < 1)$, so

$$c\pi = \tan^{-1} \frac{2y}{x^2 + y^2 - 1}$$

or

$$(x^2 + y^2 - 1) \tan c\pi = 2y.$$

Using the relations

$$\tan c\pi = \frac{1}{\cot c\pi}, \quad \text{and} \quad 1 + \cot^2 c\pi = \csc^2 c\pi$$

we have

$$x^2 + y^2 - 1 = \frac{2y}{\tan c\pi}$$

or

$$x^2 + y^2 - 2y \cot c\pi = 1.$$

Completing the square,

$$x^2 + y^2 - 2y \cot c\pi + \cot^2 c\pi = 1 + \cot^2 c\pi.$$

so

$$x^2 + (y - \cot c\pi)^2 = \csc^2 c\pi.$$

This is a circle with center on the y-axis at $y_0 = \cot c\pi$ with radius $r = \csc c\pi$.

Comment 4: Of course the isotherms are the arc of the circles in the upper half-plane. The case for $T_0 = 100$ degrees is shown in Fig. 4.4. For example, the isothermal line of temperature 75 degrees is given by the following: Since $c = \frac{75}{100} = \frac{3}{4}$, so $\cot c\pi = -1$ and $\csc^2 c\pi = 2$. This means that the center of the circle is at $y_0 = -1$ and the radius of the circle is $r = \sqrt{2}$. Thus, the circle passes the points $(\pm 1, 0)$. This means the isothermal line of 75 degrees passes the points $(\pm 1, 0)$. This is a general result in that all isotherms pass the points of $(\pm 1, 0)$ because $1 + \cot^2 c\pi = \csc^2 c\pi$, $(1 + y_0^2 = r^2)$. For $c > 1/2$, the center of the circle is on the negative y-axis and the isothermal line on the upper half-plane is the arc of the minor segment of the circle, for $c < 1/2$, the center of the circle is on the positive y-axis and the isothermal line is the arc of the major segment of the circle.

If we cannot solve a problem by a single mapping, it might be possible to solve it by a sequence of mappings. In the next example, we show that a problem can be solved by transforming it into another boundary value problem with a known solution, although the known solution is obtained by another mapping.

Example 4.1.3 Find the steady-state temperature $T(x, y)$ in a thin plate having the form of a semi-infinite strip $-\pi/2 \leq x \leq \pi/2$, $y \geq 0$ when the faces of the plate are perfectly insulated. The temperature at the bottom boundary is kept at 1 and the temperature along the vertical boundaries is zero. (Hint: Use $w = \sin z$).

Solution 4.1.3 Since the steady-state temperature $T(x, y)$ satisfies the Laplace equation and the boundary conditions are shown in Fig.4.5a, so our problem is

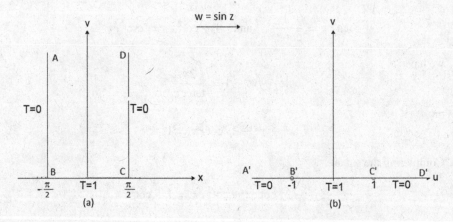

Fig. 4.5 The mapping $w = \sin z$ carries the semi-infinite strip onto the upper half plane

$$\frac{\partial^2 T\,(x,\,y)}{\partial x^2} + \frac{\partial^2 T\,(x,\,y)}{\partial y^2} = 0,$$

$$T\left(-\frac{\pi}{2},\,y\right) = T\left(\frac{\pi}{2},\,y\right) = 0 \quad for \quad y \ge 0,$$

$$T(x,\,0) = 1 \quad for \quad -\frac{\pi}{2} \le x \le \frac{\pi}{2}.$$

We wish to use the transformation $w = \sin z = u + iv$. Let us recall from Exercise 21 of Chap. 1

$$\sin z = \sin x \cosh y + i \cos x \sinh y$$

with

$$\cosh y = \frac{1}{2}(e^y + e^{-y}), \qquad \sinh y = \frac{1}{2}(e^y - e^{-y}).$$

Thus

$$u = \sin x \cosh y, \qquad v = \cos x \sinh y. \tag{4.8}$$

With this transformation, the left side vertical boundary of the semi-infinite strip in the x-y plane is at $x = -\pi/2$. Therefore, in the u-v plane, $v = \cos(-\frac{\pi}{2}) \sinh y = 0$. The bottom part of the strip is at $y = 0$, its image is also at $v = \cos(x) \sinh(0) = 0$. Similarly, the right side vertical boundary of the semi-infinite strip is at $x = \pi/2$, so it also maps on $v = \cos(\frac{\pi}{2}) \sinh y = 0$. Let the four points A, B, C, D be all on the boundary of the semi-infinite strip as shown in Fig. 4.5a, then their corresponding images A', B', C', D' are all on the u-axis in the u-v plane as shown in Fig. 4.5b. If A is at a point where $y \to \infty$, then its image A' is at $u = \sin\left(-\frac{\pi}{2}\right) \cosh(\infty) = -\infty$. Since B is at $(-\frac{\pi}{2}, 0)$, so B' is at $u = \sin(-\frac{\pi}{2}) \cosh(0) = -1$. Similarly, C' is at $(1,0)$ and D' is at $(\infty,0)$. Thus the interior of the semi-infinite strip, which is on the left-hand side of the loop ABCD, is transformed onto the upper half-plane in the u-v plane, which is also on the left-hand side of the A'B'C'D' line. This is shown in Fig. 4.5.

Thus, this problem in the u-v coordinates is exactly the same as the problem of the previous example. Although another transformation obtained the solution, here we can use the results directly and write

$$T\,(u,\,v) = \frac{1}{\pi} \tan^{-1} \frac{2v}{u^2 + v^2 - 1}.$$

To transform back to the desired solution, we can simply put $u(x, y)$ and $v(x, y)$ of Eq. (4.8) in this expression

$$T\,(x,\,y) = \frac{1}{\pi} \tan^{-1} \frac{2\cos x \sinh y}{\sin^2 x \cosh^2 y + \cos^2 x \sinh^2 y - 1}.$$

This is a perfectly legitimate result, but it so happens that it can be simplified some-what. Recall

$$\cos^2 x + \sin^2 x = 1,$$
$$\cosh^2 y - \sinh^2 y = 1,$$

so the denominator can be written as

$$\left(1 - \cos^2 x\right)\left(1 + \sinh^2 y\right) + \cos^2 x \sinh^2 y - 1 = \sinh^2 y - \cos^2 x,$$

thus

$$T(x, y) = \frac{1}{\pi} \tan^{-1} \frac{2\cos x \sinh y}{\sinh^2 y - \cos^2 x} = \frac{1}{\pi} \tan^{-1} \frac{2(\cos x / \sinh y)}{1 - (\cos x / \sinh y)^2}.$$

Let

$$\alpha = \tan^{-1}\left(\frac{\cos x}{\sinh y}\right),$$

so

$$\tan \alpha = \frac{\cos x}{\sinh y},$$

and we can write $T(x, y)$ as

$$T(x, y) = \frac{1}{\pi} \tan^{-1} \frac{2(\cos x / \sinh y)}{1 - (\cos x / \sinh y)^2} = \frac{1}{\pi} \tan^{-1} \frac{2\tan \alpha}{1 - \tan^2 \alpha}.$$

Recall

$$\tan 2\alpha = \frac{2\tan \alpha}{1 - \tan^2 \alpha},$$

thus

$$T(x, y) = \frac{1}{\pi} \tan^{-1} \frac{2\tan \alpha}{1 - \tan^2 \alpha} = \frac{1}{\pi} \tan^{-1} \tan 2\alpha.$$

or

$$T(x, y) = \frac{2}{\pi} \alpha.$$

Therefore, the final result is

$$T(x, y) = \frac{2}{\pi} \tan^{-1}\left(\frac{\cos x}{\sinh y}\right).$$

It is clear that to solve a physical problem with conformal mapping, we have to first choose a right transformation. This is a nontrivial task. There does not exist a systematic procedure that inevitably leads to a suitable mapping function. Fortunately, there exist extensive tables of conformal maps. [See Kober [26]]

While it is reasonable to rely on a conformal mapping table, we have to be sufficiently familiar with these mappings to be able to use them. In the following, we will study some of these mappings.

In fact, just with the few examples, we have seen so far, we are able to solve many similar problems. Again this is best demonstrated with examples.

Example 4.1.4 The positive and negative halves of the x-axis each represent infinite conducting plates separated from each other by insulation at the origin. The plate $x > 0$, $y = 0$ is kept at potential V_1 while the plate $x < 0$, $y = 0$ is kept at potential V_2. Find the potential $P(x, y)$ in the upper half-plane (Fig. 4.6).

Solution 4.1.4 From previous examples, we learned that logarithmic transformation could solve this type of problem. So we use

$$w = \ln(z) = u + iv$$

to transform (x, y) into (u, v). Let the potential in (x, y) coordinates be $P(x, y)$ and in (u, v) coordinates be $P(u, v)$. They are the same potential P. That is

$$\ln(z) = \ln(re^{i\theta}) = \ln r + i\theta,$$

clearly $u = \ln(r)$, $v = \theta$. The boundary conditions become very simple:

$$v = \theta = 0, \quad P(u, 0) = V_1$$
$$v = \theta = \pi, \quad P(u, \pi) = V_2.$$

Thus the Laplace equation in (u,v) becomes an ordinary differential equation in v only

Fig. 4.6 At the boundary, the potential is kept at V_1 for $x > 0$, $y = 0$; and at V_2 for $x < 0$, $y = 0$. Find the potential $P(x, y)$ in the upper half-plane

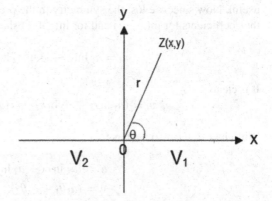

$$\frac{d^2 P}{dv^2} = 0,$$

the general solution of this equation is

$$P(u, v) = A + Bv.$$

The boundary condition at $v = 0$ leads to $A = V_1$. At $v = \pi$, it shows that

$$V_2 = V_1 + B\pi.$$

Therefore

$$P(u, v) = V_1 + \frac{V_2 - V_1}{\pi} v.$$

Transform back to (x, y) plane

$$P(x, y) = V_1 + \frac{V_2 - V_1}{\pi} \theta$$

$$= V_1 + \frac{V_2 - V_1}{\pi} \tan^{-1} \frac{y}{x}.$$

Example 4.1.5 At the boundary of x-axis, the temperature is kept at T_1 for $x > 1$, at T_2 for $-1 < x < 1$, and at T_3 for $x < -1$. Find the steady-state temperature for the upper half-plane (Fig. 4.7).

Solution 4.1.5 This problem is very similar to the Example 4.1.2 of this section, except in that case, there is symmetry in that the temperatures for $x > 1$ and for $x < -1$ are both kept at T= 0. There we find the transformation function

$$w = \ln \frac{z - 1}{z + 1} = \ln(z - 1) - \ln(z + 1)$$

useful. Now since we lost the symmetry in the present case, it stands to reason that the coefficients for $\ln(z - 1)$ and for $\ln(z + 1)$ should be different. That is, we use

$$w = a_1 \ln(z - 1) + a_2 \ln(z + 1)$$

It is clear

$$w = (a_1 \ln r_1 + a_2 \ln r_2) + i(a_1 \theta_1 + a_2 \theta_2).$$

With $w = u + iv$, so we have

$$u = (a_1 \ln r_1 + a_2 \ln r_2) \tag{4.9}$$

$$v = (a_1 \theta_1 + a_2 \theta_2). \tag{4.10}$$

The boundary conditions in v is straight forward

$$v = \begin{cases} 0 & x > 1 & \theta_1 = \theta_2 = 0 & T = T_1 \\ a_1\pi & -1 < x < 1 & \theta_1 = \pi;\ \theta_2 = 0 & T = T_2 \\ a_1\pi + a_2\pi & x < -1 & \theta_1 = \pi;\ \theta_2 = \pi & T = T_3 \end{cases} \quad (4.11)$$

The steady temperature in u, v plane satisfies Laplace's equation

$$\frac{\partial^2}{\partial u^2} T(u, v) + \frac{\partial^2}{\partial v^2} T(u, v) = 0.$$

Assuming $T(u, v)$ varies with v and not with u, this equation becomes an ordinary differential equation

$$\frac{d^2}{dv^2} T(u, v) = 0.$$

The general solution is

$$T(u, v) = A + Bv.$$

According Eq. (4.10),

$$T(u, v) = A + B(a_1\theta_1 + a_2\theta_2). \quad (4.12)$$

The boundary condition for $x > 1$ is given by the first equation of Eq. (4.11), $T = T_1$ for $\theta_1 = \theta_2 = 0$. This leads to $T_1 = A$, so

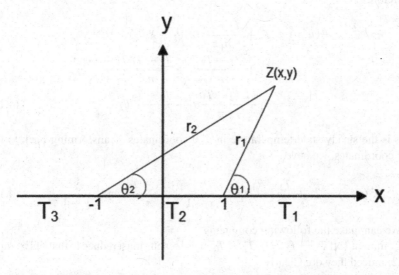

Fig. 4.7 The temperature is kept at T_1 for $x > 1$, at T_2 for $-1 < x < 1$, and at T_3 for $x < -1$. Find the steady temperature for the upper half-plane

$$T(u, v) = T_1 + B(a_1\theta_1 + a_2\theta_2).$$

The boundary condition for $-1 < x < 1$ is given by the second equation of Eq. (4.11), $T = T_2$, $\theta_1 = \pi$, $\theta_2 = 0$. So

$$T_2 = T_1 + Ba_1\pi$$

which says $B = (T_2 - T_1)/a_1\pi$. Thus

$$T(u, v) = T_1 + \frac{T_2 - T_1}{a_1\pi}(a_1\theta_1 + a_2\theta_2)$$

The boundary condition for $x < 1$ is given by the third equation of Eq. (4.11), $T = T_3$, $\theta_1 = \theta_2 = \pi$. So

$$\begin{aligned}
T_3 &= T_1 + \frac{T_2 - T_1}{a_1\pi}(a_1\pi + a_2\pi) \\
&= T_1 + T_2 - T_1 + (T_2 - T_1)\frac{a_2}{a_1}.
\end{aligned}$$

which gives

$$\frac{a_2}{a_1} = \frac{T_3 - T_2}{T_2 - T_1}.$$

Therefore

$$\begin{aligned}
T(u, v) &= T_1 + \frac{T_2 - T_1}{a_1\pi}(a_1\theta_1 + a_2\theta_2) \\
&= T_1 + \frac{T_2 - T_1}{\pi}\theta_1 + \frac{T_2 - T_1}{\pi}\frac{T_3 - T_2}{T_2 - T_1}\theta_2 \\
&= T_1 + \frac{T_2 - T_1}{\pi}\theta_1 + \frac{T_3 - T_2}{\pi}\theta_2.
\end{aligned} \tag{4.13}$$

This is the steady-state temperature in u, v coordinates. Transforming back to the x, y coordinates, we have

$$T(x, y) = T_1 + \frac{T_2 - T_1}{\pi}\tan^{-1}\frac{y}{x-1} + \frac{T_3 - T_2}{\pi}\tan^{-1}\frac{y}{x+1}. \tag{4.14}$$

We can make the following comments.

Comment 1: If $T_1 = T_3 = 0$, $T_2 = T_0$, these results must reduce to that of Example 4.1.2. Indeed they do. Clearly

$$T(u, v) = \frac{T_0}{\pi}\theta_1 - \frac{T_0}{\pi}\theta_2.$$

Since

$$\theta_1 = \tan^{-1} \frac{y}{x-1}, \quad \theta_2 = \tan^{-1} \frac{y}{x+1},$$

so

$$\tan(\theta_1 - \theta_2) = \frac{\tan\theta_1 - \tan\theta_2}{1 + \tan\theta_1 \tan\theta_2} = \frac{y/(x-1) - y/(x+1)}{1 + y/(x-1) \times y/(x+1)} = \frac{2y}{x^2 + y^2 - 1}.$$

Therefore

$$T(x, y) = \frac{T_0}{\pi} \tan^{-1} \frac{y}{x-1} - \frac{T_0}{\pi} \tan^{-1} \frac{y}{x+1}$$

$$= \frac{T_0}{\pi} \tan^{-1} \frac{2y}{x^2 + y^2 - 1}.$$

Comment 2: Clearly Eq. (4.12) can be written as

$$T(u, v) = A + B_1\theta_1 + B_2\theta_2$$

with $B_1 = Ba_1$, $B_2 = Ba_2$. Knowing this result, we can work out a solution for the case when the temperature at the x-axis is specified in n segments of constant temperatures, instead of just three segments of constant values. In that case, we can assume

$$T(u, v) = A + B_1\theta_1 + B_2\theta_2 + \cdots + B_{n-1}\theta_{n-1},$$

and use the boundary conditions to work out the coefficients quickly. But it is important and instructive to know why we can make such an assumption.

Example 4.1.6 Find the steady-state temperature $T(x, y)$ in a thin plate having the form of a semi-infinite strip $-a \le x \le a$, $y \ge 0$ when the faces of the plate are perfectly insulated. The temperature at the bottom boundary is kept at T_2 and the temperature along the right vertical boundary is T_1 and along the left vertical boundary is T_3 (Fig. 4.8).

Solution 4.1.6 This problem is almost identical to Example 4.1.3 except: (1) The location of the vertical boundaries is different; (2) The temperature at three segments of the boundary is kept at different values whereas previously $T_1 = T_3 = 0$. The first difference can be taken care of by a change of variable, that is, instead of using the transformation function $w = \sin z$, we use $w = \sin\left(\frac{\pi z}{2a}\right)$. This will transform the interior of the semi-infinite strip in the x, y plane onto the upper half-plane in the u,v coordinates. The problem in the u,v plane is exactly the same as the Example 4.1.5 in the x, y plane. Therefore, we can directly use the result by changing (x, y) into (u, v). Thus

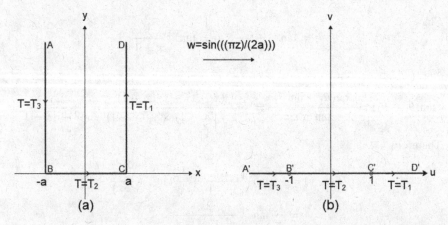

Fig. 4.8 The mapping $w = \sin(\frac{\pi z}{2a})$ carries the semi-infinite strip in (**a**) onto the upper half-plane in (**b**)

$$T(u, v) = T_1 + \frac{T_2 - T_1}{\pi} \tan^{-1} \frac{v}{u - 1} + \frac{T_3 - T_2}{\pi} \tan^{-1} \frac{v}{u + 1}.$$

Now to transform back to (x, y) coordinates, we recall

$$\sin\left(\frac{\pi z}{2a}\right) = \sin\left(\frac{\pi x}{2a}\right) \cosh\left(\frac{\pi y}{2a}\right) + i \cos\left(\frac{\pi x}{2a}\right) \sinh\left(\frac{\pi y}{2a}\right),$$

so

$$u = \sin\left(\frac{\pi x}{2a}\right) \cosh\left(\frac{\pi y}{2a}\right), \quad v = \cos\left(\frac{\pi x}{2a}\right) \sinh\left(\frac{\pi y}{2a}\right).$$

Therefore

$$T(x, y) = T_1 + \frac{T_2 - T_1}{\pi} \tan^{-1} \frac{\cos\left(\frac{\pi x}{2a}\right) \sinh\left(\frac{\pi y}{2a}\right)}{\sin\left(\frac{\pi x}{2a}\right) \cosh\left(\frac{\pi y}{2a}\right) - 1}$$

$$+ \frac{T_3 - T_2}{\pi} \tan^{-1} \frac{\cos(\frac{\pi x}{2a}) \sinh\left(\frac{\pi y}{2a}\right)}{\sin\left(\frac{\pi x}{2a}\right) \cosh\left(\frac{\pi y}{2a}\right) + 1}.$$

Example 4.1.7 Find the steady-state temperature $T(x, y)$ in a thin plate having the form of a semi-infinite strip $-\pi/2 \le x \le \pi/2$, $y \ge 0$ when the faces of the plate are perfectly insulated. The bottom boundary is insulated and the temperature along the left vertical boundary is kept at T_1 and at the right vertical boundary is kept at T_2 (Fig. 4.9).

Solution 4.1.7 The bottom boundary is insulated means that there is no heat flow between $y > 0$ and $y < 0$. Since the heat flows in proportion to the normal derivative of the temperature, this condition is equivalent to

Fig. 4.9 The bottom part of this semi-infinite strip is insulated. This condition is equivalent to $\frac{\partial}{\partial y} T = 0$ at $y = 0$

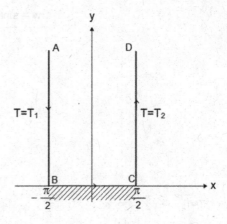

$$\frac{\partial T}{\partial y} = 0$$

at $y = 0$. With this understanding, the solution is very simple, namely

$$T(x, y) = \frac{T_2 + T_1}{2} + \frac{T_2 - T_1}{\pi} x$$

because it satisfies Laplace's equation and the boundary conditions

$$T\left(-\frac{\pi}{2}, y\right) = T_1,$$
$$T\left(\frac{\pi}{2}, y\right) = T_2;$$

and it has no y dependence, so

$$\frac{\partial}{\partial y} T(x, y) = 0.$$

Therefore it must be the solution.

Example 4.1.8 Find the temperature distribution in the upper half-plane subject to the following boundary conditions

$$T(x, 0) = \begin{cases} T_1, & x < -1 \\ T_2, & x > 1 \end{cases}$$

and the boundary between -1 and 1 is insulated, i.e. (Fig. 4.10)

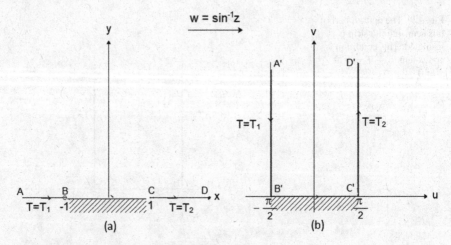

Fig. 4.10 Use $f(z) = \sin^{-1} z$ to transform the upper half-plan of the x, y plan (z plan) into the interior of the strip of the u, v plan (w plan)

$$\frac{\partial T(x, y)}{\partial y} = 0 \quad if \quad -1 < x < 1.$$

Solution 4.1.8 In view of Example 4.1.3, the inverse transformation $w = \sin^{-1} z$ will make the problem identical to the previous example, except the solution is in terms of u, i.e.

$$T(u, v) = \frac{T_2 + T_1}{2} + \frac{T_2 - T_1}{\pi} u.$$

Now we must transform it back to x and y coordinates. Since

$$w = u + iv = \sin^{-1}(x + iy),$$

so

$$\sin(u + iv) = x + iy.$$

Recall

$$\sin(u + iv) = \sin u \cosh v + i \cos u \sinh v,$$

therefore

$$x = \sin u \cosh v, \qquad y = \cos u \sinh v.$$

Hence

$$x^2 + y^2 = \sin^2 u \cosh^2 v + \cos^2 u \sinh^2 v$$
$$= \sin^2 u \cosh^2 v + (1 - \sin^2 u) \sinh^2 v$$
$$= \sin^2 u (\cosh^2 v - \sinh^2 v) + \sinh^2 v$$
$$= \sin^2 u + \cosh^2 v - 1.$$

Thus

$$x^2 + y^2 + 1 = \sin^2 u + \cosh^2 v. \tag{4.15}$$

Since

$$2x = 2 \sin u \cosh v, \tag{4.16}$$

Adding Eq. (4.16) to (4.15),

$$x^2 + y^2 + 1 + 2x = \sin^2 u + \cosh^2 v + 2 \sin u \cosh v,$$

or

$$(x + 1)^2 + y^2 = (\sin u + \cosh v)^2.$$

Taking the square root, we have

$$\sin u + \cosh v = \sqrt{(x + 1)^2 + y^2}.$$

Similarly, subtracting Eq. (4.16) from (4.15), we have

$$(x - 1)^2 + y^2 = (\sin u - \cosh v)^2.$$

Here we must be careful when we take the square root. Since the maximum value of $\sin u$ is one, and the minimum value of $\cosh v$ is also one, therefore

$$\sin u - \cosh v \le 0,$$

so

$$\sin u - \cosh v = -\sqrt{(x - 1)^2 + y^2}.$$

It follows that

$$\sin u = \frac{1}{2} \left(\sqrt{(x + 1)^2 + y^2} - \sqrt{(x - 1)^2 + y^2} \right).$$

Thus,

$$T(x, y) = \frac{T_2 + T_1}{2} + \frac{T_2 - T_1}{\pi} \sin^{-1} \frac{1}{2} \left(\sqrt{(x + 1)^2 + y^2} - \sqrt{(x - 1)^2 + y^2} \right)$$

where the inverse sine function takes values between $-\frac{\pi}{2}$ and $\frac{\pi}{2}$.

We can obtain the same result through geometry. Note that from $x = \sin u \cosh v$ and $y = \cos u \sinh v$, we have

$$\frac{x^2}{\sin^2 u} = \cosh^2 v, \quad \frac{y^2}{\cos^2 u} = \sinh^2 v.$$

Since

$$\cosh^2 v - \sinh^2 v = 1,$$

so

$$\frac{x^2}{\sin^2 u} - \frac{y^2}{\cos^2 u} = 1.$$

This is an equation of hyperbola, which has the standard form

$$\frac{x^2}{a^2} - \frac{y^2}{b^2} = 1.$$

Let us recall the definition of a hyperbola: it is a set of points in a plane whose distance from two fixed points in the plane has a constant difference.

As shown in Fig. 4.11,

$$r_1 = \sqrt{(x+c)^2 + y^2},$$
$$r_2 = \sqrt{(x-c)^2 + y^2},$$

Fig. 4.11 Definition of a hyperbola. The difference between r_1 and r_2 is a constant 2a

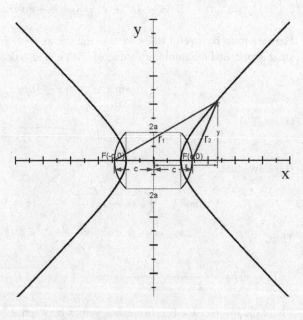

and

$$r_1 - r_2 = 2a. \tag{4.17}$$

Square both sides of the equation and simplify, we still have one radical remain. Move it to one side of the equation and square again, we have

$$\frac{x^2}{a^2} - \frac{y^2}{c^2 - a^2} = 1.$$

To have the standard form of hyperbola, we write $c^2 - a^2 = b^2$, so

$$c = \sqrt{a^2 + b^2}.$$

Compare the present equation with the standard equation of hyperbola, we can identify that

$$a = \sin u, \quad b = \cos u,$$

so

$$c = \sqrt{\sin^2 u + \cos^2 u} = 1.$$

Thus the difference equation (Eq. (4.17)) becomes

$$\sqrt{(x+1)^2 + y^2} - \sqrt{(x-1)^2 + y^2} = 2 \sin u,$$

therefore

$$u = \sin^{-1} \left[\frac{1}{2} \left(\sqrt{(x+1)^2 + y^2} - \sqrt{(x-1)^2 + y^2} \right) \right].$$

The rest follows

$$T(x, y) = \frac{T_2 + T_1}{2} + \frac{T_2 - T_1}{\pi} \sin^{-1} \frac{1}{2} \left(\sqrt{(x+1)^2 + y^2} - \sqrt{(x-1)^2 + y^2} \right).$$

For x and y in the second quadrant, the solution is obtained by making the replacement $x \to -x$. We must remember that $\sqrt{(x-1)^2}$ denotes $x - 1$ when $x > 1$ and $1 - x$ when $x < 1$. Note that the temperature at any point along the insolated part of the edge is

$$T(x, 0) = \frac{T_2 + T_1}{2} + \frac{T_2 - T_1}{\pi} \sin^{-1} x.$$

Example 4.1.9 Find the steady temperatures in a thin plate having the form of a quadrant if a segment at the end of one edge is insulated, and the rest of the edge is kept at a fixed temperature. The second edge is kept at another fixed temperature. The surface is insulated, so the problem is two-dimensional. The scales can be so

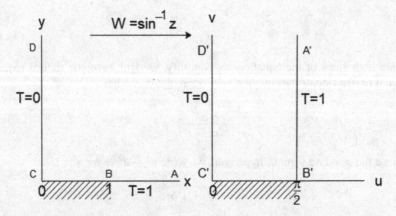

Fig. 4.12 Transform from the quadrant in the left of this figure onto the strip on the right of this figure with the transformation $w = \sin^{-1} z$

chosen that this problem can be formulated as follows:

$$\frac{\partial^2}{\partial^2 x} T(x, y) + \frac{\partial^2}{\partial^2 y} T(x, y) = 0$$

$$
\begin{aligned}
T(0, y) &= 0 & y &> 0 \\
T(x, 0) &= 1 & x &> 1 \\
\frac{\partial}{\partial y} T(x.0) &= 0 & 0 &< x < 1.
\end{aligned}
$$

The situation is shown in the left of the Fig. 4.12.

Solution 4.1.9 The transformed space is inside the strip shown in the right-hand side of the above figure. This strip is almost identical to the strip of the previous problem. Therefore, the required temperature can be written as

$$T = \frac{2}{\pi} u.$$

To transform back to the x, y space, we can take the results of the previous problem directly:

$$T = \frac{2}{\pi} \sin^{-1} \frac{1}{2} \left[\sqrt{(x+1)^2 + y^2} - \sqrt{(x-1)^2 + y^2} \right],$$

where both x and y are in the first quadrant.

4.2   Invariance of the Laplace Equation

So far in all the examples, we have assumed that the solution of Laplace's equation in
the transformed domain will automatically satisfy the Laplace equation in the original
z- plane (x, y coordinates). In this section, we will explicitly show that this is indeed
the case. A harmonic function in the transformed domain is automatically harmonic
in the original domain (the solution of Laplace's equation in the transformed w plane
(u, v coordinates) is automatically a solution of Laplace's equation in the original
z-plane (x, y coordinates)), provided the transformation $w = f(z)$ is analytic and
the derivative of the transformation function does not vanish ($f'(z) \neq 0$) inside the
domain. Note that this does not include the boundary of the domain. In other words,
we will show that

$$\frac{\partial^2 \varphi}{\partial x^2} + \frac{\partial^2 \varphi}{\partial y^2} = 0$$

if

$$\frac{\partial^2 \varphi}{\partial u^2} + \frac{\partial^2 \varphi}{\partial v^2} = 0,$$
$$f'(z) \neq 0.$$

Note that $\varphi(x, y)$ is transformed into the function $\varphi(u, v) = \varphi(x(u, v), y(u, v))$ by
the transformation $w = u + iv = f(z) = f(x + iy)$.

We will show this by straightforward differentiation.

$$\frac{\partial \varphi}{\partial x} = \frac{\partial \varphi}{\partial u}\frac{\partial u}{\partial x} + \frac{\partial \varphi}{\partial v}\frac{\partial v}{\partial x},$$
$$\frac{\partial \varphi}{\partial y} = \frac{\partial \varphi}{\partial u}\frac{\partial u}{\partial y} + \frac{\partial \varphi}{\partial v}\frac{\partial v}{\partial y}.$$

The second derivatives are

$$\frac{\partial^2 \varphi}{\partial x^2} = \frac{\partial}{\partial x}\left[\frac{\partial \varphi}{\partial x}\right] = \frac{\partial}{\partial x}\left[\frac{\partial \varphi}{\partial u}\frac{\partial u}{\partial x} + \frac{\partial \varphi}{\partial v}\frac{\partial v}{\partial x}\right]$$
$$= \frac{\partial}{\partial x}\left(\frac{\partial \varphi}{\partial u}\right)\frac{\partial u}{\partial x} + \left(\frac{\partial \varphi}{\partial u}\right)\frac{\partial^2 u}{\partial x^2} + \frac{\partial}{\partial x}\left(\frac{\partial \varphi}{\partial v}\right)\frac{\partial v}{\partial x} + \left(\frac{\partial \varphi}{\partial v}\right)\frac{\partial^2 v}{\partial x^2}$$

Since

$$\frac{\partial}{\partial x}\left(\frac{\partial \varphi}{\partial u}\right) = \frac{\partial^2 \varphi}{\partial u^2}\frac{\partial u}{\partial x} + \frac{\partial^2 \varphi}{\partial v \partial u}\frac{\partial v}{\partial x},$$

and

$$\frac{\partial}{\partial x}\left(\frac{\partial \varphi}{\partial v}\right) = \frac{\partial^2 \varphi}{\partial u \partial v}\frac{\partial u}{\partial x} + \frac{\partial^2 \varphi}{\partial v^2}\frac{\partial v}{\partial x},$$

therefore

$$\frac{\partial^2 \varphi}{\partial x^2} = \left[\frac{\partial^2 \varphi}{\partial u^2}\frac{\partial u}{\partial x} + \frac{\partial^2 \varphi}{\partial v \partial u}\frac{\partial v}{\partial x}\right]\frac{\partial u}{\partial x} + \left(\frac{\partial \varphi}{\partial u}\right)\frac{\partial^2 u}{\partial x^2}$$
$$+ \left[\frac{\partial^2 \varphi}{\partial u \partial v}\frac{\partial u}{\partial x} + \frac{\partial^2 \varphi}{\partial v^2}\frac{\partial v}{\partial x}\right]\frac{\partial v}{\partial x} + \left(\frac{\partial \varphi}{\partial v}\right)\frac{\partial^2 v}{\partial x^2}.$$

Similarly

$$\frac{\partial^2 \varphi}{\partial y^2} = \left[\frac{\partial^2 \varphi}{\partial u^2}\frac{\partial u}{\partial y} + \frac{\partial^2 \varphi}{\partial v \partial u}\frac{\partial v}{\partial y}\right]\frac{\partial u}{\partial y} + \left(\frac{\partial \varphi}{\partial u}\right)\frac{\partial^2 u}{\partial y^2}$$
$$+ \left[\frac{\partial^2 \varphi}{\partial u \partial v}\frac{\partial u}{\partial y} + \frac{\partial^2 \varphi}{\partial v^2}\frac{\partial v}{\partial y}\right]\frac{\partial v}{\partial y} + \left(\frac{\partial \varphi}{\partial v}\right)\frac{\partial^2 v}{\partial y^2}.$$

Thus

$$\frac{\partial^2 \varphi}{\partial x^2} + \frac{\partial^2 \varphi}{\partial y^2} = \frac{\partial^2 \varphi}{\partial u^2}\left[\left(\frac{\partial u}{\partial x}\right)^2 + \left(\frac{\partial u}{\partial y}\right)^2\right] + \frac{\partial^2 \varphi}{\partial v \partial u}\left(\frac{\partial v}{\partial x}\frac{\partial u}{\partial x} + \frac{\partial v}{\partial y}\frac{\partial u}{\partial y}\right)$$
$$+ \frac{\partial \varphi}{\partial u}\left(\frac{\partial^2 u}{\partial x^2} + \frac{\partial^2 u}{\partial y^2}\right) + \frac{\partial^2 \varphi}{\partial u \partial v}\left(\frac{\partial u}{\partial x}\frac{\partial v}{\partial x} + \frac{\partial u}{\partial y}\frac{\partial v}{\partial y}\right)$$
$$+ \frac{\partial^2 \varphi}{\partial v^2}\left[\left(\frac{\partial v}{\partial x}\right)^2 + \left(\frac{\partial v}{\partial y}\right)^2\right] + \frac{\partial \varphi}{\partial v}\left(\frac{\partial^2 v}{\partial x^2} + \frac{\partial^2 v}{\partial y^2}\right).$$

Now, if $f(z) = u + iv$ is analytic, then both its real part and imaginary part satisfy Laplace's equation, i.e.

$$\frac{\partial^2 u}{\partial x^2} + \frac{\partial^2 u}{\partial y^2} = \frac{\partial^2 v}{\partial x^2} + \frac{\partial^2 v}{\partial y^2} = 0.$$

Therefore, the third and the sixth terms are zero. Furthermore, the derivatives satisfy the Cauchy-Riemann equations:

$$\frac{\partial u}{\partial x} = \frac{\partial v}{\partial y}, \quad \frac{\partial v}{\partial x} = -\frac{\partial u}{\partial y},$$

thus

$$\frac{\partial v}{\partial x}\frac{\partial u}{\partial x} + \frac{\partial v}{\partial y}\frac{\partial u}{\partial y} = -\frac{\partial u}{\partial y}\frac{\partial u}{\partial x} + \frac{\partial u}{\partial x}\frac{\partial u}{\partial y} = 0,$$
$$\frac{\partial u}{\partial x}\frac{\partial v}{\partial x} + \frac{\partial u}{\partial y}\frac{\partial v}{\partial y} = -\frac{\partial v}{\partial y}\frac{\partial u}{\partial y} + \frac{\partial u}{\partial y}\frac{\partial v}{\partial y} = 0.$$

Therefore, the second and the fourth terms vanish, what remains is

$$\frac{\partial^2 \varphi}{\partial x^2} + \frac{\partial^2 \varphi}{\partial y^2} = \frac{\partial^2 \varphi}{\partial u^2}\left[\left(\frac{\partial u}{\partial x}\right)^2 + \left(\frac{\partial u}{\partial y}\right)^2\right] + \frac{\partial^2 \varphi}{\partial v^2}\left[\left(\frac{\partial v}{\partial x}\right)^2 + \left(\frac{\partial v}{\partial y}\right)^2\right].$$

In Chap. 2, we have also shown that

$$f'(z) = \frac{\partial u}{\partial x} + i\frac{\partial v}{\partial x}.$$

Because of the Cauchy-Riemann equations, we can write

$$f'(z) = \frac{\partial v}{\partial y} + i\frac{\partial v}{\partial x} = \frac{\partial u}{\partial x} - i\frac{\partial u}{\partial y},$$

therefore

$$|f'(z)|^2 = \left(\frac{\partial v}{\partial y}\right)^2 + \left(\frac{\partial v}{\partial x}\right)^2 = \left(\frac{\partial u}{\partial x}\right)^2 + \left(\frac{\partial u}{\partial y}\right)^2.$$

Thus

$$\frac{\partial^2 \varphi}{\partial x^2} + \frac{\partial^2 \varphi}{\partial y^2} = |f'(z)|^2\left(\frac{\partial^2 \varphi}{\partial u^2} + \frac{\partial^2 \varphi}{\partial v^2}\right).$$

Clearly, if φ satisfies Laplace's equation in the u, v coordinates, and the derivative is nonvanishing

$$\frac{\partial^2 \varphi}{\partial u^2} + \frac{\partial^2 \varphi}{\partial v^2} = 0,$$
$$f'(z) \neq 0,$$

then φ automatically satisfies the Laplace's equation in the x, y coordinates

$$\frac{\partial^2 \varphi}{\partial x^2} + \frac{\partial^2 \varphi}{\partial y^2} = 0.$$

This completes the proof.

However, there is another point that is worthy of mentioning. Because of the Cauchy-Riemann conditions

$$|f'(z)|^2 = \left(\frac{\partial u}{\partial x}\right)^2 + \left(\frac{\partial u}{\partial y}\right)^2 = \frac{\partial u}{\partial x}\frac{\partial v}{\partial y} - \frac{\partial v}{\partial x}\frac{\partial u}{\partial y}$$

This expression can be written in the determinant form

$$|f'(z)|^2 = \begin{vmatrix} \frac{\partial u}{\partial x} & \frac{\partial u}{\partial y} \\ \frac{\partial v}{\partial x} & \frac{\partial v}{\partial y} \end{vmatrix} = \frac{\partial(u, v)}{\partial(x, y)} = J,$$

which is known as the Jacobian. This is a very useful, or essential, quantity in coordinate transformation.

It is not always possible to have closed forms of $u(x, y)$ and $v(x, y)$. At least we can seek their Taylor expansions at a point (x_0, y_0). that is

$$u(x, y) = u(x_0, y_0) + \frac{\partial u}{\partial x}(x - x_0) + \frac{\partial u}{\partial y}(y - y_0) + \dots$$

$$v(x, y) = v(x_0, y_0) + \frac{\partial v}{\partial x}(x - x_0) + \frac{\partial v}{\partial y}(y - y_0) + \dots.$$

From linear algebra (next chapter), we know that is possible if and only if its Jacobian evaluated at (x_0, y_0) in not zero,

$$J = \begin{vmatrix} \frac{\partial u}{\partial x} & \frac{\partial u}{\partial y} \\ \frac{\partial v}{\partial x} & \frac{\partial v}{\partial y} \end{vmatrix} \neq 0.$$

For rigorous proof of this fact, we need to make use of the implicit function theorem, which can be found in advanced calculus text books [see Hildebrand [18]].

4.3   Conformal Mapping

Although the title of this chapter is Conformal Mapping, we have not yet introduced the concept of conformality. In the previous section, we have shown that in order for this method to work, the transformation function $f(z)$ must be analytic and that $f'(z) \neq 0$ in the given domain. In this section, we will show that such transformation is conformal. A mapping from the z-plane onto the w plane is said to be conformal if the angles between the tangents to the two curves in the z-plane are preserved both in magnitude and in the sense of rotation, when transformed to the corresponding curves in the w-plane.

As explained in Sect. 2.2.2, a contour in complex plan can be expressed parametrically $z = z(t)$, where t is a parameter. Let C_1 and C_2 be the two contours in the z-plane (x-y plane), they are expressed as $z_1(t)$ and $z_2(t)$, respectively, and they intersect at z_0, corresponding to t_0. Under the transformation $w = f(z)$, where f is analytic and $f'(z_0) \neq 0$, C_1 and C_2 are transformed into Γ_1 and Γ_2 in the w-plane (u-v plane), and z_0 is transformed to w_0. Conformality means that the magnitude and the sense of rotation of the angle between the tangents to C_1 and C_2 are the same as to the angle between those to Γ_1 and Γ_2.

Let us express them in the following way.

C_1: $z = z_1(t)$, $\frac{dz_1}{dt} = \left| \frac{dz_1}{dt} \right| e^{i\theta_1}$. This means that the contour C_1 is traced out by z_1 as t moves forward, and θ_1 is the inclination angle of the tangent to C_1 at t_0.

C_2: $z = z_2(t)$, $\frac{dz_2}{dt} = \left| \frac{dz_2}{dt} \right| e^{i\theta_2}$ have similar meaning to C_2.

Let the angle between the tangents to C_2 and C_1 be α, clearly

$$\alpha = \theta_2 - \theta_1.$$

With the transformation $w = f(z)$, let us write

$$\frac{df}{dz} = |f'| e^{i\varphi}.$$

From the chain rule, we have

$$\frac{dw}{dt} = \frac{df}{dz}\frac{dz}{dt}.$$

Since f is analytic and $f' \neq 0$, we can carry out this transformation. In the image w plane, we have

$$\Gamma_1 : w = w_1(z_1). \quad \frac{dw_1}{dt} = \left|\frac{dw_1}{dt}\right| e^{i\phi_1},$$

and

$$\Gamma_2 : w = w_2(z_2). \quad \frac{dw_2}{dt} = \left|\frac{dw_2}{dt}\right| e^{i\phi_2}.$$

Since

$$\frac{dw_1}{dt} = \frac{df}{dz_1}\frac{dz_1}{dt} = |f'| e^{i\varphi}\left|\frac{dz_1}{dt}\right| e^{i\theta_1} = |f'|\left|\frac{dz_1}{dt}\right| e^{i(\varphi+\theta_1)},$$

therefore

$$\phi_1 = \varphi + \theta_1.$$

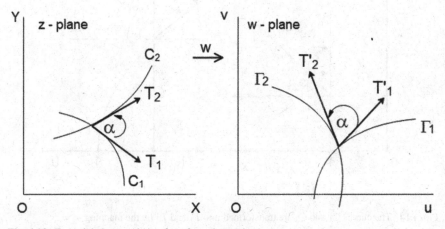

Fig. 4.13 Essential characteristic of conformal mapping

Similarly

$$\phi_2 = \varphi + \theta_2.$$

Let the angle between the tangents to Γ_2 and Γ_1 be β, and

$$\beta = \phi_2 - \phi_1 = (\varphi + \theta_2) - (\varphi + \theta_1) = \theta_2 - \theta_1 = \alpha.$$

Thus the angle and the sense of rotation are both preserved. This is the essential characteristic of a conformal mapping (Fig. 4.13).

Example 4.3.1 Let C_1 and C_2 be defined in the upper half of z-plane. Let C_1 be represented by the equation $x - y = 0$ and C_2 by the line $x = 1$, as shown in (a) of the following figure. Find the point z_0 where they intersect and the angle from C_1 to C_2 at z_0. Let Γ_1 and Γ_2 be the images of C_1 and C_2, respectively, in the w plane under the transformation of $f(z) = z^2$, as shown in (b) of the following figure. Find the equations of Γ_1 and Γ_2 in the w plane, and the point w_0 where they intersect. Show that w_0 is the image of z_0 under the same transformation. Find the angle from Γ_1 to Γ_2 at w_0 and show that this angle is the same as the angle from C_1 to C_2 in the z-plane (Fig. 4.14).

Solution 4.3.1 The equation representing C_1 is $x = y$, the one representing C_2 is $x = 1$. Clearly, $x = 1$ and $y = 1$ satisfy both these equations, therefore they intersect at

$$z_0 = x_0 + iy_0 = 1 + i.$$

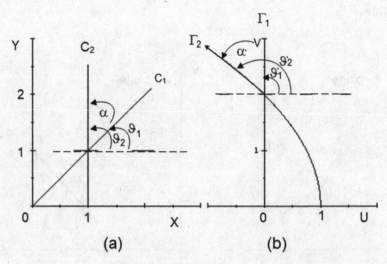

Fig. 4.14 The curves C_1 and C_2 are transformed into Γ_1 and Γ_2 by the mapping $w = z^2$

The angle from C_1 to C_2 is clearly equal to $\pi/4$. Nevertheless, it is instructive to see how formally we get this result. The angle ϑ_1 between C_1 and the x-axis is the slope of C_1, that is: $\tan \vartheta_1 = \frac{dy}{dx}$ along C_1, Since $\frac{dy}{dx} = 1$, so $\vartheta_1 = \tan^{-1}(1) = \pi/4$. The angle ϑ_2 between C_2 and the x-axis is clearly equal to $\pi/2$, since C_2 is perpendicular to the x-axis. Thus the angle α from C_1 to C_2 is

$$\alpha = \vartheta_2 - \vartheta_1 = \pi/2 - \pi/4 = \pi/4.$$

Now under the transformation

$$w = z^2 = (x + iy)^2 = (x^2 - y^2) + i2xy, \quad w = u + iv,$$

we have

$$u = x^2 - y^2, \quad v = 2xy.$$

Γ_1 is the image of C_1 on which $x = y$, therefore the equation for Γ_1 is

$$u = x^2 - y^2 = 0, \quad (v \geq 0).$$

For Γ_2, we take from C_2 that $x = 1$, so

$$u = 1 - y^2, \quad v = 2y.$$

Combine these two equations, we have the equation representing Γ_2:

$$u = 1 - \frac{1}{4}v^2.$$

Clearly, $u = 0$ and $v = 2$ satisfy the equations for both Γ_1 and Γ_2; therefore their intersection point w_0 must be

$$w_0 = 0 + i2 = 2i.$$

It can be seen that

$$f(z_0) = (1 + i)^2 = 1^2 + 2i + (i)^2 = 2i = w_0.$$

Therefore, w_0 is indeed the image of z_0 under the transformation of $w = z^2$.

Let the angle from Γ_1 to Γ_2 be α' and $\alpha' = \vartheta_2' - \vartheta_1'$. These angles are shown in (b) of the above figure. The angle ϑ_1' is clearly equal to $\pi/2$, since Γ_1 is perpendicular to u-axis. To calculate ϑ_2', we must take the slope of Γ_2 with respect to u. That is, $\tan \vartheta_2' = \frac{dv}{du}$ along Γ_2 at w_0. Take derivative with respect to u of the equation $u = 1 - v^2/4$, we have $1 = -\frac{1}{2}v\frac{dv}{du}$, or $\frac{dv}{du} = -\frac{2}{v}$. At $v = 2$, we have $\frac{dv}{du} = -1 = \tan \vartheta_2'$. Therefore $\vartheta_2' = \tan^{-1}(-1) = 3\pi/4$, as shown in (b) of the above figure. Thus

$$\alpha' = \vartheta'_2 - \vartheta'_1 = \frac{3\pi}{4} - \frac{\pi}{2} = \frac{\pi}{4}.$$

This is indeed the same as α, the angle from C_1 to C_2.

To emphasize the local nature of conformality, consider the mapping $w = f(z) = z^2$. If the domain of definition D is the first quadrant $0 \le x < \infty$, $0 \le y < \infty$, then the range D' is the upper half-plane $v \ge 0$ as shown in Fig. 4.15. Let the point z in the x-y plane be expressed in its polar coordinates $re^{i\theta}$. Under the transformation $w = z^2$, it becomes $r^2 e^{i2\theta}$ in the u-v plane. Therefore, the annulus section ABCD in the x-y plane is mapped into A'B'C'D' in the u-v plane. Since $w = f(z) = z^2$ is analytic everywhere; and $f'(z) = 2z$, so it is zero only at the origin. Therefore, the mapping is conformal everywhere except at the origin. Note that all the right angles at ABCD are preserved at their images A'B'C' and D'. However, at $z = 0$, the only point at which the conformality breaks down, the angle of $\pi/2$ is not preserved, it is doubled. Thus, conformality is local, not global.

We do not mean to imply that if conformality breaks down at a point, then angles are necessarily doubled there. That result is specific to the present example (Fig. 4.15).

4.4   Complex Potential

Let $\phi(x, y)$ be harmonic in some domain D and $\psi(x, y)$ is the conjugate of ϕ, then

$$F(z) = \phi(x, y) + i\psi(x, y)$$

is an analytic function of $z = x + iy$. This function is called the complex potential corresponding to the real potential ϕ.

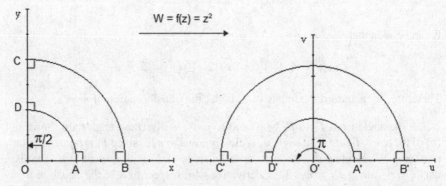

Fig. 4.15 At $z = 0$, the only point at which the conformality breaks down, the angle $\pi/2$ is not preserved, but doubled

Recall from Sect. 2.1 that for a given $\phi(x, y)$, a conjugate $\psi(x, y)$ can be uniquely determined through the Cauchy-Riemann equations, except for an additive constant. Hence, we may say the complex potential without causing misunderstandings.

The use of complex potential has a technical advantage that in complex analysis it is easier to handle a complex function than its real and imaginary parts separately. The complex potential has also a physical advantage in that its real and imaginary parts have physical meanings. For example, if $\phi(x, y) = $ constant is the equipotential in electrostatics, then $\psi(x, y) = $ constant is the profile of electrical field line (or line of force), they are the paths of moving charged particles, such as the paths of electrons in electron microscope. Or in the steady temperature distribution, $F(z)$ represents the complex heat potential. The real part equals constant represents the isotherms (lines of constant temperature) and the curves $\psi(x, y) = $ constant is the heat flow lines. Or in hydrodynamics, $\phi(x, y)$ is called velocity potential and $\psi(x, y)$ is called stream function.

This method is most powerful in dealing with fluid flow problems (see next section), But we will start with some simple examples in the electrostatics, which are more familiar to most readers.

Before we start, we must make the following point clear. Conceptually, equipotential surfaces and electrical field lines are two different quantities. If $\phi(x, y) = constant$ represent equipotential surfaces, then $\psi(x, y) = constant$ can only represent the profile of the electric field lines, but not their magnitude. Consider, for example, a capacitor consists of two parallel plates extending to infinity at a distance d apart, (no infinite extended plate exists, but our solution will approximate the situation in the section that is far away from the ends of plates). The voltage of one plate is kept at zero, and the other at V. Let these plates be perpendicular to the x-$axis$. In freshman physics, we learned that the equipotential surfaces are planes parallel to the plates. The electrical field lines are parallel to the x-$axis$ with a constant strength of V/d. Now with complex potential $F(z) = (V/d)z = (V/d)(x + iy)$, the real part $\phi(x, y) = (V/d)x$ is indeed the equipotential surface (planes perpendicular to the x-axis, everywhere on this plane the potential is equal to $(V/d)x$. However, we cannot say that the imaginary part $(V/d)y$ is the electrical field. Although $(V/d)y = constant$ are lines parallel to x-axis, as the electric field lines should be, the electrical field is, however, not equal to $(V/d)y$ (Given the equipotential surface $\phi(x, y)$, the electric field is a vector and $E_x(x, y) = -\frac{\partial}{\partial x}\phi(x, y)$ and $E_y(x, y) = -\frac{\partial}{\partial y}\phi(x, y)$. In the present case, $E_x = -V/d$, $E_y = 0$. That is, the electric field is a constant everywhere in the x direction). With this understanding, we can say the real part of the complex potential is the equipotential surface and the imaginary part represents the electrical field lines (the emphasis is on the lines).

Example 4.4.1 Sketch the equipotentials and the field lines of a long uniformly charged straight wire.

Solution 4.4.1 With the help of the Gauss law, we learned in freshman physics that in this case the electrical field is $\frac{1}{r}$ and the potential is $\ln r$, (we have ignored

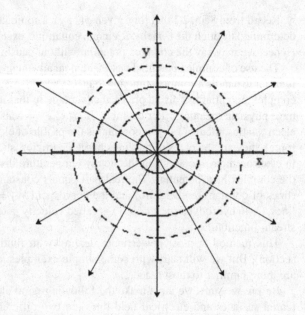

the constant coming from the sign of charges and the units we choose to do the measurements), where r is the distance between the field point and the charged wire. Now in terms of complex potential we have

$$F(z) = \ln z = \ln r e^{i\theta} = \ln r + i\theta.$$

If $r = constant$, $\ln r$ will equal to a corresponding constant. In Fig. 4.16, the concentric circles represent constant r, therefore they are equipotential surfaces. The imaginary part is just θ. Therefore, θ equal to different constant represents different field line. In Fig. 4.16, the straight lines represent $\theta = n\pi/6$, $n = 1, \ldots 12$. They are the field lines.

Comment 1: All field lines are perpendicular to all equipotential surfaces.

Comment 2: All points on the same equipotential surface have the same value. However, different points on the same field line have different values.

Example 4.4.2 Determine the potential of a pair of oppositely charged straight lines of the same strength at the $z = c$ and $z = -c$ on the real axis. Sketch the equipotential surfaces and the field lines.

Solution 4.4.2 The solution of this problem can be obtained by superposition. It follows from the last example that the potential of each of the source lines is

$$\phi_1 = \ln(z - c), \qquad \phi_2 = -\ln(z + c)$$

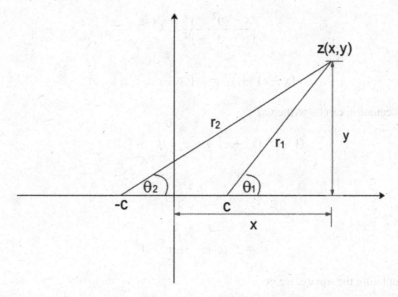

Fig. 4.17 Two uniformly charged straight wires perpendicular to xy plane with the same amount of charge density but opposite sign located at $x = c$ and $x = -c$, respectively

respectively. Hence the complex potential of the combination of the two source lines is

$$F(z) = \phi_1(z) + \phi_2(z) = \ln(z - c) - \ln(z + c) = \ln\frac{z - c}{z + c}$$

The situation is shown in Fig. 4.17.

Since

$$\ln\frac{z - c}{z + c} = \ln\frac{r_1 e^{i\theta_1}}{r_2 e^{i\theta_2}} = \ln\frac{r_1}{r_2}e^{i(\theta_1 - \theta_2)} = \ln\frac{r_1}{r_2} + i(\theta_1 - \theta_2).$$

The equipotential lines are represented by the real part of $F(z)$ equal to different constants, that is

$$\Phi = \text{Re } F(z) = \ln\frac{r_1}{r_2} = constants.$$

These conditions are satisfied if

$$\frac{r_1}{r_2} = k$$

where k is a constant. Since

$$r_1^2 = (x - c)^2 + y^2, \quad r_2^2 = (x + c)^2 + y^2,$$

therefore

$$\frac{(x-c)^2 + y^2}{(x+c)^2 + y^2} = k^2,$$

or

$$(x-c)^2 + y^2 = k^2[(x+c)^2 + y^2]. \qquad (4.18)$$

This equation can be written as

$$(1-k^2)(x^2 + c^2 + y^2) = (k^2+1)2cx.$$

Let

$$x_0 = c\frac{1+k^2}{1-k^2},$$

then

$$x^2 - 2x_0 x + c^2 + y^2 = 0.$$

Completing the square, we get

$$(x - x_0)^2 + y^2 = x_0^2 - c^2.$$

This is a series of circles with radius of $R = \left(x_0^2 - c^2\right)^{1/2}$, centered at x_0 on the x-axis. Some of them are shown in Fig. 4.18.

The field lines are represented by the imaginary part of $F(z)$ equal to different constants. That is

$$\Psi = \operatorname{Im} F(z) = \theta_1 - \theta_2 = \alpha$$

where $\alpha's$ are different fixed angles and $|\alpha| \le \pi$. Let

$$\tan \alpha = \tan(\theta_1 - \theta_2)$$

and

$$\tan(\theta_1 - \theta_2) = \frac{\tan \theta_1 - \tan \theta_2}{1 + \tan \theta_1 \tan \theta_2}.$$

Since

$$\tan \theta_1 = \frac{y}{x-c}, \quad \tan \theta_2 = \frac{y}{x+c},$$

Hence

$$\tan \alpha = \frac{\frac{y}{x-c} - \frac{y}{x+c}}{1 + \frac{y}{x-c}\frac{y}{x+c}} = \frac{2cy}{x^2 + y^2 - c^2}.$$

This equation can be written as

$$x^2 + y^2 - c^2 = \frac{2cy}{\tan \alpha}.$$

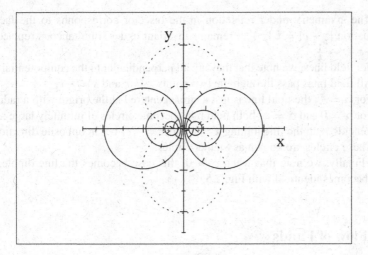

Fig. 4.18 The circles centered on the x-axis are the equipotential lines and those centered on the y-axis are the field lines

Let

$$y_0 = \frac{c}{\tan \alpha},$$

and completing the square, this equation is simplified to

$$x^2 + (y - y_0)^2 = c^2 + y_0^2.$$

These are a series of circles centered on the y-axis with radius $R = \left(c^2 + y_0^2\right)^{1/2}$. All these circles go through $\pm c$, because at $y = 0$, $x = \pm c$. Some of them are shown in Fig. 4.18.

The circles centered on the x-axis are the equipotential lines and those centered on the y-axis are the field lines.

We can make some comments about this plot.

1. For the equipotential lines, we see that from Eq. (4.18) that if $k = 1$, $x = 0$ which is the straight line bisects the line joining the two points $x = c$ and $x = -c$. It divides the plane into left and right half-planes.

2. When $k \approx 0$, again we see that from Eq. (4.18) that the equipotential line shrinks to a small circle enclosing the point $x = c$. As k increases, x_0 also increases. The radius of the circle becomes larger and larger and tends to the bisector as $k \to 1$.

3. For $k > 1$, the circle lies in the half-plane containing the point $x = -c$. As k increases, $x_0 \to c$. So the circle becomes smaller and shrinks to the point $x = -c$ when $k \to \infty$.

4. The symmetry under reflection in the bisector corresponds to the fact that the equation $|z - c| = k|z + c|$ remains invariant under simultaneous replacement $c \to -c, k \to \frac{1}{k}$.

5. For field lines, we note that they are all perpendicular to the equipotential lines.

6. All field lines pass through the two points $x = c$ and $x = -c$.

7. For $\alpha = \frac{\pi}{2}$, the field line is just a circle centered at the origin with a radius c.

8. For $\alpha = 0$ and $\alpha = \pi$, both field lines become circles of infinitely large radius, they coincide with the line joining $x = c$ and $x = -c$, but in opposite directions.

9. These circles are known as Steiner circles.

10. Finally, we note that when $c \to 0$, this case becomes the line dipole. This figure becomes identical with Fig. 2.5.

4.5   Flow of Fluids

In this section, we examine the application of analytic function theory to the study of fluid flow. In general, the real behavior of fluids is quite complicated, most of the complications are due to the viscosity of the fluids (The so-called "Wet Water Flow"). However, for many common fluids, like air and water, viscosity effects are negligible in regions distant from the boundaries of solids. Furthermore, if the velocity of the flow is below 0.3 Mach number (which is the ratio of the fluid velocity to the speed of sound), the assumption of incompressibility is acceptable, (for water and oil, but not for air). Under these conditions, harmonic functions (solutions of Laplace's equation) play an important role in hydrodynamics. Here again, we consider only the two-dimensional steady-state type of problem. That is, the motion of the fluid is assumed to be the same in all planes parallel to the xy plane, the velocity being parallel to that plane and independent of time. It is, then, sufficient to consider the motion of a sheet of fluid in the xy plane.

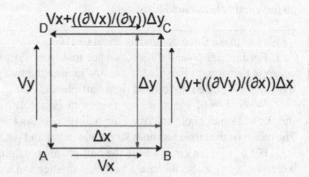

Fig. 4.19 Computing the circulation of the tangential component of the vector field V around a small square

4.5.1 *Irrotation Flow and Velocity Potential*

In a non-viscous fluid, we can assume the flow is irrotational. By which we mean the following. Consider an infinitesimal loop ABCD with width Δx and Δy shown in Fig. 4.19. Let us consider the line integral of the tangential components of V around this infinitesimal loop ABCD. Let V_x and V_y be the components of the velocity vector V in the x and y directions. Starting at the point (x, y) –at A, the lower left corner of the loop—we go around in the direction indicated by the arrows. Along AB, the tangential component is V_x and the distance is Δx. The first part of the integral is $V_x(AB)\Delta x$. Along BC, we get $V_y(BC)\Delta y$. Along CD, we get $-V_x(CD)\Delta x$, the minus sign is required because the component is in the opposite direction of travel. Along the last leg, we have $-V_y(DA)\Delta y$, again the minus sign is for the same reason. Now if we take into account of the rate of change of V_x, we can see that

$$V_x(CD) = V_x(AB) + \frac{\partial V_x}{\partial y}\Delta y.$$

Similarly

$$V_y(BC) = V_y(DA) + \frac{\partial V_y}{\partial x}\Delta x.$$

Therefore, the circulation around the square is

$$\oint V \cdot dl = V_x(AB)\Delta x + V_y(BC)\Delta y - V_x(CD)\Delta x - V_y(DA)\Delta y$$

$$= V_x(AB)\Delta x + [V_y(DA) + \frac{\partial V_y}{\partial x}\Delta x]\Delta y-$$

$$[V_x(AB) + \frac{\partial V_x}{\partial y}\Delta y]\Delta x - V_y(DA)\Delta y = \left(\frac{\partial V_y}{\partial x} - \frac{\partial V_x}{\partial y}\right)\Delta x \Delta y.$$

When we say that the flow is irrotational, we mean this circulation is equal to zero, or simply

$$\frac{\partial V_y}{\partial x} - \frac{\partial V_x}{\partial y} = 0.$$

Before proceeding further, we ask the question, why the name irrotational? (irrotational literally mean no rotation.) If a body is rotating around an axis with an angler velocity ω, then every point in the body is performing a circular motion with angler velocity ω as shown in Fig. 4.20. Let the arc distance from A to B be Δs, then by definition

$$\Delta \theta = \frac{\Delta s}{r}, \quad \omega = \frac{\Delta \theta}{\Delta t} = \frac{1}{r}\frac{\Delta s}{\Delta t}.$$

It is understood that $\Delta t \rightarrow 0$. The magnitude of the velocity v is equal to

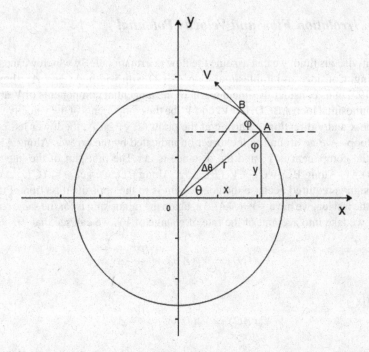

Fig. 4.20 To show that the irrotational condition means there is no rotation in the fluid

$$|v| = \frac{\Delta s}{\Delta t} = r\omega.$$

But velocity is a vector quantity as shown in the figure. Clearly

$$V_x = -|v|\cos\varphi = -r\omega\cos\varphi, \quad V_y = r\omega\sin\varphi.$$

The minus sign is because V_x is in the negative direction of x as shown in the figure.
It is also clear from the figure that

$$\cos\varphi = \frac{y}{r}, \quad \sin\varphi = \frac{x}{r}.$$

Hence

$$V_x = -\omega y, \quad V_y = \omega x.$$

Therefore

$$\frac{\partial V_y}{\partial x} - \frac{\partial V_x}{\partial y} = 2\omega$$

Thus if this expression is equal to zero, it means $\omega = 0$, no rotation.
For those who are familiar with vector analysis, it is easy to show

$$\mathbf{V} = \omega \times \mathbf{r}.$$

So

$$\nabla \times \mathbf{V} = \nabla \times (\omega \times \mathbf{r}) = 2\omega$$

(See Example 2.7.1 of vol. 2). But in two dimensions

$$\nabla \times \mathbf{V} = \begin{vmatrix} \mathbf{i} & \mathbf{j} & \mathbf{k} \\ \frac{\partial}{\partial x} & \frac{\partial}{\partial y} & 0 \\ V_x & V_y & 0 \end{vmatrix} = \left(\frac{\partial V_y}{\partial x} - \frac{\partial V_x}{\partial y} \right) \mathbf{k}$$

Thus we get the same thing

$$\left(\frac{\partial V_y}{\partial x} - \frac{\partial V_x}{\partial y} \right) \mathbf{k} = 2\omega\mathbf{k}.$$

Irrotation condition simply means no rotation. Many books simply define irrotational condition as

$$\nabla \times \mathbf{V} = 0.$$

A simple test is to put a small straw into the flow field. If the fluid is truly irrotational, the tiny straw will not rotate.

Now if we have a macroscopic closed loop Γ, which is the boundary of a simply connected domain D, and suppose that the vector velocity has continuous partial derivatives in the domain, then by Green's Lemma (Sect. 2.3.1), we have

$$\iint_D \left(\frac{\partial V_y}{\partial x} - \frac{\partial V_x}{\partial y} \right) dxdy = \oint_\Gamma \left(V_x dx + V_y dy \right)$$

To carry out the double integral, we can cut D into many small areas, each approximately a square. Then the circulation around Γ is the sum of the circulation around the little loops. If the flow is irrotational, then we have shown that the integrand of the double integral is equal to zero. Therefore, the integral around Γ is also zero

$$\oint_\Gamma \left(V_x dx + V_y dy \right) = 0.$$

Since Γ is any closed contour, this means that the integral is independent of the path. Hence if we integrate from a fixed point (a, b) to a variable point (x, y), the integral becomes a function of the point (x, y). That is

$$\int_{(a,b)}^{(x,y)} \left(V_x dx + V_y dy \right) = \Phi \left(x, y \right).$$

That is

$$\int\limits_{(a,b)}^{(x,y)} \left(V_x dx + V_y dy \right) = \int\limits_{(a,b)}^{(x,y)} d\Phi \left(x, y \right) = \Phi \left(x, y \right).$$

Since

$$d\Phi = \frac{\partial \Phi}{\partial x} dx + \frac{\partial \Phi}{\partial y} dy$$

therefore

$$V_x = \frac{\partial \Phi}{\partial x}, \qquad V_y = \frac{\partial \Phi}{\partial y}.$$

We call the function $\Phi\,(x, y)$ the velocity potential. The curves $\Phi\,(x, y) = const.$ are the equipotential lines. The vector velocity is the gradient of Φ. That is

$$\nabla \Phi = \mathbf{i} \frac{\partial \Phi}{\partial x} + \mathbf{j} \frac{\partial \Phi}{\partial x} = \mathbf{i} V_x + \mathbf{j} V_y = \mathbf{V}.$$

Furthermore, since

$$d\Phi = \frac{\partial \Phi}{\partial x} dx + \frac{\partial \Phi}{\partial y} dy = \nabla \Phi \cdot d\mathbf{r}$$

is the difference of Φ between two nearby points separated by $d\mathbf{r}$

$$d\mathbf{r} = \mathbf{i} dx + \mathbf{j} dy.$$

If the two points are on the same equipotential line, then

$$d\Phi = \nabla \Phi \cdot d\mathbf{r} = 0.$$

This means $\nabla \Phi$ is perpendicular to the equipotential line since $d\mathbf{r}$ is on that line. This also means that the velocity of the flow $\mathbf{V}$ is perpendicular to the equipotential line, Since $\nabla \Phi = \mathbf{V}$.

4.5.2   Incompressibility of the Fluid

In this section, we will show that the incompressibility of the fluid leads to the continuity equation. From this we see that the velocity potential satisfies Laplace's equation. Let us consider the the flux of fluid flowing into a given infinitesimal area ABCD with width Δx and Δy shown in Fig. 4.21. Again let V_x and V_y be the components of the velocity vector V in the x and y directions. Clearly the flux of the fluid flowing into or out of ABCD are the component of the velocity of the fluid perpendicular to its sides. Along AB, the normal component is V_y and the distance is Δx. So the flux flowing into ABCD through the side AB is $V_y \Delta x$. Along CD, the normal component is also V_y, but considering the change of V_y

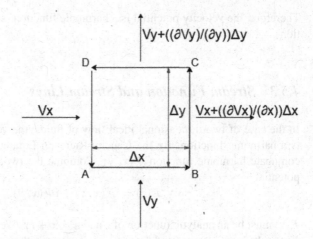

Fig. 4.21 Fluxes across the four sides of a differential area in a two-dimensional flow field

due to the displacement of Δy, we have $V_y + \frac{\partial V_y}{\partial y}\Delta y$. So the flux through CD is $-(V_y + \frac{\partial V_y}{\partial y}\Delta y)\Delta x$, the minus sign is because this is the flux outflow. Similarly the flux through DA is $V_x \Delta y$, and through BC is $-(V_x + \frac{\partial V_x}{\partial x}\Delta x)\Delta y$. Summing the four parts, we have the net rate of accumulation (Fig. 4.21)

$$Flux\ inflow = -\left(\frac{\partial V_x}{\partial x} + \frac{\partial V_y}{\partial y}\right)\Delta x)\Delta y.$$

Since the fluid is incompressible, the net rate of accumulation in any given area of the flow is zero. Therefore, the condition of incompressibility is given by

$$\frac{\partial V_x}{\partial x} + \frac{\partial V_y}{\partial y} = 0. \tag{4.19}$$

This equation is called the continuity equation. In three dimensions, this equation is written as

$$\nabla \cdot \mathbf{V} = 0.$$

In fact we can use $\nabla \cdot \mathbf{V}$ as the source or the sink of a flow. For example, if we use Fig. 4.16 to describe a flow, then it is a flow with a source at $z = 0$. In other words, $\nabla \cdot \mathbf{V} \neq 0$ at $z = 0$. Everywhere else $\nabla \cdot \mathbf{V} = 0$.

As we have seen in the last section that

$$V_x = \frac{\partial \Phi}{\partial x}, \qquad V_y = \frac{\partial \Phi}{\partial y},$$

where Φ is the velocity potential. Substituting them into Eq. (4.19), we have

$$\frac{\partial^2 \Phi}{\partial x^2} + \frac{\partial^2 \Phi}{\partial y^2} = 0.$$

Therefore, the velocity potential is a harmonic function (satisfying Laplace's equation).

4.5.3   Stream Function and Stream Lines

In the case of two-dimensional ideal flow of fluids, the velocity potential $\Phi(x, y)$ is a harmonic function. By the Cauchy-Riemann Equations, we can construct its conjugate harmonic function $\Psi(x, y)$. Combine the two, we can form a complex potential

$$F(z) = \Phi(x, y) + i\Psi(x, y).$$

$F(z)$ must be an analytic function of z and $z = x + iy$. We know that the families of curves $\Psi(x, y) = const$ and $\Phi(x, y) = const$ are orthogonal to each other. Now the velocity vector is normal to the equipotential line $\Phi(x, y) = const$, thus, they must be in the tangent direction of $\Psi(x, y) = const$. Therefore, the curve $\Psi(x, y) = const$ is called the stream lines, which represent the path followed by the fluid particles. The conjugate function $\Psi(x, y)$ is known as the stream function. Since it is also harmonic, so

$$\frac{\partial^2 \Psi}{\partial x^2} + \frac{\partial^2 \Psi}{\partial y^2} = 0.$$

In the problem of fluid flows in channels and across obstacles, the normal component of the velocity along the boundaries is zero. Since the velocity must be tangential to the boundaries, these boundaries must be stream lines. The stream function must be constant on the boundary.

Example 4.5.1 Describe the flow represented by

$$F(z) = Az,$$

where A is a positive real constant.

Solution 4.5.1 Clearly

$$F(z) = Ax + iAy,$$

is certainly analytic. Furthermore, both its real part and imaginary part satisfy Laplace's equation. Therefore, we can regard it as a complex potential. The velocity potential is Ax and equipotential lines are

$$Ax = const$$

which are lines parallel to the y-axis, and the stream function is Ay and stream lines are

$$Ay = const.$$

which are lines parallel to x-axis. Furthermore,

$$V_x = \frac{\partial}{\partial x}(Ax) = A,$$

$$V_y = \frac{\partial}{\partial y}(Ax) = 0.$$

Therefore, this represents a uniform flow from left to right with velocity A. This velocity is the same whether it is in the upper half-plane or in the lower half-plane. In other words, the flow of fluid is the same whether it is above the boundary line $y = 0$, or below this line.

In analyzing a flow in the xy plane, it is often advantageous to first make a conformal mapping w transforming the complicated boundary conditions in the x, y plane into a simpler conditions in the u, v plane. Then if Φ is a velocity potential and Ψ is a stream function for the flow in the u, v plane, we can transform them back to the x, y plane.

For example, if $f(z)$ is analytic, and the conformal mapping is

$$w = f(z) = u(x, y) + iv(x, y).$$

Now if in the u, v plane, the complex potential is

$$F(w) = F(f(z)) = Aw = Au + iAv,$$

then this represents a uniform flow from left to right in the u, v plane with velocity A, as shown in the last example. Therefore, the stream lines in the x, y plane are

$$Av(x, y) = const.$$

This is best illustrated in the following example.

Example 4.5.2 Consider a flow in the first quadrant $x > 0$, $y > 0$. It comes in downward at velocity v_0 parallel to the y-axis but is forced to turn a corner near the origin, as shown in the following figure. Find the velocity potential and stream function of the flow. Sketch the stream lines of the flow.

Solution 4.5.2 First we make a conformal mapping $w = z^2$. In polar coordinates, the y- axis $re^{i\pi/2}$ will be mapped into $r^2 e^{i\pi}$ by $w = z^2$. Therefore, this transformation will map the first quadrant of the xy plane into the upper half-plane of the uv plane as shown in Fig. 4.22. The corresponding images of ABC are A'B'C' as shown in the figure.

In the uv plane, the complex potential can be written as

$$F(w) = v_0 w = v_0(u + iv)$$

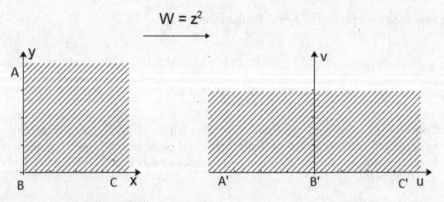

Fig. 4.22 The transformation $W = z^2$ maps the first quadrant of the xy plane into the upper half-plane of the uv plane

which represents a uniform flow from left to right with velocity v_0. Now we want to transform it back to the xy plane. Since

$$w = z^2 = (x + iy)^2 = (x^2 - y^2) + i2xy = u(x, y) + iv(x, y)$$

the velocity potential is

$$v_0 u(x, y) = v_0(x^2 - y^2),$$

the stream function is

$$v_0 v(x, y) = 2v_0 xy.$$

The stream lines are

$$xy = const,$$

which are sketched in Fig. 4.23.

Fig. 4.23 The streaming lines corresponding to the stream function $\Psi(x, y) = 2v_0 xy$

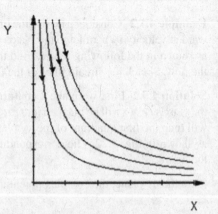

Example 4.5.3 A uniform flow of velocity v_0 is flowing over a cylinder of radius R that is perpendicular to the flow. What is the velocity potential and stream function? Determine the pattern of the flow by sketching some of its stream lines.

Solution 4.5.3 In the table of transformations, one find the following transformation $w = z + \frac{1}{z}$ which maps the upper half-plane with half of a circle of radius one removed in the xy plane onto the entire half-plane of the uv plane (see Fig. 4.24). It is not difficult to verify this transformation. It is clear that points on the x-axis with $y = 0$ will transform to points with $v = 0$ on the uv plane. Therefore, the images of A, B, D, E are A', B', D' and E'. On the circular BCD boundary, we can use the polar coordinates $z = re^{i\theta}$, where $r = 1$. Since $z + \frac{1}{z} = e^{i\theta} + e^{-i\theta} = 2\cos\theta$. Hence, there is no imaginary part. Point C is mapped to C' at $v = u = 0$.

Although this transformation is promising, it is not exactly what we need, since the radius of the cylinder is R, not one, in our problem. But we can modify this transformation to

$$ w = \frac{1}{2}\left(z + \frac{R^2}{z}\right). $$

With $z = Re^{i\theta}$, so B and D are at $x = -R$ and $x = R$, respectively, their images B' and D' are also at $u = -R$ and $u = R$, respectively. The circular boundary BCD is also transformed to $u = R\cos\theta$ with $\pi \le \theta \le 0$ and $v = 0$.

With this transformation, the fluid is flowing uniformly from left to right with velocity v_0 on the u,v plane. The complex potential is

$$ F(w) = v_0 w = v_0(u + iv). $$

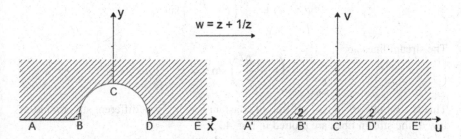

Fig. 4.24 The transformation $w = 1 + 1/z$ that maps the upper half-plane with half a circle of radius one removed in xy plane onto the entire upper half-plane of the uv plane

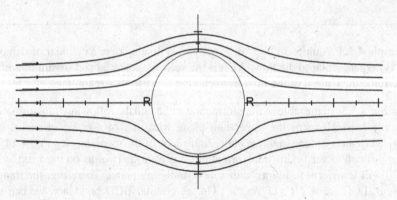

Fig. 4.25 Stream lines of a flow over a cylinder of radius R

Now we want to transform back to xy plane. Since

$$w = \frac{1}{2}\left(z + \frac{R^2}{z}\right) = \frac{1}{2}\left(re^{i\theta} + \frac{R^2}{r}e^{-i\theta}\right)$$

$$= \frac{1}{2}\left(r + \frac{R^2}{r}\right)\cos\theta + i\frac{1}{2}\left(r - \frac{R^2}{r}\right)\sin\theta,$$

so the velocity potential is

$$\Phi(r,\theta) = \frac{v_0}{2}\left(r + \frac{R^2}{r}\right)\cos\theta,$$

and the stream function is

$$\Psi(r,\theta) = \frac{v_0}{2}\left(r - \frac{R^2}{r}\right)\sin\theta.$$

The stream lines are

$$\left(r - \frac{R^2}{r}\right)\sin\theta = k$$

This is a family of curves, a different constant k will give a different stream line. A few of the stream lines are plotted in Fig. 4.25.

Note that the stream lines are symmetrical with y-axis, since $\sin\theta = \sin(\pi - \theta)$. Furthermore, they are also symmetrical with x-axis, changing k to $-k$, we will have a corresponding stream line in the other side of x-axis. Even though we mapped from one upper half-plane to another, the problem is symmetric about the x-axis and so we show stream lines both above and below the x-axis.

4.5.4 *Complex Velocity*

Velocity is a vector quantity, its x and y components are

$$\mathbf{V} = V_x \mathbf{i} + V_y \mathbf{j}$$

so the speed is

$$v = \sqrt{V_x^2 + V_y^2}. \tag{4.20}$$

Now if we define a complex velocity as

$$V(z) = V_x + i V_y,$$

the speed is also given by Eq. (4.20).

If we have a complex potential

$$F(z) = \Phi(x, y) + i\Psi(x, y), \tag{4.21}$$

then we know

$$V_x = \frac{\partial}{\partial x} \Phi(x, y), \quad V_y = \frac{\partial}{\partial y} \Phi(x, y),$$

so our complex velocity is

$$V(z) = \frac{\partial}{\partial x} \Phi(x, y) + i \frac{\partial}{\partial y} \Phi(x, y). \tag{4.22}$$

Recall Eq. (2.6) of Sect. 2.1.3 and from Eq. (4.21) we have

$$\frac{d}{dz} F(z) = \frac{\partial}{\partial x} \Phi(x, y) + i \frac{\partial}{\partial x} \Psi(x, y).$$

Since $F(z)$ is analytic, so by Cauchy-Riemann equation,

$$\frac{\partial}{\partial x} \Psi(x, y) = -\frac{\partial}{\partial y} \Phi(x, y), \tag{4.23}$$

so

$$F'(z) = \frac{\partial}{\partial x} \Phi(x, y) - i \frac{\partial}{\partial y} \Phi(x, y).$$

Thus, the complex conjugate of F',

$$(F'(z))^* = \frac{\partial}{\partial x} \Phi(x, y) + i \frac{\partial}{\partial y} \Phi(x, y) \tag{4.24}$$

is identical to the complex velocity of Eq. (4.22),

At certain points $F'(z) = 0$. This means that at those points, the velocity is zero. We call those points stagnation points.

Example 4.5.4 Find (a) the complex potential, (b) the complex velocity, (c) the stream function when a cylindrical pipe of radius "R" is placed perpendicular in a uniform horizontal flow with velocity v_0 far from the pipe. Also find the stagnation points of this flow.

Solution 4.5.4 In the last example of the previous section, we see that the transformation $w = \frac{1}{2}(z + \frac{R^2}{z})$ will give us the proper stream lines. Therefore, the complex potential can be expressed as

$$F(w) = Aw = \frac{A}{2}\left(z + \frac{R^2}{z}\right).$$

Hence the velocity of this flow is described by

$$(F'(w))^* = Complex \ \ Conjugate \ \ of \ \left\{\frac{d}{dz}\left[\frac{A}{2}(z + \frac{R^2}{z})\right]\right\}$$

$$= \frac{A}{2}\left(1 - \frac{R^2}{z^2}\right).$$

When it is far from the pipe, z is very large. The velocity is v_0, therefore $A = 2v_0$. Clearly when $z^2 = R^2$, $F'(w) = 0$. That means the stagnation points are $x = \pm R$, $y = 0$. Furthermore, the complex potential is

$$F(w) = v_0\left(z + \frac{R^2}{z}\right)$$

and

$$(F'(w))^* = v_0\left(1 - \frac{R^2}{(z^2)^*}\right) = v_0\left(1 - \frac{R^2}{r^2}e^{i2\theta}\right)$$

$$= v_0\left(1 - \frac{R^2}{r^2}\cos 2\theta\right) - iv_0\frac{R^2}{r^2}\sin 2\theta.$$

Since

$$F(w) = v_0\left[(x + iy) + \frac{R^2}{(x + iy)}\right]$$

$$= v_0\left[x\left(1 + \frac{R^2}{x^2 + y^2}\right) + iy\left(1 - \frac{R^2}{x^2 + y^2}\right)\right]$$

$$= \Phi(x, y) + i\Psi(x, y)$$

The stream function is

$$\Psi(x, y) = v_0 y \left(1 - \frac{R^2}{x^2 + y^2}\right).$$

Example 4.5.5 Find the complex potential $F(z)$ describing a uniform flow of velocity v_0 making an angle of θ with the x-axis. Let

$$F(z) = \Phi(x, y) + i\Psi(x, y).$$

Verify that

$$V_x = \frac{\partial}{\partial x}\Phi(x, y), \quad V_y = \frac{\partial}{\partial y}\Phi(x, y).$$

Solution 4.5.5 Figure 4.26 represents this flow.

It is clear that

$$V_x = V_0 \cos\theta, \quad \text{and} \quad V_y = V_0 \sin\theta.$$

The complex velocity is therefore

$$V_x + iV_y = V_0 \cos\theta + iV_0 \sin\theta = V_0 e^{i\theta}.$$

Fig. 4.26 A uniform flow with velocity V_0 making an angle of θ with the x-axis

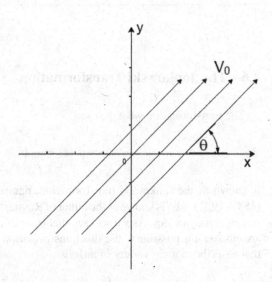

Hence

$$(F'(z))^* = V_0 e^{i\theta}.$$

Thus

$$F'(z) = V_0 e^{-i\theta}.$$

It follows that the complex potential is given by

$$F(z) = V_0 z e^{-i\theta}.$$

It can be expressed as

$$
\begin{aligned}
F(z) &= V_0(x + iy)(\cos\theta - i\sin\theta) \\
&= V_0[(x\cos\theta + y\sin\theta) + i(y\cos\theta - x\sin\theta)].
\end{aligned}
$$

Thus

$$
\begin{aligned}
\Phi(x, y) &= V_0(x\cos\theta + y\sin\theta) \\
\Psi(x, y) &= V_0(y\cos\theta - x\sin\theta).
\end{aligned}
$$

Therefore

$$
\begin{aligned}
\frac{\partial}{\partial x}\Phi(x, y) &= \frac{\partial}{\partial x}V_0(x\cos\theta + y\sin\theta) = V_0\cos\theta = V_x, \\
\frac{\partial}{\partial y}\Phi(x, y) &= \frac{\partial}{\partial y}V_0(x\cos\theta + y\sin\theta) = V_0\sin\theta = V_y.
\end{aligned}
$$

4.6 The Joukowski Transformation

The transformation we have used

$$w = f(z) = \frac{1}{2}\left(z + \frac{1}{z}\right) \tag{4.25}$$

is known as the Joukowski transformation, named in honor of Nikolai Joukowski (1847–1921), who is known as the father of Russian aviation. It has played a venerable role in aerodynamics. As we will see that by starting with figures in the z-plane that are circles, it is possible to use this transformation to produce figures in the w plane that describes a great variety of airfoils.

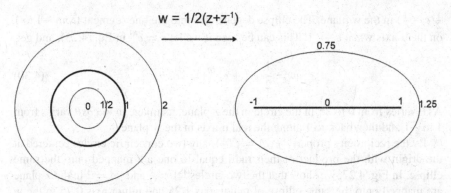

Fig. 4.27 The Joukowski transformation takes circles in the z-plane onto ellipses in the w plane. Any two circles with the product of their radii equal to one are mapped onto the same ellipse. The unit circle $|z| = 1$ is mapped onto the line segment $(-1 \text{ to } 1)$ on the u-axis

4.6.1 Properties of the Joukowski Transformation

The Joukowski transformation is conformal except at $z = 0$ and $z = \pm 1$. At $z = 0$, the transformation is singular. Since $f'(z) = \frac{1}{2}(1 - \frac{1}{z^2})$, therefore $f'(\pm 1) = 0$, thus $z = \pm 1$ are critical points. In fact, the image of a unit circle passing $z = \pm 1$ in the z-plane will collapse into a line from $u = -1$ to $u = 1$ in the w plane.

Circles to Ellipses By letting $z = ce^{i\theta}$, we see that

$$w = \frac{1}{2}\left(c + \frac{1}{c}\right)\cos\theta + \frac{i}{2}\left(c - \frac{1}{c}\right)\sin\theta \qquad (4.26)$$

Using the notation $w = u + iv$, we have

$$u = \left[\frac{1}{2}\left(c + \frac{1}{c}\right)\right]\cos\theta, \quad v = \left[\frac{1}{2}\left(c - \frac{1}{c}\right)\right]\sin\theta, \qquad (4.27)$$

or

$$\frac{u^2}{\left[\frac{1}{2}(c + \frac{1}{c})\right]^2} = \cos^2\theta, \quad \frac{v^2}{\left[\frac{1}{2}(c - \frac{1}{c})\right]^2} = \sin^2\theta. \qquad (4.28)$$

Since $\cos^2\theta + \sin^2\theta = 1$, it follows that

$$\frac{u^2}{\left[\frac{1}{2}(c + \frac{1}{c})\right]^2} + \frac{v^2}{\left[\frac{1}{2}(c - \frac{1}{c})\right]^2} = 1. \qquad (4.29)$$

This is equation of an ellipse. Therefore, Eq. (4.25) maps the circle $|z| = c$ in the z-plane onto an ellipse with semi-major axis of $\frac{1}{2}(c + \frac{1}{c})$ and semi-minor axis of

$\frac{1}{2}(c - \frac{1}{c})$ in the w plane. The ellipse degenerates into the line segment from -1 to 1 on the u-axis when $c \to 1$. This can be seen if we let $z = e^{i\theta}$ in Eq. (4.25), and get

$$w = \frac{1}{2}(e^{i\theta} + e^{-i\theta}) = \cos\theta. \tag{4.30}$$

As θ varies from 0 to 2π in the circle in the z-plane, its image $w = \cos\theta$ varies from 1 to -1 and then back to 1 along the real u-axis in the w plane.

By the reciprocity property $f(z) = f(\frac{1}{z})$, any two concentric circles centered at the origin with the product of their radii equal to one are mapped onto the same ellipse. In Fig. 4.27, we show that the two circles $|z| = \frac{1}{2}$ and $|z| = 2$ in the z-plane are mapped onto the same ellipse of major axis 1.25 and minor axis 0.75 in the w plane. Therefore, both the interior and the exterior domains of the unit circle in the z-plane are mapped onto the whole w plane minus the line segment from -1 to 1 along the real axis by the Joukowski transformation.

We can get the inverse of the Joukowski function by solving for z. Multiplying $2z$ to both sides of Eq. (4.25), we get

$$2zw = z^2 + 1.$$

Solving for z,

$$z = w \pm \sqrt{w^2 - 1} \tag{4.31}$$

we get a double valued function. Therefore, if we need the inverse map, we must select a suitable single valued branch to make it a one to one map. This can be easily done by giving it a specific value. For example, let $w = i$, then we have

$$z = i + \sqrt{i^2 - 1} = (1 + \sqrt{2})i,$$
$$z = i - \sqrt{i^2 - 1} = (1 - \sqrt{2})i.$$

Clearly, the plus sign is associated with the mapping of the exterior, and the minus sign associated with the mapping of the interior of the unit circle in the z-plane onto the whole w plane outside the line segment of $(-1$ to $1)$ in the real axis.

Rays of Constant Angles to Hyperbolas It is interesting to note that rays of constant θ in the z-plane is mapped onto the hyperbola in the w plane by the Joukowski transformation. From Eq. (4.27), we see that

$$\frac{u^2}{\cos^2\theta} = \left[\frac{1}{2}\left(c + \frac{1}{c}\right)\right]^2, \qquad \frac{v^2}{\sin^2\theta} = \left[\frac{1}{2}\left(c - \frac{1}{c}\right)\right]^2. \tag{4.32}$$

Since

$$\left[\frac{1}{2}\left(c + \frac{1}{c}\right)\right]^2 - \left[\frac{1}{2}\left(c - \frac{1}{c}\right)\right]^2 = 1,$$

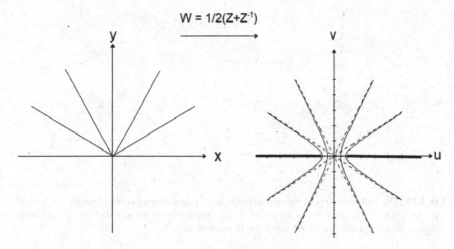

Fig. 4.28 The Joukowski transformation takes the rays of constant angles in the z plane to the hyperbolas in the w plane. It transforms (either upper or lower) half z-plane onto the whole w plane minus the slits running from 1 to ∞ and from -1 to $-\infty$ along the real axis

therefore

$$\frac{u^2}{\cos^2\theta} - \frac{v^2}{\sin^2\theta} = 1. \tag{4.33}$$

This is the equation of a hyperbola. The mapping is shown in Fig. 4.28. The ray of $\theta = 0$ is the positive real axis that passes the critical point $z = 1$. As one travels from 0 to 1 and then to ∞ along the real axis in the z-plane, the image point moves from $w = \infty$ to 1 and back to ∞ along the real axis in the w plane. Similarly, the negative x-axis, $\theta = \pi$, is mapped doubly onto the line segment running from $w = -1$ to $w = -\infty$. Thus, the family of rays with $0 < \theta < \pi$ in the z-plane is mapped onto a family of confocal hyperbolas in the w plane. The family of rays with $-\pi < \theta < 0$ is mapped onto the same family of hyperbolas.

4.6.2 Joukowski Profiles

We have discussed that circles centered at the origin are mapped into ellipse by the Joukowski transformation. Now if the circles are not centered at the origin in the z-plane, you get quite different results. The resultant figures in the w plane are known as Joukowski profiles. With the present-day computers, it is easy to draw these figures. Nevertheless, in order to develop an intuitive understanding, it is important to know the connection between the transformations and their geometric constructions.

First, let us suppose the circle in the z-plane passes through the two critical points $z = \pm 1$ and the center of the circle o' is at x distance above the origin o, as shown in Fig. 4.29. We ask the question what figure will this circle be mapped to in the w plane by the Joukowski transformation. Let us write $z = \rho e^{i\theta}$, then $w = \frac{1}{2}(z + \frac{1}{z})$

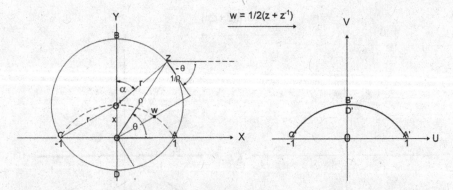

Fig. 4.29 The Joukowski transformation takes the circle passing both its critical points $z = \pm 1$ and with its center at x distance above the origin in the z plane into an arc with its highest and lowest points at the same x distance above the origin in the w plane

is simply $w = \frac{1}{2}(\rho e^{i\theta} + \frac{1}{\rho} e^{-i\theta})$. Geometrically, it is clear from the figure, that ρ is just the distance oz, so we draw a line segment of magnitude $\frac{1}{\rho}$, making an angle $-\theta$ with the horizontal, and add it to oz at z, as shown in the figure. Half of this result is the image w.

With this understanding, it is easy to find the images of A, B, C, D. At the critical point A, $\rho = 1$, $\theta = 0$, therefore $1/\rho = 1$ and $-\theta = 0$. Thus the image of A is at $w = \frac{1}{2}(1 + 1) = 1$. Hence the image A' is back to A. Similarly, at the other critical point C, its image C' coincides with C. Now at B, $\rho = r + x$ where $r^2 = 1 + x^2$. Also $\theta = \pi/2$ and $-\theta = -\pi/2$, so w is at the vertical axis, and

$$
\begin{aligned}
|w| &= \frac{1}{2}\left(\rho - \frac{1}{\rho}\right) = \frac{1}{2}\left(r + x - \frac{1}{r + x}\right) \\
&= \frac{1}{2}\left[\frac{(r + x)^2 - 1}{r + x}\right] = \frac{1}{2}\left[\frac{r^2 + 2rx + x^2 - 1}{r + x}\right] \\
&= \frac{1}{2}\frac{1 + x^2 + 2rx + x^2 - 1}{r + x} = \frac{1}{2}\frac{2x^2 + 2rx}{r + x} = x.
\end{aligned}
$$

Therefore, the image of B is B' that coincides with the center of the circle. So whatever x is, B' in the w plane is x above the origin of the w plane. It is interesting to note that the image of D is also at the vertical line, and ρ for D is $r - x$ and $\theta = 3\pi/2$. Thus $|w|$ for D' is

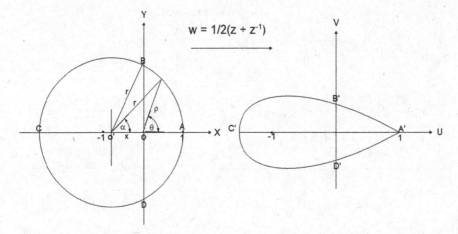

Fig. 4.30 The Joukowski transformation takes the circle passing one critical point $z = 1$ symmetrical with respect to the x-axis and with its center at $(-0.8, 0.0)$ in the z-plane onto the tear-shaped image in the w plane

$$w = \frac{1}{2}\left[-(r-x) + \frac{1}{(r-x)}\right]$$
$$= \frac{1}{2}\left[\frac{-(r-x)^2 + 1}{r-x}\right] = \frac{1}{2}\frac{-r^2 + 2rx - x^2 + 1}{r-x}$$
$$= \frac{1}{2}\frac{-1 - x^2 + 2rx - x^2 + 1}{r-x} = \frac{rx - x^2}{r-x} = x.$$

Therefore, the image of D coincides with the image of B. Thus, the whole circle in the z-plane is transformed onto an arc in the w plane. Both the major arc ABC and the minor arc ADC in the z-plane are transformed into the same arc A'B'C' in the w plane. Let the distance between the origin O and B be a, and between O and D be b, then $a = r + x$, and $b = r - x$. Clearly $x = (a - b)/2$. Therefore, the coordinates of B' and D' in the w plane are $u = 0$, $v = (a - b)/2$.

Next we look at a circle that passes only one critical point $z = 1$ at A and symmetric with respect to the x-axis. Let the distance between the origin O and the center of the circle O' be x as shown in the Fig. 4.30. Let r be the radius of this circle, and $r = 1 + x$. Clearly, this circle in z-plane will be mapped unto a figure in the w plane that is symmetrical with respect to u-axis. Let A', B', C', and D' be the images of A, B, C, and D, respectively. A' is at $u_A = 1$, $v_A = 0$ for the same reason as before. ($u_A = 1$, is the distance of A' in the transformed u coordinates, similar meaning for other symbols.) and

$$u_C = -\frac{1}{2}\left[(1+x) + x + \frac{1}{(1+x)+x}\right] = -\frac{1}{2}\left(1 + 2x + \frac{1}{1+2x}\right).$$

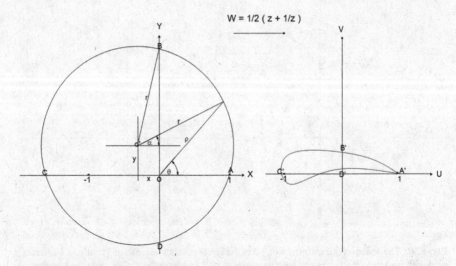

Fig. 4.31 The Joukowski transformation takes a circle passing the critical point $z = 1$ and with its center at $(-0.3, 0.4)$ in the z-plane into an airfoil in the w plane

For the point B, ρ_B is the distance of OB and $\rho_B = (r^2 - x^2)^{1/2} = [(1 + x)^2 - x^2]^{1/2} = (1 + 2x)^{1/2}$ therefore B' in the w plane is at $u_B = 0$ and

$$v_B = \frac{1}{2}\left(\rho_B - \frac{1}{\rho_B}\right) = \frac{1}{2}\left[(1 + 2x)^{1/2} - \frac{1}{(1 + 2x)^{1/2}}\right].$$

The D' in the w plane is simply at $u_D = u_B = 0$ and $v_D = -v_B$. In general, we can read from the figure that

$$\rho \sin\theta = r \sin\alpha = (1 + x)\sin\alpha = (1 + x)(1 - \cos^2\alpha)^{1/2}$$
$$\rho \cos\theta = r \cos\alpha - x = (1 + x)\cos\alpha - x$$

Thus

$$\rho = (1 + 2x - 2x(1 + x)\cos\alpha + 2x^2)^{1/2}$$
$$\theta = \tan^{-1}\frac{(1 + x)(1 - \cos^2\alpha)^{1/2}}{(1 + x)\cos\alpha - x}.$$

Let

$$w = u + iv$$

then

$$u = \frac{1}{2}\left(\rho + \frac{1}{\rho}\right)\cos\theta,$$
$$v = \frac{1}{2}\left(\rho - \frac{1}{\rho}\right)\sin\theta.$$

Since we have expressed ρ and θ in terms of $\cos \alpha$, as α moves around the circle, we can trace its image in the w plane by following the point (u, v). The tear-shaped image in the above figure is the result of $x = -0.8$.

Finally, we can look at the general case as shown in Fig. 4.31. The circle in the z-plane is passing through the critical point $z = 1$ and its center is at $x = -0.3$, $y = 0.4$. It's image in the w plane begins to resemble an airfoil. With the two previous examples, it is not difficult for us to understand that the image of this circle should be such a figure. The exact locations of A', B', C', and D' can be determined as before. In this case, $r = \left[(1 + x)^2 + y^2\right]^{1/2}$. To trace out the image, we need to express ρ and θ in terms of α (in addition to x and y). These relations can be worked out based on the conditions

$$\rho \sin \theta = r \sin \alpha + y,$$
$$\rho \cos \theta = r \cos \alpha - x.$$

Therefore

$$\rho = \left[(r \sin \alpha + y)^2 + (r \cos \alpha - x)^2\right]^{1/2}$$
$$\theta = \tan^{-1} \frac{r \sin \alpha + y}{r \cos \alpha - x}.$$

To trace out the image of this transformation, as α is moving around the circle, we can follow the point (u, v) where

$$u = \frac{1}{2}\left(\rho + \frac{1}{\rho}\right) \cos \theta,$$

$$v = \frac{1}{2}\left(\rho - \frac{1}{\rho}\right) \sin \theta.$$

To do this by hand may be tedious, but not difficult. With modern computers, we can have thousands of such plots in a matter of seconds. By starting with figures in the z-plane that are circles, it is possible to produce figures in the w plane that describe a great variety of airfoils. This is the strength of the Joukowski transformation.

4.7 The Möbius Transformation

The mapping

$$w = \frac{az + b}{cz + d}, \qquad (ad - bc \neq 0) \tag{4.34}$$

where a, b, c, and d are complex constants, is called Möbius transformation. It is called Möbius transformation because 150 years ago August Ferdinand Möbius first studied this transformation. The rich vein of knowledge which he exposed is still far from being exhausted. It lies at the heart of several exciting areas of modern mathematical research. These transformations are miraculously connected with the non-Euclidean geometry, they are also intimately connected with Einstein's theory of relativity see [Penrose and Rindler [34]]. Although we will not be able to study these subjects here, we want the reader to know that the significance of these transformations go far beyond what they are exposed here.

Multiply both sides of Eq. (4.34) by $cz + d$, we get

$$czw - az + dw - b = 0. \tag{4.35}$$

This alternative form is linear in z and linear in w. or bilinear in z and w, so it is also called bilinear transformation. This name is widely used. Any equation of this type can be put in the form of either Eq. (4.34) or (4.35).

In addition, this transformation is also called a linear fraction transformation or a homographic transformation. It is an important mapping, we shall study some of its rudimentary properties.

First if $c \neq 0$, we can factor out $\frac{a}{c}$ from Eq. (4.34) to get

$$w = \frac{a}{c} \cdot \frac{z + b/a}{z + d/c}.$$

Add and subtract $\frac{d}{c}$ in the numerator

$$\begin{aligned}
w &= \frac{a}{c} \cdot \frac{z + d/c + b/a - d/c}{z + d/c} \\
&= \frac{a}{c}\left(1 + \frac{b/a - d/c}{z + d/c}\right) = \frac{a}{c} + \frac{a}{c}\frac{(bc - ad)/ac}{z + d/c} \\
&= \frac{a}{c} + \frac{bc - ad}{c^2}\frac{1}{z + d/c}. \tag{4.36}
\end{aligned}$$

Now we can see if $ad - bc = 0$, we will have a constant function $w = a/c$. Thus the condition $ad - bc \neq 0$ is necessary.

Next if $c = 0$, then Eq. (4.35) clearly shows that the transformation becomes linear. This is a simple case, we can make a more detailed study.

4.7.1   Linear Transformation

If $c = 0$, Eq. (4.34) reduces to a linear transformation

$$w = az + b. \tag{4.37}$$

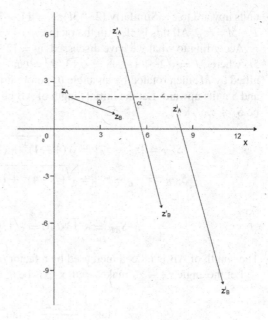

Fig. 4.32 Linear transformation through magnification, rotation, and translation

Since a and b are in general complex numbers, so a can be written as $|a|\,e^{i\theta_r}$, thus the transformation w describes the following alternations in the xy plane (z-plane):

(i) a magnification by the factor $|a|$,
(ii) a rotation through the angle θ_r,
(iii) a translation by the vector b.

These are simple processes. The details are illustrated in the following example.

Example 4.7.1 Map the straight line joining A: $(1 + 2i)$ and B: $(4 + i)$ in the z-plane onto the w plane by the transformation

$$w = (2 - 3i)z - 4 + 5i.$$

Solution 4.7.1 Let the coordinates of A be $z_A = 1 + 2i$, the coordinates of B be $z_B = 4 + i$. After the transformation

$$(2 - 3i)(1 + 2i) - 4 + 5i = (8 + i) - 4 + 5i = 4 + 6i = z'_A,$$
$$(2 - 3i)(4 + i) - 4 + 5i = (11 - 10i) - 4 + 5i = 7 - 5i = z'_B,$$

z_A is moved to z'_A and z_B is moved to z'_B. That is, the line AB is mapped to A' $(4 + 6i)$ and B' $(7 - 5i)$. This is illustrated in the Fig. 4.32

The problem is actually solved.

What went on is as follows: After the operation $(2 - 3i)z_A = 8 + i = z^i_A$, z_A is moved to an intermediate position z^i_A, then z^i_A is moved 4 units to the left and 5 units upward to z'_A. Similarly $(2 - 3i)z_B = 11 - 10i = z^i_B$ and then $z^i_B - 4 + 5i = 7 - 5i = z'_B$. All this is clear in the picture.

According to what we have discussed, $w = (2 - 3i)z - 4 + 5i = Me^{i\theta_r}z - 4 + 5i$, where $M = \sqrt{2^2 + (-3)^2} = \sqrt{13}$, $\theta_r = \tan^{-1}(-\frac{3}{2})$. That is, the length is magnified by M, then rotated by an angle θ_r before it is bodily moved 4 units to the left and 5 units upward. Let the initial length of AB be S_{AB}, and the final length of A'B' be $S_{A'B'}$:

$$S_{AB} = |z_B - z_A| = \sqrt{(4 - 1)^2 + (1 - 2)^2} = \sqrt{10},$$

$$S_{A'B'} = |z'_B - z'_A| = \sqrt{(7 - 4)^2 + (-5 - 6)^2} = \sqrt{130}.$$

So

$$M S_{AB} = \sqrt{13}\sqrt{10} = \sqrt{130} = S_{A'B'}.$$

The length of AB is indeed increased by a factor of M.

Let the angle $z_B - z_A$ makes with x-axis be θ,

$$\theta = \tan^{-1}\frac{(1 - 2)}{(4 - 1)} = \tan^{-1}\left(\frac{-1}{3}\right)$$

and the angle $z'_B - z'_A$ makes with x-axis be α,

$$\alpha = \tan^{-1}\frac{(-5 - 6)}{(7 - 4)} = \tan^{-1}\left(\frac{-11}{3}\right).$$

We want to check if α is indeed the sum of θ and θ_r. There is no problem from a table or a calculator to find θ, α and θ_r in degrees, and to see if α is indeed the sum of θ and θ_r. However, we can verify this directly in the following way.

$$\theta + \theta_r = \tan^{-1}\tan(\theta + \theta_r) = \tan^{-1}\frac{\tan\theta + \tan\theta_r}{1 - \tan\theta \cdot \tan\theta_r}$$

$$= \tan^{-1}\frac{\left(\frac{-1}{3}\right) + \left(-\frac{3}{2}\right)}{1 - \left(\frac{-1}{3}\right)\left(-\frac{3}{2}\right)} = \tan^{-1}\left(\frac{-11}{3}\right) = \alpha.$$

So the line AB is turned by angle of θ_r. Finally, it is clear from the figure that the line $z^i_B - z^i_A$ is bodily moved 4 units to the left and 5 units upward to $z'_B - z'_A$.

In this example, there is no need to analyze the transformation into intermediate step. In fact, it may even seem cumbersome. However, this is not always the case. In some other situations, this understanding can be very convenient. This is demonstrated in the following example.

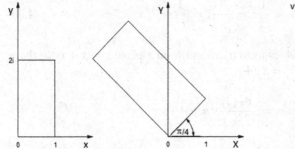

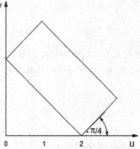

Fig. 4.33 Mapping the rectangle in z-plane to the w plane with the transformation $z = (1 + i)z + 2$

Example 4.7.2 Map the rectangle shown in the xy (z) plane of the following figure into the uv (w) plane using the transformation (Fig. 4.33)

$$w = (1 + i)z + 2.$$

Solution 4.7.2 Since $(1 + i)z = \sqrt{2}\exp(\pi/4)$, this is an expansion by a factor of $\sqrt{2}$, followed by a rotation of $\pi/4$. Finally the rectangle is bodily translated into two units to the right. This process is clearly illustrated in the figure.

4.7.2 Inversion

Another special case is so called inversion

$$w = \frac{1}{z} \tag{4.38}$$

Inversion defines a conformal mapping at all points except $z = 0$. When the point at infinity is involved our discussion, we can tacitly assume that

$$w(0) = \lim_{z \to 0} \frac{1}{z} = \infty, \quad w(\infty) = \lim_{z \to \infty} \frac{1}{z} = 0. \tag{4.39}$$

If we write $z = re^{i\theta}$, we have $w = \frac{1}{r}e^{-i\theta}$. This says that the points exterior to the unit circle $|z| = 1$ are mapped onto the points interior to it, then followed by a reflection over the real axis. This mapping can be also understood in the following way. Multiplying top and bottom of $\frac{1}{z}$ by $\bar{z}$, we have

$$w = \frac{1}{z\bar{z}}\bar{z} = \frac{1}{|z^2|}\bar{z} = \frac{1}{r^2}re^{-i\theta} \tag{4.40}$$

It is this expression that enables us to transform from z-plane $z = x + iy$ to the w plane $w = u + iv$. Since $|z^2| = x^2 + y^2$, so

$$w = \frac{x - iy}{x^2 + y^2} = u + iv,$$

therefore

$$u = \frac{x}{x^2 + y^2}, \quad v = \frac{-y}{x^2 + y^2}. \tag{4.41}$$

Also because

$$z = \frac{1}{w} = \frac{1}{|w|^2}\bar{w},$$

we have

$$x + iy = \frac{u - iv}{u^2 + v^2},$$

so

$$x = \frac{u}{u^2 + v^2}, \quad y = \frac{-v}{u^2 + v^2}. \tag{4.42}$$

At this point, we notice that a circle or a line in the z-plane can be expressed as

$$A(x^2 + y^2) + Bx + Cy + D = 0. \tag{4.43}$$

This can be shown as follows:

$$x^2 + y^2 + \frac{B}{A}x + \frac{C}{A}y + \frac{D}{A} = 0.$$

Complete the square,

$$x^2 + \frac{B}{A}x = \left(x + \frac{B}{2A}\right)^2 - \left(\frac{B}{2A}\right)^2,$$

$$y^2 + \frac{C}{A}y = \left(y + \frac{C}{2A}\right)^2 - \left(\frac{C}{2A}\right)^2,$$

so Eq. (4.43) becomes

$$\left(x + \frac{B}{2A}\right)^2 + \left(y + \frac{C}{2A}\right)^2 = \left(\frac{\sqrt{B^2 + C^2 - 4AD}}{2A}\right)^2. \tag{4.44}$$

When $A = 0$, Eq. (4.43) becomes a line, when $A \neq 0$, as seen in Eq. (4.44), it represents a circle. Equation (4.44) can be written in the form:

$$(x - x_0)^2 + (y - y_0)^2 = r_z^2 \tag{4.45}$$

where

$$x_0 = -\frac{B}{2A}, \quad y_0 = -\frac{C}{2A}, \quad r_z = \left(x_0^2 + y_0^2 - \frac{D}{A}\right)^{1/2} \tag{4.46}$$

Substituting Eq. (4.42) into (4.43), we have

$$A\left[\left(\frac{u}{u^2 + v^2}\right)^2 + \left(\frac{-v}{u^2 + v^2}\right)^2\right] + B\left(\frac{u}{u^2 + v^2}\right) + C\left(\frac{-v}{u^2 + v^2}\right) + D = 0. \tag{4.47}$$

Multiplying $(u^2 + v^2)$, it becomes

$$D\left(u^2 + v^2\right) + Bu - Cv + A = 0. \tag{4.48}$$

Therefore, its image is also either a circle or a line depending if D is equal to zero or not. Compare this equation with Eq. (4.43), we see that A and D are interchanged and the sign of C is changed. Thus, the image circle can be written as

$$(u - u_0)^2 + (v - v_0)^2 = r_w^2 \tag{4.49}$$

where

$$u_0 = -\frac{B}{2D} = -\frac{B}{2A} \cdot \frac{A}{D} = x_0\frac{A}{D}, \quad v_0 = \frac{C}{2D} = -y_0\frac{A}{D}, \tag{4.50}$$

$$r_w = \left(u_0^2 + v_0^2 - \frac{A}{D}\right)^{1/2}. \tag{4.51}$$

Notice also that if $D = 0$, Eq. (4.43) passes through the origin. If $A = 0$, its image in w plane passes through the origin. It is therefore clear that

(i) A circle ($A \neq 0$) not passing through the origin ($D \neq 0$) in the z-plane is transformed into a circle not passing through the origin in the w plane.

(ii) A circle ($A \neq 0$) passing through the origin ($D = 0$) in the z plane is transformed into a line that does not pass through the origin in the w plane.

(iii) A line ($A = 0$) not passing through the origin ($D \neq 0$) in the z plane is transformed into a circle passing through the origin in the w plane.

(iv) A line ($A = 0$) passing through the origin ($D = 0$) in the z-plane is transformed into a line through the origin in the w plane.

In Example 4.1.1, the line $x = 0$ transformed to the line $u = 0$ is a case (iv) mapping, the circle of radius $1/2$ transformed to a straight line $u = 1$ is a case (ii) mapping.

Fig. 4.34 The circle defined by $|z - 6\exp(i\pi/4)| = 4$

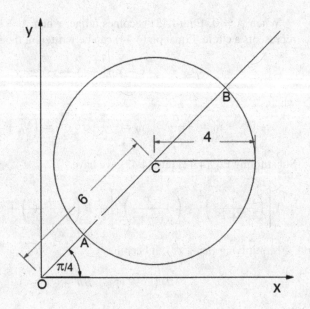

Let us see more examples.

Circle to Circle Mapping

Example 4.7.3 Find the mapping of the circle

$$\left|z - 6e^{i\pi/4}\right| = 4$$

under the transformation of

$$w = \frac{1}{z}.$$

Solution 4.7.3 This is a circle to circle case (i) mapping. The circle in the z (xy) plane is shown in Fig. 4.34.

We can solve the problem either by analysis or by geometry.
Method 1: By analysis, according to Eq. (4.45), this circle is defined by

$$(x - x_0)^2 + (y - y_0)^2 = 4.$$

According to Eq. (4.47),

$$x_0 = -\frac{B}{2A} = 6\cos\frac{\pi}{4} = 3\sqrt{2}, \quad y_0 = -\frac{C}{2A} = 6\sin\frac{\pi}{4} = 3\sqrt{2},$$

$$r_z = \left[(-3\sqrt{2})^2 + (-3\sqrt{2})^2 - \frac{D}{A}\right]^{1/2} = 4.$$

Fig. 4.35 The image circle of $\left|w - \frac{3}{10}e^{-i\pi/4}\right| = 1/5$ is the resultant into which the circle $\left|z - 6e^{i\pi/4}\right| = 4$ is mapped by the transformation of $w = \frac{1}{z}$

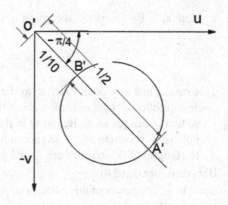

So

$$r_z^2 = \left[36 - \frac{D}{A}\right] = 16$$

Thus

$$\frac{D}{A} = 20.$$

According to Eq. (4.49), its image circle is

$$(u - u_0)^2 + (v - v_0)^2 = \left[(u_0)^2 + (-v_0)^2 - \frac{A}{D}\right] = r_w^2$$

where

$$u_0 = -\frac{B}{2D} = -\frac{B}{2A} \cdot \frac{A}{D} = x_0 \frac{A}{D} = x_0 \frac{1}{20} = \frac{3\sqrt{2}}{20},$$

$$v_0 = \frac{C}{2D} = \frac{C}{2D} \cdot \frac{A}{D} = -y_0 \frac{A}{D} = -y_0 \frac{1}{20} = -\frac{3\sqrt{2}}{20},$$

and

$$r_w = \left[\frac{18}{400} + \frac{18}{400} - \frac{1}{20}\right]^{1/2} = \frac{1}{5}.$$

Therefore, the image circle is

$$|w - w_0| = \frac{1}{5}$$

where

$$w_0 = u_0 + iv_0 = \frac{3\sqrt{2}}{20}(1 - i) = \frac{3}{10}e^{-i\pi/4}.$$

Method 2: By geometry, according to Eq. (4.40)

$$w = \frac{1}{r}e^{-i\theta}.$$

This means that any point exterior to the unit circle ($r = 1$) is mapped inside the circle and followed by a reflection about the real axis.

So let the images of A, B, and O in the above figure be A', B', and O' in the w (uv) plane, as shown in Fig. 4.35 (not in the same scale as the above figure).

The lengths O'B' and O'A' are $\frac{1}{10}$ and $\frac{1}{2}$, respectively, the distance between A' and B' is therefore equal to $\frac{1}{2} - \frac{1}{10} = \frac{2}{5}$. Since this is the diameter, the radius is therefore equal to $\frac{1}{5}$. The center of this circle is at $\frac{1}{10} + \frac{1}{5} = \frac{3}{10}$ from the origin. Therefore, the image circle is given by

$$|w - w_0| = \frac{1}{5}$$

where

$$w_0 = \frac{3}{10}e^{-i\frac{\pi}{4}}.$$

The same as obtained by method 1.

There is an important point that may have escaped from our attention. The center of the original circle is C, its image C' is not the center of the image circle, since C' is at $\frac{1}{6}e^{-i\frac{\pi}{4}}$, not $\frac{3}{10}e^{-i\frac{\pi}{4}}$. This is because the magnification is different at different points. Then how do we know that A'B' is the diameter of the image circle? In the z (xy) plane, AB is the diameter, it intersects with the circle at right angles. Since the function $\frac{1}{z}$ is analytic, so the mapping is conformal. So its images A'B' must also intersect with the image circle at right angles. Only the diameter intersects with the circle at right angles, therefore A'B' must be the diameter of the image circle. In that case, how do we know the rest of the images form a circle. Well, we have shown analytically, the circle must map into another circle. It looks that the algebra that we went through is necessary.

Example 4.7.4 Find the mapping of the following circle

$$\left| z - 2e^{i\pi/4} \right| = 4$$

under the transformation of $w = 1/z$. The original circle in z-plane is shown in Fig. 4.36.

Solution 4.7.4 The only difference of this example and the previous one is that in this problem the origin is inside of the circle. This results in quite a different map. The circle in the xy plane is

$$(x - x_0)^2 + (y - y_0)^2 = r_z^2$$

Fig. 4.36 The circle $|z - 2\exp(i\pi/4)| = 4$ in the xy plane

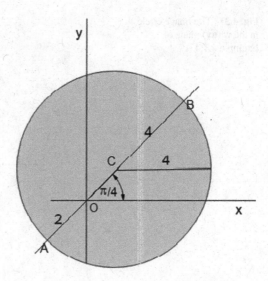

where

$$x_0 = 2\cos\frac{\pi}{4} = \sqrt{2} = -\frac{B}{2A}, \qquad y_0 = \sqrt{2} = -\frac{C}{2A}$$

$$r_z = \sqrt{2\left(\sqrt{2}\right)^2 - \frac{D}{A}} = 4. \qquad \frac{D}{A} = -12.$$

The image circle in uv plane is

$$(u - u_0)^2 + (v - v_0)^2 = r_w^2$$

where

$$u_0 = -\frac{B}{2D} = -\frac{B}{2A} \cdot \frac{A}{D} = x_0\left(-\frac{1}{12}\right) = -\frac{\sqrt{2}}{12}$$

$$v_0 = \frac{C}{2D} = \frac{D}{2A} \cdot \frac{A}{D} = -y_0\left(-\frac{1}{12}\right) = \frac{\sqrt{2}}{12}.$$

$$r_w = \sqrt{u_0^2 + v_0^2 - \frac{A}{D}} = \sqrt{2\frac{2}{144} + \frac{1}{12}} = \frac{1}{3}.$$

Thus the image circle is given by

$$\left| w - \frac{1}{6}e^{-i5\pi/4} \right| = \frac{1}{3}$$

From the geometrical construction, we see that corresponding to the A and B in the xy plane, their images are A' and B' as shown in Fig. 4.38. Therefore, the

Fig. 4.37 The image circle
in the w (uv) plane of
Example 4.7.4

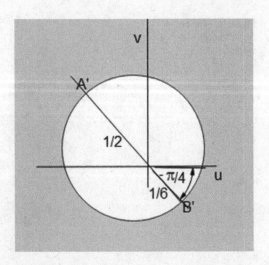

diameter is $\frac{1}{2} + \frac{1}{6} = \frac{2}{3}$, and the radius of the image circle is therefore equal to $\frac{1}{3}$. The center of this image circle is at (u_0, v_0) where $u_0 = -(\frac{1}{2} - \frac{1}{3})\cos(-\frac{5\pi}{4}) = -\frac{\sqrt{2}}{12}$, $v_0 = -(\frac{1}{2} - \frac{1}{3})\sin(-\frac{5\pi}{4}) = \frac{\sqrt{2}}{12}$. Here we have to be careful. From Fig. 4.38, we see that the angle we should use for u_0 and v_0 calculation is the angle opposite of $-\frac{\pi}{4}$, not just $-\frac{\pi}{4}$. This is where the minus sign in front of the expression from. The result is the same as obtained analytically.

Although the procedure is the same as the previous example. There is however a fundamental difference. The interior of the original circle in the xy plane is now mapped onto the exterior of the image circle. How do we know this? We can see that $z = 0$ is inside the circle in the z-plane. Its image, $w = \frac{1}{0} \to \infty$ must be outside of the image circle in the w plane.

Circle to Line Mapping If the circle in the z-plane passes through the origin, then the image of that circle becomes a line. Again this is best illustrated in the following example.

Example 4.7.5 Find the mapping of the circle

$$\left| z - \mathrm{Re}^{i(\frac{\pi}{4})} \right| = R$$

under the transformation

$$w = \frac{1}{z}.$$

Solution 4.7.5 The situation is illustrated in Fig. 4.38. This problem can be analyzed as follows. These circles can be expressed as

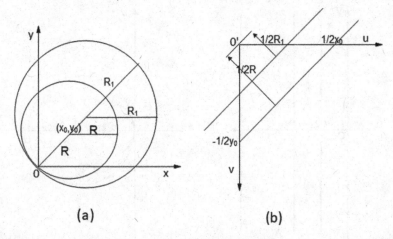

Fig. 4.38 The circle to line mapping. **a** Circles with different radius, they all pass through the origin. **b** The images of these circles become straight lines, as R becomes larger and larger, the distances between these lines and the origin o' become smaller and smaller

$$(x - x_0)^2 + (y - y_0^2) = \sqrt{x_0^2 + y_0^2}.$$

In terms of the quantities of Eq. (4.43),

$$x_0 = -\frac{B}{2A}, \qquad y_0 = -\frac{C}{2A}.$$

Since they all pass zero, D in Eq. (4.43) must vanish. This is because $x = 0$, $y = 0$ must be a point on that curve and satisfy that equation. This can be true only if $D = 0$. Therefore, the image equation, Eq. (4.48), becomes

$$Bu - Cv + A = 0$$

which is a linear equation representing a line. Furthermore when $v = 0$, $u = -\frac{A}{B} = \frac{1}{2x_0}$, and when $u = 0$, $v = \frac{A}{C} = -\frac{1}{2y_0}$. When we plot it, it is the straight line showing in (b) of Fig. 4.38 The same figure can be obtained according to geometry as we discussed before. It is interesting to note that as R becomes larger and larger, the distances d between these straight lines and the origin 0' become smaller and smaller. As $R \to \infty$, d approach zero.

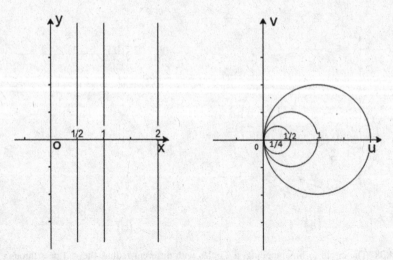

Fig. 4.39 Straight lines are mapped into circles under the transformation $w = \frac{1}{z}$

Line to Circle Transformation

Example 4.7.6 Find the mapping of the straight line

$$x = R$$

under the transformation of $w = 1/z$.

Solution 4.7.6 For the case $x = R$, in the standard equation (Eq. (4.43))

$$A(x^2 + y^2) + Bx + Cy + D = 0,$$

we have

$$A = 0, \quad B = 1, \quad C = 0, \quad D = -R.$$

Under the inversion, the standard image equation (Eq. (4.48)) is

$$D(u^2 + v^2) + Bu - Cv + A = 0.$$

Thus

$$-R(u^2 + v^2) + u = 0.$$

This equation can be written as

$$\left(u - \frac{1}{2R}\right)^2 + v^2 = \left(\frac{1}{2R}\right)^2 .$$

These are circles with radius $r = \frac{1}{2R}$, centered at $\left(\frac{1}{2R}, 0\right)$.

Figure 4.39 shows the straight lines of $x = \frac{1}{2}, 1, 2$ with their image circles centered at $1, \frac{1}{2}, \frac{1}{4}$, respectively.

Notice that as $x \to 0$, the radius $r \to \infty$. Since

$$u = \frac{x}{x^2 + y^2}, \quad v = \frac{-y}{x^2 + y^2},$$

clearly for the straight line $x = 0$, the image is also a straight line $u = 0$. Therefore we can regard the straight line as a part of the infinitely large circle.

An Extension In preparation for the truely bilinear transformation, which can be regarded as a series of transformations, we first look at a simple extension of what we have just discussed. This extension involves only two steps. Again it is best illustrated by an example.

Example 4.7.7 Find the image in the w plane of the unit circle $|z| = 1$ in the z-plane under the transformation of $w = \frac{1}{z-2}$ (Fig. 4.40).

Solution 4.7.7 We can regard this transformation as a two steps process:

$$w = \frac{1}{Z}, \quad Z = z - 2.$$

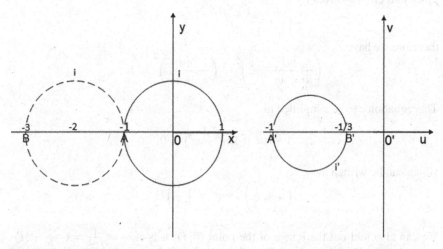

Fig. 4.40 The unit circle $|z| = 1$ is mapped into the w plane under the transformation of $w = \frac{1}{z-2}$

First, the unit circle is moved two units to the left. Then followed by an inversion. This is illustrated in Fig. 4.40. A' and B' in the w plane are the images of A and B respectively. Since

$$O'A' = \frac{1}{-OA} = \frac{1}{-1} = -1,$$

$$O'B' = \frac{1}{-OB} = -\frac{1}{3}.$$

Therefore, the diameter of the image circle is $1 - \frac{1}{3} = \frac{2}{3}$. Thus, the radius is $\frac{1}{3}$ and the center of the image circle is at $\left(-\frac{2}{3}, 0\right)$.

Of course, we should get the same result by the brute force substitution. Since

$$z - 2 = \frac{1}{w}$$

or

$$x + iy - 2 = \frac{1}{u + iv}.$$

Thus

$$(x - 2) + iy = \frac{u - iv}{u^2 + v^2},$$

so

$$x - 2 = \frac{u}{u^2 + v^2}, \qquad y = \frac{-v}{u^2 + v^2}.$$

Since unit circle is given by

$$x^2 + y^2 = 1,$$

therefore we have

$$\left(\frac{u}{u^2 + v^2} + 2\right)^2 + \left(\frac{-v}{u^2 + v^2}\right)^2 = 1.$$

This equation can be simplified to

$$1 + 4u + 3(u^2 + v^2) = 0,$$

which can be written as

$$\left(u + \frac{2}{3}\right)^2 + v^2 = \left(\frac{1}{3}\right)^2.$$

We can also find out the image of the point $(0, i)$, it is at $w = \frac{1}{i-2} = \left(-\frac{2}{5} - \frac{1}{5}i\right)$, somewhere in the lower plane. Therefore when we go through AiB in the original unit circle, it is counter-clock wise. In the image plane, we go through A'i'B', it is also counter-clock wise. So the inside of the unit circle is mapped into the inside of the image circle.

4.7.3 General Bilinear Transformation

In previous problems and examples, we mapped circular regions under the inversion transformation $w = 1/z$. In a same way, we will show how to map regions with the general bilinear transformation

$$w = \frac{az + b}{cz + d}.$$

We have shown in Eq. (4.36) that for $c \neq 0$, this equation can be written as

$$w = \frac{a}{c} + \frac{bc - ad}{c^2}\frac{1}{z + d/c}. \tag{4.52}$$

Using this equation, we can reduce this transformation to the following sequence of simple mappings:
   (i) Translate the region by the vector d/c.
   (ii) Followed by an inversion, just like the previous example.
   (iii) Magnify the region by the factor $\left|\frac{bc - ad}{c^2}\right|$.
   (iv) Rotate the region through the angle θ, where $\theta = \tan^{-1}\frac{bc - ad}{c^2}$.
   (v) Finally, translate the result of the above processes by the vector $\frac{a}{c}$.
   We see that the general bilinear transformation (Eq. (4.36)) preserves circles, in the same sense as the inversion transformation preserves circles, that is, lines as circles with $r \to \infty$. This sequence of mappings can be best illustrated by an example.

Example 4.7.8 Find the image of the unit circle $|z| = 1$ under the bilinear transformation

$$w = -i\left(\frac{z - 1}{z + 1}\right).$$

Solution 4.7.8 First we write the transformation as

$$w = -i\left(\frac{z - 1}{z + 1}\right) = -i\left(\frac{z + 1 - 2}{z + 1}\right) = -i\left(1 - \frac{2}{z + 1}\right)$$

$$= -i + i2\frac{1}{z + 1} = -i + 2e^{i(\pi/2)}\frac{1}{z + 1}$$

Therefore, this transformation can be regarded as the sum of the following steps:

1. The unit circle is moved bodily one unit to the right:

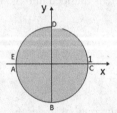 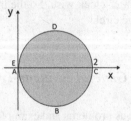

2. An inversion is shown in the following figure. Notice that A and E are the same point. Both are mapped to ∞. Although, for convenience, we put A' at $v = \infty$ and E' at $v = -\infty$. Actually both should be at $z = \infty$. We can think A' and E' are the same point on the circumference of an infinitely large circle. This includes $u \to \infty$. Thus the lower half of the circle ABC is mapped onto A'B'C', with B at $(1, -1)$ being mapped to B' where $B' = \frac{1}{1-i} = \frac{1+i}{2} = (\frac{1}{2}, \frac{1}{2})$. The upper half of the circle CDE is mapped onto C'D'E' where D' is at $(\frac{1}{2}, -\frac{1}{2})$.

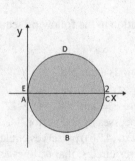 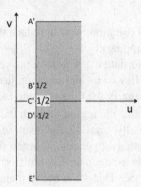

3. Magnified by a factor of two:

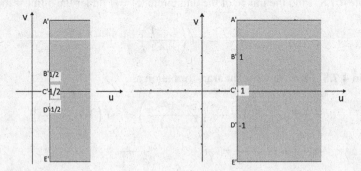

4. Rotate by $\pi/2$:

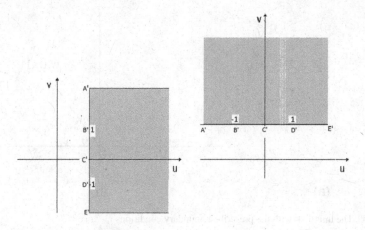

5. Bodily move downward by one unit:

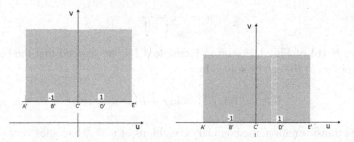

The net effect is that the disk inside the unit circle in the xy plane is transformed onto the entire upper half-plane in the uv plane and the upper half disk is transformed onto the first quadrant of the uv plane by $w = -i\frac{z-1}{z+1}$. Note that the upper circular boundary CDE of the original disk is mapped onto the positive part of the real u-axis, and the lower circular boundary ABC is mapped onto the negative part of the u-axis. The flat diameter EC that goes through zero is mapped onto the positive v-axis. As we have discussed, E' is at infinity, it can be regarded as any point on the circumference of an infinite large circle, including at $v \to \infty$, $u = 0$. For the upper half disk to be transformed onto the first quardrant of the uv plane, EC must be mapped onto the positive v-axis.

This enables us to solve some interesting problems as shown in the following examples.

Example 4.7.9 Find the electrostatic potential within the unit disk with the boundary conditions shown in (a) of Fig. 4.41.

Solution 4.7.9 From the previous example, we know that the inside of the unit disk is mapped onto the upper half-plane by the transformation

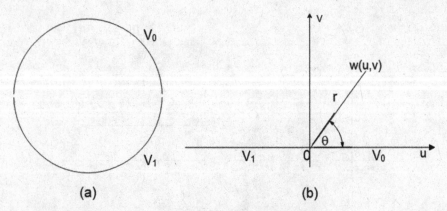

Fig. 4.41 The unit disk with the prescribed boundary conditions

$$w = -i\frac{z-1}{z+1}$$

as shown in (b) of Fig. 4.41. From Example 4.1.4, we learned that the boundary value problem of (b) can be solved by

$$\ln(z) = \ln(re^{i\theta}) = \ln r + i\theta = u' + iv'.$$

With this transformation, the boundary conditions of $v' = \theta$ becomes very simple,

$$\theta = 0, \quad V(u', 0) = V_0,$$
$$\theta = \pi, \quad V(u', \pi) = V_1.$$

Thus the Laplace equation in (u, v) becomes an ordinary differential equation in v only

$$\frac{d^2V}{dv'^2} = 0.$$

Therefore

$$V(u', v') = A + Bv'$$

The boundary conditions lead to the solution

$$V(u', v') = V_0 + \frac{V_1 - V_0}{\pi}v'$$

Since

$$v' = \theta = \tan^{-1}\frac{v}{u},$$

so we can express the solution as

$$V(u, v) = V_0 + \frac{V_1 - V_0}{\pi} \tan^{-1} \frac{v}{u}.$$

Now we must transform back to the (x, y) coordinates. Since

$$w = -i\frac{z-1}{z+1} = -i\frac{x+iy-1}{x+iy+1} = -i\frac{[(x-1)+iy][(x+1)-iy]}{(x+1)^2 + y^2}$$

$$= -i\frac{[(x-1)(x+1)+y^2]+2iy}{(x+1)^2+y^2} = \frac{2y+(1-x^2-y^2)i}{(x+1)^2+y^2} = u+iv,$$

therefore

$$u = \frac{2y}{(x+1)^2 + y^2}, \quad v = \frac{(1-x^2-y^2)}{(x+1)^2+y^2}$$

and

$$\tan^{-1}\frac{v}{u} = \tan^{-1}\frac{(1-x^2-y^2)}{2y}.$$

Thus

$$V(x, y) = V_0 + \frac{V_1 - V_0}{\pi} \tan^{-1}\frac{(1-x^2-y^2)}{2y}.$$

This is the solution of the problem. It is interesting to find the potential at the center of the disk, $V(0, 0)$. In that case

$$\tan^{-1}\frac{(1-x^2-y^2)}{2y} = \tan^{-1}\frac{1}{0} = \tan^{-1}(\infty) = \frac{\pi}{2},$$

so

$$V(0, 0) = V_0 + \frac{V_1 - V_0}{\pi} \cdot \frac{\pi}{2} = \frac{V_1 + V_0}{2},$$

an expected answer.

Example 4.7.10 Find the temperature inside the semi-circular region with the given boundary values shown in (a) of Fig. 4.42.

Solution 4.7.10 The transformation function

$$w = -i\frac{z-1}{z+1}$$

maps the inside of the semi-circular region shown in (a) unto the first quadrant of the w plane shown in (b). The upper circular boundary of (a) maps unto the positive part of the u-axis in (b). The flat horizontal part of (a) maps unto the positive part of v-axis of (b). We can solve it in exactly the same way as in the previous example

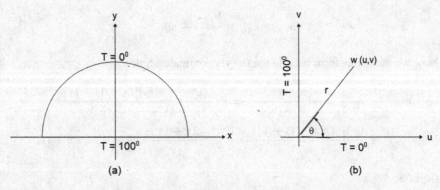

Fig. 4.42 Semi-circular region in z-plane mapped unto the first quadrant of w plane

except the boundary conditions should be changed to

$$v' = \theta = 0, \quad T(u', 0) = 0^0.$$
$$v' = \theta = \frac{\pi}{2}, \quad T(u', \frac{\pi}{2}) = 100^0.$$

Thus with

$$T(u', v') = A + Bv',$$

the first boundary condition leads to $A = 0$, the second boundary condition shows that $100 = B\frac{\pi}{2}$, or $B = \frac{200}{\pi}$, Therefore

$$T(u', v') = \frac{200}{\pi}v'$$

and

$$T(u, v) = \frac{200}{\pi} \tan^{-1}\left(\frac{v}{u}\right).$$

In terms of (x, y) coordinates, as shown in the last example,

$$\tan^{-1}\frac{v}{u} = \tan^{-1}\frac{(1 - x^2 - y^2)}{2y}.$$

Therefore

$$T(x, y) = \frac{200}{\pi} \tan^{-1}\frac{(1 - x^2 - y^2)}{2y}.$$

4.7.4 Some Properties of Bilinear Transformation

Upper Half Plane to Unit Circle We have just shown that the transformation

$$w = -i\frac{z-1}{z+1}$$

will map the inside of a unit circle in the z-plane unto the entire upper half of the
w plane. Multiply both sides of the above equation by $z + 1$, we can easily find its
inverse transformation

$$z = -\frac{w-i}{w+i} = e^{i\pi}\frac{w-i}{w+i}.$$

This equation says that the transform

$$w = e^{i\pi}\frac{z-i}{z+i} = e^{i\pi}\frac{z-i}{z-i^*}$$

will map the upper half z-plane into the inside of the unit circle in the w plane. But
this is a special transformation, it transforms $z = i$ to $w = 0$. Suppose we want to
transform any point z_0 in the upper half-plane to $w = 0$, we might try

$$w = e^{i\alpha}\frac{z-z_0}{z-z_0^*}. \tag{4.53}$$

Indeed, this is the general form that will map the entire upper half z-plane unto the
inside of a unit disk in the w plane. Now we want to prove it.

First we notice

$$|w| = \left|e^{i\alpha}\frac{z-z_0}{z-z_0^*}\right| = \left|\frac{z-z_0}{z-z_0^*}\right|,$$

and if z is on the boundary between the the upper and lower half-plane, that is $y = 0$,
$z = x$, then

$$|z - z_0| = |z - z_0^*|,$$

therefore $|w| = 1$. Next, if z is on the upper half plane (above the edge), then

$$|z - z_0| < |z - z_0^*|,$$

therefore $|w| < 1$. This completes our proof.

One to One Mapping The bilinear transformation is a one to one mapping of the
extended complex plane onto itself: that is

$$f(z_1) = f(z_2) \quad \text{if and only if} \quad z_1 = z_2.$$

To show that, we assume

$$w_1 = f(z_1) = \frac{az_1 + b}{cz_1 + d},$$

$$w_2 = f(z_2) = \frac{az_2 + b}{cz_2 + d}.$$

In Eq. (4.52), we have shown

$$w_1 = \frac{a}{c} + \frac{bc - ad}{c^2} \frac{1}{z_1 + d/c}.$$

Similarly

$$w_2 = \frac{a}{c} + \frac{bc - ad}{c^2} \frac{1}{z_2 + d/c}.$$

If $w_1 = w_2$, then

$$\frac{bc - ad}{c^2}(z_2 + d/c) = \frac{bc - ad}{c^2}(z_1 + d/c).$$

Since $bc - ad \neq 0$, this implies

$$z_2 = z_1.$$

This means the transformation maps distinct points onto distinct images. Since it is a one to one transformation, its inverse always exists. The inverse transformation is obtained by solving for z from

$$w = f(z) = \frac{az + b}{cz + d}.$$

Thus

$$w(cz + d) - (az + b) = 0,$$

or

$$(wc - a)z - (b - dw) = 0.$$

This gives

$$z = \frac{-dw + b}{cw - a}.$$

To compute the derivative of f, again we use Eq. (4.52),

$$f'(z) = \frac{bc - ad}{c^2} \frac{-1}{(z + d/c)^2} = \frac{ad - bc}{(cz + d)^2}.$$

This is well defined and never assumes the zero value in the complex plane. Therefore a bilinear transformation is conformal.

Triples to Triples There are four coefficients in the bilinear transformation, but only three of them are independent. There exists a unique bilinear transformation that maps three distinct points z_1, z_2, z_3 in the z-plane onto three distinct points w_1, w_2, w_s in the w-plane. First we assume these six points are all finite. Let

$$w_j = \frac{az_j + b}{cz_j + d}, \quad j = 1, 2, 3. \tag{4.54}$$

We have

$$w - w_j = \frac{az + b}{cz + d} - \frac{az_j + b}{cz_j + d} = \frac{(az + b)(cz_j + d) - (az_j + b)(cz + d)}{(cz + d)(cz_j + d)}$$

$$= \frac{(ad - bc)(z - z_j)}{(cz + d)(cz_j + d)}, \quad j = 1, 2. \tag{4.55}$$

It follows

$$w_3 - w_j = \frac{(ad - bc)(z_3 - z_j)}{(cz_3 + d)(cz_j + d)}, \quad j = 1, 2. \tag{4.56}$$

It is clear that

$$\frac{w - w_1}{w - w_2} = \frac{(z - z_1)}{(cz_1 + d)} \frac{(cz_2 + d)}{(z - z_2)}, \tag{4.57}$$

$$\frac{w_3 - w_1}{w_3 - w_2} = \frac{(z_3 - z_1)}{(cz_1 + d)} \frac{(cz_2 + d)}{(z_3 - z_2)}. \tag{4.58}$$

Thus

$$\frac{w - w_1}{w - w_2} \bigg/ \frac{w_3 - w_1}{w_3 - w_2} = \frac{(z - z_1)}{(z - z_2)} \bigg/ \frac{(z_3 - z_1)}{(z_3 - z_2)}. \tag{4.59}$$

What happens when some of these points are not finite? For example, when $z_1 \to \infty$, the right-hand side of Eq. (4.59) is then replaced by

$$\lim_{z_1 \to \infty} \frac{(z - z_1)}{(z - z_2)} \bigg/ \frac{(z_3 - z_1)}{(z_3 - z_2)} = \frac{z_3 - z_2}{z - z_2}. \tag{4.60}$$

This means that if any point goes to infinite, all other terms that are grouped with it can be ignored since they are dominated by the one that goes to infinite. This can be applied to all limiting cases. Again the details can be found in the following examples.

Example 4.7.11 Find the bilinear transformation that carries the points -1, ∞, i onto the points (a) i, 1, $1 + i$. (b) ∞, i, 1.

Solution 4.7.11 With $z_1 = -1$, $z_2 = \infty$, and $z_3 = i$,

(a) $w_1 = i$, $w_2 = 1$, $w_3 = 1 + i$,

$$\frac{w - w_1}{w - w_2} / \frac{w_3 - w_1}{w_3 - w_2} = \frac{w - i}{w - 1} / \frac{(1+i) - i}{(1+i) - 1} = i\frac{w - i}{w - 1}$$

$$\lim_{z_2 \to \infty} \frac{(z - z_1)}{(z - z_2)} / \frac{(z_3 - z_1)}{(z_3 - z_2)} = \frac{z - z_1}{z_3 - z_1} = \frac{z + 1}{i + 1}$$

So

$$i\frac{w - i}{w - 1} = \frac{z + 1}{i + 1}.$$

Rearranging the terms, we obtain

$$w = \frac{z + 2 + i}{z + 2 - i}.$$

(b) $w_1 = \infty$, $w_2 = i$, $w_3 = 1$,

$$\lim_{w_1 \to \infty} \frac{w - w_1}{w - w_2} / \frac{w_3 - w_1}{w_3 - w_2} = \frac{w_3 - w_2}{w - w_2} = \frac{1 - i}{w - i}$$

$$\frac{1 - i}{w - i} = \frac{z + 1}{i + 1},$$

Rearranging the terms, we obtain

$$w = \frac{iz + 2 + i}{z + 1}.$$

4.7.5 Mapping Two Distinct Circles into Concentric Circles

Symmetric Points with Respect to a Circle Given a point A in the complex plane, its symmetric point B with respect to a circle of radius R is defined as follows: B is the point on the line joining A and the center of the circle C, (on the extension of the line, if necessary). Let the distance AC be a, and the distance BC be b, the relation between a and b is given by (See Fig. 4.43).

$$a = \frac{R^2}{b}. \tag{4.61}$$

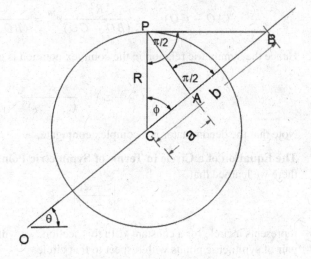

Fig. 4.43 A and B are a pair of symmetric points with respect to the circle C_R

Geometrically, it has the following meaning: From A, draw a perpendicular line that intersects the circle C_R at P. A tangent line to the circle is drawn at P. This line intersects the line CA at B.

It is clear from the triangle CAP that

$$\cos \phi = \frac{a}{R},$$

and from CPB

$$\cos \phi = \frac{R}{b},$$

therefore

$$a = \frac{R^2}{b}.$$

In terms of complex notation, we can write

$$z_c = COe^{i\theta}, \quad z_A = AOe^{i\theta}, \quad z_B = BOe^{i\theta}$$

and

$$a = AO - CO, \quad b = BO - CO.$$

Thus the relation of Eq. (4.61) can be written as

$$AO - CO = \frac{R^2}{BO - CO}.$$

Multiply both sides by $e^{i\theta}$, we have

$$(AO - CO)e^{i\theta} = \frac{R^2}{(BO - CO)}e^{i\theta} = \frac{R^2}{(BO - CO)e^{-i\theta}}.$$

Hence the symmetric relation in the complex notation is given by

$$(z_A - z_c) = \frac{R^2}{(z_B - z_c)^*} \tag{4.62}$$

Note that the denominator is a complex conjugate.

The Equation of a Circle in Terms of Symmetric Points Recall Example 1.6.14, there we learned that

$$\left| \frac{z - z_1}{z - z_2} \right| = k$$

represents a circle for a constant k. In this section, we will show that z_1 and z_2 are a pair of symmetric points with respect to that circle.

It is more clear if we write it out explicitly. Since $z = x + iy$, so $z_1 = x_1 + iy_1$ and $z_2 = x_2 + iy_2$. Thus

$$|z - z_1| = k|z - z_2|$$

can be written as

$$[(x - x_1)^2 + (y - y_1)^2]^{1/2} = k[(x - x_2)^2 + (y - y_2)^2]^{1/2}$$

Square both sides and expand it and collect terms, we have

$$x^2 - 2x\frac{x_1 - k^2 x_2}{1 - k^2} + \frac{x_1^2 - k^2 x_2^2}{1 - k^2}$$

$$+ y^2 - 2y\frac{y_1 - k^2 y_2}{1 - k^2} + \frac{y_1^2 - k^2 y_2^2}{1 - k^2} = 0.$$

Complete the square, it becomes

$$\left(x - \frac{x_1 - k^2 x_2}{1 - k^2}\right)^2 + \left(y - \frac{y_1 - k^2 y_2}{1 - k^2}\right)^2 =$$

$$\left(\frac{x_1 - k^2 x_2}{1 - k^2}\right)^2 + \left(\frac{y_1 - k^2 y_2}{1 - k^2}\right)^2 - \frac{x_1^2 - k^2 x_2^2}{1 - k^2} - \frac{y_1^2 - k^2 y_2^2}{1 - k^2}$$

$$= \frac{k^2}{(1 - k^2)^2}\left[(x_1 - x_2)^2 + (y_1 - y_2)^2\right]. \tag{4.63}$$

This is a circle since it can be recognized as

$$(x - x_c)^2 + (y - y_c)^2 = R^2.$$

Equation (4.63) is simplified considerably if we write it in terms of complex notation. Let

$$z_c = \frac{x_1 - k^2 x_2}{1 - k^2} + i \frac{y_1 - k^2 y_2}{1 - k^2} = \frac{z_1 - k^2 z_2}{1 - k^2}, \tag{4.64}$$

$$R^2 = \frac{k^2}{\left(1 - k^2\right)^2} |z_2 - z_1|^2, \tag{4.65}$$

then Eq. (4.63) can be written as

$$|z - z_c|^2 = R^2,$$

or

$$|z - z_c| = R. \tag{4.66}$$

Note that

$$z_1 - z_c = z_1 - \frac{z_1 - k^2 z_2}{1 - k^2} = \frac{k^2(z_2 - z_1)}{1 - k^2}, \tag{4.67}$$

$$z_2 - z_c = z_2 - \frac{z_1 - k^2 z_2}{1 - k^2} = \frac{z_2 - z_1}{1 - k^2}. \tag{4.68}$$

Thus

$$\frac{R^2}{(z_1 - z_c)^*} = \frac{k^2}{\left(1 - k^2\right)^2} |z_2 - z_1|^2 \cdot \frac{1 - k^2}{k^2 (z_2 - z_1)^*}$$

$$= \frac{(z_2 - z_1)(z_2 - z_1)^*}{(1 - k^2)(z_2 - z_1)^*} = z_2 - z_c, \tag{4.69}$$

satisfying Eq. (4.62). Therefore z_1 and z_2 are a pair of symmetric points with respect to the circle with a radius R centered at z_c.

Mapping Two Distinct Circles into Concentric Circles Suppose we have two separate circles, and a pair of symmetric points z_1 and z_2 with respect to both circles. Let

$$\left| \frac{z - z_1}{z - z_2} \right| = k_1 \tag{4.70}$$

be the first circle. That is, this equation in z (xy) plane describes a circle of radius R_1 with its center at z_{c_1}, where

$$R_1 = \frac{k_1}{(1 - k_1^2)} |z_2 - z_1|, \quad z_{c_1} = \frac{z_1 - k_1^2 z_2}{1 - k_1^2}. \tag{4.71}$$

And let

$$\left| \frac{z - z_1}{z - z_2} \right| = k_2 \tag{4.72}$$

be the second circle. That is, this equation in z (xy) plane describes a circle of radius R_2 with its center at z_{C_2}, where

$$R_2 = \frac{k_2}{\left(1 - k_2^2\right)} |z_2 - z_1|, \quad z_{C_2} = \frac{z_1 - k_2^2 z_2}{1 - k_2^2}. \tag{4.73}$$

Now if we make a transformation

$$w = \frac{z - z_1}{z - z_2},$$

then the first circle is mapped into

$$|w| = k_1,$$

and the second circle is mapped into

$$|w| = k_2.$$

In w (u,v) plane, these are two circles with radius k_1 and k_2 respectively. But both circles have their center at $w = 0$. Thus the two separate circles in z-plane are mapped into two concentric circles in w plane.

The details of this method are best explained in the following examples.

Example 4.7.12 Find a bilinear transformation that maps the region between two circles

$$|z| = 1, \quad |z + 1| = \frac{5}{2}$$

onto an annular domain between two concentric circles. Determine the radii of the two image circles with a common center at the origin.

Solution 4.7.12 We shall first find a pair of points z_1 and z_2, which are symmetric with respect to both circles. The common symmetric pair of points, z_1 and z_2 must lie on the line joining the center of the two circles, which is the real line. Thus z_1 and z_2 must be real and satisfy

$$z_1 z_2 = 1, \quad (z_1 + 1)(z_2 + 1) = \left(\frac{5}{2}\right)^2.$$

This gives two solutions:

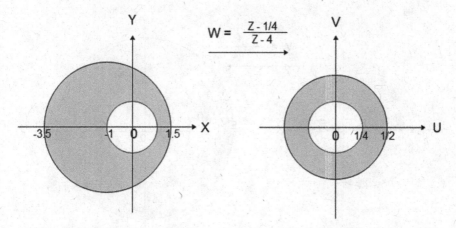

Fig. 4.44 The region bounded by the circles $|z| = 1$ and $|z + 1| = 2.5$ is conformally mapped onto the annular region $\frac{1}{4} < w < \frac{1}{2}$ by the transformation $w = \frac{z-1/4}{z-4}$

$$z_1 = \frac{1}{4}, \quad z_2 = 4 \quad \text{or} \quad z_1 = 4, \quad z_2 = \frac{1}{4}.$$

The bilinear transformation $w = \frac{z-z_1}{z-z_2}$ will transform the two separate circles into two concentric circles (Fig. 4.44).

Let us first use $z_1 = \frac{1}{4}$, $z_2 = 4$. For the case $|z| = 1$, the center is at zero, and the radius is equal to 1. Equation (4.71) gives us a set of two quadratic equations,

$$1 = \frac{k_1}{1 - k_1^2}\left(4 - \frac{1}{4}\right),$$

$$0 = \frac{\frac{1}{4} - k_1^2 4}{1 - k_1^2}.$$

Each equation gives two solutions, only one is common to both. It is easy to show that

$$k_1 = \frac{1}{4}$$

satisfies both equations. Therefore, the transformation

$$w = \frac{z - \frac{1}{4}}{z - 4}$$

will map the circle $|z| = 1$ in the z-plane onto the circle $|w| = \frac{1}{4}$ in the w plane.

Similarly, for the circle $|z + 1| = \frac{5}{2}$, the center is at -1 and the radius is 2.5. According to Eq. (4.73), $k_2 = \frac{1}{2}$. Therefore, the transformation $w = \frac{z-\frac{1}{4}}{z-4}$ will map

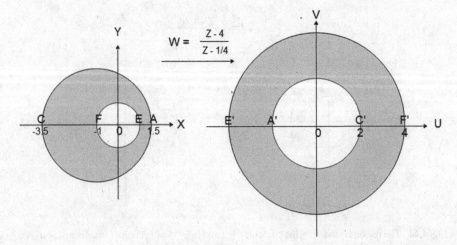

Fig. 4.45 The region bounded by the circles $|z| = 1$ and $|z + 1| = 2.5$ is conformally mapped onto the annular region $2 < w < 4$ by the transformation $w = \frac{z-4}{z-1/4}$

the circle $|z + 1| = \frac{5}{2}$ in the z plane into the circle $|w| = \frac{1}{2}$ in the w plane. Both circles in the w plane are centered at $w = 0$, so they are concentric. This is shown in Fig. 4.45. The transformation is equally valid if the two concentric circles are multiplied by the same constant.

Next we take $z_1 = 4$ and $z_2 = \frac{1}{4}$. According to Eq. (4.71), the unit circle $|z| = 1$ in z-plane corresponds to $k_1 = 4$. Thus this circle in the z-plane is transformed onto another circle in w plane with a radius 4. Similarly, the circle $|z + 1| = \frac{5}{2}$ in the z-plane corresponds to $k_2 = 2$. Therefore, the outer circle in the z-plane is transformed into the inner circle of $|w| = 2$ in the w plane, as shown in Fig. 4.45.

Example 4.7.13 Two parallel long conducting cylinders are shown in Fig. 4.46. Calculate the electric potential in the space surrounding them if the larger one is held at potential V_1 and the smaller one held at potential V_2.

Solution 4.7.13 The first step of solving this problem is to map the profile of the cross sections of the two cylinders (two circles in the figure) into two concentric circles. Therefore, we will first find a pair of points that are symmetric to both circles. This means we will solve the following equations

$$z_1 z_2 = 1, \qquad \left(z_1 - \frac{9}{2}\right)\left(z_2 - \frac{9}{2}\right) = \left(\frac{5}{2}\right)^2.$$

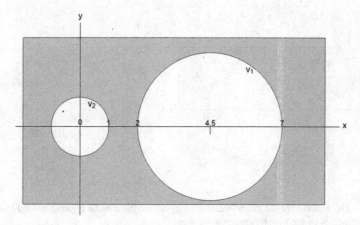

Fig. 4.46 Find the potential between the two cylinders at V_1 and V_2. The small cylinder has a radius 1 centered at the origin. The larger cylinder has a radius 2.5 centered at $x = 4.5$

The solutions are either

$$z_1 = 3, \quad z_2 = \frac{1}{3}$$

or

$$z_1 = \frac{1}{3}, \quad z_2 = 3.$$

Let us take first set of solution, and write the transformation as

$$w = \frac{z - 3}{z - \frac{1}{3}}.$$

In the z-plane, the unit circle $|z| = 1$, the center is at 0 and the radius is 1. According to Eq. (4.71) $k_1 = 3$, therefore this circle is mapped into a circle $|w| = 3$. That is a circle of radius 3 centered at 0. The second circle $\left|z - \frac{9}{2}\right| = \left(\frac{5}{2}\right)^2$, according to Eq. (4.73) corresponds to $k_2 = \frac{3}{5}$. Therefore, it is mapped to $|w| = \frac{3}{5}$ in the w plane. This mapping is sketched in the Fig. 4.46.

Now we know the potential is equal to (See Example 4.4.1)

$$V(w) = A \ln |w| + B$$

The constants A and B can be determined by the boundary conditions, that is

$$V_2 = A \ln 3 + B,$$
$$V_1 = A \ln \frac{3}{5} + B.$$

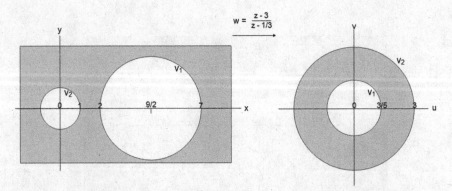

Fig. 4.47 The region between the cylinders is mapped into the annular region bounded by the two concentric circles by the transformation $w = \frac{z-3}{z-1/3}$

Thus

$$V_2 - V_1 = A \ln 3 - A \ln \frac{3}{5} = A(\ln 3 - \ln 3 + \ln 5),$$

or

$$A = \frac{V_2 - V_1}{\ln 5},$$

and

$$B = V_2 - A \ln 3 = V_2 - \frac{V_2 - V_1}{\ln 5} \ln 3.$$

Therefore, the potential between these two conductors is

$$V(z) = A \ln |w| + B = A \ln \left| \frac{z-3}{z - \frac{1}{3}} \right| + B$$

$$= A \ln 3 \left| \frac{z-3}{3z-1} \right| + B = A \ln 3 + A \ln \left| \frac{z-3}{3z-1} \right| + B$$

$$= A \ln \left| \frac{z-3}{3z-1} \right| + V_2$$

$$= \frac{V_2 - V_1}{\ln 5} \ln \left| \frac{z-3}{3z-1} \right| + V_2$$

Thus, we have

$$V(x, y) = \frac{V_2 - V_1}{\ln 5} \ln \left[\frac{(x-3)^2 + y^2}{(3x-1)^2 + 9y^2} \right]^{1/2} + V_2$$

$$= \frac{V_2 - V_1}{2 \ln 5} \ln \left[\frac{(x-3)^2 + y^2}{(3x-1)^2 + 9y^2} \right] + V_2.$$

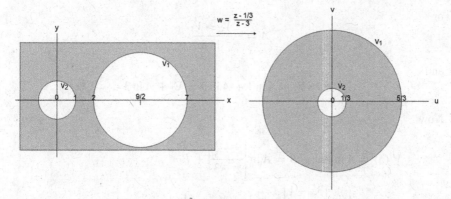

Fig. 4.48 The region between the cylinders is mapped into the annular region bounded by the two concentric circles by the transformation $w = \frac{z-1/3}{z-3}$

Now we use the other set of data and find the transformation to be

$$w = \frac{z - \frac{1}{3}}{z - 3}.$$

By the same reason, we find that

$$\frac{z - \frac{1}{3}}{z - 3} = \frac{1}{3}$$

describes the circle $|z| = 1$ in the z-plane and

$$\frac{z - \frac{1}{3}}{z - 3} = \frac{5}{3}$$

describes the other circle $\left|z - \frac{9}{2}\right| = \frac{5}{2}$ in the z-plane. The mapping is sketched in Fig. 4.48.

Now the boundary condition is

$$V_1 = A \ln \frac{5}{3} + B,$$

$$V_2 = A \ln \frac{1}{3} + B.$$

Thus

$$V_2 - V_1 = A \left(\ln \frac{1}{3} - \ln \frac{5}{3} \right) = -A \ln 5,$$

or

$$A = -\frac{V_2 - V_1}{\ln 5}$$

and

$$B = V_2 - A \ln 1 + A \ln 3 = V_2 + A \ln 3$$

Now

$$V(z) = A \ln |w| + B = A \ln \left| \frac{z - \frac{1}{3}}{z - 3} \right| + B$$

$$= A(\ln \left| \frac{3z - 1}{z - 3} \right| - \ln 3) + V_2 + A \ln 3$$

$$= -\frac{V_2 - V_1}{\ln 5} \ln \left| \frac{3z - 1}{z - 3} \right| + V_2 = \frac{V_2 - V_1}{\ln 5} \ln \left| \frac{z - 3}{3z - 1} \right| + V_2.$$

Therefore

$$V(x, y) = \frac{V_2 - V_1}{\ln 5} \ln \left[\frac{(x - 3)^2 + y^2}{(3x - 1)^2 + 9y^2} \right]^{\frac{1}{2}} + V_2$$

$$= \frac{V_2 - V_1}{2 \ln 5} \ln \frac{(x - 3)^2 + y^2}{(3x - 1)^2 + 9y^2} + V_2.$$

Exactly the same as before. It is interesting to note that the answer to the physical problem is unique, the intermediate mapping can be different.

4.8   The Schwarz-Christoffel Transformation

In this section, we construct a transformation, known as the Schwarz-Christoffel transformation, which maps the x-axis and the upper half of the z-plane onto a given simple closed polygon and its interior in the w plane. The polygon could be closed at infinity, in that case, it is practically an open polygon.

In the first section of this chapter, we have used many examples to show the usefulness of the conformal mapping. There the mapping functions are taken from tables. The selection of a particular function is guided mostly by educated guess. In this section, we will provide a systematic way to derive some of these functions.

4.8.1 *Formulation of the Schwarz-Christoffel Transformation*

Before we start, let us recall some definitions in complex variables. Any complex quantity A can be expressed as

$$A = |A| e^{i\beta_A} \tag{4.74}$$

where β_A is given by

$$\beta_A = \tan^{-1} \frac{\text{Im}(A)}{\text{Re}(A)}.$$

The argument of A is defined as β_A. That is

$$\arg A = \beta_A, \tag{4.75}$$

thus

$$A = |A| e^{i \arg A}. \tag{4.76}$$

Note that Eq. (4.74) holds even if A is a pure real number. In that case,

$$If\ A\ is\ \text{positive},\ \arg A = 0, \tag{4.77}$$

$$If\ A\ is\ \text{negative},\ \arg A = \pi. \tag{4.78}$$

Let $A = BC$, then $\arg A = \arg BC$. Since

$$B = |B| e^{i \arg B}, \qquad C = |C| e^{i \arg C},$$

and

$$BC = |B| e^{i \arg B} |C| e^{i \arg C} = |B| |C| e^{i(\arg B + \arg C)},$$

therefore

$$\arg BC = \arg B + \arg C.$$

Thus

$$\arg A = \arg BC = \arg B + \arg C. \tag{4.79}$$

Similarly, if $A = \frac{B}{C}$, then

$$\arg A = \arg \frac{B}{C} = \arg B - \arg C. \tag{4.80}$$

Our problem is to map n distinct points $x_1, x_2, ...x_n$ on the real x- axis in the z-plane unto the vertices of the n sided polygon in the w plane. Let the n points be arranged in such a way $x_1 < x_2 < \cdots < x_n$ and z moves on the x-axis from $z < x_1$ to $z > x_n$. We want to find a mapping function $w = f(z)$ so that $w_k = f(z_k), k = 1, 2, ...n$.

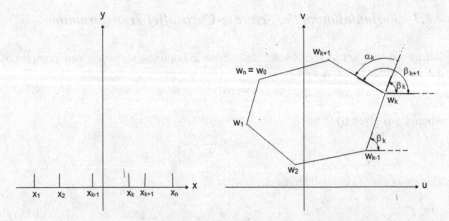

Fig. 4.49 The Schwarz-Christoffel transformation carries the upper half z-plane into the interior of an n-sided polygon. The angle of inclination of the kth side is denoted by β_k. The exterior angle α_k at the vertex w_k is given by $\beta_{k+1} - \beta_k$

By definition

$$\frac{dw}{dz} = f'(z), \tag{4.81}$$

or

$$dw = f'(z)dz.$$

According to Eq. (4.79), we have

$$\arg dw = \arg f'(z) + \arg dz.$$

Since z is on the x-axis and is moving in the positive direction, according Eq. (4.77). Therefore

$$\arg dw = \arg f'(z). \tag{4.82}$$

Now consider the closed n-sided polygon with vertices $w_1, w_2 ... w_n$ in the w plane. The angle of inclination of the kth side joining w_{k-1} and w_k is denoted by β_k, $k = 1, 2, ...n$. For convenience of notation, we also write w_n as w_0, so that the first side of the polygon refers to the line segment joining w_n and w_1, Hence β_{n+1} is taken to be β_1 (See Fig. 4.49).

When z moves along on the x-axis in the segment (x_{k-1}, x_k), its image point w moves along the kth side of the polygon. Therefore, dw is also along this side. Since $dw = |dw| e^{i\beta_k}$, so $\arg dw = \beta_k$. When z moves across the point x_k along the real axis, $\arg dw$ jumps by an amount α_k to assume a new value β_{k+1}. It is clear from the figure that $\alpha_k = \beta_{k+1} - \beta_k$. In summary: $\arg dw = \beta_k$, when z assumes real values in the interval of (x_{k-1}, x_k), $k = 2, 3, ...n$ and $\arg dw = \beta_1$ when z assumes real values in the interval of $(-\infty, x_1)$ and (x_n, ∞).

This situation can be summarized by the following expression:

$$\arg dw = \frac{\beta_1}{\pi} \arg(z - x_1) + \frac{\beta_2}{\pi} \arg \frac{z - x_2}{z - x_1} + \cdots + \frac{\beta_k}{\pi} \arg \frac{z - x_k}{z - x_{k-1}} +$$

$$\cdots + \frac{\beta_n}{\pi} \arg \frac{z - x_n}{z - x_{n-1}} + \frac{\beta_{n+1}}{\pi}(\pi - \arg(z - x_n)) \qquad (4.83)$$

Now we will show that as z moves on the real x-axis, if z is in the segment of (x_{k-1}, x_k), only the kth term in this expression is nonvanishing and the nonvanishing term is equal to β_k. All other terms are zero.

For $z < x_k$, we have $z - x_l < 0$ for $l = k + 1, k + 2, \ldots n$. According to Eq. (4.78), $\arg(z - x_l) = \pi$, since $z - x_l$ is negative. Therefore

$$\frac{\beta_l}{\pi} \arg \frac{z - x_l}{z - x_{l-1}} = \frac{\beta_l}{\pi}[\arg(z - x_l) - \arg(z - x_{l-1})] = \frac{\beta_l}{\pi}[\pi - \pi] = 0 \qquad (4.84)$$

for $l = k + 1, k + 2, \ldots n$, and the last term

$$\frac{\beta_{n+1}}{\pi} \arg[\pi - \arg(z - x_n)] = \frac{\beta_{n+1}}{\pi} \arg[\pi - \pi] = 0$$

is also equal to zero. Therefore, all terms to the right of the kth term are zero.

Similarly, for $z > x_{k-1}$, we have $z - x_m > 0$ for $m = k - 1, k - 2, \ldots 1$. According to Eq. (4.77), $\arg(z - x_m) = 0$. Therefore

$$\frac{\beta_m}{\pi} \arg \frac{z - x_m}{z - x_{m-1}} = \frac{\beta_m}{\pi}[\arg(z - x_m) - \arg(z - x_{m-1})] = \frac{\beta_m}{\pi}[0 - 0] = 0 \quad (4.85)$$

for $m = k - 1, k - 2, \ldots 1$, and the first term

$$\frac{\beta_1}{\pi} \arg(z - x_1) = \frac{\beta_1}{\pi} 0 = 0$$

is also equal to zero. Therefore, all terms to the left of the kth term are zero. For the kth term, we have

$$\frac{\beta_k}{\pi} \arg \frac{z - x_k}{z - x_{k-1}} = \frac{\beta_k}{\pi}[\arg(z - x_k) - \arg(z - x_{k-1})] = \frac{\beta_k}{\pi}[\pi - 0] = \beta_k. \qquad (4.86)$$

Thus, when z moves along the x-axis from x_{k-1} toward x_k, its image in w, moves along the kth side of the polygon between w_{k-1} and w_k, and

$$\arg dw = \beta_k.$$

Hence, the validity of Eq. (4.83) is established.

Now we use Eq. (4.80) to expand Eq. (4.83), we have

$$\arg dw = \frac{\beta_1}{\pi} \arg(z - x_1) + \frac{\beta_2}{\pi} \arg(z - x_2) - \frac{\beta_2}{\pi} \arg(z - x_1) + \frac{\beta_3}{\pi} \arg(z - x_3) - \dots$$
$$\dots \frac{\beta_n}{\pi} \arg(z - x_n) - \frac{\beta_n}{\pi} \arg(z - x_{n-1}) + \beta_{n+1} - \frac{\beta_{n+1}}{\pi} \arg(z - x_n). \quad (4.87)$$

Collecting terms

$$\arg dw = \frac{\beta_1 - \beta_2}{\pi} \arg(z - x_1) + \frac{\beta_2 - \beta_3}{\pi} \arg(z - x_2) + \dots$$
$$\dots \frac{\beta_n - \beta_{n+1}}{\pi} \arg(z - x_n) + \beta_{n+1}.$$

Using the notation $\alpha_k = \beta_{k+1} - \beta_k$, we can write this expression as

$$\arg dw = -\sum_{k=1}^{n} \frac{\alpha_k}{\pi} \arg(z - x_k) + \beta_{n+1}. \quad (4.88)$$

From Eq. (4.76), we have $A = |A| e^{i \arg A}$. If α is a real number, then

$$A^{\alpha} = |A|^{\alpha} e^{i \alpha \arg A}.$$

Therefore by definition
$$\alpha \arg A = \arg A^{\alpha}.$$

Thus
$$-\alpha_k \arg(z - x_k) = \arg(z - x_k)^{-\alpha_k}. \quad (4.89)$$

It follows that
$$-\sum_{k=1}^{n} \frac{\alpha_k}{\pi} \arg(z - x_k) = \sum_{k=1}^{n} \arg(z - x_k)^{-\frac{\alpha_k}{\pi}}.$$

Since β_{n+1} is a constant, yet to be determined. For convenience, we write it as $\beta_{n+1} = \arg A$. Thus Eq. (4.88) can be written as

$$\arg dw = \sum_{k=1}^{n} \arg(z - x_k)^{-\frac{\alpha_k}{\pi}} + \arg A.$$

By Eq. (4.79), this equation can be written as

$$\arg dw = \arg[A \Pi_{k=1}^{n} (z - x_k)^{-\frac{\alpha_k}{\pi}}]. \quad (4.90)$$

Since $\arg dw = \arg f'(z)$, therefore

$$\arg f'(z) = \arg[A\Pi_{k=1}^n (z - x_k)^{-\frac{\alpha_k}{\pi}}].$$

If $\arg(X) = \arg(Y)$, it means $X = kY$ where k is a real constant. Therefore

$$f'(z) = A'\Pi_{k=1}^n (z - x_k)^{-\frac{\alpha_k}{\pi}} \tag{4.91}$$

where A' is another constant, it is equal to kA.

Since $w = f(z)$, so $\frac{dw}{dz} = f'(z)$, therefore

$$w = A' \int^z \Pi_{k=1}^n (s - x_k)^{-\frac{\alpha_k}{\pi}} ds + B, \tag{4.92}$$

where A' and B are two constants yet to be determined. This transformation is known as the Schwarz-Christoffel transformation, named in honor of the two German mathematicians H. A. Schwarz (1843–1921) and E. B. Christoffel (1829–1900), who discovered it independently.

4.8.2 Convention Regarding the Polygon

The point w_∞ on the boundary of the polygon will map onto the point $z = \infty$. This point may or may not be a vertex of the polygon. If it is a vertex, we call it w_0. In any case, starting from w_∞ traverse the boundary of the polygon in the sense that the interior is always on the left, encounter the vertices in succession $w_1, w_2, ...w_n$. These will map onto points $x_1, x_2, ...x_n$, respectively, of the x-axis in the z-plane. Thus the upper half of the z plane will be mapped onto the inside of the polygon.

In the case of the polygon illustrated, all exterior angles α_i are positive. The general convention is that if it is a counterclockwise turn, α is positive. If it is a clockwise turn, α is negative.

Now if x_n is moved to infinity, we simply drop the factor $(s - x_n)^{-\frac{\alpha_n}{\pi}}$ from Eq. (4.92) and get only $n - 1$ factors in the integral, that is

$$w = A \int^z \Pi_{k=1}^{n-1} (s - x_k)^{-\frac{\alpha_k}{\pi}} ds + B. \tag{4.93}$$

Open Polygon. In many problems of interests, one or more vertices of the polygon are at infinity. The polygon "closed" at infinity is practically an open polygon. For example, the semi-infinite strip in the following figure is a case of a triangle with one vertex that is taken to infinity in y direction (Fig. 4.50).

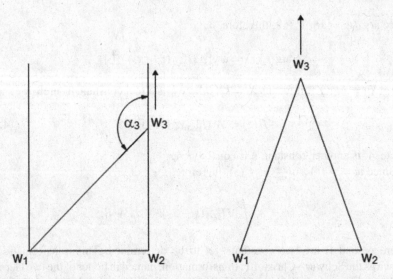

Fig. 4.50 In the triangle $w_1w_2w_3$, as w_3 moves higher and higher, α_3 becomes larger and larger, until $w_3 \to \infty$, $\alpha_3 = \pi$, the triangle becomes an open semi-infinite strip

Before we see more examples, we would like to make a few remarks:

(i)   The sum of the exterior angles of a closed polygon is equal to 2π.

$$\alpha_1 + \alpha_2 + \cdots + \alpha_n = 2\pi$$

(ii)   The constants A and B are in general complex numbers. In the following examples, we will show how they are determined.

(iii) When the vertices $w_1, w_2, \ldots, w_n$ are given points in the w plane, their images $x_1, x_2, \ldots, x_n$ on the x-axis of z-plane cannot be selected in an arbitrary manner. We can arbitrarily select at most three of these $x_i's$, the remainder must then be calculated.

4.8.3   Examples

Example 4.8.1 Find the Schwarz-Christoffel transformation that maps the upper half z-plane ($Im\ z > 0$) onto the semi-infinite strip of the w plane, such that $-\frac{\pi}{2} < u < \frac{\pi}{2}$, $v > 0$ where $w = u + iv$ (Fig. 4.51).

Solution 4.8.1 We consider the strip as the limiting form of a triangle with vertices w_1, w_2, and w_3 as the imaginary part of w_3 tends to infinity. The limiting values of the exterior angles are

$$\alpha_1 = \frac{\pi}{2}, \quad \alpha_2 = \frac{\pi}{2}.$$

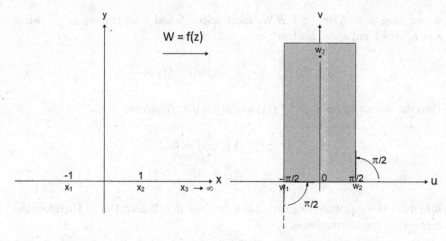

Fig. 4.51 Schwarz-Christoffel transformation of the upper half z-plane $y \geq 0$ onto the semi-finite strip of w plane $-\frac{\pi}{2} \leq u \leq \frac{\pi}{2}$, $v \geq 0$

We choose the points $x_1 = -1, x_2 = 1$, and $x_3 = \infty$ as the points whose images are $w_1 = -\frac{\pi}{2}$, $w_2 = \frac{\pi}{2}$ and $w_3 \to \infty$ as the vertices. Then according to Eq. (4.93), we have

$$w = A' \int^z (s+1)^{-\frac{1}{2}}(s-1)^{-\frac{1}{2}}ds + B$$

Since

$$(s-1)^{-\frac{1}{2}} = (-1)^{-\frac{1}{2}}(1-s)^{-\frac{1}{2}} = e^{-i\frac{\pi}{4}}(1-s)^{-\frac{1}{2}},$$

so by changing A' to A, we have

$$w = A \int^z (1+s)^{-\frac{1}{2}}(1-s)^{-\frac{1}{2}}ds + B$$

$$= A \int^z (1-s^2)^{-\frac{1}{2}}ds + B = A \int^z \frac{ds}{\sqrt{1-s^2}} + B.$$

One of the difficulty of this method is to evaluate these integrals, but this one is relative easy. Let $s = \sin x$, so $\sin^{-1} s = x$, and $ds = \cos x dx$. It follows that

$$\frac{ds}{\sqrt{1-s^2}} = \frac{\cos x dx}{\sqrt{1-\sin^2 x}} = \frac{\cos x dx}{\cos x} = dx = d\sin^{-1} s,$$

therefore

$$\int^z \frac{ds}{\sqrt{1-s^2}} = \int^z d\sin^{-1} s = \sin^{-1} z.$$

So we have $w = A \sin^{-1} z + B$. We must select A and B so that $w_1 = -\frac{\pi}{2}$ when $z = x_1 = -1$ and $w_2 = \frac{\pi}{2}$ when $z = x_2 = 1$. Since

$$\sin^{-1}(-1) = -\frac{\pi}{2} \quad \text{and} \quad \sin^{-1}(1) = \frac{\pi}{2}.$$

Thus the two equations $w_1 = f(z_1)$ and $w_2 = f(z_2)$ become

$$(-\frac{\pi}{2}) = A\left(-\frac{\pi}{2}\right) + B,$$

$$\frac{\pi}{2} = A\frac{\pi}{2} + B.$$

Add these two equations, we have $2B = 0$. Thus $B = 0$ and $A = 1$. Therefore the appropriate transformation is

$$w = \sin^{-1} z.$$

Example 4.8.2 Use the Schwarz-Christoffel method to map the upper half z-plane to the horizontal strip $0 \leq v \leq \pi$ in the w plane (Fig. 4.52).

Solution 4.8.2 Consider the strip $0 \leq v \leq \pi$ as the limiting form of a rhombus with vertices $w_1, w_2, w_3,$ and w_4 with w_2 and w_4 going to the left and right infinity, respectively, as shown in Fig. 4.52. In the limit, the exterior angles become

$$\alpha_1 = 0, \quad \alpha_2 = \pi, \quad \alpha_3 = 0.$$

Since we can arbitrary choose three x values, so we choose $x_1 = -1$, $x_2 = 0$, $x_3 = 1$, and $x_4 = \infty$. The Schwarz-Christoffel transformation is then given by

$$w = A \int^z (s+1)^0 (s)^{-\frac{\pi}{\pi}} (s-1)^0 ds + B$$

$$= A \int^z \frac{ds}{s} + B = A \ln z + B.$$

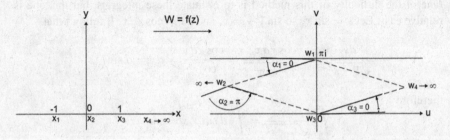

Fig. 4.52 Map the upper half z-plane onto the horizontal strip of the w plane with the Schwartz-Christoffel method

A and B can be determined by requiring $w_1 = i\pi$ at $z = x_1 = -1$, and $w_3 = 0$ at $z = x_3 = 1$. So at $x_3 = 1$, we have

$$w_3 = 0 = A \ln(1) + B.$$

Since $\ln(1) = 0$, so $B = 0$. At $x_1 = -1$, we have

$$w_1 = i\pi = A \ln(-1).$$

Since $\ln(-1) = \ln\left(e^{i\pi}\right) = i\pi$, thus $A = 1$. Therefore the Schwarz-Christoffel transformation is

$$w = \ln z.$$

Example 4.8.3 Find the Schwarz-Christoffel transformation that maps the upper-half z plane onto the unbounded region in the w plane as shown in the following figure. In particular, the point $z = x_1 = -1$ and $z = x_2 = 1$ are mapped on the points $w_1 = iH$ and $w_2 = 0$, respectively (Fig. 4.53).

Solution 4.8.3 It is clear from the above figure

$$\alpha_1 = -\frac{\pi}{2}, \qquad \alpha_2 = \frac{\pi}{2}.$$

The Schwarz-Christoffel transformation is therefore

$$w = A \int^z (s+1)^{\frac{1}{2}}(s-1)^{-\frac{1}{2}}ds + B$$

$$= A \int^z \frac{\sqrt{s+1}}{\sqrt{s-1}}ds + B.$$

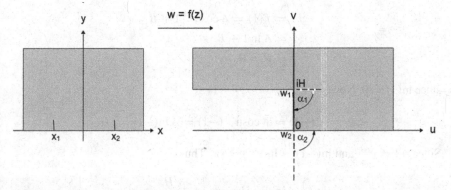

Fig. 4.53 The upper half of the z (x,y) plane (Im $z \geq 0$) is mapped onto the unbounded polygon of the w (u,v) plane as shown in this figure

Multiplying top and bottom of the integrand with $\sqrt{s+1}$, we have

$$\frac{\sqrt{s+1}}{\sqrt{s-1}}ds = \frac{s+1}{\sqrt{s^2-1}}ds = \frac{s}{\sqrt{s^2-1}}ds + \frac{1}{\sqrt{s^2-1}}ds.$$

Therefore we can split the integral into two parts

$$\int^z \frac{\sqrt{s+1}}{\sqrt{s-1}}ds = \int^z \frac{s}{\sqrt{s^2-1}}ds + \int^z \frac{1}{\sqrt{s^2-1}}ds.$$

The first part can be easily integrated out, since $d\left(s^2-1\right)^{\frac{1}{2}} = \left(s^2-1\right)^{-\frac{1}{2}}sds$. The second part can be evaluated with a substitution. Let $s = \cosh x$, so $x = \cosh^{-1}s$ and $ds = \sinh x\,dx$. Since

$$\frac{ds}{\sqrt{s^2-1}} = \frac{\sinh x\,dx}{\sqrt{\cosh^2 x - 1}} = \frac{\sinh x\,dx}{\sinh x} = dx = d\cosh^{-1}s.$$

Thus

$$\int^z \frac{\sqrt{s+1}}{\sqrt{s-1}}ds = (z^2-1)^{\frac{1}{2}} + \cosh^{-1}z.$$

Or

$$w = f(z) = A[(z^2-1)^{\frac{1}{2}} + \cosh^{-1}z] + B.$$

The two constants A and B can be determined by the requirements $f(z_i) = w_i$, that is $w_1 = f(-1) = iH$, and $w_2 = f(1) = 0$. Recall from Sect. 1.7.4 that

$$\cosh^{-1}z = \ln\left[z \pm (z^2-1)^{\frac{1}{2}}\right].$$

Thus

$$w_2 = f(1) = A\cosh^{-1}1 + B$$
$$= A\ln 1 + B$$
$$= B = 0,$$

since $\ln(1) = 0$. Now

$$w_1 = f(-1) = A\cosh^{-1}(-1) = A\ln(-1) = iH.$$

Since $-1 = e^{i\pi}$, and $\ln(-1) = \ln(e^{i\pi}) = i\pi$. Thus

$$Ai\pi = iH, \quad or \quad A = \frac{H}{\pi}.$$

Since $\cosh^{-1} z$ is a multi-valued function, for the mapping, we must select a single valued branch which is analytic in the upper half-plane. Therefore, the Schwarz-Christoffel transformation should be written as

$$f(z) = \frac{H}{\pi}\left[\sqrt{z^2 - 1} + \ln\left(z + \sqrt{z^2 - 1}\right)\right].$$

Exercises

1. Show that the transformation $w = \ln\frac{z-1}{z+1}$ maps the upper half z-plane onto a strip of $0 \le v \le \pi$ and $0 \le u$ in the w plane.

Ans. See Fig. 4.2.

2. Show that the circle $(x - \frac{1}{2})^2 + y^2 = (\frac{1}{2})^2$ in the z-plane is mapped onto $u = 1$ in the w plane by the transformation $w = \frac{1}{z}$.

Ans. See Example 4.1.1.

3. Show that the transformation $w = \sin z$ maps the semi-infinite strip $-\frac{\pi}{2} \le x \le \frac{\pi}{2}$, $y \ge 0$ in the z-plane onto the upper half $(v \ge 0)$ of the w plane.

Ans. See Example 4.1.3.

4. The function $f(z) = \sin z$ is analytic everywhere and represents a conformal mapping except at the critical points. Find these critical points.

Ans. $z = (2n + 1)\pi/2$.

5. Show that $x = a$ and $-\frac{\pi}{2} \le a \le \frac{\pi}{2}$ is mapped into a hyperbola by the transformation $w = \sin z$. Show that if $a > 0$, then the line $x = a$ is mapped into the positive branch of the hyperbola.

Hint: Recall $\sin z = \sin x \cosh y + i \cos x \sinh y$

6. Show that $y = b$ is mapped into an ellipse by the transformation of $w = \sin z$. If $y = 1$ in the z-plane, what are the major and minor axes?

Ans. Semi-major axis: $\frac{1}{2}(e + e^{-1})$; semi-minor axis: $\frac{1}{2}(e - e^{-1})$.

7. Use the function $\ln z$ to find an expression for the bounded steady temperature in a plate having the form of a quadrant $x \ge 0$, $y \ge 0$ if its faces are perfectly insulated and its edges have temperatures $T(x, 0) = 0$ and $T(0, y) = 1$.

Ans. $T = \frac{2}{\pi}\tan^{-1}(\frac{y}{x})$.

8. Find the steady temperatures in a solid whose shape is that of a long cylindrical wedge if its boundary plane $\theta = 0$ and $\theta = \theta_0$ are kept at constant temperatures zero and T_0, respectively, and if its surface $r = r_0$ is perfectly insulated.

Ans. $T = \frac{T_0}{\theta_0}\tan^{-1}(\frac{y}{x})$.

9. Find the bounded steady temperatures $T(x, y)$ in the semi-infinite solid $y \ge 0$ if $T = 0$ on the part $x < -1$ of the boundary, if $T = 1$ on the part $x > 1$ and if the strip $-1 < x < 1$ of the boundary is insulated.

Ans. $T = \frac{1}{2} + \frac{1}{\pi}\sin^{-1}\frac{1}{2}[\sqrt{(x + 1)^2 + y^2} - \sqrt{(x - 1)^2 + y^2}]$ $(-\frac{\pi}{2} \le \sin^{-1} u \le \frac{\pi}{2})$.

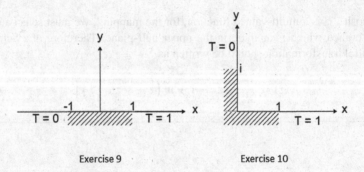

Exercise 9   Exercise 10

10. Find the bounded steady temperature in the solid $x \geq 0$, $y \geq 0$ when the boundary surfaces are kept at fixed temperature as shown in the figure except for insulated strips of equal width at the corner.

Ans. $T = \frac{1}{2} + \frac{1}{\pi}\sin^{-1}\frac{1}{2}[\sqrt{(x^2 - y^2 + 1)^2 + 4x^2y^2} - \sqrt{(x^2 - y^2 - 1)^2 + 4x^2y^2}]$

11. Find the electrostatic potential V in the space enclosed by the half cylinder $x^2 + y^2 = 1$, $y \geq 0$ and the plane $y = 0$ when $V = 0$ on the cylindrical surface and $V = 1$ on the planar surface.

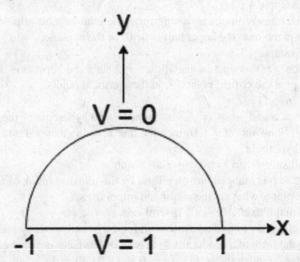

Ans. $V = \frac{2}{\pi}\tan^{-1}\left(\frac{1-x^2-y^2}{2y}\right)$.

12. Let $1 + i$ be one of the two points that are symmetric to the circle $|z| = 2$, find the other point.

Ans. $2(1 + i)$

13. Show that the transformation $w = \exp(z)$ maps (a) unto (b) of the following figure.

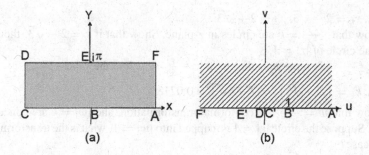

(a)   (b)

14. Find the potential V in the space between the planes $y = 0$ and $y = \pi$, when $V = 0$ on the part of each of those planes where $x > 0$, and $V = 1$ on the parts where $x < 0$.

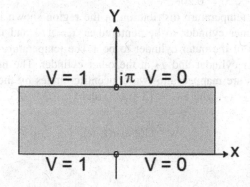

Ans. $V = \frac{1}{\pi} \tan^{-1} \frac{\sin y}{\sinh x}$.

15. Find the electrostatic potential in semi-infinite space bounded by the two half-planes and a half cylinder, when $V = 1$ on the cylindrical surface and $V = 0$ on the planar surfaces.

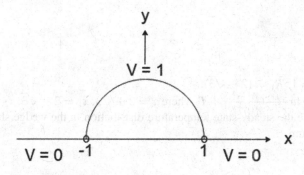

Ans. $V = \frac{2}{\pi} \tan^{-1} \frac{2y}{x^2+y^2-1}$.

16. Find the pair of points x_1, x_2 that are symmetric to both circles $\left|z - \frac{1}{4}\right| = \frac{1}{4}$ and $|z| = 1$.

Ans. $x_1 = 2 - \sqrt{3}$, $x_2 = 2 + \sqrt{3}$.

17. Show that $\frac{z-x_1}{z-x_2} = k$ are circles in z-plane. Show that if $k = 2 - \sqrt{3}$, then it is the same circle of $|z| = 1$.

18. Let $\frac{z-x_1}{z-x_2} = k_1$ represents the circle $\left|z - \frac{1}{4}\right| = \frac{1}{4}$, find the value of k_1.

Ans. $k_1 = \left(2 - \sqrt{3}\right)^2 = 7 - 4\sqrt{3} = 0.0718$

19. Show that $w = \frac{z-x_1}{z-x_2}$ is a conformal transformation, and $|w| = k$ are concentric circles. Suppose the circle $|z| = 1$ is mapped into $|w| = 1$, what is the transformation in this case?

Ans. $w = (2 + \sqrt{3})\frac{z-x_1}{z-x_2}$.

20. Under the transformation $w = (2 + \sqrt{3})\frac{z-x_1}{z-x_2}$, what is the image of the circle $\left|z - \frac{1}{4}\right| = \frac{1}{4}$?

Ans. $|w| = (2 - \sqrt{3}) = 0.268$.

21. Determine the temperature distribution in the region shown in the following figure. Take the inner cylinder to be centered at $x = 1/4$ and the radius to be $1/4$ and the radius of the outer cylinder to be 1. The temperature is kept at T_1 at the surface of inner cylinder and T_2 at the outer cylinder. The figure shows that two distinct circles are mapped into two concentric circles by the transformation $w = C(z - x_1)/(z - x_2)$ where $C = (2 + \sqrt{3})$ and $x_1 = 2 - \sqrt{3}$, $x_2 = 2 + \sqrt{3}$.

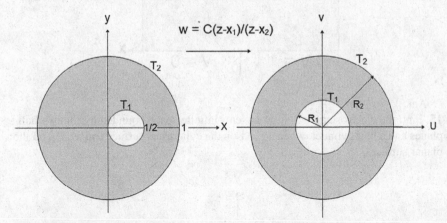

Ans. $R_2 = 1$, $R_1 = (2 - \sqrt{3})$.

$T = \frac{T_1 - T_2}{2\ln a} \ln \frac{(x-x_2)^2 + y^2}{a^2[(x-x_1)^2 + y^2]} + T_2$ where $a = 2 + \sqrt{3}$, $x_1 = 2 - \sqrt{3}$, $x_2 = 2 + \sqrt{3}$.

22. Determine the steady-state temperature distribution in the wedge shown in the following figure.

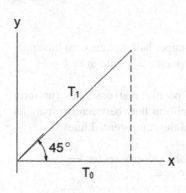

Exercise 22   Exercise 23

Ans. $T(x, y) = T_0 + \frac{4(T_1 - T_0)}{\pi} \tan^{-1} \frac{y}{x}$.

23. Determine the steady-state temperature distribution in the region shown in the above figure. The arc is insulated.

Ans. $T(x, y) = T_0 + \frac{2(T_1 - T_0)}{\pi} \tan^{-1} \frac{y}{x}$.

24. Find the bilinear transformation that maps $z_1 = -1$, $z_2 = 0$, $z_3 = 1$ onto $w_1 = -1$, $w_2 = -i$, $w_3 = 1$, respectively.

Ans. $w = \frac{z-i}{-iz+1}$.

25. Determine the bilinear transformation that maps $z_1 = 0$, $z_2 = 1$, $z_3 = \infty$ onto $w_1 = -1$, $w_2 = -i$, $w_3 = 1$, respectively.

Ans. $w = \frac{z-i}{-iz+1}$.

26. Find the bilinear transformation that maps $z_1 = -1$, $z_2 = i$, $z_3 = 1$ onto $w_1 = 0$, $w_2 = i$, $w_3 = \infty$, respectively.

Ans. $w = -\frac{z+1}{z-1}$.

27. In the Joukowski's profile, find the coordinates of the image point A', B', C', and D' of the tear-shaped image of Fig. 4.31.

Ans. A' (1,0), B' (0,0.496), C' (−1.492,0), D' (0,−0.496).

28. Find the coordinates of the image points A', B', C', and D' of the Joukowski's airfoil of Fig. 4.32.

Ans. A' (1,0), B' (0,0.574), C' (−1.113,0), D' (0,0.076).

29. In Fig. 4.39, the line joining the points $(\frac{1}{2x_0}, 0)$ and $(0, -\frac{1}{2y_0})$ is the image of the circle $|z - Re^{i\pi/4}| = R$ under the transformation of $w = \frac{1}{z}$. Show that the perpendicular distance between this line and the origin is equal to $\frac{1}{2R}$.

30. Show that $\arg[z - (1 + 2i)] = -\frac{\pi}{3}$ represents a line in the z plane. Find that line in terms of x and y.

Ans. $y + \sqrt{3}x - (2 + \sqrt{3}) = 0$.

31. Find the inverse of the bilinear transformation $w = \frac{az+b}{cz+d}$.

Ans. $z = \frac{-dw+b}{cw-a}$.

32. We have shown that a circle C_z that has z_1 and z_2 as a pair of symmetric points can be represented as $\left|\frac{z-z_1}{z-z_2}\right| = k$. Show that this equation implies that $\left|\frac{w-w_1}{w-w_2}\right| = k'$, indicating that w_1 and w_2 are a pair of symmetric points of C_w. Find k'.

Ans. $k' = k \left| \frac{cz_2+d}{cz_1+d} \right|$.

33. Find the bilinear transformation that maps the upper half z-plane onto the upper half w plane , the point $z = 0$ onto $w = 0$ and the point $z = i$ onto $w = 1 + i$.

Ans. $w = \frac{2z}{z+1}$.

34. Show that the complex potential $F(z) = Kz$ (K positive real) describes a uniform flow to the right, which can be interpreted as a uniform flow between two parallel lines. Find the velocity vector, the streamlines, and the equipotential lines."

Ans. $V = K$, $y = const.$ $x = const.$

35. Let the complex potential be $F(z) = iz$, Find the velocity vector V.

Ans. Parallel flow in the negative y-direction, $V = -i$

36. Let $F(z) = (1+i)z$, Find the velocity vector V.

Ans. Parallel flow in the direction of $y = -x$, $V = 1 - i$.

37. Let $F(z) = z^3$, Find the velocity vector V.

Ans. $V = 3(x^2 - y^2) - 6xyi$.

38. Find the streamlines of the flow corresponding to $F(z) = \frac{1}{z}$.

Ans. The streamlines are circles $x^2 + (y - k)^2 = k^2$.

39. Find the complex potential for steady flow past a cylinder of radius R if the fluid flow at infinite is uniform and if the velocity at infinity is v_0 making an angle α with the x-axis as shown in the following figure.

Ans. $F(z) = v_0(ze^{-i\alpha} + \frac{R^2}{z}e^{-i\alpha})$.

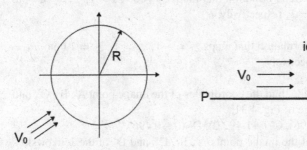

Exercise 39   Exercise 40

40. Use the Schwarz-Christoffel transformation to map the problem of flow in a river, onto the problem of flow in the upper half-plane if the river bed has a jump discontinuity as shown in the figure. Find the complex velocity if it is given that the velocity of the fluid tends to v_0 as $x \to -\infty$, and describe how the velocity varies near the discontinuity.

Ans. The complex potential is $F(w) = Az$ where $\frac{dw}{dz} = \frac{d}{\pi}\sqrt{\frac{z-1}{z+1}}$. The complex velocity is $V = v_0\sqrt{\frac{z+1}{z-1}}$.

Part II
Linear Algebra

Chapter 5
Linear Algebra and Vector Space

Ever since we learned how to solve simultaneous linear algebra equations in high school, we have encountered this topic in increasing frequency in our study of engineering and sciences. The elementary DC electrical circuit containing any number of resistors, batteries, and current loops produces such a set of equations on the unknown currents. Later we learned that many physical problems are stated in terms of differential equations. The solutions of these equations eventually lead to a system of linear algebraic equations.

The fundamental theory behind the manipulations we learned constitutes the study of linear algebra, one of the most sophisticated and beautiful fields of mathematics. The following chapters embark on a detailed analysis of determinants, matrices, and eigenvalue problems. They are tools of solving the system of linear equations. This chapter summarizes the main ideas of linear algebra and relates algebra to geometry. This leads directly to the concept of vector space.

One of the most valuable aspects of linear algebra is that concepts and results derived from two- or three-dimensional space can be extended to n-dimensional space in a straightforward way. For example, while it is difficult to see or even to imagine a 10-dimensional plane in a 11-dimensional space, in vector space, it simply means that the vector has 11 components.

As H. Poincare puts it, "mathematics is the art of giving same name to different things". Vector space puts linear algebra on an axiomatic basis. Anything that satisfies eight simple axioms can be regarded as a vector. Theoretical results developed in linear algebra are equally applicable to these vectors. Although linear algebra can be developed abstractly, it does have this connection with the geometric vector which is easier to grasp. We exploit this connection with concrete examples in two-dimensional vector space.

Defined axiomatically, we can even have a function space, such as Hilbert space, as part of vector space. Thus the fundamental assumption of quantum mechanics that the wave function is a vector in the Hilbert space becomes clear. Although we

K.-T. Tang, *Mathematical Methods for Engineers and Scientists 1*,
https://doi.org/10.1007/978-3-031-05678-9_5

are not able to go too deep in that direction, we want the readers to know that linear algebra is far more than just solving a system of equations.

5.1 Linear System of Equations

5.1.1 Cramer's Rule

Since solving a system of simultaneous equations is the central problem of linear algebra and since this is the place where we first encounter this subject, let us begin with solving the simplest system of equations, namely, two equations with two unknowns.

$$a_{11}x + a_{12}y = b_1 \tag{5.1}$$
$$a_{21}x + a_{22}y = b_2.$$

If we multiply the first of these equations by a_{22} and the second by a_{12} and then subtract, we get

$$(a_{11}a_{22} - a_{12}a_{21})x = a_{22}b_1 - a_{12}b_2$$

or

$$x = \frac{a_{22}b_1 - a_{12}b_2}{a_{11}a_{22} - a_{12}a_{21}}. \tag{5.2}$$

Similarly, if we multiply the first equation by a_{21} and the second by a_{11} and then subtract, we get

$$y = \frac{a_{21}b_1 - a_{11}b_2}{a_{12}a_{21} - a_{11}a_{22}} = \frac{a_{11}b_2 - a_{21}b_1}{a_{11}a_{22} - a_{12}a_{21}}. \tag{5.3}$$

Note that the denominators of both Eqs. (5.2) and (5.3) are the same. We represent it by

$$\begin{vmatrix} a_{11} & a_{12} \\ a_{21} & a_{22} \end{vmatrix} = a_{11}a_{22} - a_{12}a_{21}. \tag{5.4}$$

This quantity is known as a 2×2 determinant. The reason for introducing this notation is that it readily generalizes to n equations with n unknowns. An $n \times n$ determinant, called an nth-order determinant, is a square array of n^2 elements arranged in n rows and n columns. We will represent a determinant consisting of elements a_{ij} by $|A|$. Note that the element a_{ij} occurs in the ith row and jth column of $|A|$. In the next chapter, we will have a more detailed discussion of general determinants, including how to evaluate $|A|$ for $n \geq 3$.

The reader is probably familiar with this notation and know the following general properties of determinant. If not, they can be readily verified by its definition of Eq. (5.4).

1. The value of a determinant is unchanged if rows and columns are interchanged. That is

$$\begin{vmatrix} a_{11} & a_{12} \\ a_{21} & a_{22} \end{vmatrix} = \begin{vmatrix} a_{11} & a_{21} \\ a_{12} & a_{22} \end{vmatrix}.$$

2. If any two rows or any two columns are the same, the value of the determinant is zero.

3. If any two rows or any two columns are interchanged, the sign of the determinant is changed.

4. If every element in a single row or column is multiplied by a factor k, the value of the determinant is multiplied by k. For example,

$$\begin{vmatrix} 1 & 1 \times 2 \\ 3 & 2 \times 2 \end{vmatrix} = 2 \begin{vmatrix} 1 & 1 \\ 3 & 2 \end{vmatrix} = -2.$$

5. If any row or column is written as the sum or difference of two terms, the determinant can be written as the sum or difference of two determinants. For example,

$$\begin{vmatrix} 1 & 2 \\ 3 & 4 \end{vmatrix} = \begin{vmatrix} 3-2 & 2 \\ 4-1 & 4 \end{vmatrix} = \begin{vmatrix} 3 & 2 \\ 4 & 4 \end{vmatrix} + \begin{vmatrix} -2 & 2 \\ -1 & 4 \end{vmatrix} = 4 - 6 = -2.$$

6. The value of a determinant is unchanged if one row or column is added or subtracted to another. For example,

$$\begin{vmatrix} a_{11} + a_{12} & a_{12} \\ a_{21} + a_{22} & a_{22} \end{vmatrix} = \begin{vmatrix} a_{11} & a_{12} \\ a_{21} & a_{22} \end{vmatrix} + \begin{vmatrix} a_{12} & a_{12} \\ a_{22} & a_{22} \end{vmatrix} = \begin{vmatrix} a_{11} & a_{12} \\ a_{21} & a_{22} \end{vmatrix}.$$

Example 5.1.1 Show that $\begin{vmatrix} a_{11} + na_{12} & a_{12} \\ a_{21} + na_{22} & a_{22} \end{vmatrix} = \begin{vmatrix} a_{11} & a_{12} \\ a_{21} & a_{22} \end{vmatrix}.$

Solution 5.1.1

$$\begin{vmatrix} a_{11} + na_{12} & a_{12} \\ a_{21} + na_{22} & a_{22} \end{vmatrix} = \begin{vmatrix} a_{11} & a_{12} \\ a_{21} & a_{22} \end{vmatrix} + \begin{vmatrix} na_{12} & a_{12} \\ na_{22} & a_{22} \end{vmatrix}$$

$$= \begin{vmatrix} a_{11} & a_{12} \\ a_{21} & a_{22} \end{vmatrix} + n \begin{vmatrix} a_{12} & a_{12} \\ a_{22} & a_{22} \end{vmatrix} = \begin{vmatrix} a_{11} & a_{12} \\ a_{21} & a_{22} \end{vmatrix}.$$

Now let us consider the system in Equation 1 again,

$$a_{11}x + a_{12}y = b_1$$
$$a_{21}x + a_{22}y = b_2.$$

The determinant of the coefficients of x and y is

$$|A| = \begin{vmatrix} a_{11} & a_{12} \\ a_{21} & a_{22} \end{vmatrix}.$$

According to rule 4,

$$\begin{vmatrix} a_{11}x & a_{12} \\ a_{21}x & a_{22} \end{vmatrix} = x\,|A|,$$

according to rule 6,

$$\begin{vmatrix} a_{11}x + a_{12}y & a_{12} \\ a_{21}x + a_{22}y & a_{22} \end{vmatrix} = \begin{vmatrix} a_{11}x & a_{12} \\ a_{21}x & a_{22} \end{vmatrix} = x\,|A|.$$

Since $a_{11}x + a_{12}y = b_1$ and $a_{21}x + a_{22}y = b_2$, so

$$\begin{vmatrix} b_1 & a_{12} \\ b_2 & a_{22} \end{vmatrix} = x\,|A|.$$

It follows

$$x = \frac{\begin{vmatrix} b_1 & a_{12} \\ b_2 & a_{22} \end{vmatrix}}{\begin{vmatrix} a_{11} & a_{12} \\ a_{21} & a_{22} \end{vmatrix}}. \tag{5.5}$$

Similarly, we get

$$y = \frac{\begin{vmatrix} a_{11} & b_1 \\ a_{21} & b_2 \end{vmatrix}}{\begin{vmatrix} a_{11} & a_{12} \\ a_{21} & a_{22} \end{vmatrix}}. \tag{5.6}$$

Notice that Eqs. (5.5) and (5.6) are identical to Eqs. (5.2) and (5.3). Note that the determinant in the numerator is obtained by replacing the column in $|A|$ that is associated with the unknown quantity with the column associated with the right sides of the equation. This solution is called Cramer's rule. (Named after Swiss mathematician Gabriel Cramer (1704–1752).)

This method can be extended to n equations with n unknowns. Suppose we have

$$\begin{aligned}
a_{11}x_1 + a_{12}x_2 + \cdots + a_{1j}x_j + \cdots + a_{1n}x_n &= b_1 \\
a_{21}x_1 + a_{22}x_2 + \cdots + a_{2j}x_j + \cdots + a_{2n}x_n &= b_2 \\
\cdots\cdots\cdots\cdots\cdots\cdots\cdots\cdots\cdots\cdots &= \cdot \\
a_{i1}x_1 + a_{i2}x_2 + \cdots + a_{ij}x_j + \cdots + a_{in}x_n &= b_i \\
\cdots\cdots\cdots\cdots\cdots\cdots\cdots\cdots\cdots\cdots &= \cdot \\
a_{n1}x_1 + a_{n2}x_2 + \cdots + a_{nj}x_j + \cdots + a_{nn}x_n &= b_n.
\end{aligned} \tag{5.7}$$

In the next chapter, we will show that

$$x_j = \frac{D_j}{D},$$ (5.8)

where D is the coefficient determinant

$$D = \begin{vmatrix} a_{11} & a_{12} & \cdots & a_{1j} & \cdots & a_{1n} \\ a_{21} & a_{22} & \cdots & a_{2j} & \cdots & a_{2n} \\ \cdot & & \cdots & & \cdots & \cdot \\ a_{i1} & a_{i2} & \cdots & a_{ij} & \cdots & a_{in} \\ \cdot & \cdot & \cdots & \cdot & \cdots & \cdot \\ a_{n1} & a_{n2} & \cdots & a_{nj} & \cdots & a_{nn} \end{vmatrix}$$ (5.9)

just like the 2×2 system, and D_j is the determinant D with the jth column replaced by the column of the right side of the equation

$$D_j = \begin{vmatrix} a_{11} & a_{12} & \cdots & b_1 & \cdots & a_{1n} \\ a_{21} & a_{22} & \cdots & b_2 & \cdots & a_{2n} \\ \cdot & & \cdots & \cdot & \cdots & \cdot \\ a_{i1} & a_{i2} & \cdots & b_i & \cdots & a_{in} \\ \cdot & \cdot & \cdots & \cdot & \cdots & \cdot \\ a_{n1} & a_{n2} & \cdots & b_n & \cdots & a_{nn} \end{vmatrix}.$$ (5.10)

Clearly, this is valid only if $D \neq 0$. Cramer's rule is of theoretical interest and in low-dimensional engineering problems, but it is impractical in computation. In modern scientific computing, $n = 1000$ is only a moderate size. Solving 1000 equations with Cramer's rule would be a total disaster. A better way of handling such a situation is known as the Gaussian elimination.

5.1.2 Gaussian Elimination

Before we introduce the Gaussian elimination method, we must first explain the matrix notation.

A matrix is a rectangular array of numbers (or functions) enclosed in parenthesis. These numbers (or functions) are called entries or elements of the matrix. In the double-subscript notation for the entries, the first subscript always denotes the row and the second the column in which the given entry stands.

A row (column) vector is a matrix that has only one row (column). We often want to switch from one type of vector to the other. We can do this by transposition, which is indicated by the superscript T. For example,

$$\mathbf{a} = (5\ 0\ 7), \quad \mathbf{a}^T = \begin{pmatrix} 5 \\ 0 \\ 7 \end{pmatrix}.$$

A product of two matrices $\mathbf{C} = \mathbf{AB}$ is defined only if the number of rows of the second matrix is equal to the number of columns of the first matrix. In that case c_{ik} is defined as

$$c_{ik} = \sum_{l=1}^{n} a_{il}b_{lk} = a_{i1}b_{1k} + a_{i2}b_{2k} + \cdots + a_{in}b_{nk}.$$

If B is a vector, then $\mathbf{C}$ is also a vector that has only one column. Thus

$$c_i = \sum_{l=1}^{n} a_{il}b_l = a_{i1}b_1 + a_{i2}b_2 + \cdots + a_{in}b_n.$$

For example, Eq. (5.7) can be written as

$$\mathbf{Ax} = \mathbf{b},$$

where

$$\mathbf{A} = \begin{pmatrix} a_{11} & a_{12} & \cdot & a_{1j} & \cdot & a_{1n} \\ a_{21} & a_{22} & \cdot & a_{2j} & \cdot & a_{2n} \\ \cdot & \cdot & \cdot & \cdot & \cdot & \cdot \\ a_{i1} & a_{i2} & \cdot & a_{ij} & \cdot & a_{in} \\ \cdot & \cdot & \cdot & \cdot & \cdot & \cdot \\ a_{n1} & a_{n2} & \cdot & a_{nj} & \cdot & a_{nn} \end{pmatrix},$$

$$\mathbf{x} = \begin{pmatrix} x_1 \\ x_2 \\ \cdot \\ x_i \\ \cdot \\ x_n \end{pmatrix}, \qquad \mathbf{b} = \begin{pmatrix} b_1 \\ b_2 \\ \cdot \\ b_i \\ \cdot \\ b_n \end{pmatrix}.$$

The Gaussian elimination method is best illustrated by examples.

Example 5.1.2 Suppose we want to solve the following set of linear equations:

$$\begin{aligned} 3x_1 + 2x_2 - 2x_3 &= 5 \\ 2x_1 - x_2 + 3x_3 &= 5 \\ x_1 + x_2 - x_3 &= 2. \end{aligned} \qquad (5.11)$$

In terms of matrix notation, we have

$$\mathbf{Ax} = \mathbf{b}, \qquad (5.12)$$

where

$$\mathbf{A} = \begin{pmatrix} 3 & 2 & -2 \\ 2 & -1 & 3 \\ 1 & 1 & -1 \end{pmatrix}, \quad \mathbf{x} = \begin{pmatrix} x_1 \\ x_2 \\ x_3 \end{pmatrix}, \quad \mathbf{b} = \begin{pmatrix} 5 \\ 5 \\ 2 \end{pmatrix}.$$

To further simplify the writing, we use the so-called augmented matrix,

$$\mathbf{A}|\,\mathbf{b} = \begin{pmatrix} 3 & 2 & -2 & 5 \\ 2 & -1 & 3 & 5 \\ 1 & 1 & -1 & 2 \end{pmatrix}. \tag{5.13}$$

It is understood that this matrix contains all the information of Eq. (5.11). It is almost self-evident that we will create an equivalent system that has the same solution as the original system if we perform any of the following operations on this matrix:

1. interchange any pair of rows,
2. multiply any row by a non-zero constant,
3. replace any row with the sum of that row and a constant times another row.

Solution 5.1.2 Thus, the following augmented matrices are all equivalent:

$$\begin{pmatrix} 3 & 2 & -2 & 5 \\ 2 & -1 & 3 & 5 \\ 1 & 1 & -1 & 2 \end{pmatrix} \quad \begin{array}{c} \text{row 1 intact} \\ \rightarrow \text{row } 2 - \frac{2}{3} \times \text{row 1} \\ \text{row } 3 - \frac{1}{3} \times \text{row 1} \end{array} \quad \begin{pmatrix} 3 & 2 & -2 & 5 \\ 0 & -\frac{7}{3} & \frac{13}{3} & \frac{5}{3} \\ 0 & \frac{1}{3} & -\frac{1}{3} & \frac{1}{3} \end{pmatrix}$$

$$\rightarrow \quad \begin{array}{c} \text{row 1 intact} \\ \text{row 2 intact} \\ \text{row } 3 + \frac{1}{7} \times \text{row 2} \end{array} \quad \begin{pmatrix} 3 & 2 & -2 & 5 \\ 0 & -\frac{7}{3} & \frac{13}{3} & \frac{5}{3} \\ 0 & 0 & \frac{6}{21} & \frac{12}{21} \end{pmatrix}.$$

Therefore the original set of equations is equivalent to

$$3x_1 + 2x_2 - 2x_3 = 5$$
$$-\frac{7}{3}x_2 + \frac{13}{3}x_3 = \frac{5}{3}$$
$$\frac{6}{21}x_3 = \frac{12}{21}.$$

With back substitution from bottom to top, we have $x_3 = 2$, $x_2 = 3$, and $x_1 = 1$.

This method is known as Gauss elimination. Carl Friedrich Gauss (1777–1855) is a great mathematician. He made important contributions in almost all branches of mathematics. In 1801, he used this ancient method in his work on the orbit of the "planet" Ceres. Somehow his name got stuck with this method.

The system can be overdetermined (more equations than unknowns) or underdetermined (more unknowns than equations). In each case, Gaussian elimination leads us to the correct result. This is shown in the following examples.

Example 5.1.3 Solve the equations

$$x_1 + x_2 = 4$$
$$3x_1 - 4x_2 = 9$$
$$5x_1 - 2x_2 = 17.$$

Solution 5.1.3

$$\begin{pmatrix} 1 & 1 & 4 \\ 3 & -4 & 9 \\ 5 & -2 & 17 \end{pmatrix} \rightarrow \begin{matrix} \text{row 1 intact} \\ \text{row 2} - 3 \times \text{row 1} \\ \text{row 3} - 5 \times \text{row 1} \end{matrix} \begin{pmatrix} 1 & 1 & 4 \\ 0 & -7 & -3 \\ 0 & -7 & -3 \end{pmatrix}$$

$$\rightarrow \begin{matrix} \text{row 1 intact} \\ \text{row 2 intact} \\ \text{row 3} - \text{row 2} \end{matrix} \begin{pmatrix} 1 & 1 & 4 \\ 0 & -7 & -3 \\ 0 & 0 & 0 \end{pmatrix}.$$

The corresponding equations are

$$x_1 + x_2 = 4$$
$$-7x_2 = -3.$$

The solution is $x_2 = \frac{3}{7}$ and $x_1 = \frac{25}{7}$.

Example 5.1.4 Solve the equations

$$3x_1 - 2x_2 = 2$$
$$-6x_1 + 4x_2 = -4$$
$$-3x_1 + 2x_2 = 2.$$

Solution 5.1.4

$$\begin{pmatrix} 3 & -2 & 2 \\ -6 & 4 & -4 \\ -3 & 2 & 2 \end{pmatrix} \rightarrow \begin{matrix} \text{row 1 intact} \\ \text{row 2} + 2 \times \text{row 1} \\ \text{row 3} + \text{row 1} \end{matrix} \begin{pmatrix} 3 & -2 & 2 \\ 0 & 0 & 0 \\ 0 & 0 & 4 \end{pmatrix}$$

Since it is not possible for $0 = 4$, there is no solution.

Example 5.1.5 Solve the equations

$$x_1 + x_2 + x_3 + x_4 = 0$$
$$x_1 + 3x_2 + 2x_3 + 4x_4 = 0$$
$$2x_1 + x_3 - x_4 = 0.$$

Solution 5.1.5

$$\begin{pmatrix} 1 & 1 & 1 & 1 & | & 0 \\ 1 & 3 & 2 & 4 & | & 0 \\ 2 & 0 & 1 & -1 & | & 0 \end{pmatrix} \rightarrow \begin{matrix} \text{row 1 intact} \\ \text{row 2} - \text{row 1} \\ \text{row 3} - 2 \times \text{row 1} \end{matrix} \begin{pmatrix} 1 & 1 & 1 & 1 & | & 0 \\ 0 & 2 & 1 & 3 & | & 0 \\ 0 & -2 & -1 & -3 & | & 0 \end{pmatrix}$$

$$\rightarrow \begin{matrix} \text{row 1 intact} \\ \text{row 2 intact} \\ \text{row 3} + \text{row 2} \end{matrix} \begin{pmatrix} 1 & 1 & 1 & 1 & | & 0 \\ 0 & 2 & 1 & 3 & | & 0 \\ 0 & 0 & 0 & 0 & | & 0 \end{pmatrix}.$$

The corresponding equations are

$$x_1 + x_2 + x_3 + x_4 = 0$$
$$2x_2 + x_3 + 3x_4 = 0.$$

Solving for x_1 and x_2 in terms of x_3 and x_4, we have $x_2 = -\frac{1}{2}(x_3 + 3x_4)$ and $x_1 = (x_4 - x_3)/2$. Thus, there is an infinite number of solutions as long as x_1 and x_2 are given by these combinations.

Pivot The essence of the above examples is to keep the first equation intact, and to subtract a_{21}/a_{11} times the first equation from the second, and so on. The first equation is called the pivot equation, and a_{11} is called the pivot. Naturally we want $a_{11} \neq 0$. If it is zero, we have to interchange that equation with the one having a non-zero pivot. In practice, even a non-zero pivot should be rejected if it is too small. Since the smaller it is, the more susceptible is the calculation to the roundoff error as the following example shows.

Example 5.1.6 Solve the set of equations

$$0.0004x_1 + 1.402x_2 = 1.406$$
$$0.4003x_1 - 1.502x_2 = 2.501$$

with four-digit floating point arithmetic.

Solution 5.1.6 If we use the first equation as the pivot equation, then we have to multiply the equation by $0.4003/0.0004 = 1001$ and subtract it from the second equation, thus we

$$-1.502x_2 - 1001 \times 1.402x_2 = 2.501 - 1001 \times 1.406$$

or

$$-1405x_2 = -1404.$$

Thus

$$x_2 = \frac{-1404}{-1405} = 0.9993.$$

From the first equation, we have

$$x_1 = \frac{1.406 - 1.402x_2}{0.0004} = \frac{0.0050}{0.0004} = 12.5,$$

which is wrong. On the other hand, if we use the second equation as the pivot equation, and multiply that equation by $0.0004/0.4003 = 0.0010$ and subtract it from the first equation, we get

$$1.4035x_2 = 1.4035,$$

so $x_2 = 1$, and from the pivot equation, we have

$$0.4003x_1 = 2.501 + 1.502 = 4.003$$

or

$$x_1 = 10.0$$

which is the correct solution.

5.1.3 LU Decomposition

In solving the $n \times n$ matrix equation

$$Ax = b,$$

there are methods to decompose

$$A = LU,$$

where L is a lower triangular matrix and U is an upper triangular matrix. Once this is done, the original equation becomes

$$Ly = b, \quad Ux = y.$$

One may easily solve these equations with back substitution. There are several different decompositions. They are best illustrated by examples.

Doolittle's Method.

Example 5.1.7 Solve for **x**

$$\mathbf{Ax} = \begin{pmatrix} 2 & 4 & 3 \\ 1 & -2 & -2 \\ -3 & 3 & 2 \end{pmatrix} \begin{pmatrix} x_1 \\ x_2 \\ x_3 \end{pmatrix} = \begin{pmatrix} 4 \\ 0 \\ -7 \end{pmatrix} = \mathbf{b}.$$

Solution 5.1.7 Let

$$\mathbf{A} = \mathbf{LU}$$

and

$$\mathbf{L} = \begin{pmatrix} 1 & 0 & 0 \\ l_{21} & 1 & 0 \\ l_{31} & l_{32} & 1 \end{pmatrix}, \quad \mathbf{U} = \begin{pmatrix} u_{11} & u_{12} & u_{13} \\ 0 & u_{22} & u_{23} \\ 0 & 0 & u_{33} \end{pmatrix}.$$

$$\mathbf{Ax} = \mathbf{LUx} = \mathbf{b}$$

means

$$\mathbf{Ux} = \mathbf{y} \text{ and } \mathbf{Ly} = \mathbf{b}.$$

$$\mathbf{LU} = \begin{pmatrix} u_{11} & u_{12} & u_{13} \\ l_{21}u_{11} & l_{21}u_{12} + u_{22} & l_{21}u_{13} + u_{23} \\ l_{31}u_{11} & l_{31}u_{12} + l_{32}u_{22} & l_{31}u_{13} + l_{32}u_{23} + u_{33} \end{pmatrix} = \begin{pmatrix} 2 & 4 & 3 \\ 1 & -2 & -2 \\ -3 & 3 & 2 \end{pmatrix} = \mathbf{A}.$$

Clearly

$$u_{11} = 2 \qquad u_{12} = 4 \qquad u_{13} = 3$$
$$l_{21} = \tfrac{1}{2} \qquad u_{22} = -2 - \tfrac{1}{2}4 = -4 \qquad u_{23} = -2 - \tfrac{1}{2}3 = -\tfrac{7}{2}$$
$$l_{31} = -\tfrac{3}{2} \quad l_{32} = (3 - (-\tfrac{3}{2})4)/(-4) = -\tfrac{9}{4} \quad u_{33} = 2 - (-\tfrac{3}{2}3 - \tfrac{9}{4}(-\tfrac{7}{2})) = -\tfrac{11}{8}$$

$$\mathbf{Ly} = \begin{pmatrix} 1 & 0 & 0 \\ \tfrac{1}{2} & 1 & 0 \\ -\tfrac{3}{2} & -\tfrac{9}{4} & 1 \end{pmatrix} \begin{pmatrix} y_1 \\ y_2 \\ y_3 \end{pmatrix} = \begin{pmatrix} 4 \\ 0 \\ -7 \end{pmatrix} = \mathbf{b},$$

so

$$\begin{pmatrix} y_1 \\ \tfrac{1}{2}y_1 + y_2 \\ -\tfrac{3}{2}y_1 - \tfrac{9}{4}y_2 + y_3 \end{pmatrix} = \begin{pmatrix} 4 \\ 0 \\ -7 \end{pmatrix},$$

thus

$$y_1 = 4, \quad y_2 = -2, \quad y_3 = -\frac{11}{2}.$$

$$\mathbf{Ux} = \begin{pmatrix} 2 & 4 & 3 \\ 0 & -4 & -\frac{7}{2} \\ 0 & 0 & -\frac{11}{8} \end{pmatrix} \begin{pmatrix} x_1 \\ x_2 \\ x_3 \end{pmatrix} = \begin{pmatrix} 4 \\ -2 \\ -\frac{11}{2} \end{pmatrix} = \mathbf{y}.$$

Therefore

$$x_3 = 4, \quad x_2 = -3, \quad x_1 = 2.$$

Crout's Method. A similar method, known as Crout's method, is obtained if U (instead of L) is required to have 1 in its diagonal.

If A is symmetric, we can do the LU decomposition by what is known as Cholesky's method.

Cholesky's Method.

Example 5.1.8 Solve by Cholesky's method:

$$\mathbf{Ax} = \begin{pmatrix} 4 & 2 & 14 \\ 2 & 17 & -5 \\ 14 & -5 & 83 \end{pmatrix} \begin{pmatrix} x_1 \\ x_2 \\ x_3 \end{pmatrix} = \begin{pmatrix} 14 \\ -101 \\ 155 \end{pmatrix} = \mathbf{b}.$$

Solution 5.1.8 Let

$$\mathbf{A} = \mathbf{LU}$$

and

$$\mathbf{L} = \begin{pmatrix} c_{11} & 0 & 0 \\ c_{21} & c_{22} & 0 \\ c_{31} & c_{32} & c_{33} \end{pmatrix}, \quad \mathbf{U} = \begin{pmatrix} c_{11} & c_{21} & c_{31} \\ 0 & c_{22} & c_{32} \\ 0 & 0 & c_{33} \end{pmatrix}.$$

$$\mathbf{LU} = \begin{pmatrix} (c_{11})^2 & c_{11}c_{21} & c_{11}c_{31} \\ c_{21}c_{11} & (c_{21})^2 + (c_{22})^2 & c_{21}c_{31} + c_{22}c_{32} \\ c_{31}c_{11} & c_{31}c_{21} + c_{32}c_{22} & (c_{31})^2 + (c_{32})^2 + (c_{33})^2 \end{pmatrix} = \begin{pmatrix} 4 & 2 & 14 \\ 2 & 17 & -5 \\ 14 & -5 & 83 \end{pmatrix} = \mathbf{A}.$$

$$c_{11} = \sqrt{4} = 2$$
$$c_{21} = \frac{2}{2} = 1 \qquad c_{22} = \sqrt{17 - 1} = 4$$
$$c_{31} = \frac{14}{2} = 7 \quad c_{32} = \frac{1}{4}(-5 - 7) = -3 \quad c_{33} = \sqrt{83 - 49 - 9} = 5$$

$$\mathbf{L} = \begin{pmatrix} 2 & 0 & 0 \\ 1 & 4 & 0 \\ 7 & -3 & 5 \end{pmatrix}, \quad \mathbf{U} = \begin{pmatrix} 2 & 1 & 7 \\ 0 & 4 & -3 \\ 0 & 0 & 5 \end{pmatrix}.$$

$$\mathbf{Ly} = \begin{pmatrix} 2 & 0 & 0 \\ 1 & 4 & 0 \\ 7 & -3 & 5 \end{pmatrix} \begin{pmatrix} y_1 \\ y_2 \\ y_3 \end{pmatrix} = \begin{pmatrix} 14 \\ -101 \\ 155 \end{pmatrix} = \mathbf{b},$$

$$\begin{pmatrix} y_1 \\ y_2 \\ y_3 \end{pmatrix} = \begin{pmatrix} 7 \\ -27 \\ 5 \end{pmatrix}$$

$$\mathbf{Ux} = \mathbf{y}$$

$$\mathbf{Ux} = \begin{pmatrix} 2 & 1 & 7 \\ 0 & 4 & -3 \\ 0 & 0 & 5 \end{pmatrix} \begin{pmatrix} x_1 \\ x_2 \\ x_3 \end{pmatrix} = \begin{pmatrix} 7 \\ -27 \\ 5 \end{pmatrix} = \mathbf{y}.$$

Therefore $x_3 = 1$, $x_2 = -6$, $x_1 = 3$.

5.1.4 Geometry and Linear Equations

To understand the relationship between the geometry and a set of linear equations, let us return to the simplest system of two equations with two unknowns. For example, we want to solve the following set of equations

$$2x + y = 3$$
$$x - 3y = -2.$$

We can look at it from several perspectives. They should, of course, lead to the same conclusion.

1. Determinant:

$$x = \frac{\begin{vmatrix} 3 & 1 \\ -2 & -3 \end{vmatrix}}{\begin{vmatrix} 2 & 1 \\ 1 & -3 \end{vmatrix}} = \frac{-9 + 2}{-6 - 1} = 1,$$

$$y = \frac{\begin{vmatrix} 2 & 3 \\ 1 & -2 \end{vmatrix}}{\begin{vmatrix} 2 & 1 \\ 1 & -3 \end{vmatrix}} = \frac{-4 - 3}{-6 - 1} = 1.$$

Therefore the solution is $(1, 1)$.

2. Elimination:

$$\left(\begin{array}{cc|c} 2 & 1 & 3 \\ 1 & -3 & -2 \end{array} \right) \xrightarrow[\text{row 2} - \frac{1}{2}\text{ row 1}]{\text{row 1 intact}} \left(\begin{array}{cc|c} 2 & 1 & 3 \\ 0 & -\frac{7}{2} & -\frac{7}{2} \end{array} \right).$$

This leads to $y = 1$ and $x = 1$.

3. Row picture:

Figure 5.1 shows that each row represents a straight line. The line $2x + y = 3$ and the line $x - 3x = -2$ intersect at $x = 1$ and $y = 1$.

Fig. 5.1 The straight line
representing the first row and
the straight line representing
the second row intersect at
(1, 1)

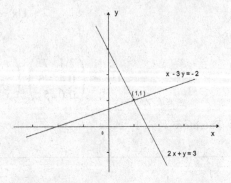

4. Column picture:

We can regard each column as a vector, so we can write this set of equations as

$$\begin{pmatrix} 2 \\ 1 \end{pmatrix} x + \begin{pmatrix} 1 \\ -3 \end{pmatrix} y = \begin{pmatrix} 3 \\ -2 \end{pmatrix}.$$

In other words,

$$\mathbf{v}_1 = \begin{pmatrix} 2 \\ 1 \end{pmatrix}, \quad \mathbf{v}_2 = \begin{pmatrix} 1 \\ -3 \end{pmatrix}, \quad \mathbf{b} = \begin{pmatrix} 3 \\ -2 \end{pmatrix}$$

and

$$x\mathbf{v}_1 + y\mathbf{v}_2 = \mathbf{b},$$

where x and y are the coefficients of this vector equation. In Fig. 5.2, we see that

$$1\mathbf{v}_1 + 1\mathbf{v}_2 = \mathbf{b}.$$

Therefore $x = 1$ and $y = 1$.

Fig. 5.2 Column picture of
a set of linear equations:
$1\begin{pmatrix} 2 \\ 1 \end{pmatrix} + 1\begin{pmatrix} 1 \\ -3 \end{pmatrix} = \begin{pmatrix} 3 \\ -2 \end{pmatrix}$

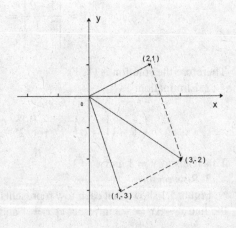

In x, y-coordinates, an equation with two unknowns represents a line, in x, y, z,-coordinates, the equation with three unknowns represents a plan. If the three-equation set has a unique solution, then in row picture the three plans intersect at a point. This is already difficult to draw, but we can at least imagine in this case. For $n > 3$, it is even difficult for us to imagine. For example, we cannot tell what a 10-dimensional plan is in a 11-dimensional space. On the other hand, in column picture, it simply means that in the n-dimensional space, the vectors have n components. This leads to the concept of vector space.

5.2   Vector Space

So far we have reviewed some computational aspects of linear algebra, the details of which will be further explained in the following three chapters. Now we want to introduce the geometrical aspect of linear algebra. The conceptual framework of this approach is the vector space.

The first question is "what is a vector?". As we have seen in the column picture, a vector can be regarded as a column of components, in two dimensions, it is $\begin{pmatrix} x \\ y \end{pmatrix}$; in three dimensions, it is $\begin{pmatrix} x \\ y \\ z \end{pmatrix}$; in n dimensions, it is a column of n components. They are equally represented by the row matrix, such as $\begin{pmatrix} x & y \end{pmatrix}$, $\begin{pmatrix} x & y & z \end{pmatrix}$ and $\begin{pmatrix} x_1 & x_2 & \cdots & x_n \end{pmatrix}$. So a computer scientist could regard a vector as an ordered list of numbers, such as one's height and weight, $\begin{pmatrix} heigt \\ weight \end{pmatrix}$. On the other hand, we learned in elementary physics that certain quantities, such as force and velocity, have both magnitude and direction. Often they are represented by a directed line segment. The length of the segment is proportional to the magnitude of the vector quantity and the direction of the vector is indicated by an arrowhead at one end of the segment, which is known as the tip of the vector. The other end is called the tail. As long as parallel to itself, this vector can be moved to anywhere in the space. We call it a free vector. A structure engineer analyzing the stress and strain of a beam knows that the forces acting on the beam cannot be moved. Vectors representing this kind of force are known as bound vectors.

5.2.1   Definition of Vector Space

The goal of mathematics is to find an axiomatic system that can be generalized to cover all these cases and more. That is where the axioms of the vector space come in. A vector space V consists of a set of objects called vectors (with notations such as **u** or **v**) and a set of rules of operations (called axioms) for the vectors. Generally we only have to remember that a vector space is closed under additivity and scaling, by which we mean

(A) If $\mathbf{u}$ and $\mathbf{v}$ are in that space, then $\mathbf{u} + \mathbf{v}$ must also be in that space.

(B) If $\mathbf{u}$ is in that space then $\alpha\mathbf{u}$ must also be in that space, where α can be any real and/or complex numbers.

To be more precise, these rules can be subdivided as follows:

Type 1: Vector addition: The sum of two vectors $\mathbf{u}$ and $\mathbf{v}$, denoted as $\mathbf{u} + \mathbf{v}$, is also a vector in V (closure rule). The sum satisfies axioms from (i) to (iv) as given below:

(i) Commutative rule: $\mathbf{u} + \mathbf{v} = \mathbf{v} + \mathbf{u}$.

(ii) Associative rule: $(\mathbf{u} + \mathbf{v}) + \mathbf{w} = \mathbf{u} + (\mathbf{v} + \mathbf{w})$.

Hence without ambiguity the sums above can be simply written as $\mathbf{u} + \mathbf{v} + \mathbf{w}$.

(iii) Existence of the zero vector $\mathbf{o}$ in V such that

$$\mathbf{u} + \mathbf{o} = \mathbf{u} \text{ (hence} = \mathbf{o} + \mathbf{u} \text{ by (i).)}$$

(iv) Existence of the (additional) inverse vector: For every vector $\mathbf{u}$,, there is a vector $-\mathbf{u}$ such that

$$\mathbf{u} + (-\mathbf{u}) = \mathbf{o} \text{ (hence } = -\mathbf{u} + \mathbf{u}).$$

Type 2: Multiplication by scalars: Let α, β be scalars (for us, a scalar is simply a complex number) multiplication of a vector by a scalar results in another vector in V (closure rule for multiplication by a scalar). Also the following axioms hold for multiplication by scalars.

(v)   Distribution rule for vector sums: $\alpha(\mathbf{u} + \mathbf{v}) = \alpha\mathbf{u} + \alpha\mathbf{v}$.

(vi) Distribution rule for scalar sums: $(\alpha + \beta)\mathbf{u} = \alpha\mathbf{u} + \beta\mathbf{u}$.

(vii) Associative rule: $\alpha(\beta\mathbf{u}) = (\alpha\beta)\mathbf{u}$.

(viii) Multiplication by the scalar "1": $1\mathbf{u} = \mathbf{u}$.

Subspace. A subspace of a vector space is a nonempty subset that satisfies the requirements for a vector space.

For example, consider all vectors whose components are positive or zero. This subset is the first quadrant of the x-y-plane; the coordinates satisfy $x \geq 0$ and $y \geq 0$. We might think that the first quadrant is a subspace because it contains zero and the sum of two vectors is a vector in this quadrant. But it is not a subspace because rule (B) is violated, since if the scalar $\alpha = -1$ and the vector $\mathbf{v}$ is $(1, 1)$, the multiple $\alpha\mathbf{v} = (-1, -1)$ is in the third quadrant instead of the first.

If we include the third quadrant along with the first, this will take care of the scaling since every multiple $\alpha\mathbf{v}$ will stay in this subset. However, rule (A) is now violated, since adding $(1, 2) + (-2, -1) = (-1, 1)$, which is not in either quadrant. The smallest subspace containing the first quadrant is the whole two-dimensional space.

5.2.2 Dot Product and Length of a Vector

We are familiar with the dot product in elementary physics and calculus. Given two vectors $\mathbf{u} = (u_1, u_2, u_3)$ and $\mathbf{v} = (v_1, v_2, v_3)$, their dot product is $\mathbf{u} \cdot \mathbf{v} = \sum_i^3 u_i v_i$.

We will generalize this definition to others in the vector spaces. What we'll do is to figure out the salient features of the definition above, and canonize them as the axioms of the scalar product in vector spaces.

The dot product (also known as scalar product or inner product) can be written as

$$
\begin{pmatrix} u_1 \\ u_2 \\ u_3 \end{pmatrix} \cdot \begin{pmatrix} v_1 \\ v_2 \\ v_3 \end{pmatrix} = u_1 v_1 + u_2 v_2 + u_3 v_3
$$

or

$$
\begin{pmatrix} u_1 & u_2 & u_3 \end{pmatrix} \begin{pmatrix} v_1 \\ v_2 \\ v_3 \end{pmatrix} = u_1 v_1 + u_2 v_2 + u_3 v_3.
$$

In two-dimensional space, the components are shown in Fig. 5.3. Therefore

$$
\mathbf{u} \cdot \mathbf{v} = u_x v_x + u_y v_y = u \cos \theta_1 v \cos \theta_2 + u \sin \theta_1 v \sin \theta_2
$$
$$
= uv(\cos \theta_1 \cos \theta_2 + \sin \theta_1 \sin \theta_2) = uv \cos(\theta_1 - \theta_2).
$$

As shown in the figure $\theta_1 - \theta_2 = \gamma$, which is the angle between $\mathbf{u}$ and $\mathbf{v}$. Thus

$$
\mathbf{u} \cdot \mathbf{v} = uv \cos \Gamma. \tag{5.14}
$$

We turn around and use this relation as the definition of the dot product. In vector space, distance and angle are sometimes defined in terms of dot (or inner) products.

Fig. 5.3 Components of vectors

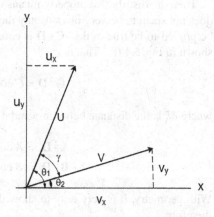

Fig. 5.4 The dot product
$(A + B) \cdot D = C \cdot D$

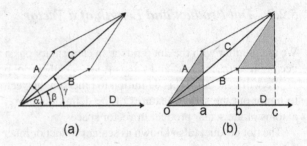

For example, the length of a vector (called norm) is defined as

$$\|\mathbf{u}\| = \sqrt{\mathbf{u} \cdot \mathbf{u}}$$

and the angle between two vectors as

$$\cos \gamma = \frac{\mathbf{u} \cdot \mathbf{v}}{uv},$$

or the dot product of two unit vectors pointing along **u** and **v**. The dot product so defined has certain properties and we elevate them to axioms that are to be satisfied by any inner products of vector space.

So we have the following axioms of dot product:

$$
\begin{aligned}
&\textit{Commutative} &&\mathbf{U} \cdot \mathbf{V} = \mathbf{V} \cdot \mathbf{U} \\
&\textit{Non negative} &&\mathbf{U} \cdot \mathbf{U} > 0 \text{ for } \mathbf{U} \neq 0 \\
& &&\mathbf{U} \cdot \mathbf{U} = 0 \text{ for } \mathbf{U} = 0 \\
&\textit{Distributive} &&(\mathbf{A} + \mathbf{B}) \cdot \mathbf{D} = \mathbf{A} \cdot \mathbf{D} + \mathbf{B} \cdot \mathbf{D}.
\end{aligned}
$$

They all seem to be self-evident except when we realize that $(\mathbf{A} + \mathbf{B}) \cdot \mathbf{D} = \mathbf{C} \cdot \mathbf{D}$ as shown in (a) of Fig. 5.4.

Then the distributive property means that $CD \cos \gamma = AD \cos \alpha + BD \cos \beta$. It does not seem to be very obvious any more. However, the distributive property can be proved to be true in that $\mathbf{C} \cdot \mathbf{D}$ is equal to the projection of $\mathbf{C}$ on $\mathbf{D}$ times D as shown in Fig. 5.4 (b). That is

$$\mathbf{C} \cdot \mathbf{D} = C \cos \gamma D = \overline{oc} D,$$

where $\overline{oc}$ is the distance between o and c in the figure. Similarly,

$$
\begin{aligned}
\mathbf{A} \cdot \mathbf{D} &= A \cos \alpha D = \overline{oa} D, \\
\mathbf{B} \cdot \mathbf{D} &= B \cos \beta D = \overline{ob} D.
\end{aligned}
$$

With geometry, it is very easy to show that the two shaded triangles are identical, therefore

Fig. 5.5 Law of Cosine:
$C^2 = A^2 + B^2 - 2AB \cos \gamma$

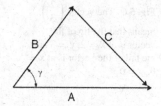

$$\overline{oa} = \overline{bc}.$$

Since

$$\overline{oc} = \overline{ob} + \overline{bc} = \overline{ob} + \overline{oa}.$$

Thus

$$(\mathbf{A} + \mathbf{B}) \cdot \mathbf{D} = \mathbf{C} \cdot \mathbf{D} = \overline{oc} D = \left(\overline{oa} + \overline{ob} \right) D = \mathbf{A} \cdot \mathbf{D} + \mathbf{B} \cdot \mathbf{D}.$$

With the distributive axiom, the law of cosine seems to be a trivial consequence. Let the three sides of the triangle be $\mathbf{A}, \mathbf{B}, \mathbf{C}$ and the angle between $\mathbf{A}$ and $\mathbf{B}$ be γ, as shown in Fig. 5.5. Let

$$\mathbf{C} = \mathbf{A} - \mathbf{B}.$$

Taking the dot product of both sides

$$\mathbf{C} \cdot \mathbf{C} = (\mathbf{A} - \mathbf{B}) \cdot (\mathbf{A} - \mathbf{B},)$$

we have

$$\mathbf{C} \cdot \mathbf{C} = \mathbf{A} \cdot \mathbf{A} - \mathbf{A} \cdot \mathbf{B} - \mathbf{B} \cdot \mathbf{A} + \mathbf{B} \cdot \mathbf{B}.$$

Thus

$$C^2 = A^2 + B^2 - 2AB \cos \gamma.$$

This is the well-known law of cosine.

5.2.3 Essence of Finite Dimensional Vector Space

Linear algebra can be developed abstractly with these axioms. However, to have a "feeling of understanding", it is helpful to have some concrete examples in mind. Therefore, in this section, we will first mostly use two-dimensional vectors, which can be easily shown on a paper, to describe an overall picture of linear algebra. Then we will show that the developed concepts can be applied to any objects that satisfy these axioms.

Fig. 5.6 The vector $\begin{pmatrix} 2 \\ 3 \end{pmatrix}$ means that the tip of the vector is at x = 2, y = 3 and the tail of the vector is at x = 0, y = 0

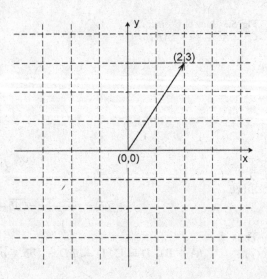

Thus whenever we encounter the word vector, we should have in our mind a two-dimensional vector that is **rooted at the origin**. This vector is written as $\begin{pmatrix} x \\ y \end{pmatrix}$, which means that the coordinates of the tip of the vector are (x, y) and the tail of the vector is at (0, 0). In other words, this vector is shown in Fig. 5.6.

Linear Dependence and Linear Independence. Let's begin with something familiar: On the xy-plane, each point (x, y) is treated as a vector $\mathbf{u} = \begin{pmatrix} x \\ y \end{pmatrix}$. Now consider these vectors

$$\mathbf{u}_1 = \begin{pmatrix} 1 \\ 2 \end{pmatrix}, \quad \mathbf{u}_2 = \begin{pmatrix} -2 \\ 1 \end{pmatrix}, \quad \mathbf{u}_3 = \begin{pmatrix} 5 \\ 6 \end{pmatrix}. \tag{5.15}$$

We ask: Is it possible to find three numbers (or scalars) $\alpha_1, \alpha_2, \alpha_3$, not all of them equal to zero, so that

$$\alpha_1 \mathbf{u}_1 + \alpha_2 \mathbf{u}_2 + \alpha_3 \mathbf{u}_3 = \mathbf{o} ? \tag{5.16}$$

or

$$\begin{pmatrix} \alpha_1 \\ 2\alpha_1 \end{pmatrix} + \begin{pmatrix} -2\alpha_2 \\ \alpha_2 \end{pmatrix} + \begin{pmatrix} 5\alpha_3 \\ 6\alpha_3 \end{pmatrix} = \begin{pmatrix} 0 \\ 0 \end{pmatrix} ?$$

In other words, is it possible that

$$\alpha_1 - 2\alpha_2 + 5\alpha_3 = 0,$$
$$2\alpha_1 + \alpha_2 + 6\alpha_3 = 0?$$

By Gaussian elimination, we have

Fig. 5.7 The unit vector **i** is along the x-axis and the unit vector **j** is along the y-axis. Both are rooted at the origin. They are the bases of the xy-space. The following equation shows that a vector can be regarded as a combination of its bases.
$$\begin{pmatrix} 2 \\ 3 \end{pmatrix} = 2 \begin{pmatrix} 1 \\ 0 \end{pmatrix} + 3 \begin{pmatrix} 0 \\ 1 \end{pmatrix} =$$
$2\mathbf{i} + 3\mathbf{j}$

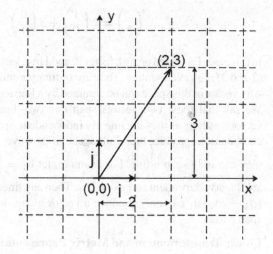

$$\begin{pmatrix} 1 & -2 & 5 & \cdot & 0 \\ 2 & 1 & 6 & \cdot & 0 \end{pmatrix} \rightarrow \begin{pmatrix} 1 & -2 & 5 & \cdot & 0 \\ 0 & 5 & -4 & \cdot & 0 \end{pmatrix}.$$

So

$$\alpha_2 = \frac{4}{5}\alpha_3, \quad \alpha_1 = -\frac{17}{5}\alpha_3,$$

where α_3 is an arbitrary number. Therefore the answer is yes, it is possible. We have found the solution. In this case, we say that vectors $\mathbf{u}_1$, $\mathbf{u}_2$, $\mathbf{u}_3$ are linearly dependent. Why is the name? The reason is this: since not all $\alpha's$ are zero, lets' assume $\alpha_1 \neq 0$, then Eq. (5.16) yields

$$\mathbf{u}_1 = -\frac{\alpha_2}{\alpha_1}\mathbf{u}_2 - \frac{\alpha_3}{\alpha_1}\mathbf{u}_3,$$

i.e., $\mathbf{u}_1$ is a linear combination of (therefore depends on) $\mathbf{u}_2$ and $\mathbf{u}_3$.

Now we are ready to give a definition of linear independence:

A set of vectors $\mathbf{v}_1, \mathbf{v}_2, \dots \mathbf{v}_n$ are said to be linearly independent if

$$\alpha_1\mathbf{v}_1 + \alpha_2\mathbf{v}_2 + \cdots + \alpha_n\mathbf{v}_3 = \mathbf{0} \qquad (5.17)$$

implies that

$$\alpha_1 = \alpha_2 = \cdots = \alpha_n = 0. \qquad (5.18)$$

Basis and Span. For the two-dimensional space we introduce the coordinate system for reference. Let **i** be a unit vector along the x-axis and **j** a unit vector along the y-axis. Both unit vectors are rooted at the origin as shown in Fig. 5.7. They are the bases of this two-dimensional vector space. Any vector can be written as a combination of these bases. The picture shows a particular case

$$\begin{pmatrix} 2 \\ 3 \end{pmatrix} = 2\begin{pmatrix} 1 \\ 0 \end{pmatrix} + 3\begin{pmatrix} 0 \\ 1 \end{pmatrix} = 2\mathbf{i} + 3\mathbf{j}.$$

In this case, $\mathbf{i}$ is scaled up by a factor 2, and $\mathbf{j}$ by a factor 3. Therefore, these coefficients (2 and 3) are called scalars. They are ordinary numbers and can vary from $-\infty$ to ∞. Any vector in this space can be obtained by a linear combination of these two vectors. We say that these two vectors span this two-dimensional space. In fact any two vectors, as long as they are linearly independent, span this space, because their linear combinations produce the whole space. If, however, they are linearly dependent, then they can only span a line. For example, let $\mathbf{u}_1 = \begin{pmatrix} 1 \\ 2 \end{pmatrix}$, $\mathbf{u}_2 = \begin{pmatrix} 2 \\ 4 \end{pmatrix}$. Clearly they are linearly dependent and $\mathbf{u}_2 = 2\mathbf{u}_1$. Then any linearly combination $\alpha_1 \mathbf{u}_1 + \alpha_2 \mathbf{u}_2 = (\alpha_1 + 2\alpha_2)\mathbf{u}_1$ can only produce a vector along the line of $\mathbf{u}_1$. Therefore they only span a line.

Linear Transformation and Matrix Representation. A transformation in the context of linear algebra is a function which takes an input vector and gives an output vector.

$$T(\mathbf{u}_{in}) = \mathbf{u}_{out}.$$

It is a linear transformation simply means

$$T(a\mathbf{u} + b\mathbf{v}) = aT(\mathbf{u}) + bT(\mathbf{v}). \tag{5.19}$$

Because of these rules, to find the transformation of a vector, we only need to find what it does to the basis vectors. For example, let

$$\mathbf{v} = x\mathbf{i} + y\mathbf{j},$$

so

$$T(\mathbf{v}) = xT(\mathbf{i}) + yT(\mathbf{j}).$$

Suppose the transformation is a rotation of 90 degrees, then

$$T(\mathbf{i}) = \begin{pmatrix} 0 \\ 1 \end{pmatrix}, \quad T(\mathbf{j}) = \begin{pmatrix} -1 \\ 0 \end{pmatrix}.$$

Therefore

$$T(\mathbf{v}) = x\begin{pmatrix} 0 \\ 1 \end{pmatrix} + y\begin{pmatrix} -1 \\ 0 \end{pmatrix} = \begin{pmatrix} 0 & -1 \\ 1 & 0 \end{pmatrix}\begin{pmatrix} x \\ y \end{pmatrix}.$$

Thus this operation of turning 90 degrees is represented by a matrix. The first column of this matrix is where the coordinates of the first basis $\mathbf{i}$ after the turn, and the second column of this matrix is where the second basis $\mathbf{j}$ lands. Let T represent the transformation. For the specific vector $\mathbf{v} = 2\mathbf{i} + 3\mathbf{j}$, $T(\mathbf{v}) = 2T(\mathbf{i}) + 3T(\mathbf{j}) = 2\mathbf{i}' +$

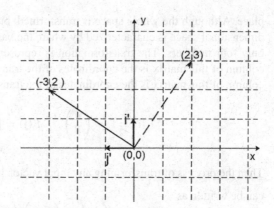

Fig. 5.8 The whole space is turned 90 degrees. The basis vector **i** lands at $\begin{pmatrix} 0 \\ 1 \end{pmatrix}$ and **j** becomes $\begin{pmatrix} -1 \\ 0 \end{pmatrix}$. The vector $\mathbf{v} = 2\mathbf{i} + 3\mathbf{j}$ turns to $2\begin{pmatrix} 0 \\ 1 \end{pmatrix}$ $+3\begin{pmatrix} -1 \\ 0 \end{pmatrix} =$ $\begin{pmatrix} 0 & -1 \\ 1 & 0 \end{pmatrix}\begin{pmatrix} 2 \\ 3 \end{pmatrix} = \begin{pmatrix} -3 \\ 2 \end{pmatrix}$

$3\mathbf{j}' = 2\begin{pmatrix} 0 \\ 1 \end{pmatrix} + 3\begin{pmatrix} -1 \\ 0 \end{pmatrix} = \begin{pmatrix} 0 & -1 \\ 1 & 0 \end{pmatrix}\begin{pmatrix} 2 \\ 3 \end{pmatrix} = \begin{pmatrix} -3 \\ 2 \end{pmatrix}$. Graphically, this is shown in Fig. 5.8.

Let us consider another transformation, under which **i** and **j** land at $\begin{pmatrix} 1 \\ -2 \end{pmatrix}$ and $\begin{pmatrix} 3 \\ 0 \end{pmatrix}$, respectively. This is shown in Fig. 5.9. The vector $\mathbf{v} = -1\mathbf{i} + 2\mathbf{j}$ is transformed to $-1\begin{pmatrix} 1 \\ -2 \end{pmatrix} + 2\begin{pmatrix} 3 \\ 0 \end{pmatrix} = \begin{pmatrix} -1 + 6 \\ 2 + 0 \end{pmatrix} = \begin{pmatrix} 5 \\ 2 \end{pmatrix}$. This equation is the same as $\begin{pmatrix} 1 & 3 \\ -2 & 0 \end{pmatrix}\begin{pmatrix} -1 \\ 2 \end{pmatrix} = \begin{pmatrix} 5 \\ 2 \end{pmatrix}$.

Under the linear transformation the coordinate grid lines are stretched and turned, but they remain evenly spaced and parallel and the origin stays at the same

Fig. 5.9 Under this transformation **i** lands at $\begin{pmatrix} 1 \\ -2 \end{pmatrix}$ and **j** at $\begin{pmatrix} 3 \\ 0 \end{pmatrix}$. This transformation is represented by the matrix $\begin{pmatrix} 1 & 3 \\ -2 & 0 \end{pmatrix}$. The vector $\mathbf{v} = -\mathbf{i} + 2\mathbf{j}$ is transformed to $\begin{pmatrix} 1 & 3 \\ -2 & 0 \end{pmatrix}\begin{pmatrix} -1 \\ 2 \end{pmatrix} = \begin{pmatrix} 5 \\ 2 \end{pmatrix}$

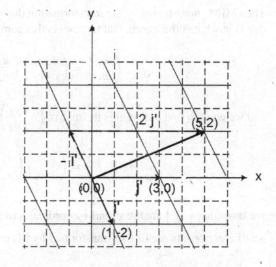

place. Although the whole space is transformed, but the transformation of the two-dimensional space is characterized by where the basis vector **i** and **j** land, altogether only four numbers. The transformation is represented by a 2×2 matrix, the first column of this matrix is the coordinates of the transformed **i** vector, and the second column of this matrix is the coordinates of the transformed **j** vector. That is, let

$$T(\mathbf{i}) = \mathbf{i}' = \begin{pmatrix} a \\ b \end{pmatrix}, \quad T(\mathbf{j}) = \mathbf{j}' = \begin{pmatrix} c \\ d \end{pmatrix}.$$

Then the process of transforming any input vector $\begin{pmatrix} x \\ y \end{pmatrix}$ into an output vector $\begin{pmatrix} x' \\ y' \end{pmatrix}$ can be written as

$$\begin{pmatrix} x' \\ y' \end{pmatrix} = \begin{pmatrix} a & c \\ b & d \end{pmatrix} \begin{pmatrix} x \\ y \end{pmatrix}. \tag{5.20}$$

It is important to understand this equation. Most of the calculations of linear algebra are based on this understanding.

Matrix Multiplication As Successive Transformations. Let

$$M_1 = \begin{pmatrix} a & c \\ b & d \end{pmatrix}, \quad M_2 = \begin{pmatrix} e & g \\ f & h \end{pmatrix}.$$

Last section shows that each matrix represents a particular transformation. Now if we have a vector **u**, and it becomes **v** under the transformation M_2. Then we apply the transformation M_1 to **v**, it becomes **w**. That is,

$$\begin{pmatrix} w_x \\ w_y \end{pmatrix} = \begin{pmatrix} a & c \\ b & d \end{pmatrix} \begin{pmatrix} v_x \\ v_y \end{pmatrix} = \begin{pmatrix} a & c \\ b & d \end{pmatrix} \begin{pmatrix} e & g \\ f & h \end{pmatrix} \begin{pmatrix} u_x \\ u_y \end{pmatrix}.$$

Then $M_1 M_2$ must represent the transformation that takes **u** directly to **w**. We ask the question what is the matrix that represents this composite transformation. That is,

$$\begin{pmatrix} w_x \\ w_y \end{pmatrix} = \begin{pmatrix} c_1 & c_3 \\ c_2 & c_3 \end{pmatrix} \begin{pmatrix} u_x \\ u_y \end{pmatrix},$$

and we want to find out what is the matrix $\begin{pmatrix} c_1 & c_3 \\ c_2 & c_3 \end{pmatrix}$. First, from

$$\begin{pmatrix} v_x \\ v_y \end{pmatrix} = \begin{pmatrix} e & g \\ f & h \end{pmatrix} \begin{pmatrix} u_x \\ u_y \end{pmatrix},$$

we know that $\begin{pmatrix} e \\ f \end{pmatrix}$ are the x- and y-coordinates of **i** for the M_2 transformation. This set of components are further transformed by M_1 to

$$e \begin{pmatrix} a \\ b \end{pmatrix} + f \begin{pmatrix} c \\ d \end{pmatrix} = \begin{pmatrix} a & c \\ b & d \end{pmatrix} \begin{pmatrix} e \\ f \end{pmatrix} = \begin{pmatrix} ae + cf \\ be + df \end{pmatrix}.$$

Similarly, for these transformations $\mathbf{j}$ is at

$$g \begin{pmatrix} a \\ b \end{pmatrix} + h \begin{pmatrix} c \\ d \end{pmatrix} = \begin{pmatrix} a & c \\ b & d \end{pmatrix} \begin{pmatrix} g \\ h \end{pmatrix} = \begin{pmatrix} ag + ch \\ bg + dh \end{pmatrix}.$$

Therefore, the matrix representing the transformation from $\mathbf{u}$ to $\mathbf{w}$ is

$$\begin{pmatrix} ae + cf & ag + ch \\ be + df & bg + dh \end{pmatrix}.$$

In other words

$$\begin{pmatrix} a & c \\ b & d \end{pmatrix} \begin{pmatrix} e & g \\ f & h \end{pmatrix} = \begin{pmatrix} ae + cf & ag + ch \\ be + df & bg + dh \end{pmatrix}. \tag{5.21}$$

We "derived" this multiplication rule instead of defining it.

Determinant and Cramer's Rule. We have defined determinants before. Now let us look at its geometrical significance. Let

$$\mathbf{u}_1 = \begin{pmatrix} a \\ b \end{pmatrix}, \quad \mathbf{u}_2 = \begin{pmatrix} c \\ d \end{pmatrix},$$

and

$$A = (\mathbf{u}_1 \mathbf{u}_2) = \begin{pmatrix} a & c \\ b & d \end{pmatrix}.$$

The determinant is defined as

$$|A| = \begin{vmatrix} a & c \\ b & d \end{vmatrix} = ad - cb.$$

It is the area of the parallelogram formed by $\mathbf{u}_1$ and $\mathbf{u}_2$ as shown in Fig. 5.10. That is,

$$|A| = ad - cb = (a + c)(b + d) - 2cb - 2\frac{1}{2}ab - 2\frac{1}{2}cd.$$

Clearly, if $\mathbf{u}_2 = \alpha \mathbf{u}_1$, then the two vectors are on top of each other, the two-dimensional space is squeezed into an one-dimensional space and $|A| = 0$. Another situation that $|A| = 0$ is either $\mathbf{u}_1$ or $\mathbf{u}_2$ is equal to zero. Here we are only interested in that $\mathbf{u}_1$ and $\mathbf{u}_2$ are still spanning the two-dimensional space and $|A| \neq 0$.

Fig. 5.10 The area of the parallelogram formed by $\begin{pmatrix} a \\ b \end{pmatrix}$ and $\begin{pmatrix} c \\ d \end{pmatrix}$ is equal to the determinant of $|A| = \begin{pmatrix} a & c \\ b & d \end{pmatrix}$

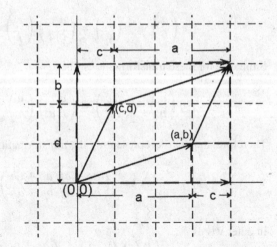

Fig. 5.11 When the space is transformed by $A = \begin{pmatrix} 3 & 0 \\ 0 & 2 \end{pmatrix}$, the unit area formed by **i** and **j** is enlarged to $|A| \times (1 \times 1) = \begin{vmatrix} 3 & 0 \\ 0 & 2 \end{vmatrix} \times 1 = 6$

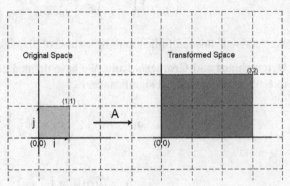

Let us consider the simple transformation $A = \begin{pmatrix} 3 & 0 \\ 0 & 2 \end{pmatrix}$. This means that the basis **i** is stretched by 3 units and **j** by 2. This is shown in Fig. 5.11. The unit area of the basic grid formed by **i** and **j** in the original space is enlarged to $\begin{vmatrix} 3 & 0 \\ 0 & 2 \end{vmatrix} = 6$ in the transformed space. When the whole space is transformed, all the square grids are transformed in the same way as the basis grid. Therefore we can use $|A|$ as a measure of how any area will change under the transformation A. In other words, if the area of certain region is S, if the space is transformed by A, then that area will become $|A| S$.

Thus the determinant has two meanings:

1. If $|A| = \begin{vmatrix} a & c \\ b & d \end{vmatrix} = ad - bc$, it is the area of the parallelogram formed by vectors $\begin{pmatrix} a \\ b \end{pmatrix}$ and $\begin{pmatrix} c \\ d \end{pmatrix}$.

2. Under the transformation A, the area S becomes $S|A|$. In other words, $|A|$ is the factor by which the area in the original space is scaled up (or down) after it is transformed by A.

The following example will make these interpretations clear.

Let

$$\mathbf{v}_1 = \begin{pmatrix} 1 \\ 0 \end{pmatrix}, \quad \mathbf{v}_2 = \begin{pmatrix} 1 \\ 3 \end{pmatrix}$$

and they are transformed and the transformation is represented by A

$$A = \begin{pmatrix} 3 & 1 \\ 1 & 2 \end{pmatrix}.$$

Thus, in the transformed space $\mathbf{v}_1$ becomes $\mathbf{v}_1'$ and $\mathbf{v}_2$ becomes $\mathbf{v}_2'$,

$$\mathbf{v}_1' = \begin{pmatrix} 3 & 1 \\ 1 & 2 \end{pmatrix} \begin{pmatrix} 1 \\ 0 \end{pmatrix} = \begin{pmatrix} 3 \\ 1 \end{pmatrix},$$

$$\mathbf{v}_2' = \begin{pmatrix} 3 & 1 \\ 1 & 2 \end{pmatrix} \begin{pmatrix} 1 \\ 3 \end{pmatrix} = \begin{pmatrix} 6 \\ 7 \end{pmatrix}.$$

In the original space, the area of the parallelogram formed by $\mathbf{v}_1$ and $\mathbf{v}_2$ is

$$|\mathbf{v}_1\mathbf{v}_2| = \begin{vmatrix} 1 & 1 \\ 0 & 3 \end{vmatrix} = 3.$$

In the transformed space, the area of the parallelogram formed by $\mathbf{v}_1'$ and $\mathbf{v}_2'$ is

$$|\mathbf{v}_1'\mathbf{v}_2'| = \begin{vmatrix} 3 & 6 \\ 1 & 7 \end{vmatrix} = 15.$$

According to what we have said, the following must hold:

$$|\mathbf{v}_1'\mathbf{v}_2'| = |A| \times |\mathbf{v}_1\mathbf{v}_2|.$$

This is indeed the case, since

$$|A| = \begin{vmatrix} 3 & 1 \\ 1 & 2 \end{vmatrix} = 6 - 1 = 5,$$

and $15 = 5 \times 3$.

Now we are ready to solve a set of linear equations. Suppose we want to solve for x, y from the set of equations:

$$3x + y = 6,$$
$$x + 2y = 7.$$

Writing it in the form of

$$Ax = b$$

with

$$\mathbf{x} = \begin{pmatrix} x \\ y \end{pmatrix}, \qquad \mathbf{b} = \begin{pmatrix} 6 \\ 7 \end{pmatrix}$$

and

$$A = \begin{pmatrix} 3 & 1 \\ 1 & 2 \end{pmatrix}.$$

To write the system of equations this way is not only a short hand, it enables us to see it as to find a vector $\mathbf{x}$ that will move to $\mathbf{b}$ under the transformation A. If we think that A is representing a transformation, then the first column is $\mathbf{i}$ expressed in the transformed space and the second column is where $\mathbf{j}$ is after it is transformed. The unknown vector $\mathbf{x}$ is in the original space, and it becomes the $\mathbf{b}$ vector in the transformed space. Let us sort out those vectors in the transformed space: $\begin{pmatrix} 3 \\ 1 \end{pmatrix}, \begin{pmatrix} 1 \\ 2 \end{pmatrix}, \begin{pmatrix} 6 \\ 7 \end{pmatrix}$. They correspond to those in the original space: $\mathbf{i} = \begin{pmatrix} 1 \\ 0 \end{pmatrix}, \mathbf{j} = \begin{pmatrix} 0 \\ 1 \end{pmatrix}$ and $\begin{pmatrix} x \\ y \end{pmatrix}$. Let us look at the area of the parallelogram formed by $\mathbf{i}$ and $\mathbf{x}$. In the transformed space, it is $\begin{vmatrix} 3 & 6 \\ 1 & 7 \end{vmatrix} = 21 - 6 = 15$. In the original space, it is $\begin{vmatrix} 1 & x \\ 0 & y \end{vmatrix} = y$. (From geometry, this area is the base ($\mathbf{i} = 1$) times the height (y), so the area is y.) The relation between these two areas is $|A| y = \begin{vmatrix} 3 & 6 \\ 1 & 7 \end{vmatrix}$, or

$$y = \frac{\begin{vmatrix} 3 & 6 \\ 1 & 7 \end{vmatrix}}{|A|} = \frac{\begin{vmatrix} 3 & 6 \\ 1 & 7 \end{vmatrix}}{\begin{vmatrix} 3 & 1 \\ 1 & 2 \end{vmatrix}} = \frac{15}{5} = 3.$$

Similarly, if we consider the parallelogram formed by $\mathbf{j}$ and $\mathbf{x}$, we have

$$x = \frac{\begin{vmatrix} 6 & 1 \\ 7 & 2 \end{vmatrix}}{\begin{vmatrix} 3 & 1 \\ 1 & 2 \end{vmatrix}} = \frac{5}{5} = 1.$$

These are identiacl to Eqs. (5.6) and (5.5) known as Cramer's rule.

Before we leave this section, we should remark that the determinant can be negative. In that case, it only means that the paper is flipped over. For example, the determinant $\begin{vmatrix} 1 & 0 \\ 0 & 1 \end{vmatrix} = 1$. If the two columns are interchanged, then the determinant

$\begin{vmatrix} 0 & 1 \\ 1 & 0 \end{vmatrix} = -1$ changes the sign. This is achieved by interchanging $\mathbf{i}$ and $\mathbf{j}$, which is equivalent to say that the paper is flipped along the 45-degree line. Other than that, everything else we have said so far is still valid.

Change of Basis and Similarity Matrix. We draw the grid lines with $\mathbf{i}$ and $\mathbf{j}$ as basis vectors for the coordinate system. Let us call this coordinate system the regular system. An vector $\mathbf{V}$ can be expressed as

$$\mathbf{V} = v_1 \mathbf{i} + v_2 \mathbf{j} = v_1 \begin{pmatrix} 1 \\ 0 \end{pmatrix} + v_2 \begin{pmatrix} 0 \\ 1 \end{pmatrix} = \begin{pmatrix} v_1 \\ v_2 \end{pmatrix} \tag{5.22}$$

in this system. As a matter of notation, when a column matrix is used to represent a vector, it is understood that the numbers in the column are the components of the vector. They are also the coordinates of the tip of the vector whose tail is at the origin. But the space has no intrinsic grids. Any set of two linearly independent vectors can be used as basis vectors. Let $\mathbf{b}_1$ and $\mathbf{b}_2$ be another set of basis. Let us call it the prime system. The same vector can be expressed in the prime system as

$$\mathbf{V} = v_1' \mathbf{b}_1 + v_2' \mathbf{b}_2.$$

As long as we know the relationship between the two sets of the basis vectors, we can find the components of any vector in one system in terms of its components of the other system. If

$$\mathbf{b}_1 = a\mathbf{i} + b\mathbf{j} = \begin{pmatrix} a \\ b \end{pmatrix} \tag{5.23}$$

$$\mathbf{b}_2 = c\mathbf{i} + d\mathbf{j} = \begin{pmatrix} c \\ d \end{pmatrix}, \tag{5.24}$$

then

$$\mathbf{V} = v_1' \mathbf{b}_1 + v_2' \mathbf{b}_2 = v_1' \begin{pmatrix} a \\ b \end{pmatrix} + v_2' \begin{pmatrix} c \\ d \end{pmatrix} = \begin{pmatrix} a & c \\ b & d \end{pmatrix} \begin{pmatrix} v_1' \\ v_2' \end{pmatrix}.$$

Therefore by Eq. (5.22)

$$\begin{pmatrix} v_1 \\ v_2 \end{pmatrix} = \begin{pmatrix} a & c \\ b & d \end{pmatrix} \begin{pmatrix} v_1' \\ v_2' \end{pmatrix}. \tag{5.25}$$

Clearly the matrix

$$S = \begin{pmatrix} a & c \\ b & d \end{pmatrix} \tag{5.26}$$

facilitates the change of coordinates. It is known as the matrix of basis change.

If we know S, we can find its inverse S^{-1}. In Chap. 7, there are several ways to find the inverse. Nowadays there are many computer routines. Usually with a given matrix, its inverse is only a stroke away on the computer. Its definition is simply

$$S^{-1}S = I,$$

where I is the identity matrix. In other words, S^{-1} is another matrix, its effect is to undo what S did.

Writing Eq. (5.25) in the form of

$$\begin{pmatrix} v_1 \\ v_2 \end{pmatrix} = S \begin{pmatrix} v_1' \\ v_2' \end{pmatrix}, \tag{5.27}$$

and applying S^{-1}, we have

$$S^{-1} \begin{pmatrix} v_1 \\ v_2 \end{pmatrix} = S^{-1} S \begin{pmatrix} v_1' \\ v_2' \end{pmatrix} = \begin{pmatrix} v_1' \\ v_2' \end{pmatrix}. \tag{5.28}$$

Thus, the components of any vector can be written either in the regular system or in the prime system.

Every linear transformation is represented by a matrix, but the contents of the matrix depend on the choice of basis. Let the matrix A represent a certain transformation written in the regular system, we want to know its form in the prime system.

First take a vector in the prime system, then transform it to the regular system as in Eq. (5.27). Now we are in the regular system, so we can apply A to it directly. After that, we can change it back to the prime system by applying S^{-1}. This sequence of operations resulted in another matrix B

$$B = S^{-1}AS \tag{5.29}$$

which is called similarity matrix. Thus, the transformation represented by A in the regular system is represented by $S^{-1}AS$ in the prime system.

Eigenbasis and Diagonalizing a Matrix. If a matrix A is representing a linear transformation, its particular form depends on the basis chosen. Usually, if not otherwise specified, it is written in the regular coordinate system, that is, with **i** and **j** as basis vectors. If **u** is a vector, $A\mathbf{u} = \mathbf{w}$ is another vector. Generally, **w** is not equal to **u**. However, sometimes $A\mathbf{u}$ stays on the same span of **u**, in other words,

$$A\mathbf{u} = \lambda\mathbf{u},$$

where λ is a scalar. To find the non-zero **u** (known as eigenvector) and λ (known as eigenvalue) for this equation to be true is known as the eigenvalue problem. The whole Chap. 8 is devoted to the details of this problem. Here we only want to mention some of its salient features.

For this equation to be true, we must solve the matrix problem

$$(A - \lambda I)\mathbf{u} = 0.$$

If A is a 2×2 matrix, this means

$$\begin{pmatrix} a_{11} - \lambda & a_{12} \\ a_{21} & a_{22} - \lambda \end{pmatrix} \begin{pmatrix} u_x \\ u_y \end{pmatrix} = 0.$$

This is possible if and only if the characteristic equation is equal to zero.

$$\begin{vmatrix} a_{11} - \lambda & a_{12} \\ a_{21} & a_{22} - \lambda \end{vmatrix} = 0.$$

In fact, for a large matrix, this is how the eigenvalue is found. The computer gradually increases λ until the right-hand side equals zero. For a 2×2 matrix, the determinant is

$$(a_{11} - \lambda)(a_{22} - \lambda) - a_{12}a_{21} = 0.$$

This quadratic equation is easily solved to give us the two eigenvalues and two eigenvectors. The procedure is demonstrated in the following example. Let

$$A = \begin{pmatrix} 19 & 10 \\ -30 & -16 \end{pmatrix}$$

then the characteristic equation is

$$\begin{vmatrix} 19 - \lambda & 10 \\ -30 & -16 - \lambda \end{vmatrix} = 0.$$

This means

$$(19 - \lambda)(-16 - \lambda) + 300 = 0$$

or

$$\lambda^2 - 3\lambda - 4 = 0.$$

Therefore

$$\lambda = \lambda_1 = 4, \quad \lambda = \lambda_2 = -1.$$

Let the eigenvector corresponding to the eigenvalue $\lambda_1 = 4$ be $\mathbf{u}_1 = \begin{pmatrix} x_1 \\ y_1 \end{pmatrix}$, and we have

$$\begin{pmatrix} 19 - 4 & 10 \\ -30 & -16 - 4 \end{pmatrix} \begin{pmatrix} x_1 \\ y_1 \end{pmatrix} = 0,$$

or

$$3x_1 + 2y_1 = 0.$$

Thus

$$\mathbf{u}_1 = \begin{pmatrix} 2 \\ -3 \end{pmatrix}$$

or any constant times it. Similarly if $\lambda = \lambda_2 = -1$, the associated eigenvector is

$$\mathbf{u}_2 = \begin{pmatrix} 1 \\ -2 \end{pmatrix}.$$

Now if we use these eigenvectors as the basis vectors, they are called eigenbasis. It should be clear that the same transformation written in the coordinate system of eigenbasis is diagonal. If it is not clear, then let us do the following. Let the first basis of this eigenbasis space be $\mathbf{e}_1 = \begin{pmatrix} 1 \\ 0 \end{pmatrix}$ and the second one be $\mathbf{e}_2 = \begin{pmatrix} 0 \\ 1 \end{pmatrix}$. (Just like $\mathbf{i} = \begin{pmatrix} 1 \\ 0 \end{pmatrix}$ and $\mathbf{j} = \begin{pmatrix} 0 \\ 1 \end{pmatrix}$ in the regular coordinate system.) Let the matrix of transformation written in this eigenbasis coordinate system be $B = \begin{pmatrix} b_1 & b_3 \\ b_2 & b_4 \end{pmatrix}$. Since $\mathbf{e}_1$ is the first eigenvector, so

$$B\mathbf{e}_1 = \lambda_1 \mathbf{e}_1,$$

or

$$\begin{pmatrix} b_1 & b_3 \\ b_2 & b_4 \end{pmatrix} \begin{pmatrix} 1 \\ 0 \end{pmatrix} = \lambda_1 \begin{pmatrix} 1 \\ 0 \end{pmatrix}.$$

That is,

$$1 \begin{pmatrix} b_1 \\ b_2 \end{pmatrix} + 0 \begin{pmatrix} b_3 \\ b_4 \end{pmatrix} = \begin{pmatrix} \lambda_1 \\ 0 \end{pmatrix}.$$

Therefore

$$b_1 = \lambda_1, \quad b_2 = 0.$$

Similarly,

$$\begin{pmatrix} b_1 & b_3 \\ b_2 & b_4 \end{pmatrix} \begin{pmatrix} 0 \\ 1 \end{pmatrix} = \lambda_2 \begin{pmatrix} 0 \\ 1 \end{pmatrix}$$

shows that

$$b_4 = \lambda_2, \quad b_3 = 0.$$

Thus, B is diagonal

$$B = \begin{pmatrix} \lambda_1 & 0 \\ 0 & \lambda_2 \end{pmatrix}.$$

Matrices A and B represent the same transformation. One is written in the space with **i** and **j** as basis, and the other in the same space with eigenbasis. They are related by a similarity transformation (Eq. (5.29))

$$B = S^{-1}AS,$$

where S is the matrix of basis change. In this case, it is the change from the eigenbasis to regular basis. Let us continue with the same example. According to Eq. (5.25), the first column of this matrix is the coordinates of the first eigenvector $\begin{pmatrix} 2 \\ -3 \end{pmatrix}$ and the second column is the coordinates of the second eigenvector $\begin{pmatrix} 1 \\ -2 \end{pmatrix}$. Therefore

$$S = \begin{pmatrix} 2 & 1 \\ -3 & -2 \end{pmatrix}.$$

In this case, S^{-1} happens to be the same

$$S^{-1} = \begin{pmatrix} 2 & 1 \\ -3 & -2 \end{pmatrix}.$$

This can be easily verified in that

$$S^{-1}S = \begin{pmatrix} 2 & 1 \\ -3 & -2 \end{pmatrix}\begin{pmatrix} 2 & 1 \\ -3 & -2 \end{pmatrix} = \begin{pmatrix} 1 & 0 \\ 0 & 1 \end{pmatrix} = I.$$

Therefore

$$B = S^{-1}AS = \begin{pmatrix} 2 & 1 \\ -3 & -2 \end{pmatrix}\begin{pmatrix} 19 & 10 \\ -30 & -16 \end{pmatrix}\begin{pmatrix} 2 & 1 \\ -3 & -2 \end{pmatrix}$$

$$= \begin{pmatrix} 2 & 1 \\ -3 & -2 \end{pmatrix}\begin{pmatrix} 8 & -1 \\ -12 & 2 \end{pmatrix} = \begin{pmatrix} 4 & 0 \\ 0 & -1 \end{pmatrix} = \begin{pmatrix} \lambda_1 & 0 \\ 0 & \lambda_2 \end{pmatrix}$$

as expected.

From Two-Dimensional Space to N-Dimensional Space. To avoid getting lost in the sea of indices, we have mostly used the two-dimensional space to develop some of the basic linear algebra concepts. Once these concepts are understood, we can easily switch back to index notation for generalization. For example, we can express the matrix A in terms of its elements,

$$A = \left(A_{ij} \right).$$

If i and j run from 1 to 2, we have a 2×2 matrix

$$A = \begin{pmatrix} A_{11} & A_{12} \\ A_{21} & A_{22} \end{pmatrix}.$$

If i and j run from 1 to n, we have a $n \times n$ matrix

$$A = \begin{pmatrix} A_{11} & A_{12} & \cdots & A_{1n} \\ A_{21} & A_{22} & \cdots & A_{2n} \\ \cdot & \cdot & \cdots & \cdot \\ \cdot & \cdot & \cdots & \cdot \\ A_{n1} & A_{n2} & \cdots & A_{nn} \end{pmatrix}.$$

Similarly, we can write B as

$$B = \left(B_{pq} \right).$$

In terms of this notation, the matrix multiplication, such as Eq. (5.21), can be written in the following form:

$$\left[(AB)_{ij} \right] = \left[C_{ij} \right]$$

$$C_{ij} = \sum_{k=1}^{m} A_{ik} B_{kj}.$$

If $m = 2$, this can be the multiplication of two 2×2 matrices, if $m = n$, this can be the multiplication of two $n \times n$ matrices.

Next, let us look at the matrix of change of basis. Let the first set of basis be $(\mathbf{a}_1, \mathbf{a}_2, \cdots \mathbf{a}_n)$ where $\mathbf{a}_1 = \mathbf{i}$, $\mathbf{a}_2 = \mathbf{j}$, and so on. Let the second set of basis be $(\mathbf{b}_1, \mathbf{b}_2, \cdots \mathbf{b}_n)$. Then Eqs. (5.23), (5.24) can be written as

$$\mathbf{b}_1 = c_{11}\mathbf{a}_1 + c_{21}\mathbf{a}_2 \tag{5.30}$$

$$\mathbf{b}_2 = c_{12}\mathbf{a}_1 + c_{22}\mathbf{a}_2 \tag{5.31}$$

with $c_{11} = a$, $c_{21} = b$, $c_{12} = c$, and $c_{22} = d$. The S matrix of Eq. (5.26) becomes

$$S = \begin{pmatrix} c_{11} & c_{12} \\ c_{21} & c_{22} \end{pmatrix}.$$

These equations can be summarized as

$$\mathbf{b}_j = \sum_i c_{ij}\mathbf{a}_i$$

$$S = \left(c_{ij} \right).$$

If i and j run from 1 to 2, we have what it was before. If it is an n-dimensional space, we simply let i and j run from 1 to n.

Invariance of the Trace. A very useful quantity is known as the trace. The trace of a square matrix A is defined as the sum of all its diagonal elements. That is, if

$$A = \left(A_{ij}\right),$$

then the trace of A is

$$Tr(A) = \sum_{i=1}^{n} A_{ii}.$$

Let A and B be two $n \times n$ matrices, although $AB \neq BA$, we can show that

$$Tr(AB) = Tr(BA).$$

Since $(AB)_{ij} = \sum_{k=1}^{n} A_{ik} B_{kj}$ and $(BA)_{ij} = \sum_{k=1}^{n} B_{ik} A_{kj}$,

$$Tr(AB) = \sum_{i=1}^{n} (AB)_{ii} = \sum_{i=1}^{n} \sum_{k=1}^{n} A_{ik} B_{ki},$$

$$Tr(BA) = \sum_{i=1}^{n} (BA)_{ii} = \sum_{i=1}^{n} \sum_{k=1}^{n} B_{ik} A_{ki} = \sum_{i=1}^{n} \sum_{k=1}^{n} A_{ki} B_{ik},$$

the indexes i and k both run from 1 to n, so their name can be interchanged, therefore $Tr(AB) = Tr(BA)$.

It also follows that if

$$B = S^{-1} A S,$$

then

$$Tr(B) = Tr(A).$$

Since

$$Tr(B) = Tr(S^{-1} A S) = Tr(A S S^{-1}) = Tr(AI).$$

Therefore $Tr(B) = Tr(A)$.

We have shown in the previous example, if

$$A = \begin{pmatrix} 19 & 10 \\ -30 & -16 \end{pmatrix},$$

then

$$B = \begin{pmatrix} 4 & 0 \\ 0 & -1 \end{pmatrix}.$$

Clearly
$$Tr(A) = 3 = Tr(B).$$

5.2.4 Infinite Dimensional Function Space

Now let the vectors be functions. Specifically, let $\mathbf{u} = u(x)$ be any continuous function defined on $0 \le x \le 1$. For addition operation, let

$$\mathbf{u} + \mathbf{v} = u(x) + v(x),$$

that is, $\mathbf{u} + \mathbf{v}$ is the function whose values are the ordinary sum $u(x) + v(x)$. For scalar multiplication let
$$\alpha\mathbf{u} = \alpha u(x),$$

and we take the ordinary 0 to be the zero vector

$$\mathbf{0} = 0.$$

For the negative vector $-\mathbf{u}$
$$-\mathbf{u} = -u(x),$$

we simply mean the negative value of the function u.

With these definitions we can verify that all the vector space requirements are satisfied, so that this collection of functions is a vector space.

A vector space need not have a dot product (also called inner product) defined. Therefore the length (or the norm) is not defined. However, to keep the geometry of ordinary Euclidean space, we should keep the familiar definition of length.

As we have discussed before, geometrical vectors in n-dimensional space have n components $\mathbf{u} = \mathbf{u}(u_1, u_2, \cdots u_n)$, $\mathbf{v} = \mathbf{v}(v_1, v_2, \cdots v_n)$. The inner product is defined as

$$\mathbf{u} \cdot \mathbf{v} = u_1 v_1 + u_2 v_2 + \cdots + u_n v_n = \sum_{i=1}^{n} u_i v_i.$$

If $\mathbf{u} \cdot \mathbf{v} = 0$ means the two vectors are orthogonal.

The length of the vector is known as its norm $\|u\|$, the square of it is defined as

$$\|\mathbf{u}\|^2 = u_1^2 + u_2^2 + \cdots + u_n^2 = \sum_{i=1}^{n} u_i^2.$$

Now vectors are continuous functions, $\mathbf{u} = u(x)$ where x is a continuum defined in the interval $0 \le x \le 1$. Thus $u(x)$ is a vector with a whole continuum of components. To find the length of such a vector, the usual rule of adding the squares of the components becomes impossible. This summation is replaced, in a natural way, by integration. Therefore we define the length of the function as

$$\|u(x)\|^2 = \int_0^1 (u(x))^2 dx.$$

The same idea of replacing summation by integration produces the inner product of two functions:

$$(f, g) = \int_0^1 f(x)g(x)dx.$$

This is exactly like the vector dot product.

An infinite dimensional vector space of functions, for which an inner product is defined is called Hilbert space. Named in honor of the great German mathematician David Hilbert (1862–1943). He was the creator of the famous Göttingen mathematical school. He made important contributions in algebra, geometry, calculus of variations, integral equations, functional analysis, and mathematical logic. Hilbert space is very important in physics. In quantum mechanics, physical observables are represented by operators in Hilbert space, and physical states are vectors in Hilbert space.

Now functions are vectors. For example, we can show that the polynomial

$$P_3(x) = 4 + 5x + 6x^2 + 7x^3 \tag{5.32}$$

satisfies all the axioms of a vector, therefore it is a vector. The most familiar transformation in the function space is to take the derivative. In fact, taking derivative is a linear transformation because it satisfies Eq. (5.19) in that

$$\frac{d}{dx}P_3(x) = 4\frac{d}{dx}1 + 5\frac{d}{dx}x + 6\frac{d}{dx}x^2 + 7\frac{d}{dx}x^3$$
$$= 5 + 12x + 21x^2.$$

To draw a parallel between taking a derivative and the linear transformation of a vector space, the polynomial needs coordinates, which requires choosing a basis set. It seems natural to use the powers of x as the basis, that is, the basis set B

$$B = (1, x, x^2, \cdots x^n \cdots).$$

In a three-dimensional space, we can only have three linearly independent vectors as basis. In the infinite dimensional function space, we can have infinite many basis, although for a finite order polynomial, the coefficients of these basis are all zero after certain order. In terms of these basis, the polynomial of Eq. (5.32) can be written as

$$P_3(x) = \begin{pmatrix} 4\ 5\ 6\ 7 \cdot \cdot \end{pmatrix} \begin{pmatrix} 1 \\ x \\ x^2 \\ x^3 \\ \cdot \\ \cdot \end{pmatrix}. \tag{5.33}$$

It is not difficult to see that taking derivative of $P_3(x)$ is equivalent to evaluate

$$\begin{pmatrix} 0\ 1\ 0\ 0\ 0 \cdot \\ 0\ 0\ 2\ 0\ 0 \cdot \\ 0\ 0\ 0\ 3\ 0 \cdot \\ 0\ 0\ 0\ 0\ 0 \cdot \\ \cdot\ \cdot\ \cdot\ \cdot\ \cdot\ \cdot \\ \cdot\ \cdot\ \cdot\ \cdot\ \cdot\ \cdot \end{pmatrix} \begin{pmatrix} 4 \\ 5 \\ 6 \\ 7 \\ \cdot \\ \cdot \end{pmatrix} = \begin{pmatrix} 5 \\ 12 \\ 21 \\ 0 \\ \cdot \\ \cdot \end{pmatrix}.$$

That is,

$$\frac{d}{dx} P_3(x) = \begin{pmatrix} 5\ 12\ 21 \end{pmatrix} \begin{pmatrix} 1 \\ x \\ x^2 \end{pmatrix} = 5 + 12x + 21x^2.$$

With this as our background, it is not so mysterious in the saying that Schrödinger's differential equation formulation of quantum mechanics is the same as Heisenberg's matrix formulation. Further details belong to more advanced courses and cannot be discussed here.

Exercises

1. Use Cramer's rule to find the inverse of a 2×2 non-singular matrix

$$A = \begin{pmatrix} a_{11}\ a_{12} \\ a_{21}\ a_{22} \end{pmatrix}.$$

Ans. $A^{-1} = \frac{1}{|A|} \begin{pmatrix} a_{22}\ -a_{12} \\ -a_{21}\ a_{11} \end{pmatrix}.$

2. Find the inverse matrix of

$$A = \begin{pmatrix} 2\ 1 \\ -3\ -2 \end{pmatrix}.$$

Ans. $A^{-1} = \begin{pmatrix} 2\ 1 \\ -3\ -2 \end{pmatrix}.$

3. In a two-dimensional space, under a transformation, the coordinates of the unit vector **i** are at $\begin{pmatrix} 0 \\ 1 \end{pmatrix}$ and **j** is at $\begin{pmatrix} 1 \\ 0 \end{pmatrix}$, so the matrix

$$A = \begin{pmatrix} 0 & 1 \\ 1 & 0 \end{pmatrix}$$

represents the interchange of **i** and **j** or a reflection along the line $x = y$. Show that this is indeed the case with a matrix equation.

Ans. $\begin{pmatrix} x' \\ y' \end{pmatrix} = \begin{pmatrix} 0 & 1 \\ 1 & 0 \end{pmatrix} \begin{pmatrix} x' \\ y' \end{pmatrix}$, $x' = y$, $y' = x$. It represents a reflection

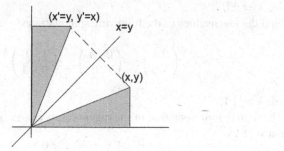

along the line $x = y$ as shown in the following figure.

4. Interpret the actions of the following matrices:

$$\begin{pmatrix} 1 & 0 \\ 0 & -1 \end{pmatrix}, \quad \begin{pmatrix} -1 & 0 \\ 0 & -1 \end{pmatrix}, \quad \begin{pmatrix} a & 0 \\ 0 & 1 \end{pmatrix}.$$

Ans. A reflection in the x-axis; a reflection over the origin; and a stretch (when $a > 1$, or a contraction when $0 < a < 1$) in the x-direction.

5. In the transformed two-dimensional space, the coordinates of **i** are $\begin{pmatrix} 2 \\ 3 \end{pmatrix}$, and **j** is $\begin{pmatrix} 1 \\ 2 \end{pmatrix}$. Find the matrix A representing this transformation. If a vector **v** is equal to $\mathbf{v} = 2\mathbf{i} + 3\mathbf{j}$, what is its coordinates after the transformation?

Ans. $A = \begin{pmatrix} 2 & 1 \\ 3 & 2 \end{pmatrix}$; $\mathbf{v} = \begin{pmatrix} 7 \\ 12 \end{pmatrix}$.

6. Find the matrix A representing a θ-degree rotation of a two-dimensional vector space.

Ans. $R(\theta) = \begin{pmatrix} \cos\theta & -\sin\theta \\ \sin\theta & \cos\theta \end{pmatrix}$.

7. Use the answer of Exercise 1 to find the inverse of the rotation matrix $R^{-1}(\theta)$ and show that

$$R^{-1}(\theta) = R(-\theta)$$

and

$$R^{-1}(\theta) = R^T(\theta).$$

The superscribe T means the transpose.

8. Show that

$$R(\theta)\, R(\theta) = R^2(\theta) = R(2\theta)$$
$$R(\theta)\, R(\varphi) = R(\theta + \varphi).$$

9. Show that the eigenvalues of a two-dimensional rotation matrix are imaginary, therefore no vector will stay put in a rotation.

Ans. $\lambda = \pm i$.

10. Find the eigenvalues of the following spin matrices:

$$\begin{pmatrix} 0 & 1 \\ 1 & 0 \end{pmatrix}, \quad \begin{pmatrix} 0 & -i \\ i & 0 \end{pmatrix}, \quad \begin{pmatrix} 1 & 0 \\ 0 & -1 \end{pmatrix}.$$

Ans. $\lambda = \pm 1$.

11. The matrix representation of the eigenvectors α and β is the two-dimensional column vectors.

$$\alpha = \begin{pmatrix} 1 \\ 0 \end{pmatrix}, \quad \beta = \begin{pmatrix} 0 \\ 1 \end{pmatrix}.$$

In this representation, show the orthonormal relations

$$(\alpha, \beta) = 0, \quad (\beta, \alpha) = 0,$$
$$(\alpha, \alpha) = 1, \quad (\beta, \beta) = 1.$$

Ans. $(\alpha, \beta) = \begin{pmatrix} 1 & 0 \end{pmatrix} \begin{pmatrix} 0 \\ 1 \end{pmatrix} = 0, \quad (\beta, \alpha) = \begin{pmatrix} 0 & 1 \end{pmatrix} \begin{pmatrix} 1 \\ 0 \end{pmatrix} = 0,$

$(\alpha, \alpha) = \begin{pmatrix} 1 & 0 \end{pmatrix} \begin{pmatrix} 1 \\ 0 \end{pmatrix} = 1, \quad (\beta, \beta) = \begin{pmatrix} 0 & 1 \end{pmatrix} \begin{pmatrix} 0 \\ 1 \end{pmatrix} = 1.$

12. Solve for x_1, x_2, and x_3 using Crout's method for

$$\begin{pmatrix} 2 & 4 & 3 \\ 1 & -2 & -2 \\ -3 & 3 & 2 \end{pmatrix} \begin{pmatrix} x_1 \\ x_2 \\ x_3 \end{pmatrix} = \begin{pmatrix} 4 \\ 0 \\ -7 \end{pmatrix}.$$

Ans. $x_1 = 2$, $x_2 = -3$, $x_3 = 4$.

13. Let $v_1 = \begin{pmatrix} 1 \\ 2 \\ 0 \end{pmatrix}$, $v_2 = \begin{pmatrix} -2 \\ 1 \\ 0 \end{pmatrix}$, $v_3 = \begin{pmatrix} 0 \\ 0 \\ 3 \end{pmatrix}$ be three-dimensional vectors, can we find some scalars β_1, β_2, β_3, not all of them zero such that

$$\beta_1 v_1 + \beta_2 v_2 + \beta_3 v_3 = \mathbf{0} \ ?$$

Are these vectors a linearly independent set?

Ans. No. The only solution of this set of equations is $\beta_1 = \beta_2 = \beta_3 = 0$. Vectors v_1, v_2, v_3 are linearly independent.

14. Let

$$\mathbf{a}_i = \begin{pmatrix} a_{1i} \\ a_{2i} \\ \cdot \\ \cdot \\ a_{di} \end{pmatrix}, \quad i = 1, 2, \cdots, d$$

and

$$A = \begin{pmatrix} a_{11} & a_{12} & \cdot & a_{1d} \\ a_{21} & a_{22} & \cdot & a_{2d} \\ \cdot & \cdot & \cdot & \cdot \\ a_{d1} & a_{d2} & \cdot & a_{dd} \end{pmatrix} = (\mathbf{a}_1, \mathbf{a}_2, \cdots \mathbf{a}_d).$$

Show that if $\mathbf{a}_1, \mathbf{a}_2, \cdots \mathbf{a}_d$ are linearly dependent then the determinant of A, $|A| = 0$.

Ans. Since $\mathbf{a}_i$ are linearly dependent, then there are numbers $\alpha_1, \alpha_2, \cdots, \alpha_d$, such that

$$\sum_i^d \alpha_i \mathbf{a}_i = 0.$$

Without loss of generality, let's assume that $\alpha_1 \neq 0$. Then

$$|A| = \left| (\frac{1}{\alpha_1} \sum_{i=2}^d \alpha_i \mathbf{a}_i), \mathbf{a}_2, \cdots, \mathbf{a}_d \right| = 0.$$

15. If

$$x_1 = \begin{pmatrix} 1 \\ 0 \end{pmatrix}, \quad Ax_1 = \begin{pmatrix} 2 \\ 3 \\ 4 \end{pmatrix}$$

and

$$x_2 = \begin{pmatrix} 0 \\ 1 \end{pmatrix}, \quad Ax_2 = \begin{pmatrix} 4 \\ 6 \\ 8 \end{pmatrix}.$$

Find the matrix A. Now with

$$b_1 = \begin{pmatrix} 1 \\ 1 \end{pmatrix}, \quad b_2 = \begin{pmatrix} 2 \\ -1 \end{pmatrix},$$

find

$$Ab_1 \text{ and } Ab_2.$$

Ans. $A = \begin{pmatrix} 2 & 4 \\ 3 & 6 \\ 4 & 8 \end{pmatrix}$, $Ab_1 = \begin{pmatrix} 6 \\ 9 \\ 12 \end{pmatrix}$, $Ab_2 = \begin{pmatrix} 0 \\ 0 \\ 0 \end{pmatrix}$.

16. Let $\mathbf{c}$ be the vector of projection of $\mathbf{a}_2$ on $\mathbf{a}_1$ in the direction of $\mathbf{a}_1$. Show that

$$\mathbf{c} = \frac{(\mathbf{a}_2 \cdot \mathbf{a}_1)\mathbf{a}_1}{\|a_1\|^2}.$$

Let $\mathbf{d}$ be the component of $\mathbf{a}_2$ perpendicular to $\mathbf{a}_1$, show that

$$\mathbf{d} = \mathbf{a}_2 - \mathbf{c}.$$

17. Continue with the previous problem. Let

$$\mathbf{a}_1 = \begin{pmatrix} 1 \\ 0 \\ 1 \end{pmatrix}, \quad \mathbf{a}_2 = \begin{pmatrix} 1 \\ 0 \\ 0 \end{pmatrix}$$

and

$$\mathbf{u}_1 = \frac{\mathbf{a}_1}{\|a_1\|}, \quad \mathbf{u}_2 = \frac{\mathbf{d}}{\|d\|}.$$

Find $\mathbf{u}_1$ and $\mathbf{u}_2$ and show that $\mathbf{u}_1 \cdot \mathbf{u}_1 = \mathbf{u}_2 \cdot \mathbf{u}_2 = 1$, and $\mathbf{u}_1 \cdot \mathbf{u}_2 = 0$.

Ans. $\mathbf{u}_1 = \begin{pmatrix} \frac{1}{\sqrt{2}} \\ 0 \\ \frac{1}{\sqrt{2}} \end{pmatrix}$ and $\mathbf{u}_2 = \begin{pmatrix} \frac{1}{\sqrt{2}} \\ 0 \\ -\frac{1}{\sqrt{2}} \end{pmatrix}$.

18. Let $y_n(x) = x^n$, $n = 0, 1, 2, \ldots$ be a linearly independent set. Construct $\phi_0(x)$, $\phi_0(x)$, $\phi_0(x)$ in the following way (known as Gram-Schmidt procedure):

$$\varphi_0 = y_0; \quad \phi_0 = \varphi_0 (\varphi_0, \varphi_0)^{-\frac{1}{2}},$$
$$\varphi_1 = y_1 - \phi_0 (\phi_0, y_1); \quad \phi_1 = \varphi_1 (\varphi_1, \varphi_1)^{-\frac{1}{2}},$$
$$\varphi_2 = y_2 - \phi_1 (\phi_1, y_2) - \phi_0 (\phi_0, y_2); \quad \phi_2 = \varphi_2 (\varphi_2, \varphi_2)^{-\frac{1}{2}}.$$

Find $\phi_0(x)$, $\phi_1(x)$, $\phi_2(x)$ and show that they are orthonormal.

Ans. $\phi_0(x) = \sqrt{\frac{1}{2}}$, $\phi_1(x) = \sqrt{\frac{3}{2}}x$, $\phi_2(x) = \sqrt{\frac{5}{2}}(\frac{3}{2}x^2 - \frac{1}{2})$.

19. We have shown that differentiation can be expressed as a matrix multiplication. The first column of that matrix comes from $\frac{d}{dx}1 = 0$. Therefore it is all zero. The second column is from $\frac{d}{dx}x = 1$ and so on, thus we have the differential matrix

$$A_{diff} = \begin{pmatrix} 0 & 1 & 0 & 0 & 0 \\ 0 & 0 & 2 & 0 & 0 \\ 0 & 0 & 0 & 3 & 0 \\ 0 & 0 & 0 & 0 & 4 \\ 0 & 0 & 0 & 0 & 0 \end{pmatrix}.$$

Now we can do the same thing for integration. Since $\int_0^x 1dx' = x$, $\int_0^x x'dx' = \frac{1}{2}x^2$ and so forth, so we have

$$A_{int} = \begin{pmatrix} 0 & 0 & 0 & 0 & 0 \\ 1 & 0 & 0 & 0 & 0 \\ 0 & \frac{1}{2} & 0 & 0 & 0 \\ 0 & 0 & \frac{1}{3} & 0 & 0 \\ 0 & 0 & 0 & \frac{1}{4} & 0 \end{pmatrix}.$$

Show that

$$\begin{pmatrix} 0 & 0 & 0 & 0 & 0 \\ 1 & 0 & 0 & 0 & 0 \\ 0 & \frac{1}{2} & 0 & 0 & 0 \\ 0 & 0 & \frac{1}{3} & 0 & 0 \\ 0 & 0 & 0 & \frac{1}{4} & 0 \end{pmatrix} \begin{pmatrix} 4 \\ 5 \\ 6 \\ 7 \\ 0 \end{pmatrix} = \begin{pmatrix} 0 \\ 4 \\ \frac{5}{2} \\ \frac{6}{3} \\ \frac{7}{4} \end{pmatrix}$$

and

$$\int (4 + 5x + 6x^2 + 7x^3)dx = \begin{pmatrix} 0 & 4 & \frac{5}{2} & \frac{6}{3} & \frac{7}{4} \end{pmatrix} \begin{pmatrix} 1 \\ x \\ x^2 \\ x^3 \\ x^4 \end{pmatrix}.$$

20. Show that

$$A_{diff} A_{int} = I$$

but

$$A_{int} A_{diff} \neq I.$$

Chapter 6
Determinants

Determinants are powerful tools for solving systems of linear equations and they are indispensable in the development of matrix theory. Most readers probably already possess the knowledge of evaluating second- and third-order determinants. After a systematic review, we introduce the formal definition of a nth-order determinant through the Levi-Civita symbol. All properties of determinants can be derived from this definition.

6.1 Systems of Linear Equations

6.1.1 Solution of Two Linear Equations

Suppose we wish to solve for x and y from the system of 2×2 linear equations (2 equations and 2 unknowns)

$$a_1 x + b_1 y = d_1 \tag{6.1}$$
$$a_2 x + b_2 y = d_2, \tag{6.2}$$

where a_1, a_2, b_1, b_2, d_1, and d_2 are known constants. We can multiply (6.1) by b_2 and (6.2) by b_1, and then take the difference. In so doing, y is eliminated, and we are left with

$$(b_2 a_1 - b_1 a_2)x = b_2 d_1 - b_1 d_2,$$

therefore

$$x = \frac{d_1 b_2 - d_2 b_1}{a_1 b_2 - a_2 b_1}, \tag{6.3}$$

© The Author(s), under exclusive license to Springer Nature Switzerland AG 2022
K.-T. Tang, *Mathematical Methods for Engineers and Scientists 1*,
https://doi.org/10.1007/978-3-031-05678-9_6

Fig. 6.1 A schematic
diagram for a second-order
determinant

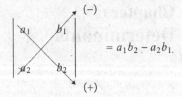

$$= a_1b_2 - a_2b_1.$$

where we have written b_2a_1 as a_1b_2, since the order is immaterial in the product of two numbers. It turns out that if we use the following notation, it is much easier to generalize this process to larger systems of $n \times n$ equations

$$a_1b_2 - a_2b_1 = \begin{vmatrix} a_1 & b_1 \\ a_2 & b_2 \end{vmatrix}. \tag{6.4}$$

The two-by-two square array of the four elements on the right-hand side of this equation is called a second-order determinant. Its meaning is just that its value is equal to the left-hand side of this equation. Explicitly, the value of a second-order determinant is defined as the difference between the two products of the diagonal elements as shown in the following schematic diagram (Fig. 6.1).

With determinants, (6.3) can be written as

$$x = \frac{\begin{vmatrix} d_1 & b_1 \\ d_2 & b_2 \end{vmatrix}}{\begin{vmatrix} a_1 & b_1 \\ a_2 & b_2 \end{vmatrix}}, \tag{6.5}$$

and with a similar procedure one can easily show that

$$y = \frac{\begin{vmatrix} a_1 & d_1 \\ a_2 & d_2 \end{vmatrix}}{\begin{vmatrix} a_1 & b_1 \\ a_2 & b_2 \end{vmatrix}}. \tag{6.6}$$

Example 6.1.1 Find the solution of

$$2x - 3y = -4,$$
$$6x - 2y = 2.$$

Solution 6.1.1

$$x = \frac{\begin{vmatrix} -4 & -3 \\ 2 & -2 \end{vmatrix}}{\begin{vmatrix} 2 & -3 \\ 6 & -2 \end{vmatrix}} = \frac{8+6}{-4+18} = 1,$$

$$y = \frac{\begin{vmatrix} 2 & -4 \\ 6 & 2 \end{vmatrix}}{\begin{vmatrix} 2 & -3 \\ 6 & -2 \end{vmatrix}} = \frac{4+24}{-4+18} = 2.$$

6.1.2 Properties of Second-Order Determinants

There are many general properties of determinants that will be discussed in later sections. At this moment, we want to list a few which we need in the following discussion of third-order determinant. For a second-order determinant, these properties are almost self-evident from its definition. Although they are generally valid for nth-order determinant, at this point, we only need them to be valid for second-order determinant to continue our discussion.

- 1. If the rows and columns are interchanged, the determinant is unaltered,

$$\begin{vmatrix} a_1 & a_2 \\ b_1 & b_2 \end{vmatrix} = a_1 b_2 - b_1 a_2 = \begin{vmatrix} a_1 & b_1 \\ a_2 & b_2 \end{vmatrix}. \tag{6.7}$$

- 2. If two columns (or two rows) are interchanged, the determinant changes sign,

$$\begin{vmatrix} b_1 & a_1 \\ b_2 & a_2 \end{vmatrix} = b_1 a_2 - b_2 a_1 = -(a_1 b_2 - a_2 b_1) = -\begin{vmatrix} a_1 & b_1 \\ a_2 & b_2 \end{vmatrix}. \tag{6.8}$$

- 3. If each element in a column (or in a row) is multiplied by m, the determinant is multiplied by m,

$$\begin{vmatrix} ma_1 & b_1 \\ ma_2 & b_2 \end{vmatrix} = ma_1 b_2 - ma_2 b_1 = m(a_1 b_2 - a_2 b_1) = m\begin{vmatrix} a_1 & b_1 \\ a_2 & b_2 \end{vmatrix}.$$

- 4. If each element of a column (or of a row) is sum of two terms, the determinant equals the sum of the two corresponding determinants,

$$\begin{vmatrix} (a_1 + c_1) & b_1 \\ (a_2 + c_2) & b_2 \end{vmatrix} = (a_1 + c_1)b_2 - (a_2 + c_2)b_1 = a_1b_2 - a_2b_1 + c_1b_2 - c_2b_1$$

$$= \begin{vmatrix} a_1 & b_1 \\ a_2 & b_2 \end{vmatrix} + \begin{vmatrix} c_1 & b_1 \\ c_2 & b_2 \end{vmatrix}.$$

6.1.3 Solution of Three Linear Equations

Now suppose we want to solve a system of three equations

$$a_1 x + b_1 y + c_1 z = d_1, \tag{6.9}$$
$$a_2 x + b_2 y + c_2 z = d_2, \tag{6.10}$$
$$a_3 x + b_3 y + c_3 z = d_3. \tag{6.11}$$

First we can solve for y and z in terms of x. Writing (6.10) and (6.11) as

$$b_2 y + c_2 z = d_2 - a_2 x$$
$$b_3 y + c_3 z = d_3 - a_3 x,$$

then in analogy to (6.5) and (6.6), we can express y and z as

$$y = \frac{\begin{vmatrix} (d_2 - a_2 x) & c_2 \\ (d_3 - a_3 x) & c_2 \end{vmatrix}}{\begin{vmatrix} b_2 & c_2 \\ b_3 & c_3 \end{vmatrix}}, \tag{6.12}$$

$$z = \frac{\begin{vmatrix} b_2 & (d_2 - a_2 x) \\ b_3 & (d_3 - a_3 x) \end{vmatrix}}{\begin{vmatrix} b_2 & c_2 \\ b_3 & c_3 \end{vmatrix}}. \tag{6.13}$$

Substituting these two expressions into (6.9) and then multiplying the entire equation by $\begin{vmatrix} b_2 & c_2 \\ b_3 & c_3 \end{vmatrix}$, we have

$$a_1 \begin{vmatrix} b_2 & c_2 \\ b_3 & c_3 \end{vmatrix} x + b_1 \begin{vmatrix} (d_2 - a_2 x) & c_2 \\ (d_3 - a_3 x) & c_3 \end{vmatrix} + c_1 \begin{vmatrix} b_2 & (d_2 - a_2 x) \\ b_3 & (d_3 - a_3 x) \end{vmatrix} = d_1 \begin{vmatrix} b_2 & c_2 \\ b_3 & c_3 \end{vmatrix}. \tag{6.14}$$

By properties 3 and 4, this equation becomes

$$a_1 \begin{vmatrix} b_2 & c_2 \\ b_3 & c_3 \end{vmatrix} x + b_1 \left\{ \begin{vmatrix} d_2 & c_2 \\ d_3 & c_3 \end{vmatrix} - \begin{vmatrix} a_2 & c_2 \\ a_3 & c_3 \end{vmatrix} x \right\}$$
$$+ c_1 \left\{ \begin{vmatrix} b_2 & d_2 \\ b_3 & d_3 \end{vmatrix} - \begin{vmatrix} b_2 & a_2 \\ b_3 & a_3 \end{vmatrix} x \right\} = d_1 \begin{vmatrix} b_2 & c_2 \\ b_3 & c_3 \end{vmatrix}. \tag{6.15}$$

It follows

$$Dx = N_x, \tag{6.16}$$

where

$$N_x = d_1 \begin{vmatrix} b_2 & c_2 \\ b_3 & c_3 \end{vmatrix} - b_1 \begin{vmatrix} d_2 & c_2 \\ d_3 & c_3 \end{vmatrix} - c_1 \begin{vmatrix} b_2 & d_2 \\ b_3 & d_3 \end{vmatrix}, \tag{6.17}$$

and

$$D = a_1 \begin{vmatrix} b_2 & c_2 \\ b_3 & c_3 \end{vmatrix} - b_1 \begin{vmatrix} a_2 & c_2 \\ a_3 & c_3 \end{vmatrix} - c_1 \begin{vmatrix} b_2 & a_2 \\ b_3 & a_3 \end{vmatrix}. \tag{6.18}$$

Expanding the second-order determinants (6.18) leads to

$$D = a_1 b_2 c_3 - a_1 b_3 c_2 - b_1 a_2 c_3 + b_1 a_3 c_2 - c_1 b_2 a_3 + c_1 b_3 a_2. \tag{6.19}$$

To express these six terms in a more systematic way, we introduce a third-order determinant as a short hand notation for (6.19)

$$D = \begin{vmatrix} a_1 & b_1 & c_1 \\ a_2 & b_2 & c_2 \\ a_3 & b_3 & c_3 \end{vmatrix}. \tag{6.20}$$

A useful device for evaluating a third-order determinant is as follows. We write down the determinant column by column, after the third column, we repeat the first, then the second column, creating a three-by-five array of numbers. We can form a product of three elements along each of the three diagonals going from upper left to lower right. These products carry a positive sign. Similarly, three products can be formed along the diagonals from lower left to upper right. These three latter products carry a minus sign. The value of the determinant is equal to the sum of these six terms. This is shown in Fig. 6.2.

This is seen to be exactly equal to the six terms in (6.19).

Using the determinant notation, one can easily show that N_x in (6.17) is equal to

$$N_x = \begin{vmatrix} d_1 & b_1 & c_1 \\ d_2 & b_2 & c_2 \\ d_3 & b_3 & c_3 \end{vmatrix}. \tag{6.21}$$

Fig. 6.2 A schematic diagram for a third-order determinant

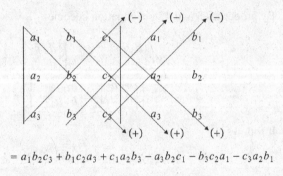

$$= a_1 b_2 c_3 + b_1 c_2 a_3 + c_1 a_2 b_3 - a_3 b_2 c_1 - b_3 c_2 a_1 - c_3 a_2 b_1$$

Therefore

$$x = \frac{\begin{vmatrix} d_1 & b_1 & c_1 \\ d_2 & b_2 & c_2 \\ d_3 & b_3 & c_3 \end{vmatrix}}{\begin{vmatrix} a_1 & b_1 & c_1 \\ a_2 & b_2 & c_2 \\ a_3 & b_3 & c_3 \end{vmatrix}}.$$

Similarly we can define

$$N_y = \begin{vmatrix} a_1 & d_1 & c_1 \\ a_2 & d_2 & c_2 \\ a_3 & d_3 & c_3 \end{vmatrix}, \qquad N_z = \begin{vmatrix} a_1 & b_1 & d_1 \\ a_2 & b_2 & d_2 \\ a_3 & b_3 & d_3 \end{vmatrix},$$

and show

$$y = \frac{N_y}{D}, \qquad z = \frac{N_z}{D}.$$

The determinant in the denominator D is called the determinant of the coefficients. It is simply formed with the array of the coefficients on the left-hand sides of Equations (6.9) to (6.11). To find the numerator determinant N_x, start with D, erase the x coefficients a_1, a_2, and a_3, and replace them with the constants d_1, d_2, and d_3 from the right-hand sides of the equations. Similarly we replace the y coefficients in D with the constant terms to find N_y, and the z coefficients in D with the constants to find N_z.

Example 6.1.2 Find the solution of

$$3x + 2y + z = 11,$$
$$2x + 3y + z = 13,$$
$$x + y + 4z = 12.$$

Solution 6.1.2

$$D = \begin{vmatrix} 3 & 2 & 1 \\ 2 & 3 & 1 \\ 1 & 1 & 4 \end{vmatrix} = 36 + 2 + 2 - 3 - 3 - 16 = 18$$

$$N_x = \begin{vmatrix} 11 & 2 & 1 \\ 13 & 3 & 1 \\ 12 & 1 & 4 \end{vmatrix} = 132 + 24 + 13 - 36 - 11 - 104 = 18$$

$$N_y = \begin{vmatrix} 3 & 11 & 1 \\ 2 & 13 & 1 \\ 1 & 12 & 4 \end{vmatrix} = 156 + 11 + 24 - 13 - 36 - 88 = 54$$

$$N_z = \begin{vmatrix} 3 & 2 & 11 \\ 2 & 3 & 13 \\ 1 & 1 & 12 \end{vmatrix} = 108 + 26 + 22 - 33 - 39 - 48 = 36.$$

Thus

$$x = \frac{18}{18} = 1, \quad y = \frac{54}{18} = 3, \quad z = \frac{36}{18} = 2.$$

Clearly, with determinant notation, the results can be given in a systematic way. While this procedure is still valid for systems of more than three equations, as we shall see in the section on Cramer's rule, but the diagonal scheme of expanding the determinants shown in this section is generally correct only for determinants of second and third orders. For determinants of higher order, we must pay attention to the formal definition of determinants.

6.2 General Definition of Determinants

6.2.1 Notations

Before we present the general definition of an arbitrary order determinant, let us write the third-order determinant in a more systematic way. Equations (6.19) and (6.20) can be written in the following form:

$$\begin{vmatrix} a_1 & b_1 & c_1 \\ a_2 & b_2 & c_2 \\ a_3 & b_3 & c_3 \end{vmatrix} = \sum_{i=1}^{3} \sum_{j=1}^{3} \sum_{k=1}^{3} \varepsilon_{ijk} a_i b_j c_k, \tag{6.22}$$

Fig. 6.3 Levi-Civita symbol ε_{ijk} where i, j, k take the value of 1, 2, or 3. If the set of indices goes clockwise, $\varepsilon_{ijk} = +1$; if counterclockwise, $\varepsilon_{ijk} = -1$

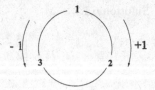

where

$$\varepsilon_{123} = \varepsilon_{231} = \varepsilon_{312} = 1,$$
$$\varepsilon_{132} = \varepsilon_{321} = \varepsilon_{213} = -1, \tag{6.23}$$
$$\varepsilon_{ijk} = 0 \ for \ all \ others.$$

Writing out term by term the right-hand side of (6.22), one can readily verify that the six nonvanishing terms are exactly the same as in (6.19).

In order to generalize this definition for a nth-order determinant, let us examine the triple sum more closely. First we note that $\varepsilon_{ijk} = 0$ if any two of the three indices i, j, k are equal, e.g., $\varepsilon_{112} = 0$, $\varepsilon_{333} = 0$. Eliminating those terms, (6.22) is a particular linear combination of six products, each product contains one and only one element from each row and from each column. Each product carries either a positive or a negative sign. The arrangements of (i, j, k) in the positive products are either in the normal order of (1, 2, 3) or are the results of an even number of interchanges between two adjacent numbers of the normal order. Those in the negative products are the results of an odd number of interchanges in the normal order. For example, it takes two interchanges to get (2, 3, 1) from (1, 2, 3) [123 (interchange 12) $\rightarrow$ 213 (interchange 13) $\rightarrow$ 231], and $a_2 b_3 c_1$ is positive ($\varepsilon_{231} = 1$); it takes only one interchange to get (1, 3, 2) from (1, 2, 3) [123 (interchange 23) $\rightarrow$ 132], and $a_1 b_3 c_2$ is negative ($\varepsilon_{132} = -1$). The following diagram can help us to find out the value of ε_{ijk} quickly. If a set of indices goes in the clockwise direction, it gives a positive one (+1); if it goes in the counterclockwise direction, it gives a negative one (−1) (Fig. 6.3).

These properties are characterized by the Levi-Civita symbol $\varepsilon_{i_1 i_2 \cdots i_n}$, which is defined as follows:

$$\varepsilon_{i_1 i_2 \cdots i_n} = \begin{cases} 1 & if \ (i_1, i_2 \ldots i_n) \ is \ an \ even \ permutation \\ & of \ the \ normal \ order \ (1, 2, \ldots n) \\ -1 & if \ (i_1, i_2 \ldots i_n) \ is \ an \ odd \ permutation \\ & of \ the \ normal \ order \ (1, 2, \ldots n) \\ 0 & if \ any \ index \ is \ repeated. \end{cases}$$

An even permutation means that an even number of pairwise interchanges of adjacent numbers is needed to obtain the given permutation from the normal order, and an odd permutation is associated with an odd number of pairwise interchanges. As we have shown, (2, 3, 1) is an even permutation and (1, 3, 2) is an odd permutation.

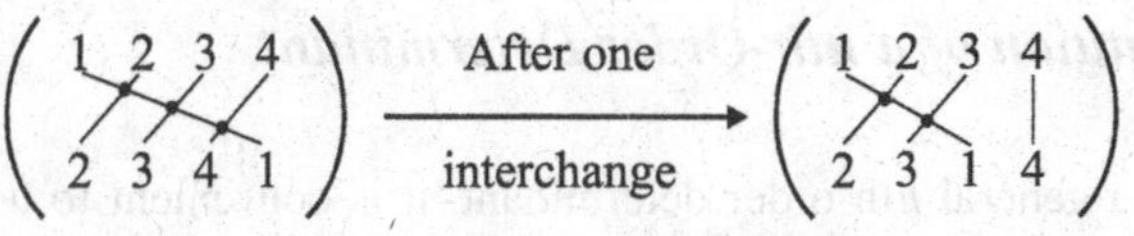

$$\begin{pmatrix} 1 & 2 & 3 & 4 \\ 2 & 3 & 4 & 1 \end{pmatrix} \xrightarrow[\text{interchange}]{\text{After one}} \begin{pmatrix} 1 & 2 & 3 & 4 \\ 2 & 3 & 1 & 4 \end{pmatrix}$$

Fig. 6.4 Permutation Diagram. The permutation is written directly below the normal order. The number of intersections between pairs of lines connecting the corresponding numbers is equal to the number of interchanges needed to obtain the permutation from the normal order. This diagram shows that one intersection point represents one interchange between two adjacent members

An easy way to determine whether a given permutation is even or odd is to write out the normal order and write the permutation directly below it. Then connect corresponding numbers in these two arrangements with line segments and count the number of intersections between pairs of these lines. If the number of intersections is even, then the given permutation is even. If the number of intersections is odd, then the permutation is odd. For example, to find the permutation $(2, 3, 4, 1)$, we write out the normal order and permutation in (we call it "permutation diagram") Fig. 6.4.

There are three intersections. Therefore, the permutation is odd and $\varepsilon_{2341} = -1$.

The reason this scheme is valid is because of the following. Starting with the smallest number that is not directly below the same number, an exchange of this number with the number to its left will eliminate one intersection. In the above example, after the interchange between 1 and 4, only two intersections remain. Clearly two more interchanges will eliminate all intersections and return the permutation to the normal order. Thus, three intersections indicate three interchanges are needed. Therefore the permutation is odd.

When we count the number of intersections, we are counting the intersections of pairs of lines. Therefore, one should avoid to have more than two lines intersecting at a point. The lines joining the corresponding numbers need not to be straight lines (Fig. 6.5).

Example 6.2.1 What is the value of the Levi-Civita symbol $\varepsilon_{1357246}$?

Solution 6.2.1 There are six intersections in the above figure, therefore the permutation is even and $\varepsilon_{1357246} = 1$.

Fig. 6.5 In this diagram, six intersections represent that six interchanges are needed to obtain the permutation 1357246 from the normal order 1234567

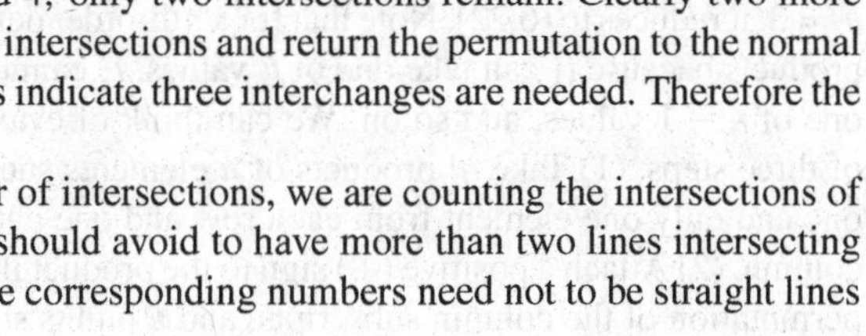

$$\begin{pmatrix} 1 & 2 & 3 & 4 & 5 & 6 & 7 \\ 1 & 3 & 5 & 7 & 2 & 4 & 6 \end{pmatrix} \rightarrow \begin{pmatrix} 1 & 2 & 3 & 4 & 5 & 6 & 7 \\ 1 & 2 & 3 & 5 & 7 & 4 & 6 \end{pmatrix}$$

$$\rightarrow \begin{pmatrix} 1 & 2 & 3 & 4 & 5 & 6 & 7 \\ 1 & 2 & 3 & 4 & 5 & 7 & 6 \end{pmatrix} \rightarrow \begin{pmatrix} 1 & 2 & 3 & 4 & 5 & 6 & 7 \\ 1 & 2 & 3 & 4 & 5 & 6 & 7 \end{pmatrix}$$

6.2.2 Definition of a nth-Order Determinant

In discussing a general nth-order determinant, it is convenient to use the double-subscript notation. Each element of the determinant is represented by the symbol a_{ij}. The subscripts ij indicate that it is the element at ith row and jth column. With this notation, $a_1b_2c_3$ becomes $a_{11}a_{22}a_{33}$; $a_2b_3c_1$ becomes $a_{21}a_{32}a_{13}$; and $a_ib_jc_k$ becomes $a_{i1}a_{j2}a_{k3}$. The determinant itself is denoted by a variety of symbols. The following notations are all equivalent:

$$\begin{vmatrix} a_{11} & a_{12} & \cdots & a_{1n} \\ a_{21} & a_{22} & \cdots & a_{2n} \\ \cdot & \cdot & \cdots & \cdot \\ a_{n1} & a_{n2} & \cdots & a_{nn} \end{vmatrix} = |a_{ij}| = |A| = \det |A| = D_n. \tag{6.24}$$

The value of the determinant is given by

$$D_n = \sum_{i_1=1}^{n} \sum_{i_2=1}^{n} \cdots \sum_{i_n=1}^{n} \varepsilon_{i_1 i_2 \cdots i_n} a_{i_1 1} a_{i_2 2} \cdots a_{i_n n}. \tag{6.25}$$

This equation is the formal definition of a nth-order determinant. Clearly, for $n = 3$, it reduces to (6.22). Note that for a nth-order determinant, there are $n!$ possible products because i_1 can take one of n values, i_2 cannot repeat i_1, so it can take only one of $n - 1$ values, and so on. We can think of evaluating a determinant in terms of three steps. (1) Take $n!$ products of n elements such that in each product there is one and only one element from each row and one and only one element from each column. (2) Attach a positive $(+)$ sign to the product if the row subscripts are an even permutation of the column subscripts, and a minus sign $(-)$ if an odd perturbation. (3) Sum over $n!$ products with these signs.

Stated in this way, it is clear that the definition of a determinant is symmetrical between the rows and columns. The determinant (6.25) can just as well be written as

$$D_n = \sum_{i_1=1}^{n} \sum_{i_2=1}^{n} \cdots \sum_{i_n=1}^{n} \varepsilon_{i_1 i_2 \cdots i_n} a_{1 i_1} a_{2 i_2} \cdots a_{n i_n}. \tag{6.26}$$

It follows that any theorem about the determinant which involves the rows is also true for the columns, and vice versa.

Another property that is clear from this definition is this. If any two rows are interchanged, the determinant changes sign. First it is easy to show that if the two rows are adjacent to each other, this is the case. This follows from the fact that an interchange of two adjacent rows corresponds to an interchange of two adjacent row indices in the Levi-Civita symbol. It changes an even permutation into an odd permutation, and vice versa. Therefore, it introduces a minus sign to all the products.

Now suppose the row indices i and j are not adjacent to each other and there are n indices between them:

$$i \; a_1 \; a_2 \; a_3 \; \cdots \; a_n \; j \, .$$

To bring j to the left requires $n + 1$ adjacent interchanges leading to

$$j \; i \; a_1 \; a_2 \; a_3 \; \cdots \; a_n \, .$$

Now bringing i to the right requires n adjacent interchanges leading to

$$j \; a_1 \; a_2 \; a_3 \; \cdots \; a_n \; i \, .$$

Therefore, all together there are $2n + 1$ number of adjacent interchanges leading to the interchange of i and j. Since $2n + 1$ is an odd integer, this brings in an overall minus sign.

Example 6.2.2 Let $D_2 = \begin{vmatrix} a_{11} & a_{12} \\ a_{21} & a_{22} \end{vmatrix}$, use (6.25) to (a) expand this second-order determinant, (b) show explicitly that the interchange of the two rows changes its sign.

Solution 6.2.2 (a) According to (6.25)

$$D_2 = \sum_{i_1=1}^{2} \sum_{i_2=1}^{2} \varepsilon_{i_1 i_2} a_{i_1 1} a_{i_2 2},$$

$$
\begin{aligned}
i_1 = 1, \; i_2 = 1: & \quad \varepsilon_{i_1 i_2} a_{i_1 1} a_{i_2 2} = \varepsilon_{11} a_{11} a_{12} \\
i_1 = 1, \; i_2 = 2: & \quad \varepsilon_{i_1 i_2} a_{i_1 1} a_{i_2 2} = \varepsilon_{12} a_{11} a_{22} \\
i_1 = 2, \; i_2 = 1: & \quad \varepsilon_{i_1 i_2} a_{i_1 1} a_{i_2 2} = \varepsilon_{21} a_{21} a_{12} \\
i_1 = 2, \; i_2 = 2: & \quad \varepsilon_{i_1 i_2} a_{i_1 1} a_{i_2 2} = \varepsilon_{22} a_{21} a_{22}.
\end{aligned}
$$

Since $\varepsilon_{11} = 0$, $\varepsilon_{12} = 1$, $\varepsilon_{21} = -1$, $\varepsilon_{22} = 0$, the double sum gives the second-order determinant as

$$\sum_{i_1=1}^{2} \sum_{i_2=1}^{2} \varepsilon_{i_1 i_2} a_{i_1 1} a_{i_2 2} = a_{11} a_{22} - a_{21} a_{12}.$$

(b) To express the interchange of two rows, we can simply replace $a_{i_1 1} a_{i_2 2}$ in the double sum with $a_{i_2 1} a_{i_1 2}$ (i_1 and i_2 are interchanged), thus

$$\begin{vmatrix} a_{21} & a_{22} \\ a_{11} & a_{12} \end{vmatrix} = \sum_{i_1=1}^{2} \sum_{i_2=1}^{2} \varepsilon_{i_1 i_2} a_{i_2 1} a_{i_1 2}.$$

Since i_1 and i_2 are running indices, we can rename i_1 as j_2 and i_2 as j_1, so

$$\sum_{i_1=1}^{2}\sum_{i_2=1}^{2} \varepsilon_{i_1 i_2} a_{i_2 1} a_{i_1 2} = \sum_{j_2=1}^{2}\sum_{j_1=1}^{2} \varepsilon_{j_2 j_1} a_{j_1 1} a_{j_2 2} = \sum_{j_1=1}^{2}\sum_{j_2=1}^{2} \varepsilon_{j_2 j_1} a_{j_1 1} a_{j_2 2}.$$

The last expression is identical with that of the original determinant except the indices of the Levi-Civita symbol are interchanged.

$$\begin{vmatrix} a_{21} & a_{22} \\ a_{11} & a_{12} \end{vmatrix} = \sum_{j_1=1}^{2}\sum_{j_2=1}^{2} \varepsilon_{j_2 j_1} a_{j_1 1} a_{j_2 2}$$

$$= \varepsilon_{11} a_{11} a_{12} + \varepsilon_{21} a_{11} a_{22} + \varepsilon_{12} a_{21} a_{12} + \varepsilon_{22} a_{21} a_{22}$$

$$= -a_{11} a_{22} + a_{21} a_{12} = - \begin{vmatrix} a_{11} & a_{12} \\ a_{21} & a_{22} \end{vmatrix}.$$

This result can, of course, be obtained by inspection. We have taken the risk of stating the obvious. Hopefully this step-by-step approach will remove any uneasy feeling of working with indices.

6.2.3 Minors, Cofactors

Let us return to (6.18), written in the double-subscript notation this equation becomes

$$D_3 = \begin{vmatrix} a_{11} & a_{12} & a_{13} \\ a_{21} & a_{22} & a_{23} \\ a_{31} & a_{32} & a_{33} \end{vmatrix}$$

$$= a_{11} \begin{vmatrix} a_{22} & a_{23} \\ a_{32} & a_{33} \end{vmatrix} - a_{12} \begin{vmatrix} a_{21} & a_{23} \\ a_{31} & a_{33} \end{vmatrix} + a_{13} \begin{vmatrix} a_{21} & a_{22} \\ a_{31} & a_{32} \end{vmatrix}, \tag{6.27}$$

where we have interchanged the two columns of the last second-order determinant of (6.18) and changed the sign. It is seen that $\begin{vmatrix} a_{22} & a_{23} \\ a_{32} & a_{33} \end{vmatrix}$ is the second-order determinant formed by removing the first row and first column from the original third-order determinant D_3. We call it M_{11} the minor complementary to a_{11}. In general, the minor M_{ij} complementary to a_{ij} is defined as the $(n-1)$th-order determinant formed by deleting the ith row and the jth column from the original n th-order determinant D_n. The cofactor C_{ij} is defined as $(-1)^{i+j} M_{ij}$.

Example 6.2.3 Find the value of the minors M_{11}, M_{23} and the cofactors C_{11}, C_{23} of the determinant

$$D_4 = \begin{vmatrix} 2 & -1 & 1 & 3 \\ -3 & 2 & 5 & 0 \\ 1 & 0 & -2 & 2 \\ 4 & 2 & 3 & 1 \end{vmatrix}.$$

Solution 6.2.3

$$M_{11} = \begin{vmatrix} * & * & * & * \\ * & 2 & 5 & 0 \\ * & 0 & -2 & 2 \\ * & 2 & 3 & 1 \end{vmatrix} = \begin{vmatrix} 2 & 5 & 0 \\ 0 & -2 & 2 \\ 2 & 3 & 1 \end{vmatrix}; \quad M_{23} = \begin{vmatrix} 2 & -1 & * & 3 \\ * & * & * & * \\ 1 & 0 & * & 2 \\ 4 & 2 & * & 1 \end{vmatrix} = \begin{vmatrix} 2 & -1 & 3 \\ 1 & 0 & 2 \\ 4 & 2 & 1 \end{vmatrix}.$$

$$C_{11} = (-1)^{1+1} \begin{vmatrix} 2 & 5 & 0 \\ 0 & -2 & 2 \\ 2 & 3 & 1 \end{vmatrix}; \quad C_{23} = (-1)^{2+3} \begin{vmatrix} 2 & -1 & 3 \\ 1 & 0 & 2 \\ 4 & 2 & 1 \end{vmatrix}.$$

6.2.4 Laplacian Development of Determinants by a Row (or a Column)

With these notations, (6.27) becomes

$$D_3 = a_{11} M_{11} - a_{12} M_{12} + a_{13} M_{13} = \sum_{j=1}^{3} (-1)^{1+j} a_{1j} M_{1j} \tag{6.28}$$

$$= a_{11} C_{11} + a_{12} C_{12} + a_{13} C_{13} = \sum_{k=1}^{3} a_{1k} C_{1k}. \tag{6.29}$$

This is known as the Laplace development of the third-order determinant on elements of the first row. It turns out this is not limited to the third-order determinant. It is a fundamental theorem that determinants of any order can be evaluated by a Laplace development on any row or column,

$$D_n = \sum_{j=1}^{n} (-1)^{i+j} a_{ij} M_{ij} = \sum_{j=1}^{n} a_{ij} C_{ij} \quad for \ any \ i, \tag{6.30}$$

$$= \sum_{i=1}^{n} (-1)^{i+j} a_{ij} M_{ij} = \sum_{i=1}^{n} a_{ij} C_{ij} \quad for \ any \ j. \tag{6.31}$$

The proof may be given by induction and is based on the definition of the determinant. According to (6.25), a determinant is the sum of all the $n!$ products which are formed by taking exactly one element from each row and each column and multiplying by 1 or -1 in accordance with the Levi-Civita rule.

Now the minor M_{ij} of a nth-order determinant is a $(n-1)$th determinant. It is a sum of $(n-1)!$ products. Each product has one element from each row and each column except the ith row and jth column. It is then clear that $\sum_{j=1}^{n} k_{ij} a_{ij} M_{ij}$ is a sum of $n(n-1)! = n!$ products, and each product is formed with exactly one element from each row and each column. It follows that, with the appropriate choice of k_{ij}, the determinant can be written in a row expansion

$$D_n = \sum_{j=1}^{n} k_{ij} a_{ij} M_{ij} \tag{6.32}$$

or in a column expansion

$$D_n = \sum_{i=1}^{n} k_{ij} a_{ij} M_{ij}. \tag{6.33}$$

The Laplace development will follow if we can show

$$k_{ij} = (-1)^{i+k}.$$

First let us consider all the terms in (6.25) containing a_{11}. In these terms $i_1 = 1$. We note that if $(1, i_2, i_3 \ldots i_n)$ is an even (or odd) permutation of $(1, 2, 3, \ldots n)$, it means $(i_2, i_3, \ldots i_n)$ is an even (or odd) permutation of $(2, 3, \ldots n)$. The number of intersections in the following two "permutation diagrams" are obviously the same,

$$\begin{pmatrix} 1 & 2 & 3 & \cdots & n \\ 1 & i_2 & i_3 & \cdots & i_n \end{pmatrix}; \quad \begin{pmatrix} 2 & 3 & \cdots & n \\ i_2 & i_3 & \cdots & i_n \end{pmatrix},$$

therefore

$$\varepsilon_{1i_2 \cdots i_n} = \varepsilon_{i_2 \cdots i_n}.$$

So terms containing a_{11} sum to

$$\sum_{i_2=1}^{n} \cdots \sum_{i_n=1}^{n} \varepsilon_{1i_2 \cdots i_n} a_{11} a_{i_2 2} \cdots a_{i_n n} = a_{11} \sum_{i_2=1}^{n} \cdots \sum_{i_n=1}^{n} \varepsilon_{i_2 \cdots i_n} a_{i_2 2} \cdots a_{i_n n},$$

which is simply $a_{11} M_{11}$, where M_{11} is the minor of a_{11}. On the other hand, according to (6.32), all the terms containing a_{11} sum to $k_{11} a_{11} M_{11}$. Therefore

$$k_{11} = +1.$$

Next consider the terms in (6.25) which contain a particular element a_{ij}. If we interchange the ith row with the one above it, the determinant changes sign. If we move the row up in this way $(i - 1)$ times, the ith row will have moved up into the first row, and the order of the other rows is not changed. The process will change the sign of the determinant $(i - 1)$ times. In a similar way, we can move the jth column to the first column without changing the order of the other columns. Then the element a_{ij} will be in the top left corner of the determinant, in the place of a_{11}, and the sign of the determinant has changed $(i - 1 + j - 1)$ times. That is

$$
\begin{vmatrix}
a_{11} & \cdots & a_{1j} & \cdots & a_{1n} \\
\cdots & \cdots & \cdots & \cdots & \cdots \\
a_{i1} & \cdots & a_{ij} & \cdots & a_{in} \\
\cdots & \cdots & \cdots & \cdots & \cdots \\
a_{n1} & \cdots & a_{nj} & \cdots & a_{nn}
\end{vmatrix}
= (-1)^{i+j-2}
\begin{vmatrix}
a_{ij} & a_{i1} & a_{i2} & \cdots & a_{in} \\
a_{1j} & a_{11} & a_{12} & \cdots & a_{1n} \\
a_{2j} & a_{21} & a_{22} & \cdots & a_{2n} \\
\cdots & \cdots & \cdots & \cdots & \cdots \\
a_{nj} & a_{n1} & a_{n2} & & a_{nn}
\end{vmatrix}.
$$

In the rearranged determinant, a_{ij} is in the place of a_{11}, thus the sum of all the terms containing a_{ij} is equal to $a_{ij} M_{ij}$. But there is a factor $(-1)^{i+j-2}$ in front of the rearranged determinant. Therefore, the terms containing a_{ij} in the right-hand side of the equation sum to $(-1)^{i+j-2} a_{ij} M_{ij}$. On the other hand, according to (6.32), all the terms containing a_{ij} in the determinant of the left-hand side of the equation sum to $k_{ij} a_{ij} M_{ij}$. Therefore

$$
k_{ij} = (-1)^{i+j-2} = (-1)^{i+j}. \tag{6.34}
$$

This completes the proof of the Laplace development, which is very important in both theory and computation of determinants. It is useful to keep in mind that k_{ij} forms a checkboard pattern:

$$
\begin{vmatrix}
+1 & -1 & +1 & & \\
-1 & 1 & -1 & & \\
+1 & -1 & +1 & & \\
& & & \ddots & \\
& & & & \begin{matrix} +1 & -1 \\ -1 & +1 \end{matrix}
\end{vmatrix}.
$$

Example 6.2.4 Find the value of the determinant $D_3 = \begin{vmatrix} 3 & -2 & 2 \\ 1 & 2 & -3 \\ 4 & 1 & 2 \end{vmatrix}$ by (a) a Laplace development on the first row; (b) a Laplace development on the second row; (c) a Laplace development on the first column.

Solution 6.2.4 (a)

$$
D_3 = a_{11} M_{11} - a_{12} M_{12} + a_{13} M_{13} = a_{11} C_{11} + a_{12} C_{12} + a_{13} C_{13}
$$
$$
= 3 \begin{vmatrix} 2 & -3 \\ 1 & 2 \end{vmatrix} - (-2) \begin{vmatrix} 1 & -3 \\ 4 & 2 \end{vmatrix} + 2 \begin{vmatrix} 1 & 2 \\ 4 & 1 \end{vmatrix}
$$
$$
= 3(4 + 3) + 2(2 + 12) + 2(1 - 8) = 35.
$$

(b)

$$D_3 = -a_{21}M_{21} + a_{22}M_{22} - a_{23}M_{23}$$

$$= -1 \begin{vmatrix} -2 & 2 \\ 1 & 2 \end{vmatrix} + 2 \begin{vmatrix} 3 & 2 \\ 4 & 2 \end{vmatrix} - (-3) \begin{vmatrix} 3 & -2 \\ 4 & 1 \end{vmatrix}$$

$$= -(-4-2) + 2(6-8) + 3(3+8) = 35.$$

(c)

$$D_3 = a_{11}M_{11} - a_{21}M_{21} + a_{31}M_{31}$$

$$= 3 \begin{vmatrix} 2 & -3 \\ 1 & 2 \end{vmatrix} - 1 \begin{vmatrix} -2 & 2 \\ 1 & 2 \end{vmatrix} + 4 \begin{vmatrix} -2 & 2 \\ 2 & -3 \end{vmatrix}$$

$$= 3(4+3) - (-4-2) + 4(6-4) = 35.$$

Example 6.2.5 Find the value of the **triangular determinant**

$$D_n = \begin{vmatrix} a_{11} & a_{12} & a_{13} & \cdots & a_{1n} \\ 0 & a_{22} & a_{23} & \cdots & a_{2n} \\ 0 & 0 & a_{33} & \cdots & a_{3n} \\ \cdots & \cdots & \cdots & \cdots & \cdots \\ 0 & 0 & 0 & \cdots & a_{nn} \end{vmatrix}.$$

Solution 6.2.5

$$D_n = a_{11} \begin{vmatrix} a_{22} & a_{23} & \cdots & a_{2n} \\ 0 & a_{33} & \cdots & a_{3n} \\ \cdots & \cdots & \cdots & \cdots \\ 0 & 0 & \cdots & a_{nn} \end{vmatrix} = a_{11}a_{22} \begin{vmatrix} a_{33} & \cdots & a_{3n} \\ \cdots & \cdots & \cdots \\ 0 & \cdots & a_{nn} \end{vmatrix} = a_{11}a_{22}a_{33}\cdots a_{nn}.$$

6.3 Properties of Determinants

By mathematical induction, we can now show that properties 1 to 4 of second-order determinants are generally valid for nth-order determinants. Based on the fact that it is true for $(n-1)$th-order determinants, we will show that it must also be true for nth-order determinants. All properties of the determinant can be derived directly from its definition of (6.25). However, in this section, we will demonstrate them with Laplace expansions.

- 1. The value of the determinant remains the same if rows and columns are interchanged.

Let the Laplace expansion of D_n on elements of the first row be

$$D_n = \sum_{j=1}^{n} (-1)^{1+j} a_{1j} M_{1j}. \tag{6.35}$$

Let D_n^T (known as the transpose of D_n) be the nth-order determinant formed by interchanging rows and columns of the determinant D_n. The Laplace expansion of D_n^T on elements of the first column (which are elements of the first row of D_n) is then given by

$$D_n^T = \sum_{j=1}^{n} (-1)^{1+j} a_{1j} M_{1j}^T, \tag{6.36}$$

where M_{1j}^T is the minor complement to a_{1j} and is equal to the determinant M_{1j} with rows and columns interchanged. In the case of $n = 3$, the minors are second-order determinants. By (6.7), $M_{1j}^T = M_{1j}$. Therefore $D_3 = D_3^T$. This process can be carried out, one step at a time, to any n. Therefore we conclude

$$D_n = D_n^T. \tag{6.37}$$

- 2. The determinant changes sign if any two columns (or any two rows) are interchanged.

First we will verify this property for the third-order determinant D_3. Let E_3 be the determinant obtained by interchanging two columns of D_3. Suppose column k is not one of those exchanged. Using Laplace development to expand D_3 and E_3 by their kth column, we have

$$D_3 = \sum_{i=1}^{3} (-1)^{i+k} a_{ik} M_{ik}; \tag{6.38}$$

$$E_3 = \sum_{i=1}^{3} (-1)^{i+k} a_{ik} M_{ik}', \tag{6.39}$$

where M_{ik}' is a second-order determinant and is equal to M_{ik} with the two columns interchanged. By (6.8) , $M_{ik}' = -M_{ik}$. Hence $E_3 = -D_3$. Now by mathematical induction, we assume this property holds for $(n-1)$th-order determinants. The same procedure will show that this property also holds for determinants of nth order.

This property is called anti-symmetric property. It is frequently used in quantum mechanics in the construction of an anti-symmetric many particle wave functions.

• 3. If each element in a column (or in a row) is multiplied by a constant m, the determinant is multiplied by m.

This property follows directly from the Laplacian expansion. If the ith column is multiplied by m, this property can be shown in the following way:

$$
\begin{vmatrix}
a_{11} & \cdots & ma_{1i} & \cdots & a_{1n} \\
a_{21} & \cdots & ma_{2i} & \cdots & a_{2n} \\
\cdot & \cdot\cdot & \cdot & \cdot\cdot\cdot & \cdot \\
a_{n1} & \cdots & ma_{ni} & \cdots & a_{nn}
\end{vmatrix}
= \sum_{j=1}^{n} ma_{ji}C_{ji} = m\sum_{j=1}^{n} a_{ji}C_{ji}
$$

$$
= m
\begin{vmatrix}
a_{11} & \cdots & a_{1i} & \cdots & a_{1n} \\
a_{21} & \cdots & a_{2i} & \cdots & a_{2n} \\
\cdot & \cdot\cdot & \cdot & \cdot\cdot\cdot & \cdot \\
a_{n1} & \cdots & a_{ni} & \cdots & a_{nn}
\end{vmatrix}.
\tag{6.40}
$$

• 4. If each element in a column (or in a row) is a sum of two terms, the determinant equals the sum of the two corresponding determinants.

If the ith column is a sum of two terms, we can expand the determinant on elements of the ith column

$$
\begin{vmatrix}
a_{11} & \cdots & a_{1i}+b_{1i} & \cdots & a_{1n} \\
a_{21} & \cdots & a_{2i}+b_{2i} & \cdots & a_{2n} \\
\cdot & \cdot\cdot & \cdot & \cdot\cdot\cdot & \cdot \\
a_{n1} & \cdots & a_{ni}+b_{ni} & \cdots & a_{nn}
\end{vmatrix}
= \sum_{j=1}^{n}(a_{ji}+b_{ji})C_{ji} = \sum_{j=1}^{n} a_{ji}C_{ji} + \sum_{j=1}^{n} b_{ji}C_{ji}
$$

$$
=
\begin{vmatrix}
a_{11} & \cdots & a_{1i} & \cdots & a_{1n} \\
a_{21} & \cdots & a_{2i} & \cdots & a_{2n} \\
\cdot & \cdot\cdot & \cdot & \cdot\cdot\cdot & \cdot \\
a_{n1} & \cdots & a_{ni} & \cdots & a_{nn}
\end{vmatrix}
+
\begin{vmatrix}
a_{11} & \cdots & b_{1i} & \cdots & a_{1n} \\
a_{21} & \cdots & b_{2i} & \cdots & a_{2n} \\
\cdot & \cdot\cdot & \cdot & \cdot\cdot\cdot & \cdot \\
a_{n1} & \cdots & b_{ni} & \cdots & a_{nn}
\end{vmatrix}.
\tag{6.41}
$$

From these four properties, one can derive many others. For example:

• 5. If two columns (or two rows) are the same, the determinant is zero.

This follows from the anti-symmetric property. If we exchange the two identical columns, the determinant will obviously remain the same. Yet the anti-symmetric property requires the determinant to change sign. The only number that is equal to its negative self is zero. Therefore, the determinant must be zero.

• 6. The value of a determinant is unchanged if a multiple of one column is added to another column (or if a multiple of one row is added to another row).

Without loss of generality, this property can be expressed as follows:

$$
\begin{vmatrix} a_{11}+ma_{12} & a_{12} & \cdots & a_{1n} \\ a_{21}+ma_{22} & a_{22} & \cdots & a_{2n} \\ & \cdot & \cdots & \cdot \\ a_{n1}+ma_{n2} & a_{n2} & \cdots & a_{nn} \end{vmatrix} = \begin{vmatrix} a_{11} & a_{12} & \cdots & a_{1n} \\ a_{21} & a_{22} & \cdots & a_{2n} \\ & \cdot & \cdots & \cdot \\ a_{n1} & a_{n2} & \cdots & a_{nn} \end{vmatrix} + \begin{vmatrix} ma_{12} & a_{12} & \cdots & a_{1n} \\ ma_{22} & a_{22} & \cdots & a_{2n} \\ & \cdot & \cdots & \cdot \\ ma_{n2} & a_{n2} & \cdots & a_{nn} \end{vmatrix}
$$

$$
= \begin{vmatrix} a_{11} & a_{12} & \cdots & a_{1n} \\ a_{21} & a_{22} & \cdots & a_{2n} \\ & \cdot & \cdots & \cdot \\ a_{n1} & a_{n2} & \cdots & a_{nn} \end{vmatrix} + m \begin{vmatrix} a_{12} & a_{12} & \cdots & a_{1n} \\ a_{22} & a_{22} & \cdots & a_{2n} \\ & \cdot & \cdots & \cdot \\ a_{n2} & a_{n2} & \cdots & a_{nn} \end{vmatrix} = \begin{vmatrix} a_{11} & a_{12} & \cdots & a_{1n} \\ a_{21} & a_{22} & \cdots & a_{2n} \\ & \cdot & \cdots & \cdot \\ a_{n1} & a_{n2} & \cdots & a_{nn} \end{vmatrix} . \tag{6.42}
$$

The first equal sign is by property 4, the second equal sign is because of property 3, and the last equal sign is due to property 5.

Example 6.3.1 Show that

$$
\begin{vmatrix} 1 & a & bc \\ 1 & b & ac \\ 1 & c & ab \end{vmatrix} = \begin{vmatrix} 1 & a & a^2 \\ 1 & b & b^2 \\ 1 & c & c^2 \end{vmatrix} .
$$

Solution 6.3.1

$$
\begin{vmatrix} 1 & a & bc \\ 1 & b & ac \\ 1 & c & ab \end{vmatrix} = \begin{vmatrix} 1 & a & (bc+a^2) \\ 1 & b & (ac+ab) \\ 1 & c & (ab+ac) \end{vmatrix} = \begin{vmatrix} 1 & a & (bc+a^2+ba) \\ 1 & b & (ac+ab+b^2) \\ 1 & c & (ab+ac+bc) \end{vmatrix}
$$

$$
= \begin{vmatrix} 1 & a & (bc+a^2+ba+ca) \\ 1 & b & (ac+ab+b^2+cb) \\ 1 & c & (ab+ac+bc+c^2) \end{vmatrix} = \begin{vmatrix} 1 & a & a^2 \\ 1 & b & b^2 \\ 1 & c & c^2 \end{vmatrix} + \begin{vmatrix} 1 & a & (bc+ba+ca) \\ 1 & b & (ac+ab+cb) \\ 1 & c & (ab+ac+bc) \end{vmatrix}
$$

$$
= \begin{vmatrix} 1 & a & a^2 \\ 1 & b & b^2 \\ 1 & c & c^2 \end{vmatrix} + (ab+bc+ca) \begin{vmatrix} 1 & a & 1 \\ 1 & b & 1 \\ 1 & c & 1 \end{vmatrix} = \begin{vmatrix} 1 & a & a^2 \\ 1 & b & b^2 \\ 1 & c & c^2 \end{vmatrix} .
$$

First, we multiply each element of the second column by a and add to the third column. For the second equal sign, we multiply the second column by b and add to the third column. Do the same thing except multiplying by c for the third equal sign. The fourth equal sign is due to property 4. The fifth equal sign is due to property 3. And lastly, the determinant with two identical column vanishes.

Example 6.3.2 Evaluate the determinant

$$
D_n = \begin{vmatrix} 1+a_1 & a_2 & a_3 & \cdot\cdot & a_n \\ a_1 & 1+a_2 & a_3 & \cdot\cdot & a_n \\ a_1 & a_2 & 1+a_3 & \cdot\cdot & a_n \\ \cdot & \cdot & \cdot & \cdot\cdot & \cdot \\ \cdot & \cdot & \cdot & \cdot\cdot & \cdot \\ a_1 & a_2 & a_3 & \cdot\cdot & 1+a_n \end{vmatrix} .
$$

Solution 6.3.2 Adding column 2, column 3, all the way to column n to column 1, we have

$$
D_n = \begin{vmatrix}
1+a_1+a_2+a_3+\cdots+a_n & a_2 & a_3 & \cdot\cdot & a_n \\
1+a_1+a_2+a_3+\cdots+a_n & 1+a_2 & a_3 & \cdot\cdot & a_n \\
1+a_1+a_2+a_3+\cdots+a_n & a_2 & 1+a_3 & \cdot\cdot & a_n \\
& \cdot & \cdot & \cdot\cdot & \cdot \\
& \cdot & \cdot & \cdot\cdot & \cdot \\
1+a_1+a_2+a_3+\cdots+a_n & a_2 & a_3 & \cdot\cdot & 1+a_n
\end{vmatrix}
$$

$$
= (1+a_1+a_2+a_3+\cdots+a_n)\begin{vmatrix}
1 & a_2 & a_3 & \cdot\cdot & a_n \\
1 & 1+a_2 & a_3 & \cdot\cdot & a_n \\
1 & a_2 & 1+a_3 & \cdot\cdot & a_n \\
\cdot & \cdot & \cdot & \cdot\cdot & \cdot \\
\cdot & \cdot & \cdot & \cdot\cdot & \cdot \\
1 & a_2 & a_3 & \cdot\cdot & 1+a_n
\end{vmatrix}.
$$

Multiplying row 1 by -1 and add it to row 2, and then add it to row 3, and so on

$$
D_n = (1+a_1+a_2+a_3+\cdots+a_n)\begin{vmatrix}
1 & a_2 & a_3 & \cdot\cdot & a_n \\
0 & 1 & 0 & \cdot\cdot & 0 \\
0 & 0 & 1 & \cdot\cdot & 0 \\
\cdot & \cdot & \cdot & \cdot\cdot & \cdot \\
\cdot & \cdot & \cdot & \cdot\cdot & \cdot \\
0 & 0 & 0 & \cdot\cdot & 1
\end{vmatrix}
$$

$$
= (1+a_1+a_2+a_3+\cdots+a_n).
$$

Example 6.3.3 Evaluate the following determinants (known as **Vandermonde determinant**):

$$
(a)\ D_3 = \begin{vmatrix}
1 & x_1 & x_1^2 \\
1 & x_2 & x_2^2 \\
1 & x_3 & x_3^2
\end{vmatrix}, \qquad
(b)\ D_n = \begin{vmatrix}
1 & x_1 & x_1^2 & \cdot\cdot & x_1^{n-1} \\
1 & x_2 & x_2^2 & \cdot\cdot & x_2^{n-1} \\
1 & x_3 & x_3^2 & \cdot\cdot & x_3^{n-1} \\
\cdot & \cdot & \cdot & \cdot\cdot & \cdot \\
\cdot & \cdot & \cdot & \cdot\cdot & \cdot \\
1 & x_n & x_n^2 & \cdot\cdot & x_n^{n-1}
\end{vmatrix}.
$$

Solution 6.3.3 (a) Method I.

$$
\begin{vmatrix}
1 & x_1 & x_1^2 \\
1 & x_2 & x_2^2 \\
1 & x_3 & x_3^2
\end{vmatrix} = \begin{vmatrix}
1 & x_1 & x_1^2 \\
0 & (x_2-x_1) & (x_2^2-x_1^2) \\
0 & (x_3-x_1) & (x_3^2-x_1^2)
\end{vmatrix} = (x_2-x_1)(x_3-x_1)\begin{vmatrix}
1 & (x_2+x_1) \\
1 & (x_3+x_1)
\end{vmatrix}
$$

$$
= (x_2-x_1)(x_3-x_1)(x_3-x_2).
$$

Method II. D_3 is a polynomial in x_1 and it vanishes when $x_1 = x_2$, since then the first two rows are the same. Hence it is divisible by $(x_1 - x_2)$. Similarly, it is divisible by $(x_2 - x_3)$ and $(x_3 - x_1)$. Therefore

$$D_3 = k(x_1 - x_2)(x_1 - x_3)(x_2 - x_3).$$

Furthermore, since D_3 is of degree 3 in x_1, x_2, x_3, k must be a constant. The coefficient of the term $x_2 x_3^2$ in this expression is $k(-1)(-1)^2$. On the other hand, the diagonal product of the D_3 is $+x_2 x_3^2$. Comparing them shows that $k(-1)(-1)^2 = 1$. Therefore $k = -1$ and

$$D_3 = -(x_1 - x_2)(x_1 - x_3)(x_2 - x_3) = (x_2 - x_1)(x_3 - x_1)(x_3 - x_2).$$

(b) With the same reason as in Method II of (a),

$$D_n = k(x_1 - x_2)(x_1 - x_3)\cdots(x_1 - x_n)(x_2 - x_3)\cdots(x_2 - x_n)\cdots(x_{n-1} - x_n).$$

The coefficient of the term $x_2 x_3^2 \cdots x_n^{n-1}$ in this expression is $k(-1)(-1)^2 \cdots (-1)^{n-1}$. Comparing this with the diagonal product of D_n, we have

$$1 = k(-1)(-1)^2 \cdots (-1)^{n-1} = k(-1)^{1+2+3+\cdots+(n-1)}.$$

Since

$$1 + 2 + 3 + \cdots + (n-1) = \frac{1}{2}n(n-1),$$

therefore

$$D_n = (-1)^{n(n-1)/2}(x_1 - x_2)(x_1 - x_3)\cdots(x_1 - x_n)(x_2 - x_3)\cdots(x_2 - x_n)\cdots(x_{n-1} - x_n).$$

Example 6.3.4 (*Pivotal Condensation*) Show that

$$D_3 = \begin{vmatrix} a_{11} & a_{12} & a_{13} \\ a_{21} & a_{22} & a_{23} \\ a_{31} & a_{32} & a_{33} \end{vmatrix} = \frac{1}{a_{11}} \begin{vmatrix} \begin{vmatrix} a_{11} & a_{12} \\ a_{21} & a_{22} \end{vmatrix} & \begin{vmatrix} a_{11} & a_{13} \\ a_{21} & a_{23} \end{vmatrix} \\ \begin{vmatrix} a_{11} & a_{12} \\ a_{31} & a_{32} \end{vmatrix} & \begin{vmatrix} a_{11} & a_{13} \\ a_{31} & a_{33} \end{vmatrix} \end{vmatrix}.$$

Clearly, a_{11} must be non-zero. If it is zero, then the first row (or first column) must be exchanged with another row (or another column), so that $a_{11} \neq 0$.

Solution 6.3.4

$$
\begin{vmatrix} a_{11} & a_{12} & a_{13} \\ a_{21} & a_{22} & a_{23} \\ a_{31} & a_{32} & a_{33} \end{vmatrix} = \frac{1}{a_{11}^2} \begin{vmatrix} a_{11} & a_{11}a_{12} & a_{11}a_{13} \\ a_{21} & a_{11}a_{22} & a_{11}a_{23} \\ a_{31} & a_{11}a_{32} & a_{11}a_{33} \end{vmatrix}
$$

$$
= \frac{1}{a_{11}^2} \begin{vmatrix} a_{11} & (a_{11}a_{12} - a_{11}a_{12}) & (a_{11}a_{13} - a_{11}a_{13}) \\ a_{21} & (a_{11}a_{22} - a_{21}a_{12}) & (a_{11}a_{23} - a_{21}a_{13}) \\ a_{31} & (a_{11}a_{32} - a_{31}a_{12}) & (a_{11}a_{33} - a_{31}a_{13}) \end{vmatrix}
$$

$$
= \frac{1}{a_{11}^2} \begin{vmatrix} a_{11} & 0 & 0 \\ a_{21} & (a_{11}a_{22} - a_{21}a_{12}) & (a_{11}a_{23} - a_{21}a_{13}) \\ a_{31} & (a_{11}a_{32} - a_{31}a_{12}) & (a_{11}a_{33} - a_{31}a_{13}) \end{vmatrix}
$$

$$
= \frac{1}{a_{11}} \begin{vmatrix} (a_{11}a_{22} - a_{21}a_{12}) & (a_{11}a_{23} - a_{21}a_{13}) \\ (a_{11}a_{32} - a_{31}a_{12}) & (a_{11}a_{33} - a_{31}a_{13}) \end{vmatrix}
$$

$$
= \frac{1}{a_{11}} \frac{\begin{vmatrix} \begin{vmatrix} a_{11} & a_{12} \\ a_{21} & a_{22} \end{vmatrix} & \begin{vmatrix} a_{11} & a_{13} \\ a_{21} & a_{23} \end{vmatrix} \\ \begin{vmatrix} a_{11} & a_{12} \\ a_{31} & a_{32} \end{vmatrix} & \begin{vmatrix} a_{11} & a_{13} \\ a_{31} & a_{33} \end{vmatrix} \end{vmatrix}}{} .
$$

This method can be applied to reduce a nth-order determinant to a $(n - 1)$th-order determinant and is known as pivotal condensation. It may not offer any advantage for hand calculation, but it is useful in evaluating determinants with computers.

6.4   Cramer's Rule

6.4.1   Nonhomogeneous Systems

Suppose we have a set of n equations and n unknowns

$$
\begin{aligned}
a_{11}x_1 + a_{12}x_2 \cdots + a_{1n}x_n &= d_1 \\
a_{21}x_1 + a_{22}x_2 \cdots + a_{2n}x_n &= d_2 \\
\cdots\cdots\cdots\cdots &= \cdot \\
a_{n1}x_1 + a_{n2}x_2 \cdots + a_{nn}x_n &= d_n.
\end{aligned} \tag{6.43}
$$

The constants $d_1, d_2, \cdots d_n$ on the right-hand side are known as nonhomogeneous terms. If they are not all equal to zero, the set of equations is known as a nonhomogeneous system. The problem is to find $x_1, x_2, \ldots x_n$ to satisfy this set of equations. We will see by using the properties of determinants, this set of equations can be readily solved for any n.

Forming the determinant of the coefficients and then multiplying by x_1, with the help of property 3, we have

$$x_1 \begin{vmatrix} a_{11} & a_{12} & \cdots & a_{1n} \\ a_{21} & a_{22} & \cdots & a_{2n} \\ \cdot & \cdot & \cdots & \cdot \\ a_{n1} & a_{n2} & \cdots & a_{nn} \end{vmatrix} = \begin{vmatrix} a_{11}x_1 & a_{12} & \cdots & a_{1n} \\ a_{21}x_1 & a_{22} & \cdots & a_{2n} \\ \cdot & \cdot & \cdots & \cdot \\ a_{n1}x_1 & a_{n2} & \cdots & a_{nn} \end{vmatrix}.$$

We multiply the second column of the right-hand side determinant by x_2 and add it to the first column, and then multiply the third column by x_3 and add it to the first column and so on. According to property 6, the determinant is unchanged

$$x_1 \begin{vmatrix} a_{11} & a_{12} & \cdots & a_{1n} \\ a_{21} & a_{22} & \cdots & a_{2n} \\ \cdot & \cdot & \cdots & \cdot \\ a_{n1} & a_{n2} & \cdots & a_{nn} \end{vmatrix} = \begin{vmatrix} a_{11}x_1 + a_{12}x_2 \cdots + a_{1n}x_n & a_{12} & \cdots & a_{1n} \\ a_{21}x_1 + a_{22}x_2 \cdots + a_{2n}x_n & a_{22} & \cdots & a_{2n} \\ & \cdots & & \\ a_{n1}x_1 + a_{n2}x_2 \cdots + a_{nn}x_n & a_{n2} & \cdots & a_{nn} \end{vmatrix}.$$

Replacing the first column of the right-hand side determinant with the constants of the right-hand side of (6.43), we obtain

$$x_1 \begin{vmatrix} a_{11} & a_{12} & \cdots & a_{1n} \\ a_{21} & a_{22} & \cdots & a_{2n} \\ \cdot & \cdot & \cdots & \cdot \\ a_{n1} & a_{n2} & \cdots & a_{nn} \end{vmatrix} = \begin{vmatrix} d_1 & a_{12} & \cdots & a_{1n} \\ d_2 & a_{22} & \cdots & a_{2n} \\ \cdot & \cdot & \cdots & \cdot \\ d_3 & a_{n2} & \cdots & a_{nn} \end{vmatrix}.$$

Clearly if we multiply the determinant of the coefficients by x_2, we can analyze the second column of the determinant in the same way. In general

$$x_i D_n = N_i, \quad 1 \le i \le n, \tag{6.44}$$

where D_n is the determinant of the coefficients

$$D_n = \begin{vmatrix} a_{11} & a_{12} & \cdots & a_{1n} \\ a_{21} & a_{22} & \cdots & a_{2n} \\ \cdot & \cdot & \cdots & \cdot \\ a_{n1} & a_{n2} & \cdots & a_{nn} \end{vmatrix}$$

and N_i is the determinant obtained by replacing the ith column of D_n by the nonhomogeneous terms

$$N_i = \begin{vmatrix} a_{11} & \cdots & a_{1i-1} & d_1 & a_{1i+1} & \cdots & a_{1n} \\ a_{21} & \cdots & a_{2i-1} & d_2 & a_{2i+1} & \cdots & a_{2n} \\ \cdot & \cdots & & & & \cdots & \cdot \\ a_{n1} & \cdots & a_{ni-1} & d_n & a_{ni+1} & \cdots & a_{nn} \end{vmatrix}. \tag{6.45}$$

Thus, if the determinant of the coefficients is not zero, the system has a unique solution

$$x_i = \frac{N_i}{D_n}, \quad 1 \le i \le n. \tag{6.46}$$

This procedure is known as Cramer's rule. For the special cases of $n = 2$ and $n = 3$, the results are, of course, identical to what we derived in the first section. Cramer's rule is very important in the development of the theory of determinants and matrices. However, to use it for solving a set of equations with large n, it is not very practical. Either because the amount of computations is so large and/or because the demand of numerical accuracy is so high with this method, even with high-speed computers it may not be possible to carry out such calculations. There are other techniques to solve that kind of problems, such as the Gauss-Jordan elimination method which we will discuss in the chapter on matrix theory.

6.4.2 Homogeneous Systems

Now if $d_1, d_2 \cdots d_n$ in the right-hand side of (6.43) are all zero, that is,

$$a_{11}x_1 + a_{12}x_2 \cdots + a_{1n}x_n = 0$$
$$a_{21}x_1 + a_{22}x_2 \cdots + a_{2n}x_n = 0$$
$$\cdots \cdots \cdots \cdots = \cdot$$
$$a_{n1}x_1 + a_{n2}x_2 \cdots + a_{nn}x_n = 0,$$

the set of equations is known as a homogeneous system. In this case, all $N_i's$ in (6.45) are equal to zero. If $D_n \ne 0$, then the only solution by (6.46) is a trivial one, namely, $x_1 = x_2 = \cdots x_n = 0$. On the other hand, if D_n is equal to zero, then it is clear from (6.44), x_i do not have to be zero. Hence, a homogeneous system can have a nontrivial solution only if the coefficient determinant is equal to zero. Conversely, one can show that if

$$\begin{vmatrix} a_{11} & a_{12} & \cdots & a_{1n} \\ a_{21} & a_{22} & \cdots & a_{2n} \\ \cdot & \cdot & \cdots & \cdot \\ a_{n1} & a_{n2} & \cdots & a_{nn} \end{vmatrix} = 0, \tag{6.47}$$

then there is always a nontrivial solution of the homogeneous equations. For a 2×2 system, the existence of a solution can be shown by direct calculation. Then one can show by mathematical induction that the statement is true for any $n \times n$ system.

This simple fact has many important applications.

Example 6.4.1 For what values of λ do the equations

$$3x + 2y = \lambda x,$$
$$4x + 5y = \lambda y$$

have a solution other than $x = y = 0$?

Solution 6.4.1 Moving the right-hand side to the left gives the homogeneous system

$$(3 - \lambda)x + 2y = 0,$$
$$4x + (5 - \lambda)y = 0.$$

For a nontrivial solution, the coefficient determinant must vanish:

$$\begin{vmatrix} 3 - \lambda & 2 \\ 4 & 5 - \lambda \end{vmatrix} = \lambda^2 - 8\lambda + 7 = (\lambda - 1)(\lambda - 7) = 0.$$

Thus the system has a nontrivial solution if and only if $\lambda = 1$ or $\lambda = 7$.

6.5 Block Diagonal Determinants

Frequently we encounter determinants with many zero elements and the non-zero elements which form square blocks along the diagonal. For example, the following fifth-order determinant is a block diagonal determinant

$$D_5 = |A| = \begin{vmatrix} a_{11} & a_{12} & 0 & 0 & 0 \\ a_{21} & a_{22} & 0 & 0 & 0 \\ * & * & a_{33} & a_{34} & a_{35} \\ * & * & a_{43} & a_{44} & a_{45} \\ * & * & a_{53} & a_{54} & a_{55} \end{vmatrix}.$$

In this section, we will show that

$$D_5 = \begin{vmatrix} a_{11} & a_{12} \\ a_{21} & a_{22} \end{vmatrix} \cdot \begin{vmatrix} a_{33} & a_{34} & a_{35} \\ a_{43} & a_{44} & a_{45} \\ a_{53} & a_{54} & a_{55} \end{vmatrix},$$

regardless the values the elements $*$ assume.

By definition

$$D_5 = \sum_{i_1=1}^{5}\sum_{i_2=1}^{5}\sum_{i_3=1}^{5}\sum_{i_4=1}^{5}\sum_{i_5=1}^{5} \varepsilon_{i_1 i_2 i_3 i_4 i_5} a_{i_1 1} a_{i_2 2} a_{i_3 3} a_{i_4 4} a_{i_5 5}.$$

Since $a_{13} = a_{14} = a_{15} = a_{23} = a_{24} = a_{25} = 0$, all terms containing these elements can be excluded from the summation. Thus

$$D_5 = \sum_{i_1=1}^{5}\sum_{i_2=1}^{5}\sum_{i_3=3}^{5}\sum_{i_4=3}^{5}\sum_{i_5=3}^{5} \varepsilon_{i_1 i_2 i_3 i_4 i_5} a_{i_1 1} a_{i_2 2} a_{i_3 3} a_{i_4 4} a_{i_5 5}.$$

Furthermore, the summation over i_1 and i_2 can be written as from 1 to 2, since 3, 4, and 5 are taken up by i_3, i_4, or i_5, and the Levi-Civita symbol is equal to zero if any index is repeated. Hence

$$D_5 = \sum_{i_1=1}^{2}\sum_{i_2=1}^{2}\sum_{i_3=3}^{5}\sum_{i_4=3}^{5}\sum_{i_5=3}^{5} \varepsilon_{i_1 i_2 i_3 i_4 i_5} a_{i_1 1} a_{i_2 2} a_{i_3 3} a_{i_4 4} a_{i_5 5}.$$

Under these circumstances, the permutation of i_1, i_2, i_3, i_4, i_5 can be separated into two permutations as schematically shown below:

$$\begin{pmatrix} 1 & 2 & 3 & 4 & 5 \\ i_1 = 1,2 & i_2 = 1,2 & i_3 = 3,4,5 & i_4 = 3,4,5 & i_5 = 3,4,5 \end{pmatrix}$$
$$= \begin{pmatrix} 1 & 2 \\ i_1 & i_2 \end{pmatrix}\begin{pmatrix} 3 & 4 & 5 \\ i_3 & i_4 & i_5 \end{pmatrix}.$$

The entire permutation is even if the two separated permutations are both even or both odd. The permutation is odd if one of the separated permutations is even and the other is odd. Therefore

$$\varepsilon_{i_1 i_2 i_3 i_4 i_5} = \varepsilon_{i_1 i_2} \cdot \varepsilon_{i_3 i_4 i_5}.$$

It follows

$$D_5 = \sum_{i_1=1}^{2} \sum_{i_2=1}^{2} \sum_{i_3=3}^{5} \sum_{i_4=3}^{5} \sum_{i_5=3}^{5} \varepsilon_{i_1 i_2} \cdot \varepsilon_{i_3 i_4 i_5} a_{i_1 1} a_{i_2 2} a_{i_3 3} a_{i_4 4} a_{i_5 5}$$

$$= \sum_{i_1=1}^{2} \sum_{i_2=1}^{2} \varepsilon_{i_1 i_2} a_{i_1 1} a_{i_2 2} \cdot \sum_{i_3=3}^{5} \sum_{i_4=3}^{5} \sum_{i_5=3}^{5} \varepsilon_{i_3 i_4 i_5} a_{i_3 3} a_{i_4 4} a_{i_5 5}$$

$$= \begin{vmatrix} a_{11} & a_{12} \\ a_{21} & a_{22} \end{vmatrix} \cdot \begin{vmatrix} a_{33} & a_{34} & a_{35} \\ a_{43} & a_{44} & a_{45} \\ a_{53} & a_{54} & a_{55} \end{vmatrix}.$$

When the blocks are along the "anti-diagonal" line, we can evaluate the determinant in a similar way, except we should be careful about its sign. For example,

$$\begin{vmatrix} 0 & 0 & a_{13} & a_{14} \\ 0 & 0 & a_{23} & a_{24} \\ a_{31} & a_{32} & * & * \\ a_{41} & a_{42} & * & * \end{vmatrix} = \begin{vmatrix} a_{31} & a_{32} \\ a_{41} & a_{42} \end{vmatrix} \cdot \begin{vmatrix} a_{13} & a_{14} \\ a_{23} & a_{24} \end{vmatrix}, \tag{6.48}$$

and

$$\begin{vmatrix} 0 & 0 & 0 & a_{14} & a_{15} & a_{16} \\ 0 & 0 & 0 & a_{24} & a_{25} & a_{26} \\ 0 & 0 & 0 & a_{34} & a_{35} & a_{36} \\ a_{41} & a_{42} & a_{43} & * & * & * \\ a_{51} & a_{52} & a_{53} & * & * & * \\ a_{61} & a_{62} & a_{63} & * & * & * \end{vmatrix} = - \begin{vmatrix} a_{41} & a_{42} & a_{43} \\ a_{51} & a_{52} & a_{53} \\ a_{61} & a_{62} & a_{63} \end{vmatrix} \cdot \begin{vmatrix} a_{14} & a_{15} & a_{16} \\ a_{24} & a_{25} & a_{26} \\ a_{34} & a_{35} & a_{36} \end{vmatrix}. \tag{6.49}$$

We can establish the result of (6.48) by changing it to a block diagonal determinant with an even number of interchanges between two rows. However, we need an odd number of interchanges between two rows to change (6.49) into a block diagonal determinant, therefore a minus sign.

Example 6.5.1 Evaluate

$$D_5 = \begin{vmatrix} 0 & 2 & 0 & 7 & 1 \\ 1 & 0 & 3 & 0 & 0 \\ 0 & 0 & 0 & 5 & 1 \\ 1 & 0 & 4 & 0 & 0 \\ 0 & 0 & 0 & 1 & 0 \end{vmatrix}.$$

Solution 6.5.1

$$D_5 = \begin{vmatrix} 0 & 2 & 0 & 7 & 1 \\ 1 & 0 & 3 & 0 & 0 \\ 0 & 0 & 0 & 5 & 1 \\ 1 & 0 & 4 & 0 & 0 \\ 0 & 0 & 0 & 1 & 0 \end{vmatrix} \rightarrow (Row4 - Row2) = \begin{vmatrix} 0 & 2 & 0 & 7 & 1 \\ 1 & 0 & 3 & 0 & 0 \\ 0 & 0 & 0 & 5 & 1 \\ 0 & 0 & 1 & 0 & 0 \\ 0 & 0 & 0 & 1 & 0 \end{vmatrix}$$

$$= \begin{vmatrix} 0 & 2 \\ 1 & 0 \end{vmatrix} \cdot \begin{vmatrix} 0 & 5 & 1 \\ 1 & 0 & 0 \\ 0 & 1 & 0 \end{vmatrix} = -2 \cdot 1 = -2.$$

6.6 Laplacian Developments by Complementary Minors

(This section can be skipped in the first reading.)

The Laplace expansion of D_3 by the elements of the third column is

$$D_3 = \begin{vmatrix} a_{11} & a_{12} & a_{13} \\ a_{21} & a_{22} & a_{23} \\ a_{31} & a_{32} & a_{33} \end{vmatrix} = a_{13} \begin{vmatrix} a_{21} & a_{22} \\ a_{31} & a_{32} \end{vmatrix} - a_{23} \begin{vmatrix} a_{11} & a_{12} \\ a_{31} & a_{32} \end{vmatrix} + a_{33} \begin{vmatrix} a_{11} & a_{12} \\ a_{21} & a_{22} \end{vmatrix}.$$

The three second-order determinants are minors complementary to their respective elements. It is also useful to think that the three elements a_{13}, a_{23}, a_{33} are complementary to their respective minors. Obviously the expansion can be written as

$$D_3 = \begin{vmatrix} a_{11} & a_{12} \\ a_{21} & a_{22} \end{vmatrix} a_{33} - \begin{vmatrix} a_{11} & a_{12} \\ a_{31} & a_{32} \end{vmatrix} a_{23} + \begin{vmatrix} a_{21} & a_{22} \\ a_{31} & a_{32} \end{vmatrix} a_{13}. \tag{6.50}$$

In this way, it is seen that the determinant D_3 is equal to the sum of the signed products of all the second-order minors contained in the first two columns, each multiplied by its complementary element. In fact, any determinant D_n, even for $n > 3$, can be expanded in the same way, except the complementary element is of course another complementary minor. For example, for a fourth-order determinant

$$D_4 = \begin{vmatrix} a_{11} & a_{12} & a_{13} & a_{14} \\ a_{21} & a_{22} & a_{23} & a_{24} \\ a_{31} & a_{32} & a_{33} & a_{34} \\ a_{41} & a_{42} & a_{43} & a_{44} \end{vmatrix}, \tag{6.51}$$

six second-order minors can be formed from the first two columns. They are

$$\begin{vmatrix} a_{11} & a_{12} \\ a_{21} & a_{22} \end{vmatrix}, \quad \begin{vmatrix} a_{11} & a_{12} \\ a_{31} & a_{32} \end{vmatrix}, \quad \begin{vmatrix} a_{11} & a_{12} \\ a_{41} & a_{42} \end{vmatrix}, \quad \begin{vmatrix} a_{21} & a_{22} \\ a_{31} & a_{32} \end{vmatrix}, \quad \begin{vmatrix} a_{21} & a_{22} \\ a_{41} & a_{42} \end{vmatrix}, \quad \begin{vmatrix} a_{31} & a_{32} \\ a_{41} & a_{42} \end{vmatrix}.$$

Let us expand D_4 in terms of these six minors. First expanding D_4 by its first column, then expanding the four minors by their first columns, we have

$$D_4 = a_{11}C_{11} + a_{21}C_{21} + a_{31}C_{31} + a_{41}C_{41}, \tag{6.52}$$

where

$$C_{11} = \begin{vmatrix} a_{22} & a_{23} & a_{24} \\ a_{32} & a_{33} & a_{34} \\ a_{42} & a_{43} & a_{44} \end{vmatrix} = a_{22}\begin{vmatrix} a_{33} & a_{34} \\ a_{43} & a_{44} \end{vmatrix} - a_{32}\begin{vmatrix} a_{23} & a_{24} \\ a_{43} & a_{44} \end{vmatrix} + a_{42}\begin{vmatrix} a_{23} & a_{24} \\ a_{33} & a_{34} \end{vmatrix}$$

$$C_{21} = -\begin{vmatrix} a_{12} & a_{13} & a_{14} \\ a_{32} & a_{33} & a_{34} \\ a_{42} & a_{43} & a_{44} \end{vmatrix} = -a_{12}\begin{vmatrix} a_{33} & a_{34} \\ a_{43} & a_{44} \end{vmatrix} + a_{32}\begin{vmatrix} a_{13} & a_{14} \\ a_{43} & a_{44} \end{vmatrix} - a_{42}\begin{vmatrix} a_{13} & a_{14} \\ a_{33} & a_{34} \end{vmatrix}$$

$$C_{31} = \begin{vmatrix} a_{12} & a_{13} & a_{14} \\ a_{22} & a_{23} & a_{24} \\ a_{42} & a_{43} & a_{44} \end{vmatrix} = a_{12}\begin{vmatrix} a_{23} & a_{24} \\ a_{43} & a_{44} \end{vmatrix} - a_{22}\begin{vmatrix} a_{13} & a_{14} \\ a_{43} & a_{44} \end{vmatrix} + a_{42}\begin{vmatrix} a_{13} & a_{14} \\ a_{23} & a_{24} \end{vmatrix}$$

$$C_{41} = -\begin{vmatrix} a_{12} & a_{13} & a_{14} \\ a_{22} & a_{23} & a_{24} \\ a_{32} & a_{33} & a_{34} \end{vmatrix} = -a_{12}\begin{vmatrix} a_{23} & a_{24} \\ a_{33} & a_{34} \end{vmatrix} + a_{22}\begin{vmatrix} a_{13} & a_{14} \\ a_{33} & a_{34} \end{vmatrix} - a_{32}\begin{vmatrix} a_{13} & a_{14} \\ a_{23} & a_{24} \end{vmatrix}.$$

Putting these cofactors back into (6.52) and collecting terms, we have

$$D_4 = (a_{11}a_{22} - a_{21}a_{12})\begin{vmatrix} a_{33} & a_{34} \\ a_{43} & a_{44} \end{vmatrix} - (a_{11}a_{32} - a_{31}a_{12})\begin{vmatrix} a_{23} & a_{24} \\ a_{43} & a_{44} \end{vmatrix}$$

$$+ (a_{11}a_{41} - a_{41}a_{12})\begin{vmatrix} a_{23} & a_{24} \\ a_{33} & a_{34} \end{vmatrix} + (a_{21}a_{32} - a_{31}a_{22})\begin{vmatrix} a_{13} & a_{14} \\ a_{43} & a_{44} \end{vmatrix}$$

$$- (a_{21}a_{42} - a_{41}a_{22})\begin{vmatrix} a_{13} & a_{14} \\ a_{33} & a_{34} \end{vmatrix} + (a_{31}a_{42} - a_{41}a_{32})\begin{vmatrix} a_{13} & a_{14} \\ a_{23} & a_{24} \end{vmatrix}. \tag{6.53}$$

Clearly,

$$D_4 = \begin{vmatrix} a_{11} & a_{12} \\ a_{21} & a_{22} \end{vmatrix} \cdot \begin{vmatrix} a_{33} & a_{34} \\ a_{43} & a_{44} \end{vmatrix} - \begin{vmatrix} a_{11} & a_{12} \\ a_{31} & a_{32} \end{vmatrix} \cdot \begin{vmatrix} a_{23} & a_{24} \\ a_{43} & a_{44} \end{vmatrix}$$
$$+ \begin{vmatrix} a_{11} & a_{12} \\ a_{41} & a_{42} \end{vmatrix} \cdot \begin{vmatrix} a_{23} & a_{24} \\ a_{33} & a_{34} \end{vmatrix} + \begin{vmatrix} a_{21} & a_{22} \\ a_{31} & a_{32} \end{vmatrix} \cdot \begin{vmatrix} a_{13} & a_{14} \\ a_{43} & a_{44} \end{vmatrix}$$
$$- \begin{vmatrix} a_{21} & a_{22} \\ a_{41} & a_{42} \end{vmatrix} \cdot \begin{vmatrix} a_{13} & a_{14} \\ a_{33} & a_{34} \end{vmatrix} + \begin{vmatrix} a_{31} & a_{32} \\ a_{41} & a_{42} \end{vmatrix} \cdot \begin{vmatrix} a_{13} & a_{14} \\ a_{23} & a_{24} \end{vmatrix} . \tag{6.54}$$

If D_4 is a block diagonal determinant,

$$D_4 = \begin{vmatrix} a_{11} & a_{12} & a_{13} & a_{14} \\ a_{21} & a_{22} & a_{23} & a_{24} \\ 0 & 0 & a_{33} & a_{34} \\ 0 & 0 & a_{43} & a_{44} \end{vmatrix} ,$$

then only the first term in (6.54) is non-zero, therefore

$$D_4 = \begin{vmatrix} a_{11} & a_{12} \\ a_{21} & a_{22} \end{vmatrix} \cdot \begin{vmatrix} a_{33} & a_{34} \\ a_{43} & a_{44} \end{vmatrix} ,$$

in agreement with the result derived in the last section.

If we adopt the following notation

$$A_{i_1 i_2, j_1 j_2} = \begin{vmatrix} a_{i_1 j_1} & a_{i_1 j_2} \\ a_{i_2 j_1} & a_{i_2 j_2} \end{vmatrix}$$

and $M_{i_1 i_2, j_1 j_2}$ as the complementary minor to $A_{i_1 i_2, j_1 j_2}$, the determinant D_4 in (6.51) can be expanded in terms of the minors formed by the elements of any two columns,

$$D_4 = \sum_{i_1=1}^{3} \sum_{i_2>i_1}^{4} (-1)^{i_1+i_2+j_1+j_2} A_{i_1 i_2, j_1 j_2} M_{i_1 i_2, j_1 j_2} . \tag{6.55}$$

With $j_1 = 1$, $j_2 = 2$, it can be readily verified, that (6.55) is, term by term, equal to (6.54). The proof of (6.55) goes the same way as in the Laplacian expansion by a row. First (6.55) is a linear combination of 4! products, each product has one element from each row and one from each column. The coefficients are either $+1$ or -1, depending on whether an even or odd number of interchange are needed to move i_1 to the first row, i_2 to the second row, and j_1 to the first column, j_2 to the second column, without changing the order of the rest of the elements. Obviously, the determinant can also be expanded in terms of the minors formed from any number of rows.

For a nth-order determinant D_n, one can expand it in a similar way, not only in terms of second-order minors but also in terms of kth-order minors with $k < n$. Of course, for $k = n - 1$, it reduces to the regular Laplacian development by a column. Following the same procedure of expanding D_4, one can show that

$$D_n = \sum_{(i)} (-1)^{i_1+i_2\cdots+i_k+j_1+j_2\cdots+j_k} A_{i_1i_2\cdots i_k, j_1 j_2\cdots j_k} M_{i_1 i_2\cdots i_k, j_1 j_2\cdots j_k},$$

where the symbol $\sum_{(i)}$ indicates that the summation is taken over all possible permutations in the following way. The first set of subscripts $i_1 i_2 \cdots i_k$ is from n indices $12 \cdots n$ taken k at a time with the restriction $i_1 < i_2 \cdots < i_k$. The second set subscripts $j_1 j_2 \cdots j_k$ are chosen arbitrarily but remain fixed for each term of the expansion. This formula is general, but is seldom needed for the evaluation of a determinant.

Example 6.6.1 Evaluate $D_4 = \begin{vmatrix} 2 & 1 & 3 & 1 \\ 1 & 0 & 2 & 5 \\ 2 & 1 & 1 & 3 \\ 1 & 3 & 0 & 2 \end{vmatrix}$ by (a) expansion with minors formed from

the first two columns, (b) expansion with minors formed from the second and fourth rows.

Solution 6.6.1 (a)

$$D_4 = \begin{vmatrix} 2 & 1 \\ 1 & 0 \end{vmatrix} \cdot \begin{vmatrix} 1 & 3 \\ 0 & 2 \end{vmatrix} - \begin{vmatrix} 2 & 1 \\ 2 & 1 \end{vmatrix} \cdot \begin{vmatrix} 2 & 5 \\ 0 & 2 \end{vmatrix} + \begin{vmatrix} 2 & 1 \\ 1 & 3 \end{vmatrix} \cdot \begin{vmatrix} 2 & 5 \\ 1 & 3 \end{vmatrix}$$

$$+ \begin{vmatrix} 1 & 0 \\ 2 & 1 \end{vmatrix} \cdot \begin{vmatrix} 3 & 1 \\ 0 & 2 \end{vmatrix} - \begin{vmatrix} 1 & 0 \\ 1 & 3 \end{vmatrix} \cdot \begin{vmatrix} 3 & 1 \\ 1 & 3 \end{vmatrix} + \begin{vmatrix} 2 & 1 \\ 1 & 3 \end{vmatrix} \cdot \begin{vmatrix} 3 & 1 \\ 2 & 5 \end{vmatrix}$$

$$= -2 - 0 + 5 + 6 - 24 + 65 = 50.$$

(b)

$$D_4 = (-1)^{2+4+1+2} \begin{vmatrix} 1 & 0 \\ 1 & 3 \end{vmatrix} \cdot \begin{vmatrix} 3 & 1 \\ 1 & 3 \end{vmatrix} + (-1)^{2+4+1+3} \begin{vmatrix} 1 & 2 \\ 1 & 0 \end{vmatrix} \cdot \begin{vmatrix} 1 & 1 \\ 1 & 3 \end{vmatrix}$$

$$+ (-1)^{2+4+1+4} \begin{vmatrix} 1 & 5 \\ 1 & 2 \end{vmatrix} \cdot \begin{vmatrix} 1 & 3 \\ 1 & 1 \end{vmatrix} + (-1)^{2+4+2+3} \begin{vmatrix} 0 & 2 \\ 3 & 0 \end{vmatrix} \cdot \begin{vmatrix} 2 & 1 \\ 2 & 3 \end{vmatrix}$$

$$+ (-1)^{2+4+2+4} \begin{vmatrix} 0 & 5 \\ 3 & 2 \end{vmatrix} \cdot \begin{vmatrix} 2 & 3 \\ 2 & 1 \end{vmatrix} + (-1)^{2+4+3+4} \begin{vmatrix} 2 & 5 \\ 0 & 2 \end{vmatrix} \cdot \begin{vmatrix} 2 & 1 \\ 2 & 1 \end{vmatrix}$$

$$= -24 - 4 - 6 + 24 + 60 - 0 = 50.$$

6.7 Multiplication of Determinants of the Same Order

If $|A|$ and $|B|$ are determinants of order n, then the product

$$|A| \cdot |B| = |C|$$

is a determinant of the same order. Its elements are given by

$$c_{ij} = \sum_{k=1}^{n} a_{ik}b_{kj}.$$

(As we shall show in the next chapter, this is the rule of multiplying two matrices.)
    For second-order determinants, this relation is expressed as

$$|A| \cdot |B| = \begin{vmatrix} a_{11} & a_{12} \\ a_{21} & a_{22} \end{vmatrix} \cdot \begin{vmatrix} b_{11} & b_{12} \\ b_{21} & b_{22} \end{vmatrix} = \begin{vmatrix} (a_{11}b_{11} + a_{12}b_{21}) & (a_{11}b_{12} + a_{12}b_{22}) \\ (a_{21}b_{11} + a_{22}b_{21}) & (a_{21}b_{12} + a_{22}b_{22}) \end{vmatrix}.$$

To prove this, we use the property of block diagonal determinants.

$$|A| \cdot |B| = \begin{vmatrix} a_{11} & a_{12} \\ a_{21} & a_{22} \end{vmatrix} \cdot \begin{vmatrix} b_{11} & b_{12} \\ b_{21} & b_{22} \end{vmatrix} = \begin{vmatrix} a_{11} & a_{12} & 0 & 0 \\ a_{21} & a_{22} & 0 & 0 \\ -1 & 0 & b_{11} & b_{12} \\ 0 & -1 & b_{21} & b_{22} \end{vmatrix}.$$

Multiplying the elements in the first column by b_{11} and the elements in the second
column by b_{21} and then adding them to the corresponding elements in the third
column, we obtain

$$|A| \cdot |B| = \begin{vmatrix} a_{11} & a_{12} & (a_{11}b_{11} + a_{12}b_{21}) & 0 \\ a_{21} & a_{22} & (a_{21}b_{11} + a_{22}b_{21}) & 0 \\ -1 & 0 & 0 & b_{12} \\ 0 & -1 & 0 & b_{22} \end{vmatrix}.$$

In the same way, we multiply the elements in the first column by b_{12} and the elements
in the second column by b_{22} and then add them to the corresponding elements in the
fourth column, it becomes

$$|A| \cdot |B| = \begin{vmatrix} a_{11} & a_{12} & (a_{11}b_{11} + a_{12}b_{21}) & (a_{11}b_{12} + a_{12}b_{22}) \\ a_{21} & a_{22} & (a_{21}b_{11} + a_{22}b_{21}) & (a_{21}b_{12} + a_{22}b_{22}) \\ -1 & 0 & 0 & 0 \\ 0 & -1 & 0 & 0 \end{vmatrix}.$$

By (6.48)

$$|A| \cdot |B| = \begin{vmatrix} -1 & 0 \\ 0 & -1 \end{vmatrix} \cdot \begin{vmatrix} (a_{11}b_{11} + a_{12}b_{21}) & (a_{11}b_{12} + a_{12}b_{22}) \\ (a_{21}b_{11} + a_{22}b_{21}) & (a_{21}b_{12} + a_{22}b_{22}) \end{vmatrix}$$

$$= \begin{vmatrix} (a_{11}b_{11} + a_{12}b_{21}) & (a_{11}b_{12} + a_{12}b_{22}) \\ (a_{21}b_{11} + a_{22}b_{21}) & (a_{21}b_{12} + a_{22}b_{22}) \end{vmatrix},$$

which is the desired result. This procedure is applicable to determinants of any order. (This property is of considerable importance, we will revisit this problem for determinant of higher order in the chapter on matrices.)

Example 6.7.1 Show that

$$\begin{vmatrix} b^2 + c^2 & ab & ca \\ ab & a^2 + b^2 & bc \\ ca & bc & a^2 + b^2 \end{vmatrix} = 4a^2b^2c^2.$$

Solution 6.7.1

$$\begin{vmatrix} b^2 + c^2 & ab & ca \\ ab & a^2 + b^2 & bc \\ ca & bc & a^2 + b^2 \end{vmatrix} = \begin{vmatrix} b & c & 0 \\ a & 0 & c \\ 0 & a & b \end{vmatrix} \cdot \begin{vmatrix} b & a & 0 \\ c & 0 & a \\ 0 & c & b \end{vmatrix} = (-2abc)^2 = 4a^2b^2c^2.$$

6.8 Differentiation of Determinants

Occasionally, we require an expression for the derivative of a determinant. If the derivative is with respect to a particular element a_{ij}, then

$$\frac{\partial D_n}{\partial a_{ij}} = C_{ij},$$

where C_{ij} is the cofactor of a_{ij}, since

$$D_n = \sum_{j=1}^{n} a_{ij} C_{ij} \quad for \ \ 1 \leq i \leq n.$$

Suppose the elements are functions of a parameter s, the derivative of D_n with respect to s is then given by

$$\frac{dD_n}{ds} = \sum_{i=1}^{n} \sum_{j=1}^{n} \frac{\partial D_n}{\partial a_{ij}} \frac{da_{ij}}{ds} = \sum_{i=1}^{n} \sum_{j=1}^{n} C_{ij} \frac{da_{ij}}{ds}.$$

For example

$$D_3 = \begin{vmatrix} a_{11} & a_{12} & a_{13} \\ a_{21} & a_{22} & a_{23} \\ a_{31} & a_{32} & a_{33} \end{vmatrix} = \sum_{j=1}^{3} a_{1j} C_{1j} = \sum_{j=1}^{3} a_{2j} C_{2j} = \sum_{j=1}^{3} a_{3j} C_{3j},$$

$$\frac{dD_3}{ds} = \sum_{j=1}^{3} \frac{da_{1j}}{ds} C_{1j} + \sum_{j=1}^{3} \frac{da_{2j}}{ds} C_{2j} + \sum_{j=1}^{3} \frac{da_{3j}}{ds} C_{3j}$$

$$= \begin{vmatrix} \dfrac{da_{11}}{ds} & \dfrac{da_{12}}{ds} & \dfrac{da_{13}}{ds} \\ a_{21} & a_{22} & a_{23} \\ a_{31} & a_{32} & a_{33} \end{vmatrix} + \begin{vmatrix} a_{11} & a_{12} & a_{13} \\ \dfrac{da_{21}}{ds} & \dfrac{da_{22}}{ds} & \dfrac{da_{23}}{ds} \\ a_{31} & a_{32} & a_{33} \end{vmatrix} + \begin{vmatrix} a_{11} & a_{12} & a_{13} \\ a_{21} & a_{22} & a_{23} \\ \dfrac{da_{31}}{ds} & \dfrac{da_{32}}{ds} & \dfrac{da_{33}}{ds} \end{vmatrix}.$$

Example 6.8.1 If $D_2 = \begin{vmatrix} \cos x & \sin x \\ -\sin x & \cos x \end{vmatrix}$, find $\dfrac{dD_2}{dx}$.

Solution 6.8.1

$$\frac{dD_2}{dx} = \begin{vmatrix} -\sin x & \cos x \\ -\sin x & \cos x \end{vmatrix} + \begin{vmatrix} \cos x & \sin x \\ -\cos x & -\sin x \end{vmatrix} = 0.$$

This is an obvious result, since $D_2 = \cos^2 x + \sin^2 x = 1$.

6.9 Determinants in Geometry

It is well known in analytic geometry that a straight line in the xy-plane is represented by the equation

$$ax + by + c = 0. \tag{6.56}$$

The line is uniquely defined by two points. If the line goes through two points (x_1, y_1) and (x_2, y_2), then both of them have to satisfy the equation

$$ax_1 + by_1 + c = 0, \tag{6.57}$$
$$ax_2 + by_2 + c = 0. \tag{6.58}$$

These equations (6.56), (6.57), and (6.58) may be regarded as a system in the unknowns a, b, c which cannot all vanish if (6.56) represents a line. Hence, the coefficient determinant must vanish:

$$\begin{vmatrix} x & y & 1 \\ x_1 & y_1 & 1 \\ x_2 & y_2 & 1 \end{vmatrix} = 0 \tag{6.59}$$

It can be easily shown that (6.59) is indeed the familiar equation of a line. Expanding (6.59) by the third column, we have

$$\begin{vmatrix} x & y & 1 \\ x_1 & y_1 & 1 \\ x_2 & y_2 & 1 \end{vmatrix} = (x_1 y_2 - x_2 y_1) - (x y_2 - x_2 y) + (x y_1 - x_1 y) = 0.$$

This equation can be readily transformed into (6.56) with $a = y_1 - y_2$, $b = x_2 - x_1$, $c = x_1 y_2 - x_2 y_1$. Or it can be put in form

$$y = mx + y_0,$$

where $m = \dfrac{y_2 - y_1}{x_2 - x_1}$ is the slope and $y_0 = y_1 - m x_1$ is the y-axis intercept.

It follows from (6.59) that a necessary and sufficient condition for three points (x_1, y_1), (x_2, y_2), and (x_3, y_3) to lie on a line is

$$\begin{vmatrix} x_1 & y_1 & 1 \\ x_2 & y_2 & 1 \\ x_3 & y_3 & 1 \end{vmatrix} = 0. \tag{6.60}$$

Now if the three points are not on a line, then they form a triangle and the determinant (6.60) is not equal to zero. In that case it is interesting to ask what does the determinant represent. Since it has the dimension of an area, this strongly suggests that the determinant is related to the area of the triangle (Fig. 6.6).

The area of the triangle formed by three points $A(x_1, y_1)$, $B(x_2, y_2)$, and $C(x_3, y_3)$ shown in the figure is seen to be

$$Area\ ABC = Area\ AA'C'C + Area\ CC'B'B - Area\ AA'B'B.$$

The area of a trapezoid is equal to half of the product of its altitude and the sum of the parallel sides:

Fig. 6.6 The area of ABC is equal to the sum of the trapezoids $AA'C'C$ and $CC'B'B$ minus the trapezoid $AA'B'B$. As a consequence, the area ABC can be represented by a determinant

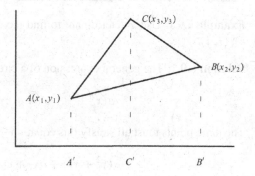

$$Area\ AA'C'C = \frac{1}{2}(x_3 - x_1)(y_1 + y_3),$$

$$Area\ CC'B'B = \frac{1}{2}(x_2 - x_3)(y_2 + y_3),$$

$$Area\ AA'B'B = \frac{1}{2}(x_2 - x_1)(y_1 + y_2).$$

Hence

$$
\begin{aligned}
Area\ ABC &= \frac{1}{2}[(x_3 - x_1)(y_1 + y_3) + (x_2 - x_3)(y_2 + y_3) \\
&\quad -(x_2 - x_1)(y_1 + y_2)] \\
&= \frac{1}{2}[(x_2 y_3 - x_3 y_2) - (x_1 y_3 - x_3 y_1) + (x_1 y_2 - x_2 y_1)] \\
&= \frac{1}{2}\begin{vmatrix} x_1 & y_1 & 1 \\ x_2 & y_2 & 1 \\ x_3 & y_3 & 1 \end{vmatrix}.
\end{aligned}
\tag{6.61}
$$

Notice the order of the points ABC in the figure is counterclockwise. If it is clockwise, the positions of B and C are interchanged. This will result in the interchange of row 2 and row 3 in the determinant. As a consequence, a minus sign will be introduced. Thus we conclude that if the three vertices of a triangle are $A(x_1, y_1)$, $B(x_2, y_2)$, $C(x_3, y_3)$, then

$$
\begin{vmatrix} x_1 & y_1 & 1 \\ x_2 & y_2 & 1 \\ x_3 & y_3 & 1 \end{vmatrix} = \pm 2 \times Area\ of\ ABC,
\tag{6.62}
$$

where the + or − sign is chosen according to the vertices being numbered consecutively in the counterclockwise or the clockwise direction.

Example 6.9.1 Use a determinant to find the circle that passes through (2, 6), (6, 4), (7, 1).

Solution 6.9.1 The general expression of a circle is

$$a(x^2 + y^2) + bx + cy + d = 0.$$

The three points must all satisfy this equation

$$a(x_1^2 + y_1^2) + bx_1 + cy_1 + d = 0,$$
$$a(x_2^2 + y_2^2) + bx_2 + cy_2 + d = 0,$$
$$a(x_3^2 + y_3^2) + bx_3 + cy_3 + d = 0.$$

These equations may be regarded as a system of equations in the unknowns a, b, c, d which cannot all be zero. Hence the coefficient determinant must vanish:

$$\begin{vmatrix} x^2 + y^2 & x & y & 1 \\ x_1^2 + y_1^2 & x_1 & y_1 & 1 \\ x_2^2 + y_2^2 & x_2 & y_2 & 1 \\ x_3^2 + y_3^2 & x_3 & y_3 & 1 \end{vmatrix} = 0.$$

Put in the specific values

$$\begin{vmatrix} x^2 + y^2 & x & y & 1 \\ 40 & 2 & 6 & 1 \\ 52 & 6 & 4 & 1 \\ 50 & 7 & 1 & 1 \end{vmatrix} = 0.$$

Replacing the first row by (row 1 − row 2), the third row by (row 3 − row 2), and the fourth row by (row 4 − Row 2), we have

$$\begin{vmatrix} x^2 + y^2 & x & y & 1 \\ 40 & 2 & 6 & 1 \\ 52 & 6 & 4 & 1 \\ 50 & 7 & 1 & 1 \end{vmatrix} = \begin{vmatrix} (x^2 + y^2 - 40) & (x - 2) & (y - 6) & 0 \\ 40 & 2 & 6 & 1 \\ 12 & 4 & -2 & 0 \\ 10 & 5 & -5 & 0 \end{vmatrix}$$

$$= \begin{vmatrix} (x^2 + y^2 - 40) & (x - 2) & (y - 6) \\ 12 & 4 & -2 \\ 10 & 5 & -5 \end{vmatrix}$$

$$= -10(x^2 + y^2 - 40) + 40 (x - 2) + 20 (y - 6) = 0,$$

or

$$x^2 + y^2 - 40 - 4(x - 2) - 2(y - 6) = 0.$$

which can be written as

$$(x - 2)^2 + (y - 1)^2 = 25.$$

So the circle is centered at $x = 2$, $y = 1$ with a radius of 5.

Example 6.9.2 What is the area of the triangle whose vertices are $(-2, 1)$, $(4, 3)$, $(0, 0)$?

Solution 6.9.2

$$Area = \begin{vmatrix} -2 & 1 & 1 \\ 4 & 3 & 1 \\ 0 & 0 & 1 \end{vmatrix} = -10.$$

The area of the triangle is 10 and the order of the vertices is clockwise.

Exercises

1. Use determinants to solve for x, y, z from the following system of equations:

$$3x + 6z = 51,$$
$$12y - 6z = -6,$$
$$x - y - z = 0.$$

Ans. $x = 7, \ y = 2, \ z = 5$.

2. By applying the Kirchhoff's rule to an electric circuit, the following equations are obtained for the currents i_1, i_2, i_3 in three branches:

$$i_1 R_1 + i_3 R_3 = V_A$$
$$i_2 R_2 + i_3 R_3 = V_C$$
$$i_1 + i_2 - i_3 = 0.$$

Express i_1, i_2, i_3 in terms of resistance R_1, R_2, R_3, and voltage source V_A, V_C.

Ans. $i_1 = \frac{(R_2 + R_3)V_A - R_3 V_C}{R_1 R_2 + R_1 R_3 + R_2 R_3}$, $i_2 = \frac{(R_1 + R_3)V_C - R_3 V_A}{R_1 R_2 + R_1 R_3 + R_2 R_3}$, $i_3 = \frac{R_2 V_A + R_1 V_C}{R_1 R_2 + R_1 R_3 + R_2 R_3}$.

3. Find the value of the following fourth-order determinant (which happens to be formed from one of the matrices appearing in Dirac's relativistic electron theory)

$$D_4 = \begin{vmatrix} 0 & 1 & 0 & 0 \\ -1 & 0 & 0 & 0 \\ 0 & 0 & 0 & 1 \\ 0 & 0 & -1 & 0 \end{vmatrix}.$$

Ans. 1.

4. Without computation, show that a skew-symmetric determinant of odd order is zero

$$D_{ss} = \begin{vmatrix} 0 & a & b & c & d \\ -a & 0 & e & f & g \\ -b & -e & 0 & h & i \\ -c & -f & -h & 0 & j \\ -d & -g & -i & -j & 0 \end{vmatrix} = 0.$$

[Hint: $D^T = D$ and $(-1)^n D_{ss} = D_{ss}^T$.]

5. Show that $\begin{vmatrix} a & d & 2a - 3d \\ b & e & 2b - 3e \\ c & f & 2c - 3f \end{vmatrix} = 0$.

6. Determine x such that $\begin{vmatrix} 1 & 2 & -3 \\ -x & 1 + 3x & 3 - x \\ 0 & -6 & 5 \end{vmatrix} = 36$.

Ans. 13.

7. The development of the determinant D_n on the ith row elements a_{ik} is $\sum_{k=1}^{n} a_{ik} C_{ik}$ where C_{ik} is the cofactor of a_{ik}. Show that

$$\sum_{k=1}^{n} a_{jk}C_{ik} = 0 \quad for \; j \neq i.$$

[Hint: The expansion is another determinant with two identical rows.]

8. Evaluate the following determinant by a development on (a) the first column, (b) the second row

$$D_4 = \begin{vmatrix} 1 & 1 & 1 & 1 \\ 1 & 2 & 3 & 4 \\ 1 & 3 & 6 & 10 \\ 1 & 4 & 10 & 20 \end{vmatrix}.$$

Ans. 1.

9. Use the properties of determinants to transform the determinant in problem 6 into a triangular form and then evaluate it as the product of the diagonal elements.

10. Evaluate the determinant in problem 6 by expanding it in terms of the 2×2 minors formed from the first two columns.

11. Evaluate the determinant

$$D_5 = \begin{vmatrix} 3 & -1 & 0 & 0 & 0 \\ 2 & 4 & 0 & 0 & 0 \\ 0 & 0 & 5 & 0 & 0 \\ 0 & 0 & 1 & 2 & 7 \\ 0 & 0 & 3 & -6 & 1 \end{vmatrix}.$$

Ans. 3080.

[The quickest way to evaluate is to expand it in terms of the 2×2 minors formed from the first two columns.]

12. Without expanding, show that

$$\begin{vmatrix} y+z & z+x & x+y \\ x & y & z \\ 1 & 1 & 1 \end{vmatrix} = 0.$$

[Hint: Add row 1 and row 2, factor out $(x + y + z)$.]

13. Show that (a)

$$\begin{vmatrix} x & y & z \\ x^2 & y^2 & z^2 \\ yz & zx & xy \end{vmatrix} = (xy + yz + zx) \begin{vmatrix} x & y & z \\ x^2 & y^2 & z^2 \\ 1 & 1 & 1 \end{vmatrix}.$$

[Hint: Replace row 3 successively by $x \cdot$ row 1 + row 3, then by $y \cdot$ row 1 + row 3, then by $z \cdot$ row 1 + row 3. Express the result as a sum of two determinants, one of them is equal to zero.]

(b) Use the result of the Vandermonde determinant to show that

$$\begin{vmatrix} x & y & z \\ x^2 & y^2 & z^2 \\ yz & zx & xy \end{vmatrix} = (xy + yz + zx)(x - y)(y - z)(z - x).$$

14. State the reason for each step of the following identity:

$$\begin{vmatrix} a & -b & -a & b \\ b & a & -b & -a \\ c & -d & c & -d \\ d & c & d & c \end{vmatrix} = \begin{vmatrix} 2a & -2b & -a & b \\ 2b & 2a & -b & -a \\ 0 & 0 & c & -d \\ 0 & 0 & d & c \end{vmatrix}$$

$$= \begin{vmatrix} 2a & -2b \\ 2b & 2a \end{vmatrix} \cdot \begin{vmatrix} c & -d \\ d & c \end{vmatrix} = 4(a^2 + b^2)(c^2 + d^2).$$

15. State the reason for each step of the following identity:

$$\begin{vmatrix} a & b & c & d \\ b & a & d & c \\ c & d & a & b \\ d & c & b & a \end{vmatrix} = \begin{vmatrix} (a+b) & b & (c+d) & d \\ (b+a) & a & (d+c) & c \\ (c+d) & d & (a+b) & b \\ (d+c) & c & (b+a) & a \end{vmatrix} = \begin{vmatrix} (a+b) & b & (c+d) & d \\ 0 & a-b & 0 & c-d \\ (c+d) & d & (a+b) & b \\ 0 & c-d & 0 & a-b \end{vmatrix}$$

$$= \begin{vmatrix} (a+b) & (c+d) & b & d \\ (c+d) & (a+b) & d & b \\ 0 & 0 & a-b & c-d \\ 0 & 0 & c-d & a-b \end{vmatrix} = [(a+b)^2 - (c+d)^2][(a-b)^2 - (c-d)^2].$$

16. Show and state the reason for each step of the following identity:

$$\begin{vmatrix} 0 & 1 & 2 & 3 & \cdots & n-1 \\ 1 & 0 & 1 & 2 & \cdots & n-2 \\ 2 & 1 & 0 & 1 & \cdots & n-3 \\ 3 & 2 & 1 & 0 & \cdots & n-4 \\ \cdot & & & & \cdots & \cdot \\ n-1 & n-2 & n-3 & n-4 & \cdots & 0 \end{vmatrix} = -(-2)^{n-2}(n-1).$$

[Hint: 1. Replace column 1 by column 1 + last column. 2. Factor out $(n - 1)$. 3. Replace row i by row $i - $ row$(i - 1)$, starting with the last row. 3. Replace row i by row $i + $ row 2. 4. Evaluating the triangular determinant.]

17. Evaluate the following determinant:

$$D_n = \begin{vmatrix} 1 & 2 & 3 & \cdots & n \\ n+1 & n+2 & n+3 & \cdots & 2n \\ 2n+1 & 2n+2 & 2n+3 & \cdots & 3n \\ \cdot & \cdot & \cdot & \cdots & \cdot \\ (n-1)n+1 & (n-1)n+2 & (n-1)n+3 & \cdots & n^2 \end{vmatrix}.$$

Ans. For $n = 1$, $D_1 = 1$; $n = 2$, $D_2 = -2$; $n \geq 3$, $D_n = 0$.

[Hint: For $n \geq 3$, replace row i by row $i - \text{row}(i - 1)$.]

18. Use the rule of product of two determinants of same order to show the

$$\begin{vmatrix} b^2 + c^2 & ab & ca \\ ab & a^2 + b^2 & bc \\ ca & bc & a^2 + b^2 \end{vmatrix} = \begin{vmatrix} b^2 + ac & bc & c^2 \\ ab & 2ac & bc \\ a^2 & ab & b^2 + ac \end{vmatrix}.$$

[Hint: $\begin{vmatrix} b & c & 0 \\ a & 0 & c \\ 0 & a & b \end{vmatrix} \cdot \begin{vmatrix} b & a & 0 \\ c & 0 & a \\ 0 & c & b \end{vmatrix} = \begin{vmatrix} b & c & 0 \\ a & 0'c \\ 0 & a & b \end{vmatrix} \cdot \begin{vmatrix} b & c & 0 \\ a & 0 & c \\ 0 & a & b \end{vmatrix}$.]

19. If $f(s)$ is given by the following determinants, without the expansion of $f(x)$ find $\dfrac{d}{dx} f(x)$

(a) $f(x) = \begin{vmatrix} e^x & e^{-x} & 1 \\ e^x & -e^{-x} & 0 \\ e^x & -e^{-x} & x \end{vmatrix}$; (b) $f(x) = \begin{vmatrix} \cos x & \sin x & \ln |x| \\ -\sin x & \cos x & \dfrac{1}{x} \\ -\cos x & -\sin x & -\dfrac{1}{x^2} \end{vmatrix}$.

Ans. (a) -2; (b) $1/x + 2/x^3$.

20. The vertices of a triangle are $(0, t)$, $(3t, 0)$, $(t, 2t)$. Find a formula for the area of the triangle.

Ans. $2t$.

21. The equation representing a plane is given by $ax + by + cz + d = 0$. Find the plane that goes through $(1, 1, 1)$, $(5, 0, 5)$, $(3, 2, 6)$.

Ans. $3x + 4y - 2z - 5 = 0$.

Chapter 7
Matrix Algebra

Matrices were introduced by British mathematician Arthur Cayley (1821–1895). The method of matrix algebra has extended far beyond mathematics into almost all disciplines of learning. In physical sciences, matrix is not only useful but essential in handling many complicated problems. These problems are mainly in three categories. First in the theory of transformation, second in the solution of systems of linear equations, and third in the solution of eigenvalue problems. In this chapter, we shall discuss various matrix operations and different situations in which they can be applied.

7.1 Matrix Notation

In this section, we shall define a matrix and discuss some of the simple operations by which two or more matrices can be combined.

7.1.1 Definition

Matrices
A rectangular array of elements is called a matrix. The array is usually enclosed within curved or square brackets. Thus, the rectangular arrays

$$\begin{pmatrix} 4 & 7 \\ 12 & 6 \\ -9 & 3 \end{pmatrix}, \quad \begin{pmatrix} x+iy \\ x-iy \end{pmatrix}, \quad \begin{pmatrix} \cos\theta & \sin\theta & 0 \\ -\sin\theta & \cos\theta & 0 \\ 0 & 0 & 1 \end{pmatrix} \tag{7.1}$$

© The Author(s), under exclusive license to Springer Nature Switzerland AG 2022   379
K.-T. Tang, *Mathematical Methods for Engineers and Scientists 1*,
https://doi.org/10.1007/978-3-031-05678-9_7

are examples of a matrix. It is convenient to think of every element of a matrix as belonging to a certain row and a certain column of the matrix. If a matrix has m rows and n columns, the matrix is said of order m by n, or $m \times n$. Every element of a matrix can be uniquely characterized by a row index and a column index. It is convenient to write a $m \times n$ matrix as

$$A = \begin{pmatrix} a_{11} & a_{12} & \cdots & a_{1n} \\ a_{21} & a_{22} & \cdots & a_{2n} \\ \cdot & & & \\ \cdot & & & \\ a_{m1} & a_{m2} & \cdots & a_{mn}, \end{pmatrix}$$

where a_{ij} is the element of ith row and jth column, it may be real or complex number or functions. The elements may even be matrices themselves, in which case the elements are called submatrices and the whole matrix is said to be partitioned.

Thus, if the first matrix in (7.1) is called matrix A, then A is a 3×2 matrix, it has three rows: (4 7), (12 6), (−9 3) and two columns: $\begin{pmatrix} 4 \\ 12 \\ -9 \end{pmatrix}$, $\begin{pmatrix} 7 \\ 6 \\ 3 \end{pmatrix}$. Its elements are $a_{11} = 4$, $a_{12} = 7$, $a_{21} = 12$, $a_{22} = 6$, $a_{31} = -9$ and $a_{32} = 3$.

Sometimes it is convenient to use the notation

$$A = (a_{ij})_{m \times n}$$

to indicate that A is a $m \times n$ matrix. The elements a_{ij} can also be expressed as

$$a_{ij} = (A)_{ij}.$$

7.1.2   Some Special Matrices

There are some special matrices, which are named after their appearances.

Zero Matrix A matrix of arbitrary order is said to be a zero matrix if and only if every element of the matrix equals zero. A zero matrix is sometimes called a **null matrix**.

Row Matrix A row matrix has only one row, such as $\begin{pmatrix} 1 & 0 & 3 \end{pmatrix}$. A row matrix is also called a **row vector.** If it is called row vector, the elements of the matrix are usually referred to as components.

Column Matrix A column matrix has only one column, such as $\begin{pmatrix} 3 \\ 4 \\ 5 \end{pmatrix}$. A column matrix is also called a **column vector.** Again if it is called column vector, the elements of the matrix are usually called the components of the vector.

Square Matrix A matrix is said to be a square matrix if the number of rows equals the number of columns. A square matrix of order n simply means it has n rows and n columns. Square matrix is of particular importance. We will be dealing mostly with square matrices together with column and row matrices.

For a square matrix A, we can calculate the determinant

$$\det(A) = |A|,$$

as defined in the last chapter. Matrix is not a determinant. Matrix is an array of numbers, determinant is a single number. The determinant of a matrix can only be defined for a square matrix.

Let $A = \left(a_{ij}\right)_n$ be a square matrix of order n. The diagonal going from the top left corner to the bottom right corner of the matrix, its elements $a_{11}, a_{22}, \cdots, a_{nn}$ are called the **diagonal elements**. All the remaining elements a_{ij} for $i \neq j$ are called the **off-diagonal elements**.

There are several special square matrices that are of interest.

Diagonal Matrix A diagonal matrix is a square matrix whose diagonal elements are not all equal to zero, but off-diagonal elements are all zero. For example,

$$\begin{pmatrix} 1 & 0 & 0 \\ 0 & 0 & 0 \\ 0 & 0 & -2 \end{pmatrix} \quad and \quad \begin{pmatrix} 3 & 0 & 0 \\ 0 & 4 & 0 \\ 0 & 0 & 5 \end{pmatrix}$$

are diagonal matrices. Therefore for a diagonal matrix

$$(A)_{ij} = a_{ii}\delta_{ij},$$

where

$$\delta_{ij} = \begin{cases} 1 & i = j \\ 0 & i \neq j \end{cases}.$$

This kind of notation may seem to be redundant, as a diagonal matrix can be easily visualized. However, this notation is useful in manipulating matrices as we shall see later.

Constant Matrix If all elements of a diagonal matrix happen to be equal to each other, it is said to be a constant matrix or a **scalar matrix**.

Unit Matrix If the elements of a constant matrix are equal to unity, then it is a unit matrix. A unit matrix is also called the **Identity matrix,** denoted by I, that is,

$$I = \begin{pmatrix} 1 & 0 & \cdots & 0 \\ 0 & 1 & \cdots & 0 \\ & \cdots & & \\ 0 & 0 & \cdots & 1 \end{pmatrix}.$$

Triangular Matrix A square matrix having only zero elements on one side of the principal diagonal is a triangular matrix. Thus

$$A = \begin{pmatrix} 1 & 2 & 3 \\ 0 & 3 & 4 \\ 0 & 0 & -2 \end{pmatrix}; \quad B = \begin{pmatrix} 1 & 0 & 0 \\ 3 & 2 & 0 \\ 4 & 5 & 0 \end{pmatrix}; \quad C = \begin{pmatrix} 0 & 0 & 0 \\ 5 & 0 & 0 \\ 4 & 3 & 0 \end{pmatrix}$$

are examples of a triangular matrix. A matrix for which $a_{ij} = 0$ for $i > j$ is called a **right-triangular matrix** or a **upper triangular matrix**, such as matrix A above. Whereas a matrix with $a_{ij} = 0$ for $i < j$ is called a **left-triangular matrix** or a **lower triangular matrix**, such as matrix B. If all the principal diagonal elements are zero, the matrix is a **strictly triangular matrix,** such as matrix C. Diagonal matrix, identity matrix as well as zero matrix are all triangular matrices.

7.1.3 Matrix Equation

Equality Two matrices A and B are equal to each other if and only if, every elements of A is equal to the corresponding element of B. Clearly A and B must be of the same order, in other words they must have the same rows and columns. Thus if

$$A = \begin{pmatrix} 1 & 2 \\ 3 & 4 \end{pmatrix}, \quad B = \begin{pmatrix} 0 & 1 & 2 \\ 0 & 3 & 4 \end{pmatrix}, \quad C = \begin{pmatrix} 1 & 2 & 0 \\ 3 & 4 & 0 \end{pmatrix},$$

we see that

$$A \neq B, \quad B \neq C, \quad C \neq A.$$

Therefore, a matrix equation $A = B$ means that A and B are of the same order and their corresponding elements are equal, that is, $a_{ij} = b_{ij}$. For example, the equation

$$\begin{pmatrix} x_1 & x_2 \\ y_1 & y_2 \end{pmatrix} = \begin{pmatrix} 3t & 1 + 2t \\ 4t^2 & 0 \end{pmatrix}$$

means $x_1 = 3t$, $y_1 = 4t^2$, $x_2 = 1 + 2t$, $y_2 = 0$.

With this understanding, often we can use a single matrix equation to replace a set of equations. This will not only simplify the writing but will also enable us to systematically manipulate these equations.

Addition and Subtraction We may now define the addition and subtraction of two matrices of the same order. The sum of two matrices A and B is another matrix C. By definition

$$A + B = C$$

means
$$c_{ij} = a_{ij} + b_{ij}.$$

For example,
$$A = \begin{pmatrix} 1 & 3 & 12 \\ -2 & 4 & -6 \end{pmatrix}, \quad B = \begin{pmatrix} -10 & 5 & -6 \\ 7 & 3 & 2 \end{pmatrix}$$

then
$$A + B = \begin{pmatrix} (1-10) & (3+5) & (12-6) \\ (-2+7) & (4+3) & (-6+2) \end{pmatrix} = \begin{pmatrix} -9 & 8 & 6 \\ 5 & 7 & -4 \end{pmatrix},$$

$$A - B = \begin{pmatrix} (1+10) & (3-5) & (12+6) \\ (-2-7) & (4-3) & (-6-2) \end{pmatrix} = \begin{pmatrix} 11 & -2 & 18 \\ -9 & 1 & -8 \end{pmatrix}.$$

The sum of several matrices is obtained by repeated addition. Since matrix addition is merely the addition of corresponding elements, it does not matter in which order we add several matrices. To be explicit, if A, B, C are three $m \times n$ matrices, then both commutative and associative laws hold

$$A + B = B + A,$$
$$A + (B + C) = (A + B) + C.$$

Multiplication by a Scalar It is possible to combine a matrix of arbitrary order and a scalar-by-scalar multiplication. If A is a matrix of order $m \times n$

$$A = (a_{ij})_{m \times n}$$

and c a scalar, we define cA to be another $m \times n$ matrix such that

$$cA = (ca_{ij})_{m \times n}.$$

For example, if
$$A = \begin{pmatrix} 1 & -3 & 5 \\ -2 & 4 & -6 \end{pmatrix},$$

then
$$-2A = \begin{pmatrix} -2 & 6 & -10 \\ 4 & -8 & 12 \end{pmatrix}.$$

The scalar can be a real number, a complex number or a function, but it cannot be a matrix quantity.

Note the difference between the scalar multiplication of a square matrix cA and the scalar multiplication of its determinant $c|A|$. For cA, c is multiplied to every elements of A, whereas for $c|A|$, c is only multiplied to the elements of a single column or a single row. Thus, if A is a square matrix of order n, then

$$\det(cA) = c^n |A|.$$

7.1.4 Transpose of a Matrix

If the rows and columns are interchanged, the resulting matrix is called the transposed matrix. The transposed matrix is denoted by $\widetilde{A}$, called A tilde, or by A^T. Usually, but not always, the transpose of a single matrix is denoted by the tilde and the transpose of the product of a number of matrices by the superscript T.

Thus, if

$$A = \begin{pmatrix} a_{11} & a_{12} & \cdots & a_{1n} \\ a_{21} & a_{22} & \cdots & a_{2n} \\ \cdot \\ \cdot \\ a_{m1} & a_{m2} & \cdots & a_{mn} \end{pmatrix},$$

then

$$\widetilde{A} = A^T = \begin{pmatrix} a_{11} & a_{21} & \cdots & a_{m1} \\ a_{12} & a_{22} & \cdots & a_{m2} \\ \cdot \\ \cdot \\ \cdot \\ a_{1n} & a_{2n} & \cdots & a_{mn} \end{pmatrix}.$$

By definition, if we transpose the matrix twice, we should get the original matrix, that is,

$$\widetilde{A}^T = A.$$

Using index notation, this means

$$\left(\widetilde{A}\right)_{ij} = (A)_{ji}, \quad \left(\widetilde{A}^T\right)_{ij} = (A)_{ij}.$$

It is clear that the transpose of $m \times n$ matrix is a $n \times m$ matrix. The transpose of a square matrix is another square matrix. The transpose of a column matrix is a row matrix, and the transpose of a row matrix is a column matrix.

Symmetric Matrix A symmetric matrix is a matrix that is equal to its transpose, that is,

$$A = \widetilde{A},$$

which means

$$a_{ij} = a_{ji}.$$

It is symmetric with respect to its diagonal. A symmetric matrix must be a square matrix.

Anti-symmetric Matrix An anti-symmetric matrix is a matrix that is equal to the negative of its transpose, that is,

$$A = -\widetilde{A},$$

which means

$$a_{ij} = -a_{ji}.$$

Thus the diagonal elements of an anti-symmetric matrix must all be zero. An anti-symmetric matrix must also be a square matrix. Anti-symmetric is also known as **skew-symmetric.**

Decomposition of a Square Matrix Any square matrix can be written as the sum of a symmetric and an anti-symmetric matrix. Clearly

$$A = \frac{1}{2}\left(A + \widetilde{A}\right) + \frac{1}{2}\left(A - \widetilde{A}\right)$$

is an identity. Furthermore, let

$$A_s = \frac{1}{2}\left(A + \widetilde{A}\right), \quad A_a = \frac{1}{2}\left(A - \widetilde{A}\right),$$

then A_s is symmetric, since

$$A_s^T = \frac{1}{2}\left(A^T + \widetilde{A}^T\right) = \frac{1}{2}\left(\widetilde{A} + A\right) = A_s,$$

and A_a is anti-symmetric, since

$$A_a^T = \frac{1}{2}\left(A^T - \widetilde{A}^T\right) = \frac{1}{2}\left(\widetilde{A} - A\right) = -A_a.$$

Therefore

$$A = A_s + A_a,$$
$$\widetilde{A} = A_s - A_a.$$

Example 7.1.1 Express the matrix

$$A = \begin{pmatrix} 2 & 0 & 1 \\ 0 & 3 & 2 \\ -1 & 4 & 2 \end{pmatrix}$$

as the sum of a symmetric matrix and an anti-symmetric matrix.

Solution 7.1.1

$$A = A_s + A_a$$

$$A_s = \frac{1}{2} \left[\begin{pmatrix} 2 & 0 & 1 \\ 0 & 3 & 2 \\ -1 & 4 & 2 \end{pmatrix} + \begin{pmatrix} 2 & 0 & -1 \\ 0 & 3 & 4 \\ 1 & 2 & 2 \end{pmatrix} \right] = \begin{pmatrix} 2 & 0 & 0 \\ 0 & 3 & 3 \\ 0 & 3 & 2 \end{pmatrix},$$

$$A_a = \frac{1}{2} \left[\begin{pmatrix} 2 & 0 & 1 \\ 0 & 3 & 2 \\ -1 & 4 & 2 \end{pmatrix} - \begin{pmatrix} 2 & 0 & -1 \\ 0 & 3 & 4 \\ 1 & 2 & 2 \end{pmatrix} \right] = \begin{pmatrix} 0 & 0 & 1 \\ 0 & 0 & -1 \\ -1 & 1 & 0 \end{pmatrix}.$$

7.2   Matrix Multiplication

7.2.1   Product of Two Matrices

The multiplication, or product, of two matrices is not a simple extension of the concept of multiplication of two numbers. The definition of matrix multiplication is motivated by the theory of linear transformation, which we will briefly discuss in the next section.

Two matrices A and B can be multiplied together only if the number of columns of A is equal to the number of rows of B. The matrix multiplication depends on the order in which the matrices occur in the product. For example, if A is of order $l \times m$ and B is of order $m \times n$, then the product matrix AB is defined but the product BA, in that order is not unless $m = l$. The multiplication is defined as follows. If

$$A = (a_{ij})_{l \times m}, \quad B = (b_{ij})_{m \times n}$$

then $AB = C$ means that C is a matrix of order $l \times n$ and

$$C = (c_{ij})_{l \times n}$$

$$c_{ij} = \sum_{k=1}^{m} a_{ik} b_{kj}.$$

$$\begin{pmatrix} a_{11} & a_{12} & \cdots & a_{1m} \\ a_{21} & a_{22} & \cdots & a_{2m} \\ \cdot & \cdot & & \cdot \\ \cdot & \cdot & & \cdot \\ \boxed{a_{i1}} & \boxed{a_{i2}} & \cdots & \boxed{a_{im}} \\ \cdot & \cdot & & \cdot \\ a_{l1} & a_{l2} & \cdots & a_{lm} \end{pmatrix} \begin{pmatrix} b_{11} & b_{12} & \cdots & \boxed{b_{1j}} & \cdots & b_{1n} \\ b_{21} & b_{22} & \cdots & \boxed{b_{2j}} & \cdots & b_{2n} \\ \cdot & \cdot & & \cdot & & \cdot \\ \cdot & \cdot & & \cdot & & \cdot \\ \cdot & \cdot & & \cdot & & \cdot \\ b_{m1} & b_{m2} & \cdots & \boxed{b_{mj}} & \cdots & b_{mn} \end{pmatrix} = \begin{pmatrix} c_{11} & c_{12} & \cdots & c_{1j} & \cdots & c_{1n} \\ c_{21} & c_{22} & \cdots & c_{2j} & \cdots & c_{2n} \\ \cdot & \cdot & & \cdot & & \cdot \\ \cdot & \cdot & & \cdot & & \cdot \\ c_{i1} & c_{i2} & \cdots & \boxed{c_{ij}} & \cdots & c_{in} \\ \cdot & \cdot & & \cdot & & \cdot \\ c_{l1} & c_{l2} & \cdots & c_{lj} & \cdots & c_{ln} \end{pmatrix}$$

Fig. 7.1 Illustration of matrix multiplication. The number of columns of A must equal the number of rows of B for the multiplication $AB = C$ to be defined. The element at ith row and jth column of C is given by $c_{ij} = a_{i1}b_{1j} + a_{i2}b_{2j} + \cdots + a_{im}b_{mj}$.

So the element of the C matrix at ith row and jth column is the sum of all the products of the elements of ith row of A and the corresponding elements of jth column of B. Thus, if

$$A = \begin{pmatrix} a_{11} & a_{12} \\ a_{21} & a_{22} \end{pmatrix}, \quad B = \begin{pmatrix} b_{11} & b_{12} & b_{13} \\ b_{21} & b_{22} & b_{23} \end{pmatrix}, \quad C = AB,$$

then

$$C = \begin{pmatrix} (a_{11}b_{11} + a_{12}b_{21}) & (a_{11}b_{12} + a_{12}b_{22}) & (a_{11}b_{13} + a_{12}b_{23}) \\ (a_{21}b_{11} + a_{22}b_{21}) & (a_{21}b_{12} + a_{22}b_{22}) & (a_{21}b_{13} + a_{22}b_{23}) \end{pmatrix}.$$

The multiplication of two matrices is illustrated in Fig. 7.1.

If the product AB is defined, A and B are said to be **comfortable** (or **compatible**). If the matrix product AB is defined, the product BA is not necessarily defined. Given two matrices A and B, both the products of AB and BA will be possible if, for example, A is of order $m \times n$ and B is of order $n \times m$. AB will be of order $m \times m$, and BA of order $n \times n$. Clearly if $m \neq n$, AB cannot equal to BA, since they are of different order. Even if $n = m$, AB is still not necessarily equal to BA. The following examples will make this clear.

Example 7.2.1 Find the product AB, if

$$A = \begin{pmatrix} 1 & 2 \\ 3 & 4 \end{pmatrix}, \quad B = \begin{pmatrix} 3 & 2 & 1 \\ 4 & 5 & 6 \end{pmatrix}.$$

Solution 7.2.1

$$AB = \begin{pmatrix} 1 & 2 \\ 3 & 4 \end{pmatrix} \begin{pmatrix} 3 & 2 & 1 \\ 4 & 5 & 6 \end{pmatrix}$$

$$= \begin{pmatrix} (1 \times 3 + 2 \times 4) & (1 \times 2 + 2 \times 5) & (1 \times 1 + 2 \times 6) \\ (3 \times 3 + 4 \times 4) & (3 \times 2 + 4 \times 5) & (3 \times 1 + 4 \times 6) \end{pmatrix}$$

$$= \begin{pmatrix} 11 & 12 & 13 \\ 25 & 26 & 27 \end{pmatrix}.$$

Here A is 2×2 and B is 2×3, so that AB comes out 2×3, whereas BA is not defined.

Example 7.2.2 Find the product AB, if

$$A = \begin{pmatrix} 1 & 2 \\ 3 & 4 \end{pmatrix}, \quad B = \begin{pmatrix} 5 \\ 6 \end{pmatrix}.$$

Solution 7.2.2

$$AB = \begin{pmatrix} 1 & 2 \\ 3 & 4 \end{pmatrix} \begin{pmatrix} 5 \\ 6 \end{pmatrix} = \begin{pmatrix} 5 + 12 \\ 15 + 24 \end{pmatrix} = \begin{pmatrix} 17 \\ 39 \end{pmatrix}.$$

Here AB is a column matrix and BA is not defined.

Example 7.2.3 Find AB and BA, if

$$A = \begin{pmatrix} 1 & 2 & 3 \end{pmatrix}, \quad B = \begin{pmatrix} 2 \\ 3 \\ 4 \end{pmatrix}.$$

Solution 7.2.3

$$AB = \begin{pmatrix} 1 & 2 & 3 \end{pmatrix} \begin{pmatrix} 2 \\ 3 \\ 4 \end{pmatrix} = (2 + 6 + 12) = (20),$$

$$BA = \begin{pmatrix} 2 \\ 3 \\ 4 \end{pmatrix} \begin{pmatrix} 1 & 2 & 3 \end{pmatrix} = \begin{pmatrix} 2 & 4 & 6 \\ 3 & 6 & 9 \\ 4 & 8 & 12 \end{pmatrix}.$$

This example dramatically shows that $AB \neq BA$.

Example 7.2.4 Find AB and BA, if

$$A = \begin{pmatrix} 1 & 2 \\ 3 & 4 \end{pmatrix}, \quad B = \begin{pmatrix} 3 & 4 \\ 5 & 6 \end{pmatrix}.$$

Solution 7.2.4

$$AB = \begin{pmatrix} 1 & 2 \\ 3 & 4 \end{pmatrix} \begin{pmatrix} 3 & 4 \\ 5 & 6 \end{pmatrix} = \begin{pmatrix} 13 & 16 \\ 29 & 36 \end{pmatrix}$$

and

$$BA = \begin{pmatrix} 3 & 4 \\ 5 & 6 \end{pmatrix} \begin{pmatrix} 1 & 2 \\ 3 & 4 \end{pmatrix} = \begin{pmatrix} 15 & 22 \\ 23 & 34 \end{pmatrix}.$$

Clearly

$$AB \neq BA.$$

Example 7.2.5 Find AB and BA, if

$$A = \begin{pmatrix} 1 & 1 \\ 2 & 2 \end{pmatrix}, \quad B = \begin{pmatrix} -1 & 1 \\ 1 & -1 \end{pmatrix}.$$

Solution 7.2.5

$$AB = \begin{pmatrix} 1 & 1 \\ 2 & 2 \end{pmatrix} \begin{pmatrix} -1 & 1 \\ 1 & -1 \end{pmatrix} = \begin{pmatrix} 0 & 0 \\ 0 & 0 \end{pmatrix},$$
$$BA = \begin{pmatrix} -1 & 1 \\ 1 & -1 \end{pmatrix} \begin{pmatrix} 1 & 1 \\ 2 & 2 \end{pmatrix} = \begin{pmatrix} 1 & 1 \\ -1 & -1 \end{pmatrix}.$$

Not only $AB \neq BA$, $AB = 0$ does not necessarily imply $A = 0$ or $B = 0$ or $BA = 0$.

Example 7.2.6 Let

$$A = \begin{pmatrix} 2 & 0 \\ 0 & 0 \end{pmatrix}, \quad B = \begin{pmatrix} 0 & 3 \\ 0 & 4 \end{pmatrix}, \quad C = \begin{pmatrix} 0 & 3 \\ 0 & 2 \end{pmatrix},$$

show that

$$AB = AC.$$

Solution 7.2.6

$$AB = \begin{pmatrix} 2 & 0 \\ 0 & 0 \end{pmatrix} \begin{pmatrix} 0 & 3 \\ 0 & 4 \end{pmatrix} = \begin{pmatrix} 0 & 6 \\ 0 & 0 \end{pmatrix},$$
$$AC = \begin{pmatrix} 2 & 0 \\ 0 & 0 \end{pmatrix} \begin{pmatrix} 0 & 3 \\ 0 & 2 \end{pmatrix} = \begin{pmatrix} 0 & 6 \\ 0 & 0 \end{pmatrix}.$$

This example shows that $AB = AC$ can hold without $B = C$ or $A = 0$.

7.2.2 Motivation of Matrix Multiplication

Much of the usefulness of matrix algebra is due to its multiplication property. The definition of matrix multiplication, as we have seen, seems to be "unnatural" and somewhat complicated. The motivation of this definition comes from the "linear transformations". It provides a simple mechanism for changing variables. For example, suppose

$$y_1 = a_{11}x_1 + a_{12}x_2 + a_{13}x_3 \qquad (7.2a)$$
$$y_2 = a_{21}x_1 + a_{22}x_2 + a_{23}x_3 \qquad (7.2b)$$

and further

$$z_1 = b_{11}y_1 + b_{12}y_2 \qquad (7.3a)$$
$$z_2 = b_{21}y_1 + b_{22}y_2. \qquad (7.3b)$$

In these equations, the x's and the y's are variables, while the a's and the b's are constants. The x's are related to the y's by the first set of equations, and the y's are related to the z's by the second set of equations. To find out how the x's are related to the z's, we must substitute the values of the y's given by the first set of equations into the second set of equations:

$$z_1 = b_{11}(a_{11}x_1 + a_{12}x_2 + a_{13}x_3) + b_{12}(a_{21}x_1 + a_{22}x_2 + a_{23}x_3) \qquad (7.4a)$$
$$z_2 = b_{21}(a_{11}x_1 + a_{12}x_2 + a_{13}x_3) + b_{22}(a_{21}x_1 + a_{22}x_2 + a_{23}x_3). \qquad (7.4b)$$

By multiplying them out and collecting coefficients, they become

$$z_1 = (b_{11}a_{11} + b_{12}a_{21})x_1$$
$$+(b_{11}a_{12} + b_{12}a_{22})x_2 + (b_{11}a_{13} + b_{12}a_{23})x_3 \qquad (7.5a)$$
$$z_2 = (b_{21}a_{11} + b_{22}a_{21})x_1$$
$$+(b_{21}a_{12} + b_{22}a_{22})x_2 + (b_{21}a_{13} + b_{22}a_{23})x_3. \qquad (7.5b)$$

Now, using matrix notation, (7.2) can be written simply as

$$\begin{pmatrix} y_1 \\ y_2 \end{pmatrix} = \begin{pmatrix} a_{11} & a_{12} & a_{13} \\ a_{21} & a_{22} & a_{23} \end{pmatrix} \begin{pmatrix} x_1 \\ x_2 \\ x_3 \end{pmatrix}, \qquad (7.6)$$

and (7.3) as

$$\begin{pmatrix} z_1 \\ z_2 \end{pmatrix} = \begin{pmatrix} b_{11} & b_{12} \\ b_{21} & b_{22} \end{pmatrix} \begin{pmatrix} y_1 \\ y_2 \end{pmatrix}. \qquad (7.7)$$

The coefficients of x_1, x_2, and x_3 in (7.5) are precisely the elements of the matrix product:

$$\begin{pmatrix} b_{11} & b_{12} \\ b_{21} & b_{22} \end{pmatrix} \begin{pmatrix} a_{11} & a_{12} & a_{13} \\ a_{21} & a_{22} & a_{23} \end{pmatrix}.$$

Not only that, they are also located in the proper position. In other words, (7.5) can be obtained by simply substituting $\begin{pmatrix} y_1 \\ y_2 \end{pmatrix}$ from (7.6) into (7.7):

$$\begin{pmatrix} z_1 \\ z_2 \end{pmatrix} = \begin{pmatrix} b_{11} & b_{12} \\ b_{21} & b_{22} \end{pmatrix} \begin{pmatrix} a_{11} & a_{12} & a_{13} \\ a_{21} & a_{22} & a_{23} \end{pmatrix} \begin{pmatrix} x_1 \\ x_2 \\ x_3 \end{pmatrix}. \tag{7.8}$$

What we have shown here is essentially two things. First, matrix multiplication is defined in such a way that linear transformation can be written in compact forms. Second, if we substitute linear transformations into each other, we can obtain the composite transformation simply by multiplying coefficient matrices in the right order. This kind of transformation is not only common in mathematics, but is also extremely important in physics. We will discuss some of them in later sections.

7.2.3 Properties of Product Matrices

Transpose of a Product Matrix A result of considerable importance in matrix algebra is that the transpose of the product of two matrices equals the product of the transposed matrices taken in reverse order,

$$(AB)^T = \tilde{B}\tilde{A}. \tag{7.9}$$

To prove this we must show that every element of the left-hand side is equal to the corresponding element in the right-hand side. The ijth element of the left-hand side of Eq. (7.9) is given by

$$\left((AB)^T\right)_{ij} = (AB)_{ji} = \sum_k (A)_{jk}(B)_{ki}. \tag{7.10}$$

The ijth element of the left-hand side of Eq. (7.9) is

$$\left(\tilde{B}\tilde{A}\right)_{ij} = \sum_k \left(\tilde{B}\right)_{ik}\left(\tilde{A}\right)_{kj} = \sum_k (B)_{ki}(A)_{jk}$$
$$= \sum_k (A)_{jk}(B)_{ki}, \tag{7.11}$$

where, in the last step, we have interchanged $(B)_{kj}$ and $(A)_{jk}$ because they are just numbers. Thus Eq. (7.9) follows.

Example 7.2.7 Let

$$A = \begin{pmatrix} 2 & 3 \\ 0 & -1 \end{pmatrix} \quad \text{and} \quad B = \begin{pmatrix} 1 & 5 \\ 2 & 4 \end{pmatrix},$$

show that

$$(AB)^T = \tilde{B}\tilde{A}.$$

Solution 7.2.7

$$AB = \begin{pmatrix} 2 & 3 \\ 0 & -1 \end{pmatrix}\begin{pmatrix} 1 & 5 \\ 2 & 4 \end{pmatrix} = \begin{pmatrix} 8 & 22 \\ -2 & -4 \end{pmatrix}; \quad (AB)^T = \begin{pmatrix} 8 & -2 \\ 22 & -4 \end{pmatrix};$$

$$\tilde{B} = \begin{pmatrix} 1 & 2 \\ 5 & 4 \end{pmatrix}, \quad \tilde{A} = \begin{pmatrix} 2 & 0 \\ 3 & -1 \end{pmatrix}; \quad \tilde{B}\tilde{A} = \begin{pmatrix} 1 & 2 \\ 5 & 4 \end{pmatrix}\begin{pmatrix} 2 & 0 \\ 3 & -1 \end{pmatrix} = \begin{pmatrix} 8 & -2 \\ 22 & -4 \end{pmatrix}.$$

Thus, $(AB)^T = \tilde{B}\tilde{A}$.

Trace of a Matrix The trace of square matrix $A = (a_{ij})$ is defined as the sum of its diagonal elements and is denoted by Tr A

$$Tr A = \sum_{i=1}^{n} a_{ii}.$$

An important theorem about trace is that the trace of the product of a finite number of matrices is invariant under any cyclic permutation of the matrices. We can first prove this theorem for the product of two matrices, and then the rest automatically follow.

Let A be a $n \times m$ matrix and B be a $m \times n$ matrix, then

$$Tr(AB) = \sum_{i=1}^{m}(AB)_{ii} = \sum_{i=1}^{m}\sum_{j=1}^{n} a_{ij}b_{ji},$$

$$Tr(BA) = \sum_{j=1}^{n}(BA)_{jj} = \sum_{j=1}^{n}\sum_{i=1}^{m} b_{ji}a_{ij}.$$

Since a_{ij} and b_{ij} are just numbers, their order can be reversed. Thus

$$Tr(AB) = Tr(BA).$$

Notice that the trace is defined only for a square matrix, but A and B do not have to be square matrices as long as their product is a square matrix. The order of AB may be different from the order of BA, yet their traces are the same.

Now

$$Tr(ABC) = Tr(A(BC)) = Tr((BC)A)$$
$$= Tr(BCA) = Tr(CAB) \tag{7.12}$$

It is important to note that the trace of the product of a number of matrices is not invariant under any permutation, but only a cyclic permutation of the matrices.

Example 7.2.8 Let

$$A = \begin{pmatrix} 4 & 0 & 6 \\ 5 & 2 & 1 \\ 7 & 8 & 3 \end{pmatrix}, \qquad B = \begin{pmatrix} 1 & 0 & 1 \\ 9 & 1 & 2 \\ 0 & 4 & 1 \end{pmatrix},$$

show that (a) $Tr(A + B) = Tr(A) + Tr(B)$; (b) $Tr(AB) = Tr(BA)$.

Solution 7.2.8 (a)

$$Tr(A + B) = Tr\left\{ \begin{pmatrix} 4 & 0 & 6 \\ 5 & 2 & 1 \\ 7 & 8 & 3 \end{pmatrix} + \begin{pmatrix} 1 & 0 & 1 \\ 9 & 1 & 2 \\ 0 & 4 & 1 \end{pmatrix} \right\} = Tr \begin{pmatrix} 5 & 0 & 7 \\ 14 & 3 & 3 \\ 7 & 12 & 4 \end{pmatrix}$$
$$= 5 + 3 + 4 = 12;$$

$$Tr(A) + Tr(B) = Tr \begin{pmatrix} 4 & 0 & 6 \\ 5 & 2 & 1 \\ 7 & 8 & 3 \end{pmatrix} + Tr \begin{pmatrix} 1 & 0 & 1 \\ 9 & 1 & 2 \\ 0 & 4 & 1 \end{pmatrix}$$
$$= (4 + 2 + 3) + (1 + 1 + 1) = 12.$$

(b)

$$Tr(AB) = Tr \begin{pmatrix} 4 & 0 & 6 \\ 5 & 2 & 1 \\ 7 & 8 & 3 \end{pmatrix} \begin{pmatrix} 1 & 0 & 1 \\ 9 & 1 & 2 \\ 0 & 4 & 1 \end{pmatrix} = Tr \begin{pmatrix} 4 & 24 & 10 \\ 23 & 6 & 10 \\ 79 & 20 & 26 \end{pmatrix} = 36;$$

$$Tr(BA) = Tr \begin{pmatrix} 1 & 0 & 1 \\ 9 & 1 & 2 \\ 0 & 4 & 1 \end{pmatrix} \begin{pmatrix} 4 & 0 & 6 \\ 5 & 2 & 1 \\ 7 & 8 & 3 \end{pmatrix} = Tr \begin{pmatrix} 11 & 8 & 9 \\ 55 & 18 & 61 \\ 27 & 16 & 7 \end{pmatrix} = 36.$$

Associative Law of Matrix Multiplication If A, B, and C are three matrices such that the matrix products AB and BC are defined, then

$$(AB) C = A (BC). \tag{7.13}$$

In other words, it is immaterial which two matrices are multiplied together first. To prove this, let

$$A = (a_{ij})_{m \times n}, \quad B = (b_{ij})_{n \times o}, \quad C = (c_{ij})_{o \times p}.$$

The ijth element of the left-hand side of Eq. (7.13) is then

$$((AB)C)_{ij} = \sum_{k=1}^{o}(AB)_{ik}(C)_{kj} = \sum_{k=1}^{o}\left(\sum_{l=1}^{n}(A)_{il}(B)_{lk}\right)(C)_{kj}$$

$$= \sum_{k=1}^{o}\sum_{l=1}^{n}a_{il}b_{lk}c_{kj},$$

while the ijth element of the right-hand side Eq. (7.13) is

$$(A(BC))_{ij} = \sum_{l=1}^{n}(A)_{il}(BC)_{lj} = \sum_{l=1}^{n}(A)_{il}\sum_{k=1}^{0}(B)_{lk}(C)_{kj}$$

$$= \sum_{l=1}^{n}\sum_{k=1}^{o}a_{il}b_{lk}c_{kj}.$$

Clearly $((AB)C)_{ij} = (A(BC))_{ij}$.

Example 7.2.9 Let

$$A = \begin{pmatrix} 1 & 2 \\ -1 & 3 \end{pmatrix}, \quad B = \begin{pmatrix} 1 & 0 & -1 \\ 2 & 1 & 0 \end{pmatrix}, \quad C = \begin{pmatrix} 1 & -1 \\ 3 & 2 \\ 2 & 1 \end{pmatrix},$$

show that

$$A(BC) = (AB)C.$$

Solution 7.2.9

$$A(BC) = \begin{pmatrix} 1 & 2 \\ -1 & 3 \end{pmatrix}\begin{pmatrix} 1 & 0 & -1 \\ 2 & 1 & 0 \end{pmatrix}\begin{pmatrix} 1 & -1 \\ 3 & 2 \\ 2 & 1 \end{pmatrix}$$

$$= \begin{pmatrix} 1 & 2 \\ -1 & 3 \end{pmatrix}\begin{pmatrix} -1 & -2 \\ 5 & 0 \end{pmatrix} = \begin{pmatrix} 9 & -2 \\ 16 & 2 \end{pmatrix},$$

$$(AB)C = \left[\begin{pmatrix} 1 & 2 \\ -1 & 3 \end{pmatrix}\begin{pmatrix} 1 & 0 & -1 \\ 2 & 1 & 0 \end{pmatrix}\right]\begin{pmatrix} 1 & -1 \\ 3 & 2 \\ 2 & 1 \end{pmatrix}$$

$$= \begin{pmatrix} 5 & 2 & -1 \\ 5 & 3 & 1 \end{pmatrix}\begin{pmatrix} 1 & -1 \\ 3 & 2 \\ 2 & 1 \end{pmatrix} = \begin{pmatrix} 9 & -2 \\ 16 & 2 \end{pmatrix}.$$

Clearly $A(BC) = (AB)C$. This is one of the most important properties of matrix algebra.

Distributive Law of Matrix Multiplication If A, B, and C are three matrices such that the addition $B + C$ and the product AB and BC are defined, then

$$A(B + C) = AB + AC. \tag{7.14}$$

To prove this, let

$$A = \left(a_{ij}\right)_{m \times n}, \quad B = \left(b_{ij}\right)_{n \times p}, \quad C = \left(c_{ij}\right)_{n \times p}, \tag{7.15}$$

so that the addition $B + C$ and the products AB and AC are defined. The ijth element of the left-hand side of Eq. (7.14) is then

$$(A(B + C))_{ij} = \sum_{k=1}^{n} (A)_{ik}(B + C)_{kj} = \sum_{k=1}^{n} (A)_{ik}(B_{kj} + C_{kj})$$

$$= \sum_{k=1}^{n} a_{ik}(b_{kj} + c_{kj}).$$

The ijth element of the right-hand side of Eq. (7.14) is

$$(AB + AC)_{ij} = (AB)_{ij} + (AC)_{ij}$$

$$= \sum_{k=1}^{n} (A)_{ik}(B)_{kj} + \sum_{k=1}^{n} (A)_{ik}(C)_{kj}$$

$$= \sum_{k=1}^{n} a_{ik}(b_{kj} + c_{kj}).$$

Thus, Eq. (7.14) follows.

Example 7.2.10 Let

$$A = \begin{pmatrix} 1 & 2 \\ 3 & 0 \end{pmatrix}, \quad B = \begin{pmatrix} 2 & -1 \\ 3 & 4 \end{pmatrix}, \quad C = \begin{pmatrix} 2 & -2 \\ 1 & 3 \\ 4 & -1 \end{pmatrix},$$

show that

$$C(A + B) = CA + CB.$$

Solution 7.2.10

$$C(A+B) = \begin{pmatrix} 2 & -2 \\ 1 & 3 \\ 4 & -1 \end{pmatrix} \left[\begin{pmatrix} 1 & 2 \\ 3 & 0 \end{pmatrix} + \begin{pmatrix} 2 & -1 \\ 3 & 4 \end{pmatrix} \right]$$

$$= \begin{pmatrix} 2 & -2 \\ 1 & 3 \\ 4 & -1 \end{pmatrix} \begin{pmatrix} 3 & 1 \\ 6 & 4 \end{pmatrix} = \begin{pmatrix} -6 & -6 \\ 21 & 13 \\ 6 & 0 \end{pmatrix},$$

$$CA + CB = \begin{pmatrix} 2 & -2 \\ 1 & 3 \\ 4 & -1 \end{pmatrix} \begin{pmatrix} 1 & 2 \\ 3 & 0 \end{pmatrix} + \begin{pmatrix} 2 & -2 \\ 1 & 3 \\ 4 & -1 \end{pmatrix} \begin{pmatrix} 2 & -1 \\ 3 & 4 \end{pmatrix}$$

$$= \begin{pmatrix} -4 & 4 \\ 10 & 2 \\ 1 & 8 \end{pmatrix} + \begin{pmatrix} -2 & -10 \\ 11 & 11 \\ 5 & -8 \end{pmatrix} = \begin{pmatrix} -6 & -6 \\ 21 & 13 \\ 6 & 0 \end{pmatrix}.$$

Hence, $C(A+B) = CA + CB$.

7.2.4 *Determinant of Matrix Product*

We have shown in the chapter on determinants that the value of the determinant of the product of two matrices is equal to the product of two determinants. That is, if A and B are square matrices of the same order, then

$$|AB| = |A||B|.$$

This relation is of considerable interests. It is instructive to prove it with the properties of matrix products. We will use 2×2 matrices to illustrate the steps of the proof, but it will be obvious that the process is generally valid for all orders.

1. If D is a diagonal matrix, it is easy to show $|DA| = |D||A|$.
For example, let $D = \begin{pmatrix} d_{11} & 0 \\ 0 & d_{22} \end{pmatrix}$, then

$$DA = \begin{pmatrix} d_{11} & 0 \\ 0 & d_{22} \end{pmatrix} \begin{pmatrix} a_{11} & a_{12} \\ a_{21} & a_{22} \end{pmatrix} = \begin{pmatrix} d_{11}a_{11} & d_{11}a_{12} \\ d_{22}a_{21} & d_{22}a_{22} \end{pmatrix},$$

$$|D| = \begin{vmatrix} d_{11} & 0 \\ 0 & d_{22} \end{vmatrix} = d_{11}d_{22},$$

$$|DA| = \begin{vmatrix} d_{11}a_{11} & d_{11}a_{12} \\ d_{22}a_{21} & d_{22}a_{22} \end{vmatrix} = d_{11}d_{22} \begin{vmatrix} a_{11} & a_{12} \\ a_{21} & a_{22} \end{vmatrix} = |D||A|.$$

2. Any square matrix can be diagonalized by a series of row operations which add a multiple of a row to another row.

For example, let $B = \begin{pmatrix} 1 & 2 \\ 3 & 4 \end{pmatrix}$. Multiply row 1 by -3 and add it to row 3, the matrix becomes $\begin{pmatrix} 1 & 2 \\ 0 & -2 \end{pmatrix}$. Then add row 2 to row 1, we have the diagonal matrix $\begin{pmatrix} 1 & 0 \\ 0 & -2 \end{pmatrix}$.

3. Each row operation is equivalent to premultiplying the matrix by an elementary matrix obtained from applying the same operation to the identity matrix.

For example, multiply row 1 by -3 and add it to row 2 of $\begin{pmatrix} 1 & 0 \\ 0 & 1 \end{pmatrix}$, we obtain the elementary matrix $\begin{pmatrix} 1 & 0 \\ -3 & 1 \end{pmatrix}$. Multiply this matrix to the left of B,

$$\begin{pmatrix} 1 & 0 \\ -3 & 1 \end{pmatrix}(B) = \begin{pmatrix} 1 & 0 \\ -3 & 1 \end{pmatrix}\begin{pmatrix} 1 & 2 \\ 3 & 4 \end{pmatrix} = \begin{pmatrix} 1 & 2 \\ 0 & -2 \end{pmatrix},$$

we get the same result as operating directly on B. The elementary matrix for adding row 2 to row 1 is $\begin{pmatrix} 1 & 1 \\ 0 & 1 \end{pmatrix}$. Multiplying this matrix to the left of $\begin{pmatrix} 1 & 2 \\ 0 & -2 \end{pmatrix}$, we have the diagonal matrix

$$\begin{pmatrix} 1 & 1 \\ 0 & 1 \end{pmatrix}\begin{pmatrix} 1 & 2 \\ 0 & -2 \end{pmatrix} = \begin{pmatrix} 1 & 0 \\ 0 & -2 \end{pmatrix}.$$

4. Combine the last equations,

$$\begin{pmatrix} 1 & 1 \\ 0 & 1 \end{pmatrix}\begin{pmatrix} 1 & 2 \\ 0 & -2 \end{pmatrix} = \begin{pmatrix} 1 & 1 \\ 0 & 1 \end{pmatrix}\begin{pmatrix} 1 & 0 \\ -3 & 1 \end{pmatrix}\begin{pmatrix} 1 & 2 \\ 3 & 4 \end{pmatrix} = \begin{pmatrix} 1 & 0 \\ 0 & -2 \end{pmatrix}.$$

Let

$$E = \begin{pmatrix} 1 & 1 \\ 0 & 1 \end{pmatrix}\begin{pmatrix} 1 & 0 \\ -3 & 1 \end{pmatrix} = \begin{pmatrix} -2 & 1 \\ -3 & 1 \end{pmatrix}, \qquad D = \begin{pmatrix} 1 & 0 \\ 0 & -2 \end{pmatrix},$$

we can write the equation as

$$EB = \begin{pmatrix} -2 & 1 \\ -3 & 1 \end{pmatrix}\begin{pmatrix} 1 & 2 \\ 3 & 4 \end{pmatrix} = \begin{pmatrix} 1 & 0 \\ 0 & -2 \end{pmatrix} = D.$$

This equation says that the matrix B is diagonalized by the matrix E, which is the product of a series of elementary matrices.

5. Because of the way E is constructed, multiplying E to the left of any matrix M is equivalent to repeatedly adding a multiple of a row to another row of M. From the theory of determinants, we know that these operations do not change the value of the determinant. For example,

$$|EB| = |D| = \begin{vmatrix} 1 & 0 \\ 0 & -2 \end{vmatrix} = -2,$$

$$|B| = \begin{vmatrix} 1 & 2 \\ 3 & 4 \end{vmatrix} = 4 - 6 = -2.$$

Therefore, the determinant of the diagonalized matrix D is equal to the determinant of the original matrix B,

$$|D| = |B|.$$

In fact M can be any matrix, as long as it is compatible,

$$|EM| = |M|.$$

6. Now let $M = BA$,

$$|E(BA)| = |BA|.$$

But

$$|E(BA)| = |(EB)A| = |DA| = |D||A|,$$

since D is diagonal. On the other hand, $|D| = |B|$, therefore

$$|BA| = |B||A|.$$

Since $|B||A| = |A||B|$, it follows $|BA| = |AB|$, even though BA may not be equal to AB.

7.2.5 The Commutator

The difference between the two products AB and BA is known as the commutator

$$[A, B] = AB - BA.$$

If, in particular, AB is equal to BA, then

$$[A, B] = 0,$$

the two matrices A and B are said to commute with each other.
   It follows directly from the definition that

- $[A, A] = 0,$
- $[A, I] = [I, A] = 0,$

- $[A, B] = -[B, A]$,
- $[A, (B + C)] = [A, B] + [A, C]$,
- $[A, [B, C]] + [B, [C, A]] + [C, [A, B]] = 0$.

Example 7.2.11 Let

$$\sigma_x = \begin{pmatrix} 0 & 1 \\ 1 & 0 \end{pmatrix}, \quad \sigma_y = \begin{pmatrix} 0 & -i \\ i & 0 \end{pmatrix}, \quad \sigma_z = \begin{pmatrix} 1 & 0 \\ 0 & -1 \end{pmatrix},$$

show that

$$[\sigma_x, \sigma_y] = 2i\sigma_z, \quad [\sigma_y, \sigma_z] = 2i\sigma_x, \quad [\sigma_z, \sigma_x] = 2i\sigma_y.$$

Solution 7.2.11

$$\sigma_x \sigma_y = \begin{pmatrix} 0 & 1 \\ 1 & 0 \end{pmatrix} \begin{pmatrix} 0 & -i \\ i & 0 \end{pmatrix} = i \begin{pmatrix} 1 & 0 \\ 0 & -1 \end{pmatrix},$$

$$\sigma_y \sigma_x = \begin{pmatrix} 0 & -i \\ i & 0 \end{pmatrix} \begin{pmatrix} 0 & 1 \\ 1 & 0 \end{pmatrix} = -i \begin{pmatrix} 1 & 0 \\ 0 & -1 \end{pmatrix},$$

$$[\sigma_x, \sigma_y] = \sigma_x \sigma_y - \sigma_y \sigma_x = 2i \begin{pmatrix} 1 & 0 \\ 0 & -1 \end{pmatrix} = 2i\sigma_z.$$

Similarly, $[\sigma_y, \sigma_z] = 2i\sigma_x$, and $[\sigma_z, \sigma_x] = 2i\sigma_y$.

Example 7.2.12 If a matrix B commutes with a diagonal matrix with no two elements equal to each other, then B must also be a diagonal matrix.

Solution 7.2.12 To prove this, let B commute with a diagonal matrix A of order n, whose elements are

$$(A)_{ij} = a_i \delta_{ij} \tag{7.16}$$
$$a_i \neq a_j \ \ if \ \ i \neq j.$$

We are given that

$$AB = BA.$$

Let the elements of B be b_{ij}, we wish to show that $b_{ij} = 0$, unless $i = j$. Taking the ijth element of both sides, we have

$$\sum_{k=1}^{n} (A)_{ik}(B)_{kj} = \sum_{k=1}^{n} (B)_{ik}(A)_{kj}.$$

On using (7.16), this becomes

$$\sum_{k=1}^{n} a_i \delta_{ik} b_{kj} = \sum_{k=1}^{n} b_{ik} a_k \delta_{kj},$$

with the definition of delta function

$$a_i b_{ij} = b_{ij} a_j.$$

This shows

$$(a_i - a_j) b_{ij} = 0.$$

Thus b_{ij} must be all equal to zero for $i \neq j$, since for those cases $a_i \neq a_j$. The only elements of B which can be different from zero are the diagonal elements b_{ii}, proving that B must be a diagonal matrix.

7.3 Systems of Linear Equations

The method of matrix algebra is very useful in solving a system of linear equations. Let $x_1, x_2, \cdots, x_n$ be a set of n unknown variables. An equation which contains first degree of x_i and no products of two or more variables is called a linear equation. The most general system of m linear equations in n unknowns can be written in the form

$$
\begin{aligned}
a_{11}x_1 + a_{12}x_2 + \cdots \cdot a_{1n}x_n &= d_1, \\
a_{21}x_1 + a_{22}x_2 + \cdots \cdot a_{2n}x_n &= d_2, \\
&\cdots \cdots \cdots \cdots \cdots \cdots \\
a_{m1}x_1 + a_{m2}x_2 + \cdots \cdot a_{mn}x_n &= d_m.
\end{aligned}
\tag{7.17}
$$

Here the coefficients a_{ij} and the right-hand side terms d_i are supposed to be known constants.

We can regard the variables $x_1, x_2, \cdots, x_n$ as components of the $n \times 1$ column vector **x**

$$\mathbf{x} = \begin{pmatrix} x_1 \\ x_2 \\ \cdot \\ \cdot \\ x_n \end{pmatrix}$$

and the constants $d_1, d_2, \cdots, d_m$ as components of the $m \times 1$ column vector **d**

$$\mathbf{d} = \begin{pmatrix} d_1 \\ d_2 \\ \cdot \\ d_m \end{pmatrix}.$$

The coefficients a_{ij} can be written as elements of the $m \times n$ matrix A

$$A = \begin{pmatrix} a_{11} & a_{12} & \cdots & a_{1n} \\ a_{21} & a_{22} & \cdots & a_{2n} \\ \cdot & \cdot & \cdots & \cdot \\ a_{m1} & a_{m2} & \cdots & a_{mn} \end{pmatrix}$$

With the matrix multiplication defined in the previous section, Eq. 7.17) can be written as

$$\begin{pmatrix} a_{11} & a_{12} & \cdots & a_{1n} \\ a_{21} & a_{22} & \cdots & a_{2n} \\ \cdot & \cdot & \cdots & \cdot \\ a_{m1} & a_{m2} & \cdots & a_{mn} \end{pmatrix} \begin{pmatrix} x_1 \\ x_2 \\ \cdot \\ \cdot \\ x_n \end{pmatrix} = \begin{pmatrix} d_1 \\ d_2 \\ \cdot \\ \cdot \\ d_m \end{pmatrix}.$$

If all the components of d are equal to zero, the system is called homogeneous. If at least one component of d is not zero, the system is called nonhomogeneous. If the system of linear equations is such that the equations are all satisfied simultaneously by at least one set of values of x_i, then it is said to be consistent. The system is said to be inconsistent if the equations are not satisfied simultaneously by any set of values. An inconsistent system has no solution. A consistent system may have an unique solution, or an infinite number of solutions. In the following sections, we will discuss practical ways of finding these solutions, as well as answer the question of existence and uniqueness of the solutions.

7.3.1 Gauss Elimination Method

Two linear systems are equivalent if every solution of either system is a solution of the other. There are three elementary operations that will transform a linear system into another equivalent system:

1. Interchanging two equations.
2. Multiplying an equation through by a non-zero number.
3. Adding to one equation a multiple of some other equation.

That a system is transformed into an equivalent system by the first operation is quite apparent. The reason that the second and third kinds of operations have the same effect is that when the same operations are done on both sides of an equal sign, the equation should remain valid. In fact, these are just the techniques we learned in elementary algebra to solve a set of simultaneous equations. The goal is to transform the set of equations into a simple form so that the solution is obvious. A practical procedure is suggested by the observation that a linear system, whose coefficient matrix is either upper triangular or diagonal, is easy to solve.

For example, the system of equations

$$-2x_2 + x_3 = 8$$
$$2x_1 - x_2 + 4x_3 = -3 \qquad (7.18)$$
$$x_1 - x_2 + x_3 = -2$$

can be written as

$$\begin{pmatrix} 0 & -2 & 1 \\ 2 & -1 & 4 \\ 1 & -1 & 1 \end{pmatrix} \begin{pmatrix} x_1 \\ x_2 \\ x_3 \end{pmatrix} = \begin{pmatrix} 8 \\ -3 \\ -2 \end{pmatrix}.$$

Interchange Eqs. 7.1 and 7.3, the system becomes

$$\begin{aligned} x_1 - x_2 + x_3 &= -2 \\ 2x_1 - x_2 + 4x_3 &= -3 \\ -2x_2 + x_3 &= 8 \end{aligned} \qquad \begin{pmatrix} 1 & -1 & 1 \\ 2 & -1 & 4 \\ 0 & -2 & 1 \end{pmatrix} \begin{pmatrix} x_1 \\ x_2 \\ x_3 \end{pmatrix} = \begin{pmatrix} -2 \\ -3 \\ 8 \end{pmatrix},$$

where we have put the matrix equation representing the system right next to it. Multiply Eq. 7.1 of the rearranged system by -2 and add to Eq. 7.2, we have

$$\begin{aligned} x_1 - x_2 + x_3 &= -2 \\ x_2 + 2x_3 &= 1 \\ -2x_2 + x_3 &= 8 \end{aligned} \qquad \begin{pmatrix} 1 & -1 & 1 \\ 0 & 1 & 2 \\ 0 & -2 & 1 \end{pmatrix} \begin{pmatrix} x_1 \\ x_2 \\ x_3 \end{pmatrix} = \begin{pmatrix} -2 \\ 1 \\ 8 \end{pmatrix}.$$

Multiply Eq. 7.2 of the last system by 2 and add to Eq. 7.3:

$$\begin{aligned} x_1 - x_2 + x_3 &= -2 \\ x_2 + 2x_3 &= 1 \\ 5x_3 &= 10 \end{aligned} \qquad \begin{pmatrix} 1 & -1 & 1 \\ 0 & 1 & 2 \\ 0 & 0 & 5 \end{pmatrix} \begin{pmatrix} x_1 \\ x_2 \\ x_3 \end{pmatrix} = \begin{pmatrix} -2 \\ 1 \\ 10 \end{pmatrix}. \qquad (7.19)$$

These four systems of equations are equivalent because they have the same solution. From the last set of equations, it is clear that $x_3 = 2$, $x_2 = 1 - 2x_3 = -3$ and $x_1 = -2 + x_2 - x_3 = -7$.

This procedure is often referred to as the **Gauss elimination method**, the **echelon method**, or **triangularization**.

Augmented Matrix To simplify the writing further, we introduce the augmented matrix. The matrix composed of the coefficient matrix plus an additional column whose elements are the nonhomogeneous constants d_i is called the **augmented matrix** of the system. Thus

$$\left(\begin{array}{cccc|c} a_{11} & a_{12} & \cdots & a_{1n} & d_1 \\ a_{21} & a_{22} & \cdots & a_{2n} & d_2 \\ \cdot & \cdot & \cdots & \cdot & \cdot \\ a_{m1} & a_{m2} & \cdots & a_{mn} & d_n \end{array} \right)$$

is the augmented matrix of (7.17). The portion in front of the vertical line is the coefficient matrix. The entire matrix, disregarding the vertical line, is the augmented matrix of the system. Clearly the augmented matrix is just a succinct expression of the linear system.

Instead of operating on the equations of the system, we can just operate on the rows of the augmented matrix with the three **elementary row operations** which consists of

- 1. Interchanging of any two rows.
- 2. Multiplying of any row by a non-zero scalar.
- 3. Adding a multiple of a row to another row.

Thus, we can summarize the Gauss elimination method as using the elementary row operations to reduce the augmented matrix of the original system to an echelon form. A matrix is said to be in echelon form if

1. The first element in the first row is non-zero.
2. The first $(n - 1)$ elements of the nth row are zero, the rest elements may or may not be zero.
3. The first non-zero element of any row appears to the right of the first non-zero element in the row above.
4. As a consequence, if there are rows whose elements are all zero, then they must be at the bottom of the matrix.

Thus, we can think of solving the linear system of (7.18) in the above example as reducing the augmented matrix from

$$\begin{pmatrix} 0 & -2 & 1 & 8 \\ 2 & -1 & 4 & -3 \\ 1 & -1 & 1 & -2 \end{pmatrix}$$

to the echelon form

$$\begin{pmatrix} 1 & -1 & 1 & -2 \\ 0 & 1 & 2 & 1 \\ 0 & 0 & 5 & 10 \end{pmatrix},$$

from which the solution is easily obtained.

Gauss-Jordan Elimination Method For a large set of linear equations, it is sometimes advantageous to continue the process to reduce the coefficient matrix from the triangular form to a diagonal form. For example, multiply the 3rd row of the last matrix by 1/5:

$$\begin{pmatrix} 1 & -1 & 1 & -2 \\ 0 & 1 & 2 & 1 \\ 0 & 0 & 1 & 2 \end{pmatrix}. \tag{7.20}$$

Multiply row 3 by -2 and add to row 2:

$$\begin{pmatrix} 1 & -1 & 1 & -2 \\ 0 & 1 & 0 & -3 \\ 0 & 0 & 1 & 2 \end{pmatrix}.$$

Multiply row 3 by -1 and add to row 1:

$$\begin{pmatrix} 1 & -1 & 0 & -4 \\ 0 & 1 & 0 & -3 \\ 0 & 0 & 1 & 2 \end{pmatrix}.$$

Add row 2 to row 1:

$$\begin{pmatrix} 1 & 0 & 0 & -7 \\ 0 & 1 & 0 & -3 \\ 0 & 0 & 1 & 2 \end{pmatrix},$$

which corresponds to $x_1 = -7$, $x_2 = -3$, and $x_3 = 2$. This process is known as the **Gauss-Jordan elimination method.**

7.3.2 Existence and Uniqueness of Solutions of Linear Systems

For a linear system of m equations and n unknowns, the order of the coefficient matrix is $m \times n$ and that of the augmented matrix is $m \times (n + 1)$. If $m < n$, the system is underdetermined. If $m > n$, the system is overdetermined. The most interesting case is $m = n$. In all three cases, we can use Gauss elimination method to reduce the augmented matrix into an echelon form. Once in the echelon form, the problem is either solved, or else shown to be inconsistent. A few examples will make this clear.

Example 7.3.1 Solve the following system of equations:

$$\begin{aligned} x_1 + x_2 - x_3 &= 2 \\ 2x_1 - x_2 + x_3 &= 1 \\ 3x_1 - x_2 + x_3 &= 4. \end{aligned}$$

Solution 7.3.1 The augmented matrix is $\begin{pmatrix} 1 & 1 & -1 & 2 \\ 2 & -1 & 1 & 1 \\ 3 & -1 & 1 & 4 \end{pmatrix}.$

Multiply row 1 by -2 and add to row 2: $\begin{pmatrix} 1 & 1 & -1 & 2 \\ 0 & -3 & 3 & -3 \\ 3 & -1 & 1 & 4 \end{pmatrix};$

multiply row 1 by -3 and add to row 3: $\begin{pmatrix} 1 & 1 & -1 & 2 \\ 0 & -3 & 3 & -3 \\ 0 & -4 & 4 & -2 \end{pmatrix}$;

multiply row 2 by $-\dfrac{1}{3}$: $\begin{pmatrix} 1 & 1 & -1 & 2 \\ 0 & 1 & -1 & 1 \\ 0 & -4 & 4 & -2 \end{pmatrix}$;

multiply row 2 by 4 and add to row 3: $\begin{pmatrix} 1 & 1 & -1 & 2 \\ 0 & 1 & -1 & 1 \\ 0 & 0 & 0 & 2 \end{pmatrix}$.

This represents the system of equations

$$x_1 + x_2 - x_3 = 2$$
$$x_2 - x_3 = 1$$
$$0 = 2.$$

Since no values of x_1, x_2 and x_3 can make $0 = 2$, the system is inconsistent and has no solution.

Example 7.3.2 Solve the following system of equations:

$$x_1 + 3x_2 + x_3 = 6$$
$$3x_1 - 2x_2 - 8x_3 = 7$$
$$4x_1 + 5x_2 - 3x_3 = 17.$$

Solution 7.3.2 The augmented matrix is $\begin{pmatrix} 1 & 3 & 1 & 6 \\ 3 & -2 & -8 & 7 \\ 4 & 5 & -3 & 17 \end{pmatrix}$.

Multiply row 1 by -3 and add to row 2: $\begin{pmatrix} 1 & 3 & 1 & 6 \\ 0 & -11 & -11 & -11 \\ 4 & 5 & -3 & 17 \end{pmatrix}$;

multiply row 1 by -4 and add to row 3: $\begin{pmatrix} 1 & 3 & 1 & 6 \\ 0 & -11 & -11 & -11 \\ 0 & -7 & -7 & -7 \end{pmatrix}$;

multiply row 2 by $-\dfrac{1}{11}$: $\begin{pmatrix} 1 & 3 & 1 & 6 \\ 0 & 1 & 1 & 1 \\ 0 & -7 & -7 & -7 \end{pmatrix}$;

multiply row 2 by 7 and add to row 3: $\begin{pmatrix} 1 & 3 & 1 & 6 \\ 0 & 1 & 1 & 1 \\ 0 & 0 & 0 & 0 \end{pmatrix}$.

This represents the system

$$x_1 + 3x_2 + x_3 = 6$$
$$x_2 + x_3 = 1$$
$$0 = 0.$$

This says $x_2 = 1 - x_3$, and $x_1 = 6 - 3x_2 - x_3 = 3 + 2x_3$. The value of x_3 may be assigned arbitrarily, and therefore the system has an infinite number of solutions.

Example 7.3.3 Solve the following system of equations:

$$x_1 + x_2 = 2$$
$$x_1 + 2x_2 = 3$$
$$2x_1 + x_2 = 3.$$

Solution 7.3.3 The augmented matrix is

$$\begin{pmatrix} 1 & 1 & | & 2 \\ 1 & 2 & | & 3 \\ 2 & 1 & | & 3 \end{pmatrix}.$$

Multiply row 1 by -1 and add to row 2: $\begin{pmatrix} 1 & 1 & | & 2 \\ 0 & 1 & | & 1 \\ 2 & 1 & | & 3 \end{pmatrix}$;

multiply row 1 by -2 and add to row 3: $\begin{pmatrix} 1 & 1 & | & 2 \\ 0 & 1 & | & 1 \\ 0 & -1 & | & -1 \end{pmatrix}$;

add row 2 to row 3: $\begin{pmatrix} 1 & 1 & | & 2 \\ 0 & 1 & | & 1 \\ 0 & 0 & | & 0 \end{pmatrix}$.

The last augmented matrix says

$$x_1 + x_2 = 2$$
$$x_2 = 1$$
$$0 = 0,$$

clearly $x_2 = 1$ and $x_1 = 1$. Therefore this system has an unique solution.

To answer questions of existence and uniqueness of solutions of linear systems, it is useful to introduce the concept of the rank of a matrix.

Rank of a Matrix There are several equivalent definitions for the rank of a matrix. For our purpose, it is most convenient to define the rank of a matrix as the number of non-zero rows in the matrix after it has been transformed into a echelon form by elementary row operations.

In Example 7.3.1, the echelon forms of the coefficient matrix C_e and of the augmented matrix A_e are, respectively,

$$C_e = \begin{pmatrix} 1 & 1 & -1 \\ 0 & 1 & -1 \\ 0 & 0 & 0 \end{pmatrix}, \quad A_e = \begin{pmatrix} 1 & 1 & -1 & 2 \\ 0 & 1 & -1 & 1 \\ 0 & 0 & 0 & 2 \end{pmatrix}.$$

In C_e, there are two non-zero rows, and therefore the rank of the coefficient matrix is 2. In A_e, there are three non-zero rows, and therefore the rank of the augmented matrix is 3. As we have shown, this system has no solution.

In Example 7.3.2, the echelon forms of these two matrices are

$$C_e = \begin{pmatrix} 1 & 3 & 1 \\ 0 & 1 & 1 \\ 0 & 0 & 0 \end{pmatrix}, \quad A_e = \begin{pmatrix} 1 & 3 & 1 & 6 \\ 0 & 1 & 1 & 1 \\ 0 & 0 & 0 & 0 \end{pmatrix}.$$

They both have only two non-zero rows. Therefore, the rank of the coefficient matrix equals the rank of the augmented matrix. They both equal to 2. As we have seen, this system has infinite number of solutions.

In Example 7.3.3, the two echelon forms are

$$C_e = \begin{pmatrix} 1 & 1 \\ 0 & 1 \\ 0 & 0 \end{pmatrix}, \quad A_e = \begin{pmatrix} 1 & 1 & 2 \\ 0 & 1 & 1 \\ 0 & 0 & 0 \end{pmatrix}.$$

Both of them have two non-zero rows, and therefore the coefficient matrix and the augmented matrix have the same rank of 2. As we have shown, this system has an unique solution.

From the results of these examples, we can make the following observations:

- 1. A linear system of m equations and n unknowns has solutions if and only if the coefficient matrix and the augmented matrix have the same rank.
- 2. If the rank of both matrices is r, and $r < n$, the system has infinitely many solutions.
- 3. If $r = n$, the system has only one solution.

Actually these statements are generally valid for all linear systems regardless of whether $m < n$, $m = n$, or $m > n$.

The most interesting case is $m = n = r$. In that case, the coefficient matrix is a square matrix. The solution of such systems can be obtained from (1) Cramer's rule discussed in the chapter of determinants, (2) the Gauss elimination method discussed in this section, and (3) the inverse matrix which we will discuss in the next section.

7.4 Inverse Matrix

7.4.1 Non-singular Matrix

TheindexInverse matrix square matrix A is said to be non-singular if there exists a matrix B such that

$$BA = I,$$

where I is the identity (unit) matrix. If no matrix B exists, then A is said to be singular. The matrix B is the inverse of A and vice versa. The inverse matrix is denoted by A^{-1}

$$A^{-1} = B.$$

The relationship is reciprocal. If B is the inverse of A, then A is the inverse of B. Since

$$BA = A^{-1}A = I, \tag{7.21}$$

applying B^- from the left

$$B^{-1}BA = B^{-1}I.$$

It follows

$$A = B^{-1}. \tag{7.22}$$

Existence If A is non-singular, then determinant $|A| \neq 0$.

Proof If A is non-singular, then by definition A^{-1} exists and $AA^{-1} = I$. Thus

$$\left|AA^{-1}\right| = |A| \cdot \left|A^{-1}\right| = |I|.$$

Since $|I| = 1$, neither $|A|$ nor $\left|A^{-1}\right|$ can be zero.

If $|A| \neq 0$, we will show in following sections that A^{-1} can always be found.

Uniqueness The inverse of a matrix, if it exists, is unique. That is, if

$$AB = I;$$
$$AC = I,$$

then

$$B = C.$$

This can be seen as follows. Since $AC = I$, by definition $C = A^{-1}$. It follows that

$$CA = AC = I.$$

Multiplying this equation from the right by B, we have

$$(CA)B = IB = B.$$

But

$$(CA)B = C(AB) = CI = C.$$

It is clear from the last two equations that $B = C$.

Inverse of Matrix Products The inverse of the product of a number of matrices, none of which is singular, equals the product of the inverses taken in the reverse order.

Proof Consider three non-singular matrices A, B, and C. We will show

$$(ABC)^{-1} = C^{-1}B^{-1}A^{-1}.$$

By definition

$$ABC(ABC)^{-1} = I.$$

Now

$$ABC(C^{-1}B^{-1}A^{-1}) = AB(CC^{-1})B^{-1}A^{-1}$$
$$= ABB^{-1}A^{-1} = AIA^{-1} = AA^{-1} = I.$$

Since the inverse is unique, it follows that

$$(ABC)^{-1} = C^{-1}B^{-1}A^{-1}.$$

7.4.2 Inverse Matrix by Cramer's Rule

To find A^{-1}, let us consider the set of nonhomogeneous linear equation

$$\begin{pmatrix} a_{11} & a_{12} & \cdots & a_{1n} \\ a_{21} & a_{22} & \cdots & a_{2n} \\ \cdot & \cdot & \cdots & \cdot \\ a_{n1} & a_{n2} & \cdots & a_{nn} \end{pmatrix} \begin{pmatrix} x_1 \\ x_2 \\ \cdot \\ x_n \end{pmatrix} = \begin{pmatrix} d_1 \\ d_2 \\ \cdot \\ d_n \end{pmatrix} \tag{7.23}$$

written as

$$(A)(x) = (d). \tag{7.24}$$

According to Cramer's rule discussed in the chapter on determinants,

$$x_i = \frac{N_i}{|A|},$$

where $|A|$ is the determinant of A and N_i is the determinant

$$N_i = \begin{vmatrix} a_{11} & \cdots & a_{1i-1} & d_1 & a_{1i+1} & \cdots & a_{1n} \\ a_{21} & \cdots & a_{2i-1} & d_2 & a_{2i+1} & \cdots & a_{2n} \\ \cdot & \cdots & \cdot & \cdot & \cdot & \cdots & \cdot \\ a_{n1} & \cdots & a_{ni-1} & d_n & a_{ni+1} & \cdots & a_{nn} \end{vmatrix}.$$

Expanding N_i over the ith column, we have

$$x_i = \frac{1}{|A|} \sum_{j=1}^{n} d_j C_{ji}, \tag{7.25}$$

where C_{ji} is the cofactor of jth row and ith column of A.

Now let $A^{-1} = B$, that is,

$$A^{-1} = B = \begin{pmatrix} b_{11} & b_{12} & \cdots & b_{1n} \\ b_{21} & b_{22} & \cdots & b_{2n} \\ \cdot & \cdot & \cdots & \cdot \\ b_{n1} & b_{n2} & \cdots & b_{nn} \end{pmatrix}.$$

Applying A^{-1} to (7.24) from the left

$$(A^{-1})(A)(x) = (A^{-1})(d),$$

so

$$(x) = (A^{-1})(d),$$

or

$$\begin{pmatrix} x_1 \\ x_2 \\ \cdot \\ x_n \end{pmatrix} = \begin{pmatrix} b_{11} & b_{12} & \cdots & b_{1n} \\ b_{21} & b_{22} & \cdots & b_{2n} \\ \cdot & \cdot & \cdots & \cdot \\ b_{n1} & b_{n2} & \cdots & b_{nn} \end{pmatrix} \begin{pmatrix} d_1 \\ d_2 \\ \cdot \\ d_n \end{pmatrix}.$$

Thus

$$x_i = \sum_{j=1}^{n} b_{ij} d_j. \tag{7.26}$$

Comparing (7.25) and (7.26), it is clear

$$b_{ij} = \frac{1}{|A|} C_{ji} = \frac{1}{|A|} \tilde{C}_{ij}.$$

Thus the process of obtaining the inverse of a non-singular matrix involves the following steps:

(a) Obtain the cofactor of every element of the matrix A and write the matrix of cofactors in the form

$$C = \begin{pmatrix} C_{11} & C_{12} & \cdots & C_{1n} \\ C_{21} & C_{22} & \cdots & C_{2n} \\ & \cdot & \cdots & \cdot \\ C_{n1} & C_{n2} & \cdots & C_{nn} \end{pmatrix}.$$

(b) Transpose the matrix of cofactors to obtain

$$\tilde{C} = \begin{pmatrix} C_{11} & C_{21} & \cdots & C_{n1} \\ C_{12} & C_{22} & \cdots & C_{n2} \\ & \cdot & \cdots & \cdot \\ C_{1n} & C_{2n} & \cdots & C_{nn} \end{pmatrix}.$$

(c) Divide this by det A to obtain the inverse of A

$$A^{-1} = \frac{1}{|A|} \begin{pmatrix} C_{11} & C_{21} & \cdots & C_{n1} \\ C_{12} & C_{22} & \cdots & C_{n2} \\ & \cdot & \cdots & \cdot \\ C_{1n} & C_{2n} & \cdots & C_{nn} \end{pmatrix}.$$

Example 7.4.1 Find the inverse of the following matrix by Cramer's rule:

$$A = \begin{pmatrix} -3 & 1 & -1 \\ 15 & -6 & 5 \\ -5 & 2 & -2 \end{pmatrix}.$$

Solution 7.4.1 The nine cofactors of A are

$$C_{11} = \begin{vmatrix} -6 & 5 \\ 2 & -2 \end{vmatrix} = 2, \quad C_{12} = -\begin{vmatrix} 15 & 5 \\ -5 & -2 \end{vmatrix} = 5, \quad C_{13} = \begin{vmatrix} 15 & -6 \\ -5 & 2 \end{vmatrix} = 0,$$

$$C_{21} = -\begin{vmatrix} 1 & -1 \\ 2 & -2 \end{vmatrix} = 0, \quad C_{22} = \begin{vmatrix} -3 & -1 \\ -5 & -2 \end{vmatrix} = 1, \quad C_{23} = -\begin{vmatrix} -3 & 1 \\ -5 & 2 \end{vmatrix} = 1,$$

$$C_{31} = \begin{vmatrix} 1 & -1 \\ -6 & 5 \end{vmatrix} = -1, \quad C_{32} = -\begin{vmatrix} -3 & -1 \\ 15 & 5 \end{vmatrix} = 0, \quad C_{33} = \begin{vmatrix} -3 & 1 \\ 15 & -6 \end{vmatrix} = 3.$$

The value of the determinant of A can be obtained from the Laplacian expansion over any row or any column. For example, over the first column

$$|A| = -3C_{11} + 15C_{21} - 5C_{31} = -6 + 0 + 5 = -1.$$

So the inverse exists. The matrix of cofactors C is

$$C = \begin{pmatrix} 2 & 5 & 0 \\ 0 & 1 & 1 \\ -1 & 0 & 3 \end{pmatrix}.$$

The inverse of A is then obtained by transposing C and dividing it by $\det A$. Therefore

$$A^{-1} = \frac{1}{|A|}\tilde{C} = \frac{1}{-1}\begin{pmatrix} 2 & 5 & 0 \\ 0 & 1 & 1 \\ -1 & 0 & 3 \end{pmatrix}^T = \begin{pmatrix} -2 & 0 & 1 \\ -5 & -1 & 0 \\ 0 & -1 & -3 \end{pmatrix}. \tag{7.27}$$

It can be directly verified that

$$A^{-1}A = \begin{pmatrix} -2 & 0 & 1 \\ -5 & -1 & 0 \\ 0 & -1 & -3 \end{pmatrix}\begin{pmatrix} -3 & 1 & -1 \\ 15 & -6 & 5 \\ -5 & 2 & -2 \end{pmatrix} = \begin{pmatrix} 1 & 0 & 0 \\ 0 & 1 & 0 \\ 0 & 0 & 1 \end{pmatrix}.$$

In literature, the transpose of the cofactor matrix of A is sometimes defined as the adjoint of A, that is, $adj\ A = \tilde{C}$. However, the name adjoint has another meaning, especially in quantum mechanics. It is often defined as the Hermitian conjugate $A^\dagger$, that is, $adj\ A = A^\dagger$. We will discuss Hermitian matrix in the next chapter.

For a large matrix, there are more efficient techniques to find the inverse matrix. However, for a 2×2 non-singular matrix

$$A = \begin{pmatrix} a_{11} & a_{12} \\ a_{21} & a_{22} \end{pmatrix},$$

one readily obtain from this method

$$A^{-1} = \frac{1}{|A|}\begin{pmatrix} a_{22} & -a_{12} \\ -a_{21} & a_{11} \end{pmatrix}.$$

This result is simple and useful. It may even be worthwhile to memorize it.

Example 7.4.2 Find the inverse matrices for

$$A = \begin{pmatrix} 1 & 2 \\ 3 & 4 \end{pmatrix}, \qquad R = \begin{pmatrix} \cos\theta & \sin\theta \\ -\sin\theta & \cos\theta \end{pmatrix}.$$

Solution 7.4.2

$$A^{-1} = \frac{1}{(4-6)}\begin{pmatrix} 4 & -2 \\ -3 & 1 \end{pmatrix} = \begin{pmatrix} -2 & 1 \\ 3/2 & -1/2 \end{pmatrix},$$

$$R^{-1} = \frac{1}{(\cos^2\theta + \sin^2\theta)}\begin{pmatrix} \cos\theta & -\sin\theta \\ \sin\theta & \cos\theta \end{pmatrix} = \begin{pmatrix} \cos\theta & -\sin\theta \\ \sin\theta & \cos\theta \end{pmatrix}.$$

One can readily verify

$$\begin{pmatrix} -2 & 1 \\ 3/2 & -1/2 \end{pmatrix} \begin{pmatrix} 1 & 2 \\ 3 & 4 \end{pmatrix} = \begin{pmatrix} 1 & 0 \\ 0 & 1 \end{pmatrix},$$

$$\begin{pmatrix} \cos\theta & -\sin\theta \\ \sin\theta & \cos\theta \end{pmatrix} \begin{pmatrix} \cos\theta & \sin\theta \\ -\sin\theta & \cos\theta \end{pmatrix} = \begin{pmatrix} 1 & 0 \\ 0 & 1 \end{pmatrix}.$$

7.4.3 Inverse of Elementary Matrices

Elementary Matrices An elementary matrix is a matrix that can be obtained from the identity matrix I by an elementary operation. For example, the elementary matrix E_1 obtained from interchanging row 1 and row 2 of the identity matrix of third order is

$$E_1 = \begin{pmatrix} 0 & 1 & 0 \\ 1 & 0 & 0 \\ 0 & 0 & 1 \end{pmatrix}.$$

On the other hand, the elementary row operation of interchanging row 1 and row 2 of any matrix A of order $3 \times n$ can be accomplished by premultiplying A by the elementary matrix E_1 :

$$\begin{pmatrix} 0 & 1 & 0 \\ 1 & 0 & 0 \\ 0 & 0 & 1 \end{pmatrix} \begin{pmatrix} a_{11} & a_{12} \\ a_{21} & a_{22} \\ a_{31} & a_{32} \end{pmatrix} = \begin{pmatrix} a_{21} & a_{22} \\ a_{11} & a_{12} \\ a_{31} & a_{32} \end{pmatrix}.$$

The second elementary operation, namely, multiplying a row, say row 2, by a scalar k can be accomplished as follows:

$$\begin{pmatrix} 1 & 0 & 0 \\ 0 & k & 0 \\ 0 & 0 & 1 \end{pmatrix} \begin{pmatrix} a_{11} & a_{12} \\ a_{21} & a_{22} \\ a_{31} & a_{32} \end{pmatrix} = \begin{pmatrix} a_{11} & a_{12} \\ ka_{21} & ka_{22} \\ a_{31} & a_{32} \end{pmatrix}.$$

Finally, to add the third row k times to the second row, we can proceed in the following way:

$$\begin{pmatrix} 1 & 0 & 0 \\ 0 & 1 & k \\ 0 & 0 & 1 \end{pmatrix} \begin{pmatrix} a_{11} & a_{12} \\ a_{21} & a_{22} \\ a_{31} & a_{32} \end{pmatrix} = \begin{pmatrix} a_{11} & a_{12} \\ a_{21}+ka_{31} & a_{22}+ka_{32} \\ a_{31} & a_{32} \end{pmatrix}.$$

   Thus, to effect any elementary operation on a matrix A, one may first perform the same elementary operation on an identity matrix to obtain the corresponding elementary matrix. Then premultiply A by the elementary matrix.

Inverse of an Elementary Matrix Since the elementary matrix is obtained from the elementary operation on the identity matrix, its inverse simply represents the reverse operation. For example, E_1 is obtained from interchanging row 1 and row 2 of the identity matrix I :

$$E_1 I = E_1 = \begin{pmatrix} 0 & 1 & 0 \\ 1 & 0 & 0 \\ 0 & 0 & 1 \end{pmatrix}.$$

Since

$$E_1^{-1} E_1 = I = \begin{pmatrix} 1 & 0 & 0 \\ 0 & 1 & 0 \\ 0 & 0 & 1 \end{pmatrix},$$

E_1^{-1} represents the operation of interchanging row 1 and row 2 of E_1, Thus E_1^{-1} is also given by

$$E_1^{-1} = \begin{pmatrix} 0 & 1 & 0 \\ 1 & 0 & 0 \\ 0 & 0 & 1 \end{pmatrix} = E_1.$$

   The inverses of the two other kinds of elementary matrices can be obtained in a similar way, namely,

$$E_2 = \begin{pmatrix} 1 & 0 & 0 \\ 0 & k & 0 \\ 0 & 0 & 1 \end{pmatrix}, \qquad E_2^{-1} = \begin{pmatrix} 1 & 0 & 0 \\ 0 & 1/k & 0 \\ 0 & 0 & 1 \end{pmatrix};$$

$$E_3 = \begin{pmatrix} 1 & 0 & 0 \\ 0 & 1 & k \\ 0 & 0 & 1 \end{pmatrix}, \qquad E_3^{-1} = \begin{pmatrix} 1 & 0 & 0 \\ 0 & 1 & -k \\ 0 & 0 & 1 \end{pmatrix}.$$

   It can be readily shown by successive elementary operations that

$$E_4 = \begin{pmatrix} a & 0 & 0 \\ 0 & b & 0 \\ 0 & 0 & c \end{pmatrix}, \qquad E_4^{-1} = \begin{pmatrix} 1/a & 0 & 0 \\ 0 & 1/b & 0 \\ 0 & 0 & 1/c \end{pmatrix}$$

and

$$E_5 = \begin{pmatrix} 1 & 0 & 0 & 0 \\ 0 & 1 & n & m \\ 0 & 0 & 1 & 0 \\ 0 & 0 & 0 & 1 \end{pmatrix}, \qquad E_5^{-1} = \begin{pmatrix} 1 & 0 & 0 & 0 \\ 0 & 1 & -n & -m \\ 0 & 0 & 1 & 0 \\ 0 & 0 & 0 & 1 \end{pmatrix}.$$

7.4.4 Inverse Matrix by Gauss-Jordan Elimination

For a matrix of large order, Cramer's rule is of little practical use. One of the most commonly used methods for inverting a large matrix is the Gauss-Jordan method.

Equation (7.23) can be written in the form

$$
\begin{pmatrix} a_{11} & a_{12} & \cdots & a_{1n} \\ a_{21} & a_{22} & \cdots & a_{2n} \\ \cdot & \cdot & \cdots & \cdot \\ a_{n1} & a_{n2} & \cdots & a_{nn} \end{pmatrix} \begin{pmatrix} x_1 \\ x_2 \\ \cdot \\ x_n \end{pmatrix} = \begin{pmatrix} 1 & 0 & \cdots & 0 \\ 0 & 1 & \cdots & 0 \\ \cdot & \cdot & \cdots & \cdot \\ 0 & 0 & \cdots & 1 \end{pmatrix} \begin{pmatrix} d_1 \\ d_2 \\ \cdot \\ d_n \end{pmatrix}, \tag{7.28}
$$

or symbolically as

$$
(A)(x) = (I)(d). \tag{7.29}
$$

If both sides of this equation is under the same operation, the equation will remain to be valid. We will operate them with the Gauss-Jordan procedure. Each step is an elementary row operation which can be thought as premultiplying (multiplying from the left) both sides by the elementary matrix representing that operation. Thus, the entire Gauss-Jordan process is equivalent to multiplying (7.29) by a matrix B which is a product of all the elementary matrices representing the steps of the Gauss-Jordan procedure

$$
(B)(A)(x) = (B)(I)(d). \tag{7.30}
$$

Since the process reduces the coefficient matrix A to the identity matrix I, so

$$
BA = I.
$$

Post-multiplying both sides by A^{-1}

$$
BAA^{-1} = IA^{-1},
$$

we have

$$
B = A^{-1}.
$$

Therefore, when the left-hand side of (7.30) becomes a unit matrix times the column matrix x, the right-hand side of the equation must be equal to the inverse matrix times the column matrix d.

Thus, if we want to find the inverse of A, we can first augment A by the identity matrix I, and then use elementary operations to transform this matrix. When the submatrix A is in the form of I, the form assumed of the original identity matrix I must be A^{-1}.

We have found the inverse of

$$A = \begin{pmatrix} -3 & 1 & -1 \\ 15 & -6 & 5 \\ -5 & 2 & -2 \end{pmatrix}$$

in Example 7.4.1 by Cramer's rule. Now let us do the same problem by Gauss-Jordan elimination. First we augment A by the identity matrix I

$$\left(\begin{array}{ccc|ccc} -3 & 1 & -1 & 1 & 0 & 0 \\ 15 & -6 & 5 & 0 & 1 & 0 \\ -5 & 2 & -2 & 0 & 0 & 1 \end{array} \right).$$

Divide the first row by -3, second row by 15, and third row by -5 :

$$\left(\begin{array}{ccc|ccc} 1 & -\frac{1}{3} & \frac{1}{3} & -\frac{1}{3} & 0 & 0 \\ 1 & -\frac{6}{15} & \frac{5}{15} & 0 & \frac{1}{15} & 0 \\ 1 & -\frac{2}{5} & \frac{2}{5} & 0 & 0 & -\frac{1}{5} \end{array} \right),$$

leave the first row as it is, subtract the first row from the second row and put it in the second row, and subtract the first row from the third row and put it back in the third row:

$$\left(\begin{array}{ccc|ccc} 1 & -\frac{1}{3} & \frac{1}{3} & -\frac{1}{3} & 0 & 0 \\ 0 & -\frac{1}{15} & 0 & \frac{1}{3} & \frac{1}{15} & 0 \\ 0 & -\frac{1}{15} & \frac{1}{15} & \frac{1}{3} & 0 & -\frac{1}{5} \end{array} \right),$$

multiply the second and third row by -15 :

$$\left(\begin{array}{ccc|ccc} 1 & -\frac{1}{3} & \frac{1}{3} & -\frac{1}{3} & 0 & 0 \\ 0 & 1 & 0 & -5 & -1 & 0 \\ 0 & 1 & -1 & -5 & 0 & 3 \end{array} \right),$$

leave the second row where it is, subtract it from the third row and put the result back to the third row, and then add $\frac{1}{3}$ of the second row to the first row:

$$\left(\begin{array}{ccc|ccc} 1 & 0 & \frac{1}{3} & -2 & -\frac{1}{3} & 0 \\ 0 & 1 & 0 & -5 & -1 & 0 \\ 0 & 0 & -1 & 0 & 1 & 3 \end{array} \right),$$

multiply the third row by -1 and then subtract $\frac{1}{3}$ of it from the first row:

$$\left(\begin{array}{ccc|ccc} 1 & 0 & 0 & -2 & 0 & 1 \\ 0 & 1 & 0 & -5 & -1 & 0 \\ 0 & 0 & 1 & 0 & -1 & -3 \end{array} \right).$$

Finally, we have changed matrix A to the unit matrix I, the original unit matrix on the right side must have changed to A^{-1}, thus

$$A^{-1} = \begin{pmatrix} -2 & 0 & 1 \\ -5 & -1 & 0 \\ 0 & -1 & -3 \end{pmatrix},$$

which is the same as (7.27) obtained in the last section.

This technique is actually more adapted to modern computers. Computer Codes and extensive literature for the Gauss-Jordan elimination method are given in W.H. Press, B.P. Flannery, S.A. Teukolsky, and W.T. Vetterling, *Numerical Recipes*, 2nd ed. Cambridge University Press (1992).

Exercises

1.

Given two matrices

$$A = \begin{pmatrix} 2 & 5 \\ -2 & 1 \end{pmatrix}, \quad B = \begin{pmatrix} 2 & 0 \\ 2 & 1 \end{pmatrix},$$

find $B - 5A$.

Ans. $\begin{pmatrix} -8 & -25 \\ 12 & -4 \end{pmatrix}$.

2.

If A and B are the 2×2 matrices

$$A = \begin{pmatrix} 2 & 4 \\ -3 & 1 \end{pmatrix}, \quad B = \begin{pmatrix} 3 & -1 \\ 4 & 2 \end{pmatrix},$$

find the products AB and BA.

Ans. $AB = \begin{pmatrix} 22 & 6 \\ -5 & 5 \end{pmatrix}$, $BA = \begin{pmatrix} 9 & 11 \\ 2 & 18 \end{pmatrix}$.

3.

If

$$A = \begin{pmatrix} 2 & -1 & 4 \\ -3 & 2 & 1 \end{pmatrix}, \quad B = \begin{pmatrix} 1 & -4 \\ 3 & -2 \\ -1 & 1 \end{pmatrix},$$

find AB and BA if they exist.

Ans. $AB = \begin{pmatrix} -5 & -2 \\ 2 & 9 \end{pmatrix}$, $BA = \begin{pmatrix} 14 & -9 & 0 \\ 12 & -7 & 10 \\ -5 & 3 & -3 \end{pmatrix}$.

4.

If

$$A = \begin{pmatrix} 2 & 1 & -3 \\ 0 & 2 & -2 \\ -1 & -1 & 3 \\ 2 & 0 & 1 \end{pmatrix}, \quad B = \begin{pmatrix} 3 & 0 \\ 2 & 4 \\ 2 & -1 \end{pmatrix},$$

Find AB and BA if they exist.

Ans. $AB = \begin{pmatrix} 2 & 7 \\ 0 & 10 \\ 1 & -7 \\ 8 & -1 \end{pmatrix}, \quad BA$ does not exist.

5.

If

$$A = \begin{pmatrix} 1 & 2 \\ -2 & 3 \end{pmatrix}, \quad B = \begin{pmatrix} 3 & 1 \\ 2 & 0 \end{pmatrix}, \quad C = \begin{pmatrix} 4 & 3 \\ 2 & 1 \end{pmatrix},$$

verify the associative law by showing that

$$(AB)C = A(BC).$$

6.

Show that if

$$A = \begin{pmatrix} 1 & 1 \\ 0 & 1 \end{pmatrix},$$

then

$$A^n = \begin{pmatrix} 1 & n \\ 0 & 1 \end{pmatrix}.$$

Hint: $A^n = \left[\begin{pmatrix} 1 & 0 \\ 0 & 1 \end{pmatrix} + \begin{pmatrix} 0 & 1 \\ 0 & 0 \end{pmatrix} \right]^n.$

7.

Given

$$A = \begin{pmatrix} -1 & 0 \\ 0 & -1 \end{pmatrix}, \quad B = \begin{pmatrix} 0 & 1 \\ 1 & 0 \end{pmatrix}, \quad C = \begin{pmatrix} 0 & -1 \\ -1 & 0 \end{pmatrix}, \quad I = \begin{pmatrix} 1 & 0 \\ 0 & 1 \end{pmatrix}.$$

Find all possible products of A, B, C, and I, two at a time including squares.

(Note that the product of any two matrices is another matrix in this group. These four matrices form a representation of a mathematical group, known as viergruppe. [vier is the German word four]).

8.

If
$$A = \begin{pmatrix} ab & b^2 \\ -a^2 & -ab \end{pmatrix},$$

show that $A^2 = 0$.

9.

Find the value of
$$(1\ 2) \begin{pmatrix} -1 & 0 \\ 2 & 1 \end{pmatrix} \begin{pmatrix} 2 \\ 1 \end{pmatrix}.$$

Ans. 8.

10.

Explicitly verify that $(AB)^T = \tilde{B}\tilde{A}$, if
$$A = \begin{pmatrix} 1 & 2 \\ 3 & 4 \end{pmatrix}, \quad B = \begin{pmatrix} -1 & 0 \\ 1 & 2 \end{pmatrix}.$$

11.

Show that matrix A is symmetric, if
$$A = B\tilde{B}.$$

Hint: $a_{ij} = \sum_k b_{ik} \tilde{b}_{kj}$.

12.

Let $A = \begin{pmatrix} 1 & 3 \\ 5 & 12 \end{pmatrix}$, find a matrix E such that EA is diagonal and $|EA| = |A|$.

Ans. $\begin{pmatrix} -4 & 1 \\ -5 & 1 \end{pmatrix}$.

13.

Let
$$A = \begin{pmatrix} 1 & 0 \\ 2 & -1 \end{pmatrix}, \quad B = \begin{pmatrix} 2 & 1 \\ 0 & 3 \end{pmatrix},$$

explicitly show that
$$AB \neq BA \quad \text{but} \quad |AB| = |BA|.$$

14.

Show that if
$$[A, B] \neq 0,$$

then

$$(A - B)(A + B) \neq A^2 - B^2,$$
$$(A + B)^2 \neq A^2 + 2AB + B^2.$$

15.

Show that

$$[A, [B, C]] + [B, [C, A]] + [C, [A, B]] = 0.$$

16.

Show that

$$\left|A^{-1}\right| = |A|^{-1}.$$

Hint: $AA^{-1} = I$, $|AB| = |A| |B|$.

17.

Let

$$A = \begin{pmatrix} 1 & 3 \\ 5 & 7 \end{pmatrix}, \qquad B = \begin{pmatrix} 2 & 4 \\ 6 & 8 \end{pmatrix},$$

find A^{-1}, B^{-1} and $(AB)^{-1}$ by Cramer's rule and verify that $(AB)^{-1} = B^{-1}A^{-1}$.

18.

Reduce the augmented matrix of the following system to an echelon form and show that this system has no solution.

$$x_1 + x_2 + 2x_3 + x_4 = 5$$
$$2x_1 + 3x_2 - x_3 - 2x_4 = 2$$
$$4x_1 + 5x_2 + 3x_3 = 7.$$

Ans. $\begin{pmatrix} 1 & 1 & 2 & 1 & 5 \\ 0 & 1 & -5 & -4 & -8 \\ 0 & 0 & 0 & 0 & -5 \end{pmatrix}$.

19.

Solve the following equations by Gauss' elimination.

$$x_1 + 2x_2 - 3x_3 = -1$$
$$3x_1 - 2x_2 + 2x_3 = 10$$
$$4x_1 + x_2 + 2x_3 = 3.$$

Ans. $x_3 = -1, x_2 = -3, x_1 = 2.$

20.

Let

$$A = \begin{pmatrix} 1 & 2 & -3 \\ 3 & -2 & 2 \\ 4 & 1 & 2 \end{pmatrix},$$

find A^{-1} by Gauss-Jordan elimination. Find x_1, x_2, x_3 from

$$\begin{pmatrix} x_1 \\ x_2 \\ x_3 \end{pmatrix} = A^{-1} \begin{pmatrix} -1 \\ 10 \\ 3 \end{pmatrix},$$

and show that

$$A \begin{pmatrix} x_1 \\ x_2 \\ x_3 \end{pmatrix} = \begin{pmatrix} -1 \\ 10 \\ 3 \end{pmatrix}.$$

21.

Determine the rank of the following matrices:

$$(a) \begin{pmatrix} 4 & 2 & -1 & 3 \\ 0 & 5 & -1 & 2 \\ 12 & -4 & -1 & 5 \end{pmatrix}, \quad (b) \begin{pmatrix} 3 & -1 & 4 & -2 \\ 0 & 2 & 4 & 6 \\ 6 & -1 & 10 & -1 \end{pmatrix}.$$

Ans. (a) 2, (b) 2.

22.

Determine if the following systems are consistent. If consistent, is the solution unique?

$$(a) \quad \begin{aligned} x_1 - x_2 + 3x_3 &= -5 \\ -x_1 + 3x_3 &= 0 \\ 2x_1 + x_2 &= 1. \end{aligned} \qquad (b) \quad \begin{aligned} x_1 - 2x_2 + 3x_3 &= 0 \\ 2x_1 + 3x_2 - x_3 &= 0 \\ 4x_1 - x_2 + 5x_3 &= 0. \end{aligned}$$

Ans. (a) unique solution, (b) infinite number of solutions.

23.

Find the value of λ so that the following linear system has a solution:

$$\begin{aligned} x_1 + 2x_2 + 3x_3 &= 2 \\ 3x_1 + 2x_2 + x_3 &= 0 \\ x_1 + x_2 + x_3 &= \lambda. \end{aligned}$$

Ans. $\lambda = 0.5$.

24.

   Let

$$L^+ = \begin{pmatrix} 0 & 1 & 0 \\ 0 & 0 & 1 \\ 0 & 0 & 0 \end{pmatrix}, \qquad L^- = \begin{pmatrix} 0 & 0 & 0 \\ 1 & 0 & 0 \\ 0 & 1 & 0 \end{pmatrix},$$

$$|-1\rangle = \begin{pmatrix} 0 \\ 0 \\ 1 \end{pmatrix}, \quad |0\rangle = \begin{pmatrix} 0 \\ 1 \\ 0 \end{pmatrix}, \quad |1\rangle = \begin{pmatrix} 1 \\ 0 \\ 0 \end{pmatrix}, \quad |null\rangle = \begin{pmatrix} 0 \\ 0 \\ 0 \end{pmatrix}.$$

Show that

$$L^+ |-1\rangle = |0\rangle, \quad L^+ |0\rangle = |1\rangle, \quad L^+ |1\rangle = |null\rangle,$$
$$L^- |1\rangle = |0\rangle, \quad L^- |0\rangle = |-1\rangle, \quad L^- |-1\rangle = |null\rangle.$$

Chapter 8
Eigenvalue Problems of Matrices

Given a square matrix A, to determine the scalars λ and the non-zero column matrix $\mathbf{x}$ which simultaneously satisfy the equation

$$A\mathbf{x} = \lambda\mathbf{x}, \tag{8.1}$$

is known as the **eigenvalue** problem (eigen in German means proper). The solution of this problem is intimately connected to the question of whether the matrix can be transformed into a diagonal form.

The eigenvalue problem is of great interest in many engineering applications, such as mechanical vibrations, alternating currents, and rigid body dynamics. It is of crucial importance in modern physics. The whole structure of quantum mechanics is based on the diagonalization of a certain type of matrices.

8.1 Eigenvalues and Eigenvectors

8.1.1 Secular Equation

In the **eigenvalue** problem, the value λ is called the **eigenvalue (character-istic value)** and the corresponding column matrix $\mathbf{x}$ is called the **eigenvector (characteristic vector)**. If A is a $n \times n$ matrix, (8.1) is given by

$$\begin{pmatrix} a_{11} & a_{12} & \cdots & a_{1n} \\ a_{21} & a_{22} & \cdots & a_{2n} \\ \cdot & \cdot & \cdots & \cdot \\ a_{n1} & a_{n2} & \cdots & a_{nn} \end{pmatrix} \begin{pmatrix} x_1 \\ x_2 \\ \cdot \\ x_n \end{pmatrix} = \lambda \begin{pmatrix} x_1 \\ x_2 \\ \cdot \\ x_n \end{pmatrix}.$$

© The Author(s), under exclusive license to Springer Nature Switzerland AG 2022
K.-T. Tang, *Mathematical Methods for Engineers and Scientists 1*,
https://doi.org/10.1007/978-3-031-05678-9_8

Since

$$\lambda \begin{pmatrix} x_1 \\ x_2 \\ \cdot \\ x_n \end{pmatrix} = \lambda \begin{pmatrix} 1\,0\cdots 0 \\ 0\,1\cdots 0 \\ \cdot\cdot\cdot\cdot\cdot\cdot \\ 0\,0\cdots 1 \end{pmatrix} \begin{pmatrix} x_1 \\ x_2 \\ \cdot \\ x_n \end{pmatrix} = \lambda I \mathbf{x},$$

where I is the unit matrix, we can write (8.1) as

$$(A - \lambda I)\,\mathbf{x} = 0. \tag{8.2}$$

This system has nontrivial solutions if and only if the determinant of the coefficient matrix vanishes

$$\begin{vmatrix} a_{11} - \lambda & a_{12} & \cdots\cdots & a_{1n} \\ a_{21} & a_{22} - \lambda & \cdots\cdots & a_{2n} \\ \cdot\cdot & \cdot\cdot & \cdots\cdots & \cdot\cdot \\ a_{n1} & a_{n2} & \cdots\cdots & a_{nn} - \lambda \end{vmatrix} = 0. \tag{8.3}$$

The expansion of this determinant yields a polynomial of degree n in λ, which is called **characteristic polynomial** $P(\lambda)$. The equation

$$P(\lambda) = |A - \lambda I| = 0$$

is known as the **characteristic equation** (or **secular equation**). Its n roots are the eigenvalues and will be denoted $\lambda_1, \lambda_2, \cdots \lambda_n$. They may be real or complex. When one of the eigenvalues is substituted back into (8.2), the corresponding eigenvector $\mathbf{x}(x_1, x_2, \ldots x_n)$ may be determined. Note that the eigenvectors may be multiplied by any constant and remain a solution of the equation.

We will denote $\mathbf{x}_i$ as the the eigenvector belonging to the eigenvalue λ_i. That is, if

$$P(\lambda_i) = 0,$$

then

$$\mathbf{A}\mathbf{x}_i = \lambda_i \mathbf{x}_i.$$

If n eigenvalues are all different, we will have n distinct eigenvectors. If two or more eigenvalues are the same, we say that they are degenerate. In some problems, a degenerate eigenvalue may produce only one eigenvector, in other problems a degenerate eigenvalue may produce more than one distinct eigenvectors.

Example 8.1.1 Find the eigenvalues and eigenvectors of A, if

$$A = \begin{pmatrix} 1 & 2 \\ 2 & 1 \end{pmatrix}.$$

Solution 8.1.1 The characteristic polynomial of A is

$$P(\lambda) = \begin{vmatrix} 1-\lambda & 2 \\ 2 & 1-\lambda \end{vmatrix} = \lambda^2 - 2\lambda - 3,$$

and the secular equation is

$$\lambda^2 - 2\lambda - 3 = (\lambda + 1)(\lambda - 3) = 0.$$

Thus, the eigenvalues are

$$\lambda_1 = -1; \qquad \lambda_2 = 3.$$

Let the eigenvector $\mathbf{x}_1$ corresponding to $\lambda_1 = -1$ be $\begin{pmatrix} x_{11} \\ x_{12} \end{pmatrix}$, then $\mathbf{x}_1$ must satisfy

$$\begin{pmatrix} 1-\lambda_1 & 2 \\ 2 & 1-\lambda_1 \end{pmatrix} \begin{pmatrix} x_{11} \\ x_{12} \end{pmatrix} = 0, \implies \begin{pmatrix} 2 & 2 \\ 2 & 2 \end{pmatrix} \begin{pmatrix} x_{11} \\ x_{12} \end{pmatrix} = 0$$

This reduces to

$$2x_{11} + 2x_{12} = 0.$$

Thus, for this eigenvector, $x_{11} = -x_{12}$. That is, $x_{11} : x_{12} = -1 : 1$. Therefore the eigenvector can be written as

$$\mathbf{x}_1 = \begin{pmatrix} -1 \\ 1 \end{pmatrix}.$$

Any constant, positive or negative, times it will also be a solution, but it will not be regarded as another distinct eigenvector. With a similar procedure, we find the eigenvector corresponding to $\lambda_2 = 3$ to be

$$\mathbf{x}_2 = \begin{pmatrix} x_{21} \\ x_{22} \end{pmatrix} = \begin{pmatrix} 1 \\ 1 \end{pmatrix}.$$

Example 8.1.2 Find the eigenvalues and eigenvectors of A if

$$A = \begin{pmatrix} 3 & -5 \\ 1 & -1 \end{pmatrix}.$$

Solution 8.1.2 The characteristic polynomial of A is

$$P(\lambda) = \begin{vmatrix} 3-\lambda & -5 \\ 1 & -1-\lambda \end{vmatrix} = \lambda^2 - 2\lambda + 2,$$

so the secular equation is

$$\lambda^2 - 2\lambda + 2 = 0.$$

Thus, the eigenvalues are

$$\lambda = 1 \pm i.$$

Let $\lambda_1 = 1 + i$, and the corresponding eigenvector $\mathbf{x}_1$ be $\begin{pmatrix} x_{11} \\ x_{12} \end{pmatrix}$, then $\mathbf{x}_1$ must satisfy

$$\begin{pmatrix} 3 - (1+i) & -5 \\ 1 & -1 - (1+i) \end{pmatrix} \begin{pmatrix} x_{11} \\ x_{12} \end{pmatrix} = 0,$$

which gives

$$(2-i)x_{11} - 5x_{12} = 0$$
$$x_{11} - (2+i)x_{12} = 0.$$

The first equation gives

$$x_{11} = \frac{5}{2-i}x_{12} = \frac{5(2+i)}{4+1}x_{12} = \frac{2+i}{1}x_{12},$$

which is the same result from the second equation, as it should be. Therefore, $\mathbf{x}_1$ can be written as

$$\mathbf{x}_1 = \begin{pmatrix} 2+i \\ 1 \end{pmatrix}.$$

Similarly, for $\lambda = \lambda_2 = 1 - i$, the corresponding eigenvector is

$$\mathbf{x}_2 = \begin{pmatrix} 2-i \\ 1 \end{pmatrix}.$$

So, we have an example of a real matrix with complex eigenvalues and complex eigenvectors.

Example 8.1.3 Find the eigenvalues and eigenvectors of A if

$$A = \begin{pmatrix} -2 & 2 & -3 \\ 2 & 1 & -6 \\ -1 & -2 & 0 \end{pmatrix}.$$

Solution 8.1.3 The characteristic polynomial of A is

$$P(\lambda) = \begin{vmatrix} -2-\lambda & 2 & -3 \\ 2 & 1-\lambda & -6 \\ -1 & -2 & -\lambda \end{vmatrix} = -\lambda^3 - \lambda^2 + 21\lambda + 45.$$

The secular equation can be written as

$$\lambda^3 + \lambda^2 - 21\lambda - 45 = (\lambda - 5)(\lambda + 3)^2 = 0.$$

This equation has a single root of 5 and a double root of -3. Let

$$\lambda_1 = 5, \quad \lambda_2 = -3, \quad \lambda_3 = -3.$$

The eigenvector belonging to the eigenvalue of λ_1 must satisfy the equation

$$\begin{pmatrix} -2-5 & 2 & -3 \\ 2 & 1-5 & -6 \\ -1 & -2 & 0-5 \end{pmatrix} \begin{pmatrix} x_{11} \\ x_{12} \\ x_{13} \end{pmatrix} = 0.$$

With Gauss' elimination method, this equation can be shown to be equivalent to

$$\begin{pmatrix} -7 & 2 & -3 \\ 0 & 1 & 2 \\ 0 & 0 & 0 \end{pmatrix} \begin{pmatrix} x_{11} \\ x_{12} \\ x_{13} \end{pmatrix} = 0,$$

which means

$$-7x_{11} + 2x_{12} - 3x_{13} = 0$$
$$x_{12} + 2x_{13} = 0.$$

Assign $x_{13} = 1$, then $x_{12} = -2$, $x_{11} = -1$. So, corresponding to $\lambda_1 = 5$, the eigenvector $\mathbf{x}_1$ can be written as

$$\mathbf{x}_1 = \begin{pmatrix} -1 \\ -2 \\ 1 \end{pmatrix}.$$

Since the eigenvalue of -3 is two-fold degenerate, corresponding to this eigenvalue, we may have one or two eigenvectors. Let us express the eigenvector corresponding to the eigenvalue of -3 as $\begin{pmatrix} x_1 \\ x_2 \\ x_3 \end{pmatrix}$. It must satisfy the equation

$$\begin{pmatrix} -2+3 & 2 & -3 \\ 2 & 1+3 & -6 \\ -1 & -2 & 0+3 \end{pmatrix} \begin{pmatrix} x_1 \\ x_2 \\ x_3 \end{pmatrix} = 0.$$

With Gauss' elimination method, this equation can be shown to be equivalent to

$$\begin{pmatrix} 1 & 2 & -3 \\ 0 & 0 & 0 \\ 0 & 0 & 0 \end{pmatrix} \begin{pmatrix} x_1 \\ x_2 \\ x_3 \end{pmatrix} = 0,$$

which means

$$x_1 + 2x_2 - 3x_3 = 0.$$

We can express x_1 in terms of x_2 and x_3, and there is no restriction on x_2 and x_3. Let $x_2 = c_2$ and $x_3 = c_3$, then $x_1 = -2c_2 + 3c_3$. So, we can write

$$\begin{pmatrix} x_1 \\ x_2 \\ x_3 \end{pmatrix} = \begin{pmatrix} -2c_2 + 3c_3 \\ c_2 \\ c_3 \end{pmatrix} = c_2 \begin{pmatrix} -2 \\ 1 \\ 0 \end{pmatrix} + c_3 \begin{pmatrix} 3 \\ 0 \\ 1 \end{pmatrix}.$$

Since c_2 and c_3 are arbitrary, we can first assign $c_3 = 0$ and get an eigenvector, and then assign $c_2 = 0$ and get another eigenvector. So, corresponding to the degenerate eigenvalue $\lambda = -3$, there are two distinct eigenvectors

$$\mathbf{x}_2 = \begin{pmatrix} -2 \\ 1 \\ 0 \end{pmatrix}, \quad \mathbf{x}_3 = \begin{pmatrix} 3 \\ 0 \\ 1 \end{pmatrix}.$$

In this example, we have only two distinct eigenvalues, but we still have three distinct eigenvectors.

Example 8.1.4 Find the eigenvalues and eigenvectors of A if

$$A = \begin{pmatrix} 4 & 6 & 6 \\ 1 & 3 & 2 \\ -1 & -5 & -2 \end{pmatrix}.$$

Solution 8.1.4 The characteristic polynomial of A is

$$P(\lambda) = \begin{vmatrix} 4-\lambda & 6 & 6 \\ 1 & 3-\lambda & 2 \\ -1 & -5 & -2-\lambda \end{vmatrix} = -\lambda^3 + 5\lambda^2 - 8\lambda + 4.$$

The secular equation can be written as

$$\lambda^3 - 5\lambda^2 + 8\lambda - 4 = (\lambda - 1)(\lambda - 2)^2 = 0.$$

The three eigenvalues are

$$\lambda_1 = 1, \quad \lambda_2 = \lambda_3 = 2.$$

From the equation for the eigenvector $\mathbf{x}_1$ belonging to the eigenvalue of λ_1

$$\begin{pmatrix} 4-1 & 6 & 6 \\ 1 & 3-1 & 2 \\ -1 & -5 & -2-1 \end{pmatrix} \begin{pmatrix} x_{11} \\ x_{12} \\ x_{13} \end{pmatrix} = 0,$$

we obtain the solution

$$\mathbf{x}_1 = \begin{pmatrix} 4 \\ 1 \\ -3 \end{pmatrix}.$$

The eigenvector $\begin{pmatrix} x_1 \\ x_2 \\ x_3 \end{pmatrix}$, corresponding to the two-fold degenerate eigenvalue 2, satisfies the equation

$$\begin{pmatrix} 4-2 & 6 & 6 \\ 1 & 3-2 & 2 \\ -1 & -5 & -2-2 \end{pmatrix} \begin{pmatrix} x_1 \\ x_2 \\ x_3 \end{pmatrix} = 0.$$

With Gauss' elimination method, this equation can be shown to be equivalent to

$$\begin{pmatrix} 1 & 1 & 2 \\ 0 & 2 & 1 \\ 0 & 0 & 0 \end{pmatrix} \begin{pmatrix} x_1 \\ x_2 \\ x_3 \end{pmatrix} = 0,$$

which means

$$x_1 + x_2 + 2x_3 = 0$$
$$2x_2 + x_3 = 0.$$

If we assign $x_3 = -2$, then $x_2 = 1$ and $x_1 = 3$. So

$$\mathbf{x}_2 = \begin{pmatrix} 3 \\ 1 \\ -2 \end{pmatrix}.$$

The two equations above do not allow any other eigenvector which is not just a constant time $\mathbf{x}_2$. Therefore, for this 3×3 matrix, there are only two distinct eigenvectors.

Computer Code. It should be noted that for large systems, the eigenvalues and eigenvectors would usually be found with specialized numerical methods (See, for example, G.H. Golub and C.F. Van Loan, *Matrix Computations,* John Hopkins University Press, 1983). There are excellent general-purpose computer programs for the efficient and accurate determination of eigensystems. (See, for example, B.T. Smith, J.M. Boyle, J. Dongarra, B. Garbow, Y. Ikebe, V.C. Klema, and C.B. Moler, *Matrix Eigensystem Routines: EISPACK Guide,* 2nd edition, Springer Verlag, 1976).

In addition, eigenvalues and eigenvectors can be found with a simple command in computer packages such as , Maple, Mathematica, MathCad and MuPAD.. These packages are known as Computer Algebraic Systems.

This book is written with the software "Scientific WorkPlace", which also provides an interface to MuPAD. (Before version 5, it also came with Maple). Instead of requiring the user to adhere to a rigid syntax, the user can use natural mathematical notations. For example, to find the eigenvalues and eigenvectors of

$$
\begin{pmatrix} 5 & -6 & -6 \\ -1 & 4 & 2 \\ 3 & -6 & -4 \end{pmatrix},
$$

all you have to do is (1) type the expression in the math-mode, and (2) click on the "Compute" button, and (3) click on the "Matrices" button in the pull-down menu, and (4) click on the "Eigenvectors" button in the submenu. The program will return with

$$
\text{eigenvectors}: \left\{ \begin{pmatrix} 1 \\ -\frac{1}{3} \\ 1 \end{pmatrix} \right\} \leftrightarrow 1, \left\{ \begin{pmatrix} 2 \\ 0 \\ 1 \end{pmatrix}, \begin{pmatrix} 2 \\ 1 \\ 0 \end{pmatrix} \right\} \leftrightarrow 2.
$$

You can ask the program to check the results. For example, you can type

$$
\begin{pmatrix} 5 & -6 & -6 \\ -1 & 4 & 2 \\ 3 & -6 & -4 \end{pmatrix} \begin{pmatrix} 2 \\ 1 \\ 0 \end{pmatrix},
$$

and click on the "Compute" button, and then click on the "Evaluate" button. The program will return with

$$
\begin{pmatrix} 5 & -6 & -6 \\ -1 & 4 & 2 \\ 3 & -6 & -4 \end{pmatrix} \begin{pmatrix} 2 \\ 1 \\ 0 \end{pmatrix} = \begin{pmatrix} 4 \\ 2 \\ 0 \end{pmatrix},
$$

which is of course equal to $2 \begin{pmatrix} 2 \\ 1 \\ 0 \end{pmatrix}$, showing $\begin{pmatrix} 2 \\ 1 \\ 0 \end{pmatrix}$ is indeed an eigenvector belonging to eigenvalue 2. The other two eigenvectors can be similarly checked.

Computer Algebraic Systems are wonderful as they are, they must be used with caution. It is not infrequent that the system will return with an answer to a wrong problem without the user knowing it. Therefore, answers from these systems should be checked. Computer Algebraic Systems are useful supplements, but they are no substitute for the knowledge of the subject matter.

8.1.2 Properties of Characteristic Polynomial

The characteristic polynomial has several useful properties. To elaborate on them, let us first consider the case of $n = 3$.

$$
P(\lambda) =
\begin{vmatrix}
a_{11} - \lambda & a_{12} & a_{13} \\
a_{21} & a_{22} - \lambda & a_{23} \\
a_{31} & a_{32} & a_{33} - \lambda
\end{vmatrix}
$$

$$
=
\begin{vmatrix}
a_{11} & a_{12} & a_{13} \\
a_{21} & a_{22} - \lambda & a_{23} \\
a_{31} & a_{32} & a_{33} - \lambda
\end{vmatrix}
+
\begin{vmatrix}
-\lambda & a_{12} & a_{13} \\
0 & a_{22} - \lambda & a_{23} \\
0 & a_{32} & a_{33} - \lambda
\end{vmatrix}
$$

$$
=
\begin{vmatrix}
a_{11} & a_{12} & a_{13} \\
a_{21} & a_{22} & a_{23} \\
a_{31} & a_{32} & a_{33} - \lambda
\end{vmatrix}
+
\begin{vmatrix}
a_{11} & 0 & a_{13} \\
a_{21} & -\lambda & a_{23} \\
a_{31} & 0 & a_{33} - \lambda
\end{vmatrix}
$$

$$
+
\begin{vmatrix}
-\lambda & a_{12} & a_{13} \\
0 & a_{22} & a_{23} \\
0 & a_{32} & a_{33} - \lambda
\end{vmatrix}
+
\begin{vmatrix}
-\lambda & 0 & a_{13} \\
0 & -\lambda & a_{23} \\
0 & 0 & a_{33} - \lambda
\end{vmatrix}
$$

$$
=
\begin{vmatrix}
a_{11} & a_{12} & a_{13} \\
a_{21} & a_{22} & a_{23} \\
a_{31} & a_{32} & a_{33}
\end{vmatrix}
+
\begin{vmatrix}
a_{11} & a_{12} & 0 \\
a_{21} & a_{22} & 0 \\
a_{31} & a_{32} & -\lambda
\end{vmatrix}
+
\begin{vmatrix}
a_{11} & 0 & a_{13} \\
a_{21} & -\lambda & a_{23} \\
a_{31} & 0 & a_{33}
\end{vmatrix}
+
\begin{vmatrix}
a_{11} & 0 & 0 \\
a_{21} & -\lambda & 0 \\
a_{31} & 0 & -\lambda
\end{vmatrix}
$$

$$
+
\begin{vmatrix}
-\lambda & a_{12} & a_{13} \\
0 & a_{22} & a_{23} \\
0 & a_{32} & a_{33}
\end{vmatrix}
+
\begin{vmatrix}
-\lambda & a_{12} & 0 \\
0 & a_{22} & 0 \\
0 & a_{32} & -\lambda
\end{vmatrix}
+
\begin{vmatrix}
-\lambda & 0 & a_{13} \\
0 & -\lambda & a_{23} \\
0 & 0 & a_{33}
\end{vmatrix}
+
\begin{vmatrix}
-\lambda & 0 & 0 \\
0 & -\lambda & 0 \\
0 & 0 & -\lambda
\end{vmatrix}
$$

$$
= |A| + \left(
\begin{vmatrix}
a_{11} & a_{12} \\
a_{21} & a_{22}
\end{vmatrix}
+
\begin{vmatrix}
a_{11} & a_{13} \\
a_{31} & a_{33}
\end{vmatrix}
+
\begin{vmatrix}
a_{22} & a_{23} \\
a_{32} & a_{33}
\end{vmatrix}
\right)(-\lambda)
$$

$$
+ (a_{11} + a_{22} + a_{33})(-\lambda)^2 + (-\lambda)^3. \tag{8.4}
$$

Now let $\lambda_1, \lambda_2, \lambda_3$ be the eigenvalues, so $P(\lambda_1) = P(\lambda_2) = P(\lambda_3) = 0$. Since $P(\lambda)$ is a polynomial of degree 3, it follows that

$$
P(\lambda) = (\lambda_1 - \lambda)(\lambda_2 - \lambda)(\lambda_3 - \lambda) = 0.
$$

Expanding the characteristic polynomial,

$$
P(\lambda) = \lambda_1\lambda_2\lambda_3 + (\lambda_1\lambda_2 + \lambda_2\lambda_3 + \lambda_3\lambda_1)(-\lambda) + (\lambda_1 + \lambda_2 + \lambda_3)(-\lambda)^2 + (-\lambda)^3.
$$

Comparison with (8.4) shows

$$
\lambda_1 + \lambda_2 + \lambda_3 = a_{11} + a_{22} + a_{33} = Tr\ A.
$$

This means that the **sum of the eigenvalues is equal to the trace of** A. This is a very useful relation to check if the eigenvalues are calculated correctly. Furthermore,

$$\lambda_1\lambda_2 + \lambda_2\lambda_3 + \lambda_3\lambda_1 = \begin{vmatrix} a_{11} & a_{12} \\ a_{21} & a_{22} \end{vmatrix} + \begin{vmatrix} a_{11} & a_{13} \\ a_{31} & a_{33} \end{vmatrix} + \begin{vmatrix} a_{22} & a_{23} \\ a_{32} & a_{33} \end{vmatrix}.$$

which is the sum of principal minors (minors of the diagonal elements), and

$$\lambda_1\lambda_2\lambda_3 = |A|.$$

That the product of all eigenvalues is equal to the determinant of A is also a very useful relation. If A is singular $|A| = 0$, at least one of the eigenvalue must be zero. It follows that the inverse of A exists if and only if none of the eigenvalues of A is zero.

Similar calculations can generalize these relationships for matrices of higher orders.

Example 8.1.5 Find the eigenvalues and the corresponding eigenvectors of the matrix A if

$$A = \begin{pmatrix} 5 & 7 & -5 \\ 0 & 4 & -1 \\ 2 & 8 & -3 \end{pmatrix}.$$

Solution 8.1.5

$$P(\lambda) = \begin{pmatrix} 5-\lambda & 7 & -5 \\ 0 & 4-\lambda & -1 \\ 2 & 8 & -3-\lambda \end{pmatrix}$$

$$= \begin{vmatrix} 5 & 7 & -5 \\ 0 & 4 & -1 \\ 2 & 8 & -3 \end{vmatrix} - \left(\begin{vmatrix} 4 & -1 \\ 8 & -3 \end{vmatrix} + \begin{vmatrix} 5 & -5 \\ 2 & -3 \end{vmatrix} + \begin{vmatrix} 5 & 7 \\ 0 & 4 \end{vmatrix} \right) \lambda + (5+4-3)\lambda^2 - \lambda^3$$

$$= 6 - 11\lambda + 6\lambda^2 - \lambda^3 = (1-\lambda)(2-\lambda)(3-\lambda) = 0.$$

Thus, the three eigenvalues are

$$\lambda_1 = 1, \quad \lambda_2 = 2, \quad \lambda_3 = 3.$$

As a check, the sum of the eigenvalues

$$\lambda_1 + \lambda_2 + \lambda_3 = 1 + 2 + 3 = 6$$

is indeed equal to the trace of A

$$Tr\ A = 5 + 4 - 3 = 6.$$

Furthermore, the product of three eigenvalues

$$\lambda_1 \, \lambda_2 \, \lambda_3 = 6$$

is indeed equal to the determinant

$$\begin{vmatrix} 5 & 7 & -5 \\ 0 & 4 & -1 \\ 2 & 8 & -3 \end{vmatrix} = 6.$$

Let the eigenvector $\mathbf{x}_1$ corresponding to λ_1 be $\begin{pmatrix} x_{11} \\ x_{12} \\ x_{13} \end{pmatrix}$, then

$$\begin{pmatrix} 5-\lambda_1 & 7 & -5 \\ 0 & 4-\lambda_1 & -1 \\ 2 & 8 & -3-\lambda_1 \end{pmatrix} \begin{pmatrix} x_{11} \\ x_{12} \\ x_{13} \end{pmatrix} = \begin{pmatrix} 4 & 7 & -5 \\ 0 & 3 & -1 \\ 2 & 8 & -4 \end{pmatrix} \begin{pmatrix} x_{11} \\ x_{12} \\ x_{13} \end{pmatrix} = 0.$$

By Gauss' elimination method, one can readily show that

$$\begin{pmatrix} 4 & 7 & -5 \\ 0 & 3 & -1 \\ 2 & 8 & -4 \end{pmatrix} \Longrightarrow \begin{pmatrix} 4 & 7 & -5 \\ 0 & 3 & -1 \\ 0 & 4.5 & -1.5 \end{pmatrix} \Longrightarrow \begin{pmatrix} 4 & 7 & -5 \\ 0 & 3 & -1 \\ 0 & 0 & 0 \end{pmatrix}.$$

Thus, the set of equations is reduced to

$$4x_{11} + 7x_{12} - 5x_{13} = 0$$
$$3x_{12} - x_{13} = 0.$$

Only one of the three unknowns can be assigned arbitrary. For example, let $x_{13} = 3$, then $x_{12} = 1$ and $x_{11} = 2$. Therefore, corresponding to the eigenvalue $\lambda_1 = 1$, the eigenvector can be written as

$$\mathbf{x}_1 = \begin{pmatrix} 2 \\ 1 \\ 3 \end{pmatrix}.$$

Similarly, corresponding to $\lambda_2 = 2$ and $\lambda_3 = 3$, the respective eigenvectors are

$$\mathbf{x}_2 = \begin{pmatrix} 1 \\ 1 \\ 2 \end{pmatrix} \quad \text{and} \quad \mathbf{x}_3 = \begin{pmatrix} -1 \\ 1 \\ 1 \end{pmatrix}.$$

8.1.3 Properties of Eigenvalues

There are other properties related to eigenvalue problems. Taken individually, they are almost self-evident, but collectively they are useful in matrix applications.

- The transpose $\widetilde{A}\left(A^{T}\right)$ has the same eigenvalues as A.

  The eigenvalues of A and A^{T} are, respectively, the solutions of $|A - \lambda I| = 0$ and $\left|A^{T} - \lambda I\right| = 0$. Since $A^{T} - \lambda I = (A - \lambda I)^{T}$ and the determinant of a matrix is equal to the determinant of its transpose

$$|A - \lambda I| = \left|(A - \lambda I)^{T}\right| = \left|A^{T} - \lambda I\right|,$$

the secular equations of A and A^{T} are identical. Therefore, they have the same set of eigenvalues.

- If A is either upper or lower triangular, then the eigenvalues are the diagonal elements.

  Let $|A - \lambda I| = 0$ be

$$\begin{vmatrix} a_{11} - \lambda & a_{12} & \cdots & a_{1n} \\ 0 & a_{22} - \lambda & \cdots & a_{2n} \\ 0 & 0 & \cdots & \cdots \\ 0 & 0 & 0 & a_{nn} - \lambda \end{vmatrix} = (a_{11} - \lambda)(a_{22} - \lambda) \cdots (a_{nn} - \lambda) = 0.$$

Clearly $\lambda = a_{11}, \ \lambda = a_{22}, \ \cdots \lambda = a_{nn}$.

- If $\lambda_1, \ \lambda_2, \ \lambda_3 \cdots \cdot \lambda_n$ are the eigenvalues of A, then the eigenvalues of the inverse A^{-1} are $1/\lambda_1, \ 1/\lambda_2, \ 1/\lambda_3 \cdots \cdot 1/\lambda_n$.

  Multiplying the equation $A\mathbf{x} = \lambda \mathbf{x}$ from the left by A^{-1}

$$A^{-1}A\mathbf{x} = A^{-1}\lambda \mathbf{x} = \lambda A^{-1}\mathbf{x}$$

and using $A^{-1}A\mathbf{x} = I\mathbf{x} = \mathbf{x}$, we have $\mathbf{x} = \lambda A^{-1}\mathbf{x}$. Thus,

$$A^{-1}\mathbf{x} = \frac{1}{\lambda}\mathbf{x}.$$

- If $\lambda_1, \ \lambda_2, \ \lambda_3 \cdots \cdot \lambda_n$ are the eigenvalues of A, then the eigenvalues of A^m are $\lambda_1^m, \ \lambda_2^m, \ \lambda_3^m \cdots \cdot \lambda_n^m$.

  Since $A\mathbf{x} = \lambda \mathbf{x}$, it follows:

$$A^{2}\mathbf{x} = A(A\mathbf{x}) = A\lambda \mathbf{x} = \lambda A\mathbf{x} = \lambda^{2}\mathbf{x}.$$

Similarly,

$$A^3\mathbf{x} = \lambda^3\mathbf{x}, \cdots\cdots\cdots A^m\mathbf{x} = \lambda^m\mathbf{x}.$$

8.2 Some Terminology

As we have seen that for a $n \times n$ square matrix, the eigenvalues may or may not be real numbers. If the eigenvalues are degenerate, we may or may not have n distinct eigenvectors.

However, there is a class of matrices, known as hermitian matrices, the eigenvalues of which are always real. A $n \times n$ hermitian matrix will always have n distinct eigenvectors.

To facilitate the discussion of these and other properties of matrices, we will first introduce the following terminology.

8.2.1 Hermitian Conjugation

Complex Conjugation. If $A = (a_{ij})_{m \times n}$ is an arbitrary matrix whose elements may be complex numbers, the complex conjugate matrix denoted by A^* is also a matrix of order $m \times n$ with every element of which is the complex conjugate of the corresponding element of A, that is

$$(A^*)_{ij} = a_{ij}^*.$$

It is clear that

$$(cA)^* = c^* A^*.$$

Hermitian Conjugation. When the two operations of complex conjugation and transposition are carried out one after another on a matrix, the resulting matrix is called the **hermitian conjugate** of the original matrix and is denoted by $A^\dagger$, called A dagger. Mathematicians also refer to $A^\dagger$ as the adjoint matrix. The order of the two operations is immaterial. Thus,

$$A^\dagger = (A^*)^T = (\widetilde{A})^*. \tag{8.5}$$

For example, if

$$A = \begin{pmatrix} (6+i) & (1-6i) & 1 \\ (3+i) & 4 & 3i \end{pmatrix}, \tag{8.6}$$

then

$$A^\dagger = (A^*)^T = \begin{pmatrix} (6-i) & (1+6i) & 1 \\ (3-i) & 4 & -3i \end{pmatrix}^T = \begin{pmatrix} (6-i) & (3-i) \\ (1+6i) & 4 \\ 1 & -3i \end{pmatrix}, \qquad (8.7)$$

$$A^\dagger = (\widetilde{A})^* = \begin{pmatrix} (6+i) & (3+i) \\ (1-6i) & 4 \\ 1 & 3i \end{pmatrix}^* = \begin{pmatrix} (6-i) & (3-i) \\ (1+6i) & 4 \\ 1 & -3i \end{pmatrix}. \qquad (8.8)$$

Hermitian Conjugate of Matrix Products. We have shown in previous chapter that the transpose of the product of two matrices is equal to the product of the transposed matrices taken in reverse order, This leads directly to the fact that

$$(AB)^\dagger = B^\dagger A^\dagger,$$

since

$$(AB)^\dagger = (A^* B^*)^T = \widetilde{B}^* \widetilde{A}^* = B^\dagger A^\dagger. \qquad (8.9)$$

8.2.2   Orthogonality

Inner Product. If **a** and **b** are two column vectors of the same order n, the inner product (or **scalar product**) is defined as $\mathbf{a}^\dagger \mathbf{b}$. The hermitian conjugate of the column vector is a row vector

$$\mathbf{a}^\dagger = \begin{pmatrix} a_1 \\ a_2 \\ \cdot \\ \cdot \\ a_n \end{pmatrix}^\dagger = \begin{pmatrix} a_1^* & a_2^* & \cdots & a_n^* \end{pmatrix},$$

therefore the inner product is one number

$$\mathbf{a}^\dagger \mathbf{b} = \begin{pmatrix} a_1^* & a_2^* & \cdots & a_n^* \end{pmatrix} \begin{pmatrix} b_1 \\ b_2 \\ \cdot \\ \cdot \\ b_n \end{pmatrix} = \sum_{k=1}^{n} a_k^* b_k.$$

There are two other commonly used notations for the inner product.

The notation most often used in quantum mechanics is the bra-ket notation of Dirac. The row and column vectors are, respectively, defined as the bra and ket vectors. Thus, we may write the column vector

$$\mathbf{b} = |\mathbf{b}\rangle$$

as the ket vector, and the row vector

$$\mathbf{a}^\dagger = \langle \mathbf{a} |$$

as the bra vector. The inner product of two vectors is then represented by

$$\langle \mathbf{a} | \mathbf{b} \rangle = \mathbf{a}^\dagger \mathbf{b}.$$

Notice that for any scalar c,

$$\langle \mathbf{a} | c\mathbf{b} \rangle = c \langle \mathbf{a} | \mathbf{b} \rangle,$$

whereas

$$\langle c\mathbf{a} | \mathbf{b} \rangle = c^* \langle \mathbf{a} | \mathbf{b} \rangle.$$

Another notation that is often used is the parenthesis notation:

$$(\mathbf{a}, \mathbf{b}) = \mathbf{a}^\dagger \mathbf{b} = \langle \mathbf{a} | \mathbf{b} \rangle.$$

If A is a matrix, then

$$(\mathbf{a}, A\mathbf{b}) = (A^\dagger \mathbf{a}, \mathbf{b})$$

is an identity, since

$$(A^\dagger \mathbf{a}, \mathbf{b}) = (A^\dagger \mathbf{a})^\dagger \mathbf{b} = \mathbf{a}^\dagger (A^\dagger)^\dagger \mathbf{b} = \mathbf{a}^\dagger A\mathbf{b} = (\mathbf{a}, A\mathbf{b}).$$

Thus, if

$$(\mathbf{a}, A\mathbf{b}) = (A\mathbf{a}, \mathbf{b}),$$

then A is hermitian. Mathematicians refer to the relation $A^\dagger = A$ as **self-adjoint**.

Orthogonality. Two vectors a and b are said to be orthogonal if and only if

$$\mathbf{a}^\dagger \mathbf{b} = 0.$$

Note that in three-dimensional real space

$$\mathbf{a}^\dagger \mathbf{b} = \sum_{k=1}^{n} a_k^* b_k^* = a_1 b_1 + a_2 b_2 + a_3 b_3$$

is just the dot product of **a** and **b**. It is well known in vector analysis that if the dot product of two vectors is equal to zero, then they are perpendicular.

Length of a Complex Vector. If we adopt this definition of the scalar product of two complex vectors, then we have a natural definition of the length of a complex vector in a n dimensional space. The length $\|\mathbf{x}\|$ of a complex vector $\mathbf{x}$ is taken to be

$$\|\mathbf{x}\|^2 = \mathbf{x}^\dagger \mathbf{x} = \sum_{k=1}^{n} a_k^* a_k = \sum_{k=1}^{n} |a_k|^2.$$

8.2.3 Gram-Schmidt Process

Linear Independence. The set of vectors $\mathbf{x}_1, \mathbf{x}_2, \cdots, \mathbf{x}_n$ is linearly independent if and only if

$$\sum_{i=1}^{n} a_i \mathbf{x}_i = 0$$

implies every $a_i = 0$. Otherwise the set is linearly dependent.

Let us test the three vectors

$$\mathbf{x}_1 = \begin{pmatrix} 1 \\ 0 \\ 1 \end{pmatrix}, \quad \mathbf{x}_2 = \begin{pmatrix} 0 \\ 1 \\ 0 \end{pmatrix}, \quad \mathbf{x}_3 = \begin{pmatrix} 1 \\ 0 \\ 0 \end{pmatrix},$$

for linear independence. The question is if we can find a set of a_i, not all zero such that

$$\sum_{i=1}^{3} a_i \mathbf{x}_i = a_1 \begin{pmatrix} 1 \\ 0 \\ 1 \end{pmatrix} + a_2 \begin{pmatrix} 0 \\ 1 \\ 0 \end{pmatrix} + a_3 \begin{pmatrix} 1 \\ 0 \\ 0 \end{pmatrix} = \begin{pmatrix} a_1 + a_3 \\ a_2 \\ a_1 \end{pmatrix} = \begin{pmatrix} 0 \\ 0 \\ 0 \end{pmatrix}.$$

Clearly, this requires $a_1 = 0$, $a_2 = 0$ and $a_3 = 0$. Therefore, these three vectors are linearly in depend.

Note that linear independence or dependence is a property of the set as a whole, not of the individual vectors.

It is obvious that if $\mathbf{x}_1, \mathbf{x}_2, \mathbf{x}_3$ represent three non-coplanar three-dimensional vectors, they are linearly independent.

Gram-Schmidt Process. Given any n linearly independent vectors, one can construct from their linear combinations a set of n mutually orthogonal unit vectors.

Let the given linearly independent vectors be $\mathbf{x}_1, \mathbf{x}_2, \cdots, \mathbf{x}_n$. Define

$$\mathbf{u}_1 = \frac{\mathbf{x}_1}{\|\mathbf{x}_1\|}$$

to be the first unit vector. Now define

$$\mathbf{u}_2' = \mathbf{x}_2 - (\mathbf{x}_2, \mathbf{u}_1)\,\mathbf{u}_1.$$

The inner product of $\mathbf{u}_2'$ and $\mathbf{u}_1$ is equal to zero,

$$\left(\mathbf{u}_2', \mathbf{u}_1\right) = (\mathbf{x}_2, \mathbf{u}_1) - (\mathbf{x}_2, \mathbf{u}_1)\,(\mathbf{u}_1, \mathbf{u}_1) = 0,$$

since $(\mathbf{u}_1, \mathbf{u}_1) = 1$. This shows $\mathbf{u}_2'$ is orthogonal to $\mathbf{u}_1$.

We can normalize $\mathbf{u}_2'$

$$\mathbf{u}_2 = \frac{\mathbf{u}_2'}{\left\|\mathbf{u}_2'\right\|}$$

to obtain the second unit vector $\mathbf{u}_2$ which is orthogonal to $\mathbf{u}_1$.

We can continue this process by defining

$$\mathbf{u}_k' = \mathbf{x}_k - \sum_{i=1}^{k-1} (\mathbf{x}_k, \mathbf{u}_i)\,\mathbf{u}_i,$$

and

$$\mathbf{u}_k = \frac{\mathbf{u}_k'}{\left\|\mathbf{u}_k'\right\|}.$$

When all $\mathbf{x}_k$ are used up, we will have n unit vectors $u_1, u_2, \cdots, u_k$ orthogonal to each other. They are called an **orthonormal set**. This procedure is known as Gram-Schmidt process.

8.3 Unitary Matrix and Orthogonal Matrix

8.3.1 Unitary Matrix

If the square matrix U satisfies the condition

$$U^\dagger U = I,$$

then U is called unitary. The n columns in a unitary matrix can be considered as n column vectors in an orthonormal set.

In other words, if

$$\mathbf{u}_1 = \begin{pmatrix} u_{11} \\ u_{12} \\ \cdot \\ \cdot \\ u_{1n} \end{pmatrix}, \quad \mathbf{u}_2 = \begin{pmatrix} u_{21} \\ u_{22} \\ \cdot \\ \cdot \\ u_{2n} \end{pmatrix}, \quad \cdots\cdots \quad \mathbf{u}_n = \begin{pmatrix} u_{n1} \\ u_{n2} \\ \cdot \\ \cdot \\ u_{nn} \end{pmatrix},$$

and

$$\mathbf{u}_i^\dagger \mathbf{u}_j = \begin{pmatrix} u_{i1}^* & u_{i2}^* & \cdots & u_{in}^* \end{pmatrix} \begin{pmatrix} u_{j1} \\ u_{j2} \\ \cdot \\ \cdot \\ \cdot \\ u_{jn} \end{pmatrix} = \begin{cases} 1 & if \ \ i = j \\ 0 & if \ \ i \neq j \end{cases},$$

then

$$U = \begin{pmatrix} u_{11} & u_{21} & \cdot\cdot & u_{n1} \\ u_{12} & u_{22} & \cdot\cdot & u_{n2} \\ \cdot & & \cdot\cdot\cdot & \cdot \\ \cdot & & \cdot\cdot\cdot & \cdot \\ u_{1n} & u_{2n} & \cdot\cdot & u_{nn} \end{pmatrix}$$

is unitary. This is because

$$U^\dagger = \begin{pmatrix} u_{11}^* & u_{12}^* & \cdot\cdot & u_{1n}^* \\ u_{21}^* & u_{22}^* & \cdot\cdot & u_{2n}^* \\ \cdot & & \cdot\cdot\cdot & \cdot \\ \cdot & & \cdot\cdot\cdot & \cdot \\ u_{n1}^* & u_{n2}^* & \cdot\cdot & u_{nn}^* \end{pmatrix}$$

therefore

$$U^\dagger U = \begin{pmatrix} u_{11}^* & u_{12}^* & \cdot\cdot & u_{1n}^* \\ u_{21}^* & u_{22}^* & \cdot\cdot & u_{2n}^* \\ \cdot & & \cdot\cdot\cdot & \cdot \\ \cdot & & \cdot\cdot\cdot & \cdot \\ u_{n1}^* & u_{n2}^* & \cdot\cdot & u_{nn}^* \end{pmatrix} \begin{pmatrix} u_{11} & u_{21} & \cdot\cdot & u_{n1} \\ u_{12} & u_{22} & \cdot\cdot & u_{n2} \\ \cdot & & \cdot\cdot\cdot & \cdot \\ \cdot & & \cdot\cdot\cdot & \cdot \\ u_{1n} & u_{2n} & \cdot\cdot & u_{nn} \end{pmatrix} = \begin{pmatrix} 1 & 0 & \cdot\cdot & 0 \\ 0 & 1 & \cdot\cdot & 0 \\ \cdot & & \cdot\cdot\cdot & \cdot \\ \cdot & & \cdot\cdot\cdot & \cdot \\ 0 & 0 & \cdot\cdot & 1 \end{pmatrix}.$$

Multiply U^{-1} from the right, we have

$$U^\dagger U U^{-1} = I U^{-1}.$$

It follows that **hermitian conjugate of a unitary matrix is its inverse**, that is

$$U^\dagger = U^{-1}.$$

8.3.2 *Properties of Unitary Matrix*

• **Unitary Transformations leave lengths of vectors invariant.**

Let

$$\mathbf{a} = U\mathbf{b}, \quad so \ \ \mathbf{a}^\dagger = \mathbf{b}^\dagger U^\dagger,$$

and
$$\|\mathbf{a}\|^2 = \mathbf{a}^\dagger \mathbf{a} = \mathbf{b}^\dagger U^\dagger U \mathbf{b}.$$

Since
$$U^\dagger U = U^{-1} U = I,$$

it follows
$$\|\mathbf{a}\|^2 = \mathbf{a}^\dagger \mathbf{a} = \mathbf{b}^\dagger \mathbf{b} = \|\mathbf{b}\|^2.$$

Thus, the length of the initial vector is equal to the length of the transformed vector.

- **The absolute value of the eigenvalues of a unitary matrix is equal to one.**

Let $\mathbf{x}$ be a nontrivial eigenvector of the unitary matrix U belonging to the eigenvalue λ,
$$U\mathbf{x} = \lambda \mathbf{x}.$$

Take the hermitian conjugate of both sides
$$\mathbf{x}^\dagger U^\dagger = \lambda^* \mathbf{x}^\dagger.$$

Multiply the last two equations:
$$\mathbf{x}^\dagger U^\dagger U \mathbf{x} = \lambda^* \mathbf{x}^\dagger \lambda \mathbf{x}.$$

Since $U^\dagger U = I$ and $\lambda^* \lambda = |\lambda|^2$, it follows:
$$\mathbf{x}^\dagger \mathbf{x} = |\lambda|^2 \mathbf{x}^\dagger \mathbf{x}.$$

Therefore,
$$|\lambda|^2 = 1.$$

In other words, the **eigenvalues of a unitary matrix must be on the unit circle in the complex plane** centered at the origin.

8.3.3 Orthogonal Matrix

If the elements of a unitary matrix are all real, the matrix is known as an **orthogonal matrix**. Thus, the properties of unitary matrices are also properties of orthogonal matrices. In addition,

- **The determinant of an orthogonal matrix is equal to either positive one or negative one.**

If A is a real square matrix, then by definition

$$A^\dagger = \widetilde{A}^* = \widetilde{A}.$$

If, in addition, A is unitary, $A^\dagger = A^{-1}$, then

$$\widetilde{A} = A^{-1}.$$

Thus,

$$A\widetilde{A} = I. \tag{8.10}$$

Since the determinant of A is equal to the determinant of $\widetilde{A}$, so

$$\left|A\widetilde{A}\right| = |A|\left|\widetilde{A}\right| = |A|^2.$$

But

$$\left|A\widetilde{A}\right| = |I| = 1,$$

therefore

$$|A|^2 = 1.$$

Thus, the determinant of an orthogonal matrix is either $+1$ or -1.

Very often (8.10) is used as the definition of an orthogonal matrix. That is, a square real matrix A satisfying the relation expressed in (8.10) is called an orthogonal matrix. This is equivalent to the statement that **the inverse of an orthogonal matrix is equal to its transpose.**

Written in terms of its elements, (8.10) is given by

$$\sum_{j=1} a_{ij}\widetilde{a}_{jk} = \sum_{j=1} a_{ij}a_{kj} = \delta_{ik} \tag{8.11}$$

for any i and any j. Similarly, $\widetilde{A}A = I$ can be expressed as

$$\sum_{j=1} \widetilde{a}_{ij}a_{jk} = \sum_{j=1} a_{ji}a_{jk} = \delta_{ik}. \tag{8.12}$$

However, (8.12) is not independent of Eq. (8.11), since $A\widetilde{A} = \widetilde{A}A$. If one set of conditions is valid, the other set must also be valid.

Put in words, these conditions mean that the sum of the products of the corresponding elements of two distinct columns (or rows) of an orthogonal matrix is zero, while the sum of the squares of the elements of any column (or row) is equal to unity. If we regard the n columns of the matrix as n real vectors, this means that these n column vectors are orthogonal and normalized. Similarly, all the rows of an orthogonal matrix are orthonormal.

8.3.4 Independent Elements of an Orthogonal Matrix

An nth-order square matrix has n^2 elements. For an orthogonal matrix, not all these elements are independent of each other, because there are certain conditions they must satisfy. First, there are n conditions for each column to be normalized. Then there are $n(n-1)/2$ conditions for each column to be orthogonal to any other column. Therefore, the number of independent parameters of an orthogonal matrix is

$$n^2 - [n + n(n-1)/2] = n(n-1)/2.$$

In other words, an nth-order orthogonal matrix can be fully characterized by $n(n-1)/2$ independent parameters.

For $n = 2$, the number of independent parameters is one. This is illustrated as follows.

Consider an arbitrary orthogonal matrix of order 2

$$A = \begin{pmatrix} a & c \\ b & d \end{pmatrix}.$$

The fact that each column is normalized leads to

$$a^2 + b^2 = 1, \tag{8.13}$$
$$c^2 + d^2 = 1. \tag{8.14}$$

Furthermore, the two columns are orthogonal

$$\begin{pmatrix} a & b \end{pmatrix} \begin{pmatrix} c \\ d \end{pmatrix} = ac + bd = 0. \tag{8.15}$$

The general solution of (8.13) is $a = \cos\theta$, $b = \sin\theta$, where θ is a scalar. Similarly, (8.14) can be satisfied, if we choose $c = \cos\phi$, $d = \sin\phi$, where ϕ is another scalar. On the other hand, (8.15) requires

$$\cos\theta\cos\phi + \sin\theta\sin\phi = \cos(\theta - \phi) = 0,$$

therefore

$$\phi = \theta \pm \frac{\pi}{2}.$$

Thus, the most general orthogonal matrix of order 2 is

$$A_1 = \begin{pmatrix} \cos\theta & -\sin\theta \\ \sin\theta & \cos\theta \end{pmatrix} \quad or \quad A_2 = \begin{pmatrix} \cos\theta & \sin\theta \\ \sin\theta & -\cos\theta \end{pmatrix}. \tag{8.16}$$

Every orthogonal matrix of order 2 can be expressed in this form with some value of θ. Clearly, the determinant of A_1 is equal to $+1$ and that of A_2, -1.

8.3.5   *Orthogonal Transformation and Rotation Matrix*

The fact that in real space, orthogonal transformation preserves the length of a vector suggests that the orthogonal matrix is associated with rotation of vectors. In fact, the orthogonal matrix is related to two kinds of rotations in space. First, it can be thought as an operator which rotates a vector. This is often called active transformation. Secondly, it can be thought as the transformation matrix when the coordinate axes of the reference system are rotated. This is also referred to as passive transformation.

First let us consider the vectors shown in Fig. 8.1a. The x and y components of the vector $\mathbf{r}_1$ are given by $x_1 = r\cos\varphi$ and $y_1 = r\sin\varphi$, where r is the length of the vector. Now let us rotate the vector counterclockwise by an angle θ, so that $x_2 = r\cos(\varphi + \theta)$ and $y_2 = r\sin(\varphi + \theta)$. Using trigonometry, we can write

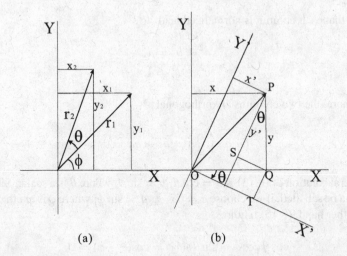

(a)   (b)

Fig. 8.1 Two interpretations of the orthogonal matrix A_1 whose determinant is equal to $+1$. **a** As a operator, it rotates the vector $\mathbf{r}_1$ to $\mathbf{r}_2$ without changing its length. **b** As the transformation matrix between the coordinates of the tip of a fixed vector when the coordinate axes are rotated. Note that the rotation in **b** is in the opposite direction as in **a**

$$x_2 = r\cos(\varphi + \theta) = r\cos\varphi\cos\theta - r\sin\varphi\sin\theta = x_1\cos\theta - y_1\sin\theta$$
$$y_2 = r\sin(\varphi + \theta) = r\sin\varphi\cos\theta + r\cos\varphi\sin\theta = y_1\cos\theta + x_1\sin\theta.$$

We can display the set of coefficients in the form of

$$\begin{pmatrix} x_2 \\ y_2 \end{pmatrix} = \begin{pmatrix} \cos\theta & -\sin\theta \\ \sin\theta & \cos\theta \end{pmatrix} \begin{pmatrix} x_1 \\ y_1 \end{pmatrix}.$$

It is seen that the coefficient matrix is the orthogonal matrix A_1 of (8.16). Therefore, the orthogonal matrix with a determinant equal to $+1$ is also called rotation matrix. It rotates the vector from $\mathbf{r}_1$ to $\mathbf{r}_2$ without changing the magnitude.

The second interpretation of rotation matrix is as follows. Let P be the tip of a fixed vector. The coordinates of P is (x, y) in a particular rectangular coordinate system. Now the coordinate axes are rotated clockwise by an angle θ as shown in Fig. 8.1b. The coordinates of P in the rotated system become (x', y'). From the geometry in Fig. 8.1b, it is clear that

$$x' = OT - SQ = OQ\cos\theta - PQ\sin\theta = x\cos\theta - y\sin\theta$$
$$y' = QT + PS = OQ\sin\theta + PQ\cos\theta = x\sin\theta + y\cos\theta,$$

or

$$\begin{pmatrix} x' \\ y' \end{pmatrix} = \begin{pmatrix} \cos\theta & -\sin\theta \\ \sin\theta & \cos\theta \end{pmatrix} \begin{pmatrix} x \\ y \end{pmatrix}.$$

Note that the matrix involved is again the orthogonal matrix A_1. However, this time A_1 is the transformation matrix between the coordinates of the tip of a fixed vector when the coordinate axes are rotated.

The equivalence between these two interpretations might be expected, since the relative orientation between the vector and coordinate axes is the same whether the vector is rotated counterclockwise by an angle θ, or the coordinate axes are rotated clockwise by the same angle.

Next, let us consider the orthogonal matrix A_2, the determinant of which is equal to -1. The matrix A_2 can be expressed as

$$A_2 = \begin{pmatrix} \cos\theta & \sin\theta \\ \sin\theta & -\cos\theta \end{pmatrix} = \begin{pmatrix} \cos\theta & -\sin\theta \\ \sin\theta & \cos\theta \end{pmatrix} \begin{pmatrix} 1 & 0 \\ 0 & -1 \end{pmatrix}.$$

The transformation

$$\begin{pmatrix} x_2 \\ y_2 \end{pmatrix} = \begin{pmatrix} 1 & 0 \\ 0 & -1 \end{pmatrix} \begin{pmatrix} x_1 \\ y_1 \end{pmatrix}$$

gives

$$x_2 = x_1, \qquad y_2 = -y_1.$$

Fig. 8.2 Two interpretations of the orthogonal matrix A_2 whose determinant is -1. **a** As an operator, it flips the vector $\mathbf{r}_1$ to $\mathbf{r}_2$ symmetrically with respect to x-axis, and then rotates $\mathbf{r}_2$ to $\mathbf{r}_3$. **b** As the transformation matrix between the tip of a fixed vector when the y-axis is inverted and then the coordinate axes are rotated. Note that the rotation in **b** seems to be in the same direction as in **a**

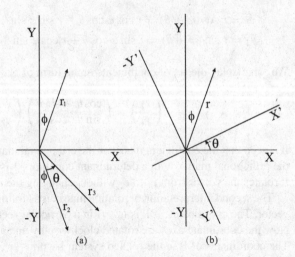

(a)   (b)

Clearly, this corresponds to the reflection of the vector with respect to the x-axis. Therefore, A_2 can be considered as an operator which first symmetrically flips the vector $\mathbf{r}_1$ over the x-axis and then rotates it to $\mathbf{r}_3$ as shown in Fig. 8.2a.

In terms of coordinate transformation, one can show that (x', y') in the equation

$$\begin{pmatrix} x' \\ y' \end{pmatrix} = \begin{pmatrix} \cos\theta & -\sin\theta \\ \sin\theta & \cos\theta \end{pmatrix} \begin{pmatrix} x \\ y \end{pmatrix}$$

represent the new coordinates of the tip of a fixed vector after the y-axis is inverted and the whole coordinate axes are rotated by an angle θ, as shown in Fig. 8.2b. In this case, one has to be careful about the sign of the angle. The sign convention is that a counterclockwise rotation is positive and a clockwise rotation is negative. However, after the y-axis is inverted as in Fig. 8.2b, a negative rotation (rotating from the direction of the positive x-axis towards the negative of the y-axis) appears to be counterclockwise. This is why in Fig. 8.1a, b, the vector and the coordinate axes are rotating in the opposite direction, whereas in Fig. 8.2a, b, they seem to rotate in the same direction.

So far, we have used rotations in two dimensions as examples. However, the conclusions that orthogonal matrix whose determinant equals to $+1$ represents pure rotation and orthogonal matrix whose determinant is equal to -1 represents a reflection followed by rotations that are generally valid in higher-dimensional space. In the chapter on vector transformation, we will have a more detailed discussion.

8.4 Diagonalization

8.4.1 Similarity Transformation

If A is a $n \times n$ matrix and $\mathbf{u}$ is a $n \times 1$ column vector, then $A\mathbf{u}$ is another $n \times 1$ column vector. The equation

$$A\mathbf{u} = \mathbf{v} \qquad (8.17)$$

represents a linear transformation. Matrix A acts as a linear operator sending vector $\mathbf{u}$ to vector $\mathbf{v}$. Let

$$\mathbf{u} = \begin{pmatrix} u_1 \\ u_2 \\ \cdot \\ \cdot \\ u_n \end{pmatrix}, \quad \mathbf{v} = \begin{pmatrix} v_1 \\ v_2 \\ \cdot \\ \cdot \\ v_n \end{pmatrix}$$

where u_i and v_i are, respectively, the ith components of $\mathbf{u}$ and $\mathbf{v}$ in the n-dimensional space. These components are, of course, measured in a certain coordinate system (reference frame). Let the unit vectors, $\mathbf{e}_i$, known as bases, along the coordinate axes of this system be

$$\mathbf{e}_1 = \begin{pmatrix} 1 \\ 0 \\ \cdot \\ \cdot \\ 0 \end{pmatrix}, \quad \mathbf{e}_2 = \begin{pmatrix} 0 \\ 1 \\ \cdot \\ \cdot \\ 0 \end{pmatrix}, \cdots \cdot \mathbf{e}_n = \begin{pmatrix} 0 \\ 0 \\ \cdot \\ \cdot \\ 1 \end{pmatrix},$$

then

$$\mathbf{u} = \begin{pmatrix} u_1 \\ u_2 \\ \cdot \\ \cdot \\ u_n \end{pmatrix} = u_1 \begin{pmatrix} 1 \\ 0 \\ \cdot \\ \cdot \\ 0 \end{pmatrix} + u_2 \begin{pmatrix} 0 \\ 1 \\ \cdot \\ \cdot \\ 0 \end{pmatrix} + \cdots + u_n \begin{pmatrix} 0 \\ 0 \\ \cdot \\ \cdot \\ 1 \end{pmatrix} = \sum_{i=1}^{n} u_i \mathbf{e}_i. \qquad (8.18)$$

Suppose there is another coordinate system, which we designate as the prime system. Measured in that system, the components of $\mathbf{u}$ and $\mathbf{v}$ become

$$\begin{pmatrix} u_1' \\ u_2' \\ \cdot \\ \cdot \\ u_n' \end{pmatrix} = \mathbf{u}', \quad \begin{pmatrix} v_1' \\ v_2' \\ \cdot \\ \cdot \\ v_n' \end{pmatrix} = \mathbf{v}'. \qquad (8.19)$$

We emphasize that $\mathbf{u}$ and $\mathbf{u}'$ are the same vector except measured in two different coordinate systems. The symbol $\mathbf{u}'$ does not mean a vector different from $\mathbf{u}$, it simply represents the collection of components of $\mathbf{u}$ in the prime system as shown in (8.19). Similarly, $\mathbf{v}$ and $\mathbf{v}'$ are the same vector. We can find these components if we know the components of $\mathbf{e}_i$ in the prime system.

In (8.18)

$$\mathbf{u} = u_1\mathbf{e}_1 + u_1\mathbf{e}_1 + \cdots + u_n\mathbf{e}_n,$$

the u_i' are just numbers which are independent of the coordinate system. To find the components of $\mathbf{u}$ in the prime system, we only need to express $\mathbf{e}_i$ in the prime system.

Let $\mathbf{e}_i$ measured in the prime system be

$$\mathbf{e}_1 = \begin{pmatrix} s_{11} \\ s_{21} \\ \cdot \\ \cdot \\ s_{n1} \end{pmatrix}, \quad \mathbf{e}_2 = \begin{pmatrix} s_{12} \\ s_{22} \\ \cdot \\ \cdot \\ s_{n2} \end{pmatrix}, \quad \cdots \quad \mathbf{e}_n = \begin{pmatrix} s_{1n} \\ s_{2n} \\ \cdot \\ \cdot \\ s_{nn} \end{pmatrix},$$

then the components of $\mathbf{u}$ measured in the prime system can be written as

$$\begin{pmatrix} u_1' \\ u_2' \\ \cdot \\ \cdot \\ u_n' \end{pmatrix} = u_1 \begin{pmatrix} s_{11} \\ s_{21} \\ \cdot \\ \cdot \\ s_{n1} \end{pmatrix} + u_2 \begin{pmatrix} s_{12} \\ s_{22} \\ \cdot \\ \cdot \\ s_{n2} \end{pmatrix} + \cdots + u_n \begin{pmatrix} s_{1n} \\ s_{2n} \\ \cdot \\ \cdot \\ s_{nn} \end{pmatrix}$$

$$= \begin{pmatrix} u_1 s_{11} + u_2 s_{12} + \cdots + u_n s_{1n} \\ u_1 s_{21} + u_2 s_{22} + \cdots + u_n s_{2n} \\ \cdot \\ \cdot \\ u_1 s_{n1} + u_2 s_{n2} + \cdots + u_n s_{nn} \end{pmatrix} = \begin{pmatrix} s_{11} & s_{12} & \cdot\cdot & s_{1n} \\ s_{21} & s_{22} & & s_{2n} \\ \cdot & & & \\ \cdot & & \cdot & \\ s_{n1} & s_{n2} & & s_{nn} \end{pmatrix} \begin{pmatrix} u_1 \\ u_2 \\ \cdot \\ \cdot \\ u_n \end{pmatrix}.$$

This equation can be written in the form

$$\mathbf{u}' = T\mathbf{u}, \tag{8.20}$$

where

$$T = \begin{pmatrix} s_{11} & s_{12} & \cdot\cdot & s_{1n} \\ s_{21} & s_{22} & & s_{2n} \\ \cdot & & & \\ \cdot & & \cdot & \\ s_{n1} & s_{n2} & & s_{nn} \end{pmatrix}.$$

It is clear from this analysis that the transformation matrix between the vector components in two coordinate systems is the same for all vectors, since it depends only on the transformation of the basis vectors in the two reference frames. Therefore, $\mathbf{v}'$ and $\mathbf{v}$ are also related through the same transformation matrix T,

$$\mathbf{v}' = T\mathbf{v}. \tag{8.21}$$

The operation of sending $\mathbf{u}$ to $\mathbf{v}$, expressed in the original system is given by $A\mathbf{u} = \mathbf{v}$. Let the same operation expressed in the prime system be

$$A'\mathbf{u}' = \mathbf{v}'.$$

Since $\mathbf{u}' = T\mathbf{u}$ and $\mathbf{v}' = T\mathbf{v}$,

$$A'T\mathbf{u} = T\mathbf{v}.$$

Multiply both sides by the inverse of T from the left,

$$T^{-1}A'T\mathbf{u} = T^{-1}T\mathbf{v} = \mathbf{v}.$$

Since $A\mathbf{u} = \mathbf{v}$, if follows that

$$A = T^{-1}A'T. \tag{8.22}$$

If we multiply this equation by T from the left and by T^{-1} from the right, we have

$$TAT^{-1} = A'.$$

What we have found is that as long as we know the relationship between the coordinate axes of two reference frames, not only can we transform a vector from one reference frame to the other, we can also transform a matrix representing a linear operator from one reference frame to the other.

In general, if there exits a non-singular matrix T such that $T^{-1}AT = B$ for any two square matrices A and B of the same order, then A and B are called **similar matrices**, and the transformation from A to B is called **similarity transformation**.

If two matrices are related by a similarity transformation, then they represent the same linear transformation in two different coordinate systems.

If the rectangular coordinate axes in the prime system are generated through a rotation from the original system, then T is an orthogonal matrix as discussed in the last section. In that case $T^{-1} = \widetilde{T}$, and the similarity transformation can be written as $\widetilde{T}AT$. If we are working in a complex space, the transformation matrix is unitary, and the similarity transformation can be written as $T^{\dagger}AT$. Both of these transformations are known as unitary similarity transformation.

A matrix that can be brought to diagonal form by a similarity transformation is said to be diagonalizable. Whether a matrix is diagonalizable and how to diagonalize it are very important questions in the theory of linear transformation. Not only because it is much more convenient to work with diagonal matrix but also because it is of fundamental importance in the structure of quantum mechanics. In the following sections, we will answer these questions.

8.4.2 *Diagonalizing a Square Matrix*

The eigenvectors of the matrix A can be used to form another matrix S in such a way that $S^{-1}AS$ becomes a diagonal matrix. This process often greatly simplifies a physical problem by a better choice of variables.

If A is a square matrix of order n, the eigenvalues λ_i and eigenvectors $\mathbf{x}_i$ satisfy the equation

$$A\mathbf{x}_i = \lambda_i \mathbf{x}_i \tag{8.23}$$

for $i = 1, 2, \cdots n$. Each eigenvector is a column matrix with n elements

$$\mathbf{x}_1 = \begin{pmatrix} x_{11} \\ x_{12} \\ \cdot \\ x_{1n} \end{pmatrix}, \quad \mathbf{x}_2 = \begin{pmatrix} x_{21} \\ x_{22} \\ \cdot \\ x_{2n} \end{pmatrix}, \cdots \cdots \mathbf{x}_n = \begin{pmatrix} x_{n1} \\ x_{n2} \\ \cdot \\ x_{nn} \end{pmatrix}.$$

Each of the n equations of (8.23) is of the form

$$\begin{pmatrix} a_{11} & a_{12} & \cdots & a_{1n} \\ a_{21} & a_{22} & \cdots & a_{2n} \\ \cdot & \cdot & \cdots & \cdot \\ a_{n1} & a_{n2} & \cdots & a_{nn} \end{pmatrix} \begin{pmatrix} x_{i1} \\ x_{i2} \\ \cdot \\ x_{in} \end{pmatrix} = \begin{pmatrix} \lambda_i x_{i1} \\ \lambda_i x_{i2} \\ \cdot \\ \lambda_i x_{in} \end{pmatrix}. \tag{8.24}$$

Collectively, they can be written as

$$\begin{pmatrix} a_{11} & a_{12} & \cdots & a_{1n} \\ a_{21} & a_{22} & \cdots & a_{2n} \\ \cdot & \cdot & \cdots & \cdot \\ a_{n1} & a_{n2} & \cdots & a_{nn} \end{pmatrix} \begin{pmatrix} x_{11} & x_{21} & \cdots & x_{n1} \\ x_{12} & x_{22} & \cdots & x_{n2} \\ \cdot & \cdot & \cdots & \cdot \\ x_{1n} & x_{2n} & \cdots & x_{nn} \end{pmatrix} = \begin{pmatrix} \lambda_1 x_{11} & \lambda_2 x_{21} & \cdots & \lambda_n x_{n1} \\ \lambda_1 x_{12} & \lambda_2 x_{22} & \cdots & \lambda_n x_{n2} \\ \cdot & \cdot & \cdots & \cdot \\ \lambda_1 x_{1n} & \lambda_2 x_{2n} & \cdots & \lambda_n x_{nn} \end{pmatrix}$$

$$= \begin{pmatrix} x_{11} & x_{21} & \cdots & x_{n1} \\ x_{12} & x_{22} & \cdots & x_{n2} \\ \cdot & \cdot & \cdots & \cdot \\ x_{1n} & x_{2n} & \cdots & x_{nn} \end{pmatrix} \begin{pmatrix} \lambda_1 & 0 & \cdots & 0 \\ 0 & \lambda_2 & \cdots & 0 \\ \cdot & \cdot & \cdots & \cdot \\ 0 & 0 & \cdots & \lambda_n \end{pmatrix}. \tag{8.25}$$

To simplify the writing, let

$$S = \begin{pmatrix} x_{11} & x_{21} & \cdots & x_{n1} \\ x_{12} & x_{22} & \cdots & x_{n2} \\ \cdot & \cdot & \cdots & \cdot \\ x_{1n} & x_{2n} & \cdots & x_{nn} \end{pmatrix}, \tag{8.26}$$

$$
\Lambda = \begin{pmatrix} \lambda_1 & 0 & \cdots & 0 \\ 0 & \lambda_2 & \cdots & 0 \\ \cdot & \cdot & \cdots & \cdot \\ 0 & 0 & \cdots & \lambda_n \end{pmatrix}, \tag{8.27}
$$

and write (8.25) as

$$
AS = S\Lambda. \tag{8.28}
$$

Multiplying both sides of this equation by S^{-1} from the left, we obtain

$$
S^{-1}AS = \Lambda. \tag{8.29}
$$

Thus, by using the matrix of eigenvectors and its inverse, it is possible to transform a matrix A into a diagonal matrix whose elements are the eigenvalues of A. The transformation expressed by (8.29) is referred to as the **diagonalization** of matrix A.

Example 8.4.1 If $A = \begin{pmatrix} 1 & 2 \\ 2 & 1 \end{pmatrix}$, find S such that $S^{-1}AS$ is a diagonal matrix. Show that the elements of $S^{-1}AS$ are the eigenvalues of A.

Solution 8.4.1 Since the secular equation is

$$
\begin{vmatrix} 1-\lambda & 2 \\ 2 & 1-\lambda \end{vmatrix} = (\lambda+1)(\lambda-3) = 0,
$$

the eigenvalues are $\lambda_1 = -1$, $\lambda_2 = 3$. The eigenvectors are, respectively, $\mathbf{x}_1 = \begin{pmatrix} 1 \\ -1 \end{pmatrix}$, $\mathbf{x}_2 = \begin{pmatrix} 1 \\ 1 \end{pmatrix}$. therefore

$$
S = \begin{pmatrix} 1 & 1 \\ -1 & 1 \end{pmatrix}.
$$

It can be readily checked that $S^{-1} = \dfrac{1}{2}\begin{pmatrix} 1 & -1 \\ 1 & 1 \end{pmatrix}$ and

$$
S^{-1}AS = \frac{1}{2}\begin{pmatrix} 1 & -1 \\ 1 & 1 \end{pmatrix}\begin{pmatrix} 1 & 2 \\ 2 & 1 \end{pmatrix}\begin{pmatrix} 1 & 1 \\ -1 & 1 \end{pmatrix} = \begin{pmatrix} -1 & 0 \\ 0 & 3 \end{pmatrix} = \begin{pmatrix} \lambda_1 & 0 \\ 0 & \lambda_2 \end{pmatrix}.
$$

Note that the diagonalizing matrix S is not necessarily unitary. However, if the eigenvectors are orthogonal, then we can normalize the eigenvectors and form an orthonormal set. The matrix with members of this orthonormal set as columns is a unitary matrix. The diagonalization process becomes a unitary similarity transformation which is much more convenient and useful.

The two eigenvectors in the above example are orthogonal, since

$$(1 \; -1) \begin{pmatrix} 1 \\ 1 \end{pmatrix} = 0.$$

Normalizing them, we get

$$\mathbf{u}_1 = \frac{1}{\sqrt{2}} \begin{pmatrix} 1 \\ -1 \end{pmatrix}, \quad \mathbf{u}_2 = \frac{1}{\sqrt{2}} \begin{pmatrix} 1 \\ 1 \end{pmatrix}.$$

The matrix constructed with these two normalized eigenvectors is

$$U = (\mathbf{u}_1 \; \mathbf{u}_2) = \frac{1}{\sqrt{2}} \begin{pmatrix} 1 & 1 \\ -1 & 1 \end{pmatrix}$$

which is an orthogonal matrix. The transformation

$$\tilde{U} A U = \begin{pmatrix} -1 & 0 \\ 0 & 3 \end{pmatrix}$$

is a unitary similarity transformation.

First, we have eliminated the step of finding the inverse of U, since U is an orthogonal matrix, the inverse of U is simply its transpose. More importantly, U represents a rotation as discussed in the last section. If we rotate the two original coordinate axes to coincide with $\mathbf{u}_1$ and $\mathbf{u}_2$, then with respect to these rotated axes, A is diagonal.

The coordinate axes of a reference system, in which the matrix is diagonal are known as **principal axes**. In this example, $\mathbf{u}_1$ and $\mathbf{u}_2$ are the unit vectors along the principal axes. From the components of $\mathbf{u}_1$, we can easily find the orientation of the principal axes. Let θ_1 be the angle between $\mathbf{u}_1$ and the original horizontal axis, then

$$\mathbf{u}_1 = \begin{pmatrix} \cos \theta_1 \\ \sin \theta_1 \end{pmatrix} = \frac{1}{\sqrt{2}} \begin{pmatrix} 1 \\ -1 \end{pmatrix}$$

which gives $\theta_1 = -\pi/4$. This means that to get the principal axes, we have to rotate the original coordinate axes 45 degrees clockwise. For consistence check, we can calculate θ_2, the angle between $\mathbf{u}_2$ and the original horizontal axis. Since

$$\mathbf{u}_2 = \begin{pmatrix} \cos \theta_2 \\ \sin \theta_2 \end{pmatrix} = \frac{1}{\sqrt{2}} \begin{pmatrix} 1 \\ 1 \end{pmatrix},$$

$\theta_2 = +\pi/4$. Therefore, the angle between θ_1 and θ_2 is $\pi/2$. This shows that $\mathbf{u}_1$ and $\mathbf{u}_2$ are perpendicular to each other, as they must.

Since $\theta_2 = \pi/2 + \theta_1$, $\cos\theta_2 = -\sin\theta_1$ and $\sin\theta_2 = \cos\theta_1$, the unitary matrix U can be written as

$$U = (\mathbf{u}_1 \ \mathbf{u}_2) = \begin{pmatrix} \cos\theta_1 & \cos\theta_2 \\ \sin\theta_1 & \sin\theta_2 \end{pmatrix} = \begin{pmatrix} \cos\theta_1 & -\sin\theta_1 \\ \sin\theta_1 & \cos\theta_1 \end{pmatrix},$$

which, as seen in (8.16) , is indeed a rotation matrix.

8.4.3 Quadratic Forms

A quadratic form is a homogeneous second-degree expression in n variables. For example,

$$Q(x_1, x_2) = 5x_1^2 - 4x_1x_2 + 2x_2^2$$

is a quadratic form in x_1 and x_2. Through a change of variables, this expression can be transformed into a form in which there is no cross-product term. Such a form is known as canonical form. Quadratic forms are important because they occur in a wide variety of applications.

The first step to change it into a canonical form is to separate the cross-product term into two equal terms, $(4x_1x_2 = 2x_1x_2 + 2x_2x_1)$, so that $Q(x_1, x_2)$ can be written as

$$Q(x_1, x_2) = (x_1 \ x_2) \begin{pmatrix} 5 & -2 \\ -2 & 2 \end{pmatrix} \begin{pmatrix} x_1 \\ x_2 \end{pmatrix} \tag{8.30}$$

where the coefficient matrix

$$C = \begin{pmatrix} 5 & -2 \\ -2 & 2 \end{pmatrix}$$

is symmetric. As we shall see in the next section that symmetric matrices can always be diagonalized. In this particular case, we can first find the eigenvalues and eigenvectors of C.

$$\begin{vmatrix} 5 - \lambda & -2 \\ -2 & 2 - \lambda \end{vmatrix} = (\lambda - 1)(\lambda - 6) = 0.$$

Corresponding to $\lambda_1 = 1$ and $\lambda_2 = 6$, the two normalized eigenvectors are found to be, respectively,

$$\mathbf{v}_1 = \frac{1}{\sqrt{5}} \begin{pmatrix} 1 \\ 2 \end{pmatrix}, \quad \mathbf{v}_2 = \frac{1}{\sqrt{5}} \begin{pmatrix} -2 \\ 1 \end{pmatrix}.$$

Therefore, the orthogonal matrix

$$U = (\mathbf{v}_1 \ \mathbf{v}_2) = \frac{1}{\sqrt{5}} \begin{pmatrix} 1 & -2 \\ 2 & 1 \end{pmatrix}$$

will diagonalize the coefficient matrix

$$\tilde{U}CU = \frac{1}{5}\begin{pmatrix} 1 & 2 \\ -2 & 1 \end{pmatrix}\begin{pmatrix} 5 & -2 \\ -2 & 2 \end{pmatrix}\begin{pmatrix} 1 & -2 \\ 2 & 1 \end{pmatrix} = \begin{pmatrix} 1 & 0 \\ 0 & 6 \end{pmatrix}.$$

If we make a change of variables

$$\begin{pmatrix} x_1 \\ x_2 \end{pmatrix} = U\begin{pmatrix} u_1 \\ u_2 \end{pmatrix}$$

and take the transpose of both sides

$$\begin{pmatrix} x_1 & x_2 \end{pmatrix} = \begin{pmatrix} u_1 & u_2 \end{pmatrix}\tilde{U},$$

we can write (8.30) as

$$\begin{pmatrix} u_1 & u_2 \end{pmatrix}\tilde{U}CU\begin{pmatrix} u_1 \\ u_2 \end{pmatrix} = \begin{pmatrix} u_1 & u_2 \end{pmatrix}\begin{pmatrix} 1 & 0 \\ 0 & 6 \end{pmatrix}\begin{pmatrix} u_1 \\ u_2 \end{pmatrix} = u_1^2 + 6u_2^2, \qquad (8.31)$$

which is in a canonical form, that is, it has no cross-term.

Note that the transformation matrix T defined in (8.20) is equal to $\tilde{U}$.

Example 8.4.2 Show that the following equation

$$9x^2 - 4xy + 6y^2 - 2\sqrt{5}x - 4\sqrt{6}y = 15$$

describes an ellipse by transforming it into a standard conic section form. Where is the center and what are the lengths of its major and minor axes?

Solution 8.4.2 The quadratic part of the equation can be written as

$$9x^2 - 4xy + 6y^2 = \begin{pmatrix} x & y \end{pmatrix}\begin{pmatrix} 9 & -2 \\ -2 & 6 \end{pmatrix}\begin{pmatrix} x \\ y \end{pmatrix}.$$

The eigenvalues of the coefficient matrix are given by

$$\begin{vmatrix} 9 - \lambda & -2 \\ -2 & 6 - \lambda \end{vmatrix} = (\lambda - 5)(\lambda - 10) = 0.$$

The normalized eigenvectors corresponding to $\lambda = 5$ and $\lambda = 10$ are found to be, respectively,

$$\mathbf{v}_1 = \frac{1}{\sqrt{5}}\begin{pmatrix} 1 \\ 2 \end{pmatrix}, \quad \mathbf{v}_2 = \frac{1}{\sqrt{5}}\begin{pmatrix} -2 \\ 1 \end{pmatrix}.$$

Therefore, the orthogonal matrix

$$U = (\mathbf{v}_1 \ \mathbf{v}_2) = \frac{1}{\sqrt{5}} \begin{pmatrix} 1 & -2 \\ 2 & 1 \end{pmatrix}$$

diagonalizes the coefficient matrix

$$\tilde{U}CU = \frac{1}{5} \begin{pmatrix} 1 & 2 \\ -2 & 1 \end{pmatrix} \begin{pmatrix} 9 & -2 \\ -2 & 6 \end{pmatrix} \begin{pmatrix} 1 & -2 \\ 2 & 1 \end{pmatrix} = \begin{pmatrix} 5 & 0 \\ 0 & 10 \end{pmatrix}.$$

Let

$$\begin{pmatrix} x \\ y \end{pmatrix} = U \begin{pmatrix} x' \\ y' \end{pmatrix} = \frac{1}{\sqrt{5}} \begin{pmatrix} 1 & -2 \\ 2 & 1 \end{pmatrix} \begin{pmatrix} x' \\ y' \end{pmatrix},$$

which is equivalent to

$$x = \frac{1}{\sqrt{5}} (x' - 2y'), \quad y = \frac{1}{\sqrt{5}} (2x' + y'),$$

then the equation can be written as

$$(x' \ y') \tilde{U}CU \begin{pmatrix} x' \\ y' \end{pmatrix} - 2\sqrt{5} \frac{1}{\sqrt{5}} (x' - 2y') - 4\sqrt{5} \frac{1}{\sqrt{5}} (2x' + y') = 15,$$

or

$$5x'^2 + 10y'^2 - 10x' = 15,$$

$$x'^2 + 2y'^2 - 2x' = 3.$$

Using $(x' - 1)^2 = x'^2 - 2x' + 1$, the last equation becomes

$$(x' - 1)^2 + 2y'^2 = 4,$$

or

$$\frac{(x' - 1)^2}{4} + \frac{y'^2}{2} = 1,$$

which is the standard form of an ellipse. The center of the ellipse is at $x = \frac{1}{\sqrt{5}}$, $y = \frac{2}{\sqrt{5}}$ (corresponding to $x' = 1$, $y' = 0$). The length of the major axis is $2\sqrt{4} = 4$, that of the minor axis is $2\sqrt{2}$.

To transform the equation into this standard form, we have rotated the coordinate axes. The major axis of the ellipse is along the vector $\mathbf{v}_1$ and the minor axis is along $\mathbf{v}_2$. Since

$$\mathbf{v}_1 = \begin{pmatrix} \cos \theta \\ \sin \theta \end{pmatrix} = \frac{1}{\sqrt{5}} \begin{pmatrix} 1 \\ 2 \end{pmatrix},$$

the major axis of the ellipse makes an angle θ with respect to the horizontal coordinate axis and $\theta = \cos^{-1} \frac{1}{\sqrt{5}}$.

8.5 Hermitian Matrix and Symmetric Matrix

8.5.1 Definitions

Real Matrix. If $A^* = A$, then $a_{ij} = a_{ij}^*$. Since every element of this matrix is real, it is called a real matrix.

Imaginary Matrix. If $A^* = -A$, this implies that $a_{ij} = -a_{ij}^*$. Every element of this matrix is purely imaginary or zero, so it is called a imaginary matrix.

Hermitian Matrix. A square matrix is called hermitian if $A^\dagger = A$. It is easy to show that the elements of a hermitian matrix satisfy the relation $a_{ij}^* = a_{ji}$. Hermitian matrix is very important in quantum mechanics.

Symmetric Matrix. If the elements of the matrix are all real, a hermitian matrix is just a symmetric matrix. A symmetric matrix is of great importance in classical physics, hermitian matrix is essential in quantum mechanics.

Anti-hermitian Matrix and Anti-symmetric Matrix. Finally, a matrix is called anti-hermitian or **skew-hemitian** if

$$A^\dagger = -A, \tag{8.32}$$

which implies $a_{ij}^* = -a_{ji}$.

Again, if the elements of the anti-hermitian matrix are all real, then the matrix is just an anti-symmetric matrix.

8.5.2 Eigenvalues of Hermitian Matrix

- **The eigenvalues of a hermitian (or real symmetric) matrix are all real.**

Let A be a hermitian matrix, and $\mathbf{x}$ be the nontrivial eigenvector belonging to eigenvalue λ

$$A\mathbf{x} = \lambda\mathbf{x}. \tag{8.33}$$

Take the hermitian conjugate of the equation,

$$\mathbf{x}^\dagger A^\dagger = \lambda^* \mathbf{x}^\dagger. \tag{8.34}$$

Note that λ is only a number (real or complex), its hermitian conjugate is just the complex conjugate, it can be multiplied either from left or from the right.

Multiply (8.33) by $\mathbf{x}^\dagger$ from the left

$$\mathbf{x}^\dagger A \mathbf{x} = \lambda \mathbf{x}^\dagger \mathbf{x}.$$

Multiply (8.34) by $\mathbf{x}$ from the right

$$\mathbf{x}^\dagger A^\dagger \mathbf{x} = \lambda^* \mathbf{x}^\dagger \mathbf{x}.$$

Subtract it from the preceding equation:

$$\left(\lambda - \lambda^*\right) \mathbf{x}^\dagger \mathbf{x} = \mathbf{x}^\dagger (A - A^\dagger) \mathbf{x}.$$

But A is hermitian, $A = A^\dagger$, so

$$\left(\lambda - \lambda^*\right) \mathbf{x}^\dagger \mathbf{x} = 0.$$

Since $\mathbf{x}^\dagger \mathbf{x} \neq 0$, it follows that $\lambda = \lambda^*$. That is, λ is real.

For real symmetric matrices, the proof is identical, since if the matrix is real, a hermitian matrix is a symmetric matrix.

- **If two eigenvalues of a hermitian (or a real symmetric) matrix are different, the corresponding eigenvectors are orthogonal.**

Let

$$A\mathbf{x}_1 = \lambda_1 \mathbf{x}_1,$$
$$A\mathbf{x}_2 = \lambda_2 \mathbf{x}_2.$$

Multiply the first equation by $\mathbf{x}_2^\dagger$ from the left:

$$\mathbf{x}_2^\dagger A \mathbf{x}_1 = \lambda_1 \mathbf{x}_2^\dagger \mathbf{x}_1.$$

Take the hermitian conjugate of the second equation and multiply by $\mathbf{x}_1$ from the right:

$$\mathbf{x}_2^\dagger A \mathbf{x}_1 = \lambda_2 \mathbf{x}_2^\dagger \mathbf{x}_1,$$

where we have used the facts that $(A\mathbf{x}_2)^\dagger = \mathbf{x}_2^\dagger A^\dagger$, $A^\dagger = A$, and $\lambda_2 = \lambda_2^*$. Subtracting these two equations, we have

$$(\lambda_1 - \lambda_2)\mathbf{x}_2^\dagger \mathbf{x}_1 = 0.$$

Since $\lambda_1 \neq \lambda_2$, it follows

$$\mathbf{x}_2^\dagger \mathbf{x}_1 = 0.$$

Therefore, $\mathbf{x}_1$ and $\mathbf{x}_2$ are orthogonal. For real symmetric matrices, the proof is the same.

8.5.3 Diagonalizing a Hermitian Matrix

- **A hermitian (or a real symmetric) matrix can be diagonalized by a unitary (or a real orthogonal) matrix.**

If the eigenvalues of the matrix are all distinct, the matrix can be diagonalized by a similarity transformation as we discussed before. Here, we only need to show that even if the eigenvalues are degenerate, as long as the matrix is hermitian, it can always be diagonalized. We will prove it by actually constructing a unitary matrix that will diagonalize a degenerate hermitian matrix.

Let λ_1 be a repeated eigenvalue of the $n \times n$ hermitian matrix H, let $\mathbf{x}_1$ be a normalized eigenvector corresponding to λ_1. We can take any n linearly independ vectors with the only condition that the first one is $\mathbf{x}_1$ and construct with the Gram-Schmidt process an orthonormal set of n vectors $\mathbf{x}_1, \mathbf{x}_2, \cdots, \mathbf{x}_n$, each of them has n elements.

Let U_1 be the matrix with $\mathbf{x}_i$ as its ith column,

$$U_1 = \begin{pmatrix} x_{11} & x_{21} & \cdot\cdot & x_{n1} \\ x_{12} & x_{22} & \cdot\cdot & x_{n2} \\ \cdot & \cdot & \cdot\cdot\cdot & \cdot \\ \cdot & \cdot & \cdot\cdot\cdot & \cdot \\ x_{1n} & x_{2n} & \cdot\cdot & x_{nn} \end{pmatrix},$$

as we have shown this automatically makes U_1 an unitary matrix. The unitary transformation $U_1^\dagger H U_1$ has exactly the same set of eigenvalues as H, since they have the same characteristic polynomial,

$$\left| U_1^\dagger H U_1 - \lambda I \right| = \left| U_1^{-1} H U_1 - \lambda U_1^{-1} U_1 \right| = \left| U_1^{-1}(H - \lambda I)U_1 \right|$$
$$= \left| U_1^{-1} \right| |H - \lambda I| |U_1| = |H - \lambda I|.$$

Furthermore, since H is hermitian, $U_1^\dagger H U_1$ is also hermitian, since

$$\left(U_1^\dagger H U_1 \right)^\dagger = (H U_1)^\dagger \left(U_1^\dagger \right)^\dagger = U_1^\dagger H^\dagger U_1 = U_1^\dagger H U_1.$$

Now

$$U_1^\dagger H U_1 = \begin{pmatrix} x_{11}^* & x_{12}^* & \cdots & x_{1n}^* \\ x_{21}^* & x_{22}^* & \cdots & x_{2n}^* \\ \cdot & \cdot & \cdots & \cdot \\ \cdot & \cdot & \cdots & \cdot \\ x_{n1}^* & x_{n2}^* & \cdots & x_{nn}^* \end{pmatrix} H \begin{pmatrix} x_{11} & x_{21} & \cdots & x_{n1} \\ x_{12} & x_{22} & \cdots & x_{n2} \\ \cdot & \cdot & \cdots & \cdot \\ \cdot & \cdot & \cdots & \cdot \\ x_{1n} & x_{2n} & \cdots & x_{nn} \end{pmatrix}$$

$$= \begin{pmatrix} x_{11}^* & x_{12}^* & \cdots & x_{1n}^* \\ x_{21}^* & x_{22}^* & \cdots & x_{2n}^* \\ \cdot & \cdot & \cdots & \cdot \\ \cdot & \cdot & \cdots & \cdot \\ x_{n1}^* & x_{n2}^* & \cdots & x_{nn}^* \end{pmatrix} \begin{pmatrix} \lambda_1 x_{11} & h_{21} & \cdots & h_{n1} \\ \lambda_1 x_{12} & h_{22} & \cdots & h_{n2} \\ \cdot & \cdot & \cdots & \cdot \\ \cdot & \cdot & \cdots & \cdot \\ \lambda_1 x_{1n} & h_{2n} & \cdots & h_{nn} \end{pmatrix},$$

where we have used the fact that $\mathbf{x}_1$ is an eigenvector of H belonging to the eigenvalue λ_1,

$$H \begin{pmatrix} x_{11} \\ x_{12} \\ \cdot \\ \cdot \\ x_{1n} \end{pmatrix} = \lambda_1 \begin{pmatrix} x_{11} \\ x_{12} \\ \cdot \\ \cdot \\ x_{1n} \end{pmatrix},$$

and have written

$$H \begin{pmatrix} x_{i1} \\ x_{i2} \\ \cdot \\ \cdot \\ x_{in} \end{pmatrix} = \begin{pmatrix} h_{i1} \\ h_{i2} \\ \cdot \\ \cdot \\ h_{in} \end{pmatrix}$$

for $i \neq 1$. Furthermore,

$$U_1^\dagger H U_1 = \begin{pmatrix} x_{11}^* & x_{12}^* & \cdots & x_{1n}^* \\ x_{21}^* & x_{22}^* & \cdots & x_{2n}^* \\ \cdot & \cdot & \cdots & \cdot \\ \cdot & \cdot & \cdots & \cdot \\ x_{n1}^* & x_{n2}^* & \cdots & x_{nn}^* \end{pmatrix} \begin{pmatrix} \lambda_1 x_{11} & h_{21} & \cdots & h_{n1} \\ \lambda_1 x_{12} & h_{22} & \cdots & h_{n2} \\ \cdot & \cdot & \cdots & \cdot \\ \cdot & \cdot & \cdots & \cdot \\ \lambda_1 x_{1n} & h_{2n} & \cdots & h_{nn} \end{pmatrix} = \begin{pmatrix} \lambda_1 & 0 & 0 & \cdot & 0 \\ 0 & \alpha_{22} & \alpha_{32} & \cdot & \alpha_{n2} \\ 0 & \cdot & \cdot & \cdots & \cdot \\ \cdot & \cdot & \cdot & \cdots & \cdot \\ 0 & \alpha_{2n} & \alpha_{3n} & \cdot & \alpha_{nn} \end{pmatrix}.$$

The first column is determined by the orthonormal condition

$$\begin{pmatrix} x_{i1}^* & x_{i2}^* & \cdots & x_{in}^* \end{pmatrix} \begin{pmatrix} x_{11} \\ x_{12} \\ \cdot \\ \cdot \\ x_{1n} \end{pmatrix} = \begin{cases} 1 & if \ i = 1 \\ 0 & if \ i \neq 1 \end{cases}.$$

The first row must be the transpose of the first column because $U_1^\dagger H U_1$ is hermitian (or real symmetric) and λ_1 is real and the complex conjugate of zero is itself. The crucial fact in this process is that the last $n - 1$ elements of the first row are all zero. This is what distinguishes hermitian (or real symmetric) matrices from other square matrices.

If λ_1 is a two-fold degenerate eigenvalue of H, then in the characteristic polynomial $p(\lambda) = |H - \lambda I|$, there must be a factor $(\lambda_1 - \lambda)^2$. Since

$$p(\lambda) = |H - \lambda I| = \left| U_1^\dagger H U_1 - \lambda I \right|$$

$$= \begin{vmatrix} \lambda_1 - \lambda & 0 & 0 & \cdot\cdot & 0 \\ 0 & \alpha_{22} - \lambda & \alpha_{32} & \cdot\cdot & \alpha_{n2} \\ 0 & \alpha_{23} & \alpha_{33} - \lambda & \cdot\cdot & \alpha_{n3} \\ \cdot & & \cdot & \cdot\cdot & \cdot \\ 0 & \alpha_{2n} & \alpha_{3n} & \cdot\cdot & \alpha_{nn} - \lambda \end{vmatrix}$$

$$= (\lambda_1 - \lambda) \begin{vmatrix} \alpha_{22} - \lambda & \alpha_{32} & \cdot\cdot & \alpha_{n2} \\ \alpha_{23} & \alpha_{33} - \lambda & \cdot\cdot & \alpha_{n3} \\ \cdot & & \cdot\cdot & \cdot \\ \alpha_{2n} & \alpha_{3n} & \cdot\cdot & \alpha_{nn} - \lambda \end{vmatrix},$$

the part

$$\begin{vmatrix} \alpha_{22} - \lambda & \alpha_{32} & \cdot\cdot & \alpha_{n2} \\ \alpha_{23} & \alpha_{33} - \lambda & \cdot\cdot & \alpha_{n3} \\ \cdot & & \cdot\cdot & \cdot \\ \alpha_{2n} & \alpha_{3n} & \cdot\cdot & \alpha_{nn} - \lambda \end{vmatrix}$$

must contain another factor of $(\lambda_1 - \lambda)$. In other words, if we define H_1 as the $(n - 1) \times (n - 1)$ submatrix

$$\begin{pmatrix} \alpha_{22} & \alpha_{32} & \cdot\cdot & \alpha_{n2} \\ \alpha_{23} & \alpha_{33} & \cdot\cdot & \alpha_{n3} \\ \cdot & \cdot & \cdot & \cdot \\ \alpha_{2n} & \alpha_{3n} & \cdot\cdot & \alpha_{nn} \end{pmatrix} = H_1,$$

then λ_1 must be an eigenvalue of H_1. Thus, we can repeat the process and construct an orthonormal set of $n - 1$ column vectors with the first one being the eigenvector of H_1 belonging to the eigenvalue λ_1. Let this orthonormal set be

$$\mathbf{y}_1 = \begin{pmatrix} y_{22} \\ y_{23} \\ \cdot \\ y_{2n} \end{pmatrix}, \quad \mathbf{y}_2 = \begin{pmatrix} y_{32} \\ y_{33} \\ \cdot \\ y_{3n} \end{pmatrix}, \quad \cdots\cdots, \quad \mathbf{y}_{n-1} = \begin{pmatrix} y_{n2} \\ y_{n3} \\ \cdot \\ y_{nn} \end{pmatrix},$$

and let U_2 be another unitary matrix defined as

$$U_2 = \begin{pmatrix} 1 & 0 & 0 & \cdots & 0 \\ 0 & y_{22} & y_{32} & \cdots & y_{n2} \\ 0 & y_{23} & y_{33} & \cdots & y_{n3} \\ \cdot & \cdot & \cdot & \cdots & \cdot \\ 0 & y_{2n} & y_{3n} & \cdots & y_{nn} \end{pmatrix},$$

then the unitary transformation $U_2^\dagger(U_1^\dagger H U_1)U_2$ can be shown as

$$U_2^\dagger(U_1^\dagger H U_1)U_2 = \begin{pmatrix} 1 & 0 & 0 & \cdots & 0 \\ 0 & y_{22}^* & y_{23}^* & \cdots & y_{2n}^* \\ 0 & y_{32}^* & y_{33}^* & \cdots & y_{3n}^* \\ \cdot & \cdot & \cdot & \cdots & \cdot \\ 0 & y_{2n}^* & y_{3n}^* & \cdots & y_{nn}^* \end{pmatrix} \begin{pmatrix} \lambda_1 & 0 & 0 & \cdots & 0 \\ 0 & \alpha_{22} & \alpha_{32} & \cdots & \alpha_{n2} \\ 0 & \alpha_{23} & \alpha_{33} & \cdots & \alpha_{n3} \\ \cdot & \cdot & \cdot & \cdots & \cdot \\ 0 & \alpha_{2n} & \alpha_{3n} & \cdots & \alpha_{nn} \end{pmatrix} \begin{pmatrix} 1 & 0 & 0 & \cdots & 0 \\ 0 & y_{22} & y_{32} & \cdots & y_{n2} \\ 0 & y_{23} & y_{33} & \cdots & y_{n3} \\ \cdot & \cdot & \cdot & \cdots & \cdot \\ 0 & y_{2n} & y_{3n} & \cdots & y_{nn} \end{pmatrix}$$

$$= \begin{pmatrix} \lambda_1 & 0 & 0 & \cdots & 0 \\ 0 & \lambda_1 & 0 & \cdots & 0 \\ 0 & 0 & \beta_{33} & \cdots & \beta_{n3} \\ \cdot & \cdot & \cdot & \cdots & \cdot \\ 0 & 0 & \beta_{3n} & \cdots & \beta_{nn} \end{pmatrix}.$$

If λ_1 is m-fold degenerate, we repeat this process m times. The rest can be diagonalized by the eigenvectors belonging to different eigenvalues. After the $n \times n$ matrix is so transformed $n - 1$ times, it will become diagonal.

Let us define

$$U = U_1 U_2 \cdots U_{n-1},$$

then U is unitary because all U_i are unitary. Consequently, the hermitian matrix H is diagonalized by the unitary transformation $U^\dagger H U$ and the theorem is established.

This construction leads to the following important corollary.

• Every $n \times n$ **hermitian (or real symmetric) matrix has** n **orthogonal eigenvectors regardless of the degeneracy of its eigenvalues.**

This is because $U^\dagger H U = \Lambda$, where the elements of the diagonal matrix Λ are the eigenvalues of H. Since $U^\dagger = U^{-1}$, it follows from the equation $U(U^\dagger H U) = U\Lambda$ that $HU = U\Lambda$ which shows that each column of U is an normalized eigenvector of H.

The following example illustrates this process.

Example 8.5.1 Find an unitary matrix that will diagonalize the hermitian matrix

$$H = \begin{pmatrix} 2 & i & 1 \\ -i & 2 & i \\ 1 & -i & 2 \end{pmatrix}.$$

Solution 8.5.1 The eigenvalues of H are the roots of the characteristic equation

$$p(\lambda) = \begin{vmatrix} 2-\lambda & i & 1 \\ -i & 2-\lambda & i \\ 1 & -i & 2-\lambda \end{vmatrix} = -\lambda^3 + 6\lambda^2 - 9\lambda = -\lambda(\lambda-3)^2 = 0.$$

Therefore, the eigenvalues $\lambda_1, \lambda_2, \lambda_3$ are

$$\lambda_1 = 3, \quad \lambda_2 = 3, \quad \lambda_3 = 0.$$

It is seen that $\lambda = \lambda_1 = \lambda_2 = 3$ is two-fold degenerate. Let one of the eigenvectors corresponding to λ_1 be

$$E_1 = \begin{pmatrix} x_1 \\ x_2 \\ x_3 \end{pmatrix},$$

so

$$\begin{pmatrix} 2-\lambda_1 & i & 1 \\ -i & 2-\lambda_1 & i \\ 1 & -i & 2-\lambda_1 \end{pmatrix} \begin{pmatrix} x_1 \\ x_2 \\ x_3 \end{pmatrix} = \begin{pmatrix} -1 & i & 1 \\ -i & -1 & i \\ 1 & -i & -1 \end{pmatrix} \begin{pmatrix} x_1 \\ x_2 \\ x_3 \end{pmatrix} = 0.$$

The three equations

$$-x_1 + ix_2 + x_3 = 0$$
$$-ix_1 - x_2 + ix_3 = 0$$
$$x_1 - ix_2 - x_3 = 0$$

are identical to each other. For example, multiply the middle one by i will change it to the last one. The equation

$$x_1 - ix_2 - x_3 = 0. \tag{8.35}$$

has an infinite number of solutions. A simple choice is to set $x_2 = 0$, then $x_1 = x_3$. Therefore,

$$E_1 = \begin{pmatrix} 1 \\ 0 \\ 1 \end{pmatrix}$$

is an eigenvector. Certainly

$$E_1 = \begin{pmatrix} 1 \\ 0 \\ 1 \end{pmatrix}, \quad E_2 = \begin{pmatrix} 0 \\ 1 \\ 0 \end{pmatrix}, \quad E_3 = \begin{pmatrix} 1 \\ 0 \\ 0 \end{pmatrix}$$

are linearly independent. Now let us use the Gram-Schmidt process to find an orthonormal set $\mathbf{x}_1, \mathbf{x}_2, \mathbf{x}_3$.

$$\mathbf{x}_1 = \frac{\mathbf{E}_1}{\|\mathbf{E}_1\|} = \frac{\sqrt{2}}{2} \begin{pmatrix} 1 \\ 0 \\ 1 \end{pmatrix},$$

$\mathbf{E}_2$ is already normalized and it is orthogonal to $\mathbf{E}_1$, so

$$\mathbf{x}_2 = \mathbf{E}_2 = \begin{pmatrix} 0 \\ 1 \\ 0 \end{pmatrix},$$

$$\mathbf{x}_3' = \mathbf{E}_3 - (\mathbf{E}_3, \mathbf{x}_1)\,\mathbf{x}_1 - (\mathbf{E}_3, \mathbf{x}_2)\,\mathbf{x}_2$$

$$= \begin{pmatrix} 1 \\ 0 \\ 0 \end{pmatrix} - \left[(1\,0\,0)\frac{\sqrt{2}}{2}\begin{pmatrix} 1 \\ 0 \\ 1 \end{pmatrix} \right] \frac{\sqrt{2}}{2}\begin{pmatrix} 1 \\ 0 \\ 1 \end{pmatrix} - \left[(1\,0\,0)\begin{pmatrix} 0 \\ 1 \\ 0 \end{pmatrix} \right]\begin{pmatrix} 0 \\ 1 \\ 0 \end{pmatrix}$$

$$= \begin{pmatrix} 1 \\ 0 \\ 0 \end{pmatrix} - \frac{1}{2}\begin{pmatrix} 1 \\ 0 \\ 1 \end{pmatrix} = \frac{1}{2}\begin{pmatrix} 1 \\ 0 \\ -1 \end{pmatrix},$$

$$\mathbf{x}_3 = \frac{\mathbf{x}_3'}{\|\mathbf{x}_3'\|} = \frac{\sqrt{2}}{2}\begin{pmatrix} 1 \\ 0 \\ -1 \end{pmatrix}.$$

Form a unitary matrix with $\mathbf{x}_1, \mathbf{x}_2, \mathbf{x}_3$:

$$U_1 = (\mathbf{x}_1\ \mathbf{x}_2\ \mathbf{x}_3) = \begin{pmatrix} \frac{\sqrt{2}}{2} & 0 & \frac{\sqrt{2}}{2} \\ 0 & 1 & 0 \\ \frac{\sqrt{2}}{2} & 0 & -\frac{\sqrt{2}}{2} \end{pmatrix}.$$

The unitary similarity transformation of H by U_1 is

$$U_1^\dagger H U_1 = \begin{pmatrix} \frac{\sqrt{2}}{2} & 0 & \frac{\sqrt{2}}{2} \\ 0 & 1 & 0 \\ \frac{\sqrt{2}}{2} & 0 & -\frac{\sqrt{2}}{2} \end{pmatrix}\begin{pmatrix} 2 & i & 1 \\ -i & 2 & i \\ 1 & -i & 2 \end{pmatrix}\begin{pmatrix} \frac{\sqrt{2}}{2} & 0 & \frac{\sqrt{2}}{2} \\ 0 & 1 & 0 \\ \frac{\sqrt{2}}{2} & 0 & -\frac{\sqrt{2}}{2} \end{pmatrix}$$

$$= \begin{pmatrix} 3 & 0 & 0 \\ 0 & 2 & -\sqrt{2}i \\ 0 & \sqrt{2}i & 1 \end{pmatrix}$$

Since H and $U_1^\dagger H U_1$ have the same set of eigenvalues, therefore, $\lambda = 3$ and $\lambda = 0$ must be the eigenvalue of the submatrix

$$H_1 = \begin{pmatrix} 2 & -\sqrt{2}i \\ \sqrt{2}i & 1 \end{pmatrix}.$$

This can also be shown directly. The two normalized eigenvector of H_1 corresponding to $\lambda = 3$ and $\lambda = 0$ are found, respectively, to be

$$\mathbf{y}_1 = \begin{pmatrix} -\frac{\sqrt{6}}{3}i \\ \frac{\sqrt{3}}{3} \end{pmatrix}, \quad \mathbf{y}_2 = \begin{pmatrix} \frac{\sqrt{3}}{3}i \\ \frac{\sqrt{6}}{3} \end{pmatrix}.$$

Therefore,

$$U_2 = \begin{pmatrix} 1 & 0 & 0 \\ 0 & -\frac{\sqrt{6}}{3}i & \frac{\sqrt{3}}{3}i \\ 0 & \frac{\sqrt{3}}{3} & \frac{\sqrt{6}}{3} \end{pmatrix},$$

and

$$U = U_1 U_2 = \begin{pmatrix} \frac{\sqrt{2}}{2} & 0 & \frac{\sqrt{2}}{2} \\ 0 & 1 & 0 \\ \frac{\sqrt{2}}{2} & 0 & -\frac{\sqrt{2}}{2} \end{pmatrix} \begin{pmatrix} 1 & 0 & 0 \\ 0 & -\frac{\sqrt{6}}{3}i & \frac{\sqrt{3}}{3}i \\ 0 & \frac{\sqrt{3}}{3} & \frac{\sqrt{6}}{3} \end{pmatrix}$$

$$= \begin{pmatrix} \frac{\sqrt{2}}{2} & \frac{\sqrt{6}}{6} & \frac{\sqrt{3}}{3} \\ 0 & -\frac{\sqrt{6}}{3}i & \frac{\sqrt{3}}{3}i \\ \frac{\sqrt{2}}{2} & -\frac{\sqrt{6}}{6} & -\frac{\sqrt{3}}{3} \end{pmatrix}.$$

It can be easily checked that

$$U^\dagger H U = \begin{pmatrix} \frac{\sqrt{2}}{2} & 0 & \frac{\sqrt{2}}{2} \\ \frac{\sqrt{6}}{6} & \frac{\sqrt{6}}{3}i & -\frac{\sqrt{6}}{6} \\ \frac{\sqrt{3}}{3} & -\frac{\sqrt{3}}{3}i & -\frac{\sqrt{3}}{3} \end{pmatrix} \begin{pmatrix} 2 & i & 1 \\ -i & 2 & i \\ 1 & -i & 2 \end{pmatrix} \begin{pmatrix} \frac{\sqrt{2}}{2} & \frac{\sqrt{6}}{6} & \frac{\sqrt{3}}{3} \\ 0 & -\frac{\sqrt{6}}{3}i & \frac{\sqrt{3}}{3}i \\ \frac{\sqrt{2}}{2} & -\frac{\sqrt{6}}{6} & -\frac{\sqrt{3}}{3} \end{pmatrix}$$

$$= \begin{pmatrix} 3 & 0 & 0 \\ 0 & 3 & 0 \\ 0 & 0 & 0 \end{pmatrix}$$

is indeed diagonal and the diagonal elements are the eigenvalues. Furthermore, the three columns of U are indeed three orthogonal eigenvectors of H,

$$H\mathbf{u}_1 = \lambda_1 \mathbf{u}_1 : \quad \begin{pmatrix} 2 & i & 1 \\ -i & 2 & i \\ 1 & -i & 2 \end{pmatrix} \begin{pmatrix} \frac{\sqrt{2}}{2} \\ 0 \\ \frac{\sqrt{2}}{2} \end{pmatrix} = 3 \begin{pmatrix} \frac{\sqrt{2}}{2} \\ 0 \\ \frac{\sqrt{2}}{2} \end{pmatrix},$$

$$H\mathbf{u}_2 = \lambda_2\mathbf{u}_2 : \quad \begin{pmatrix} 2 & i & 1 \\ -i & 2 & i \\ 1 & -i & 2 \end{pmatrix} \begin{pmatrix} \frac{\sqrt{6}}{6} \\ -\frac{\sqrt{6}}{3}i \\ -\frac{\sqrt{6}}{6} \end{pmatrix} = 3 \begin{pmatrix} \frac{\sqrt{6}}{6} \\ -\frac{\sqrt{6}}{3}i \\ -\frac{\sqrt{6}}{6} \end{pmatrix},$$

$$H\mathbf{u}_3 = \lambda_3\mathbf{u}_3 : \quad \begin{pmatrix} 2 & i & 1 \\ -i & 2 & i \\ 1 & -i & 2 \end{pmatrix} \begin{pmatrix} \frac{\sqrt{3}}{3} \\ \frac{\sqrt{3}}{3}i \\ -\frac{\sqrt{3}}{3} \end{pmatrix} = 0, \begin{pmatrix} \frac{\sqrt{3}}{3} \\ \frac{\sqrt{3}}{3}i \\ -\frac{\sqrt{3}}{3} \end{pmatrix}.$$

We have followed the steps of the proof in order to illustrate the procedure. Once it is established, we can make use of the theorem and the process of finding the eigenvectors can be simplified considerably.

In this example, one can find the eigenvector for the non-degenerate eigenvalue the usual way. For the degenerate eigenvalue $\lambda = 3$, the components (x_1, x_2, x_3) of the corresponding eigenvectors must satisfy

$$x_1 - ix_2 - x_3 = 0,$$

as shown in (8.35). This equation can be written as $x_2 = i(x_3 - x_1)$. So, in general

$$\mathbf{u} = \begin{pmatrix} x_1 \\ i(x_3 - x_1) \\ x_3 \end{pmatrix},$$

where x_1 and x_3 are arbitrary. It is seen that $\mathbf{u}_1$ is just the normalized eigenvector obtained by choosing $x_1 = x_3$,

$$\mathbf{u}_1 = \frac{\sqrt{2}}{2} \begin{pmatrix} 1 \\ 0 \\ 1 \end{pmatrix}.$$

The other eigenvector must also satisfy the same equation and be orthogonal to $\mathbf{u}_1$. Thus,

$$(1\ 0\ 1) \begin{pmatrix} x_1 \\ i(x_3 - x_1) \\ x_3 \end{pmatrix} = 0,$$

which gives $x_1 + x_3 = 0$, or $x_3 = -x_1$. Normalizing the vector $\begin{pmatrix} x_1 \\ -2x_1 \\ -x_1 \end{pmatrix}$, one

obtains the other eigenvector belonging to $\lambda = 3$:

$$\mathbf{u}_2 = \frac{\sqrt{6}}{6} \begin{pmatrix} 1 \\ -2i \\ -1 \end{pmatrix}.$$

8.5.4 Simultaneous Diagonalization

If A and B are two hermitian matrices of the same order, the following important question often arises. Can they be simultaneously diagonalized by the same matrix S? That is to say, does there exist a basis in which they are both diagonal? The answer is yes if and only if they commute.

First, we will show that if they can be simultaneously diagonalized, then they must commute. That is, if

$$D_1 = S^{-1}AS \quad \text{and} \quad D_2 = S^{-1}BS,$$

where D_1 and D_2 are both diagonal matrices, then $AB = BA$.

This follows from the fact

$$D_1 D_2 = S^{-1}ASS^{-1}BS = S^{-1}ABS,$$
$$D_2 D_1 = S^{-1}BSS^{-1}AS = S^{-1}BAS.$$

Since diagonal matrices of the same order always commute $(D_1 D_2 = D_2 D_1)$, so we have

$$S^{-1}ABS = S^{-1}BAS..$$

Multiplying S from the left and S^{-1} from the right, we obtain $AB = BA$.

Now we will show that the converse is also true. That is, if they commute, then they can be simultaneously diagonalized. First, let A and B be 2×2 matrices. Since hermitian matrix is always diagonalizable, let S be the unitary matrix that diagonalizes A,

$$S^{-1}AS = \begin{pmatrix} \lambda_1 & 0 \\ 0 & \lambda_2 \end{pmatrix}$$

where λ_1 and λ_2 are the eigenvalues of A. Let

$$S^{-1}BS = \begin{pmatrix} b_{11} & b_{12} \\ b_{21} & b_{22} \end{pmatrix}$$

Now

$$S^{-1}ABS = S^{-1}ASS^{-1}BS = \begin{pmatrix} \lambda_1 & 0 \\ 0 & \lambda_2 \end{pmatrix} \begin{pmatrix} b_{11} & b_{12} \\ b_{21} & b_{22} \end{pmatrix} = \begin{pmatrix} b_{11}\lambda_1 & b_{12}\lambda_1 \\ b_{21}\lambda_2 & b_{22}\lambda_2 \end{pmatrix}$$
$$S^{-1}BAS = S^{-1}BSS^{-1}AS = \begin{pmatrix} b_{11} & b_{12} \\ b_{21} & b_{22} \end{pmatrix} \begin{pmatrix} \lambda_1 & 0 \\ 0 & \lambda_2 \end{pmatrix} = \begin{pmatrix} b_{11}\lambda_1 & b_{12}\lambda_2 \\ b_{21}\lambda_1 & b_{22}\lambda_2 \end{pmatrix}.$$

Since $AB = BA$, so $S^{-1}ABS = S^{-1}BAS$,

$$\begin{pmatrix} b_{11}\lambda_1 & b_{12}\lambda_1 \\ b_{21}\lambda_2 & b_{22}\lambda_2 \end{pmatrix} = \begin{pmatrix} b_{11}\lambda_1 & b_{12}\lambda_2 \\ b_{21}\lambda_1 & b_{22}\lambda_2 \end{pmatrix}.$$

It follows that

$$b_{21}\lambda_2 = b_{21}\lambda_1, \qquad b_{12}\lambda_1 = b_{12}\lambda_2.$$

If $\lambda_2 \neq \lambda_1$, then $b_{12} = 0$ and $b_{21} = 0$. In other words,

$$S^{-1}BS = \begin{pmatrix} b_{11} & 0 \\ 0 & b_{22} \end{pmatrix}.$$

Therefore, A and B are simultaneously diagonalized.

If $\lambda_2 = \lambda_1 = \lambda$, we cannot conclude that $S^{-1}BS$ is diagonal. However, in this case,

$$S^{-1}AS = \begin{pmatrix} \lambda & 0 \\ 0 & \lambda \end{pmatrix}$$

Moveover, since B is hermitian, so the unitary similarity transform $S^{-1}BS$ is also hermitian, therefore, $S^{-1}BS$ is diagonalizable. Let T be the unitary matrix that diagonalize $S^{-1}BS$,

$$T^{-1}\left(S^{-1}BS\right)T = \begin{pmatrix} \lambda_1' & 0 \\ 0 & \lambda_2' \end{pmatrix}.$$

On the other hand,

$$T^{-1}\left(S^{-1}AS\right)T = T^{-1}\begin{pmatrix} \lambda & 0 \\ 0 & \lambda \end{pmatrix}T = \begin{pmatrix} \lambda & 0 \\ 0 & \lambda \end{pmatrix}T^{-1}T = \begin{pmatrix} \lambda & 0 \\ 0 & \lambda \end{pmatrix}.$$

Thus, the product matrix $U = ST$ diagonalizes both A and B. Therefore, with or without degeneracy, as long as A and B commute, they can be simultaneously diagonalized.

Although we have used 2×2 matrices for illustration, the same "proof" can obviously be applied to matrices of higher order.

Example 8.5.2 Let

$$A = \begin{pmatrix} 2 & 1 \\ 1 & 2 \end{pmatrix}, \qquad B = \begin{pmatrix} 3 & 2 \\ 2 & 3 \end{pmatrix}.$$

Can A and B be simultaneously diagonalized? If so, find the unitary matrix that diagonalized them.

Solution 8.5.2

$$AB = \begin{pmatrix} 2 & 1 \\ 1 & 2 \end{pmatrix} \begin{pmatrix} 3 & 2 \\ 2 & 3 \end{pmatrix} = \begin{pmatrix} 8 & 7 \\ 7 & 8 \end{pmatrix},$$

$$BA = \begin{pmatrix} 3 & 2 \\ 2 & 3 \end{pmatrix} \begin{pmatrix} 2 & 1 \\ 1 & 2 \end{pmatrix} = \begin{pmatrix} 8 & 7 \\ 7 & 8 \end{pmatrix}.$$

Thus, $[A, B] = 0$, therefore, they can be simultaneously diagonalized.

$$\begin{vmatrix} 2 - \lambda & 1 \\ 1 & 2 - \lambda \end{vmatrix} = (\lambda - 1)(\lambda - 3) = 0,$$

The normalized eigenvectors corresponding to $\lambda = 1, 3$ are, respectively

$$\mathbf{x}_1 = \frac{1}{\sqrt{2}} \begin{pmatrix} 1 \\ -1 \end{pmatrix}, \quad \mathbf{x}_2 = \frac{1}{\sqrt{2}} \begin{pmatrix} 1 \\ 1 \end{pmatrix}.$$

Therefore,

$$S = \frac{1}{\sqrt{2}} \begin{pmatrix} 1 & 1 \\ -1 & 1 \end{pmatrix}, \quad S^{-1} = \frac{1}{\sqrt{2}} \begin{pmatrix} 1 & -1 \\ 1 & 1 \end{pmatrix}.$$

$$S^{-1}AS = \frac{1}{\sqrt{2}} \begin{pmatrix} 1 & -1 \\ 1 & 1 \end{pmatrix} \begin{pmatrix} 2 & 1 \\ 1 & 2 \end{pmatrix} \frac{1}{\sqrt{2}} \begin{pmatrix} 1 & 1 \\ -1 & 1 \end{pmatrix} = \begin{pmatrix} 1 & 0 \\ 0 & 3 \end{pmatrix},$$

$$S^{-1}BS = \frac{1}{\sqrt{2}} \begin{pmatrix} 1 & -1 \\ 1 & 1 \end{pmatrix} \begin{pmatrix} 3 & 2 \\ 2 & 3 \end{pmatrix} \frac{1}{\sqrt{2}} \begin{pmatrix} 1 & 1 \\ -1 & 1 \end{pmatrix} = \begin{pmatrix} 1 & 0 \\ 0 & 5 \end{pmatrix}.$$

Thus, they are simultaneously diagonalized. It also shows that 1 and 5 are the eigenvalues of B. This can be easily verified, since

$$\begin{vmatrix} 3 - \lambda & 2 \\ 2 & 3 - \lambda \end{vmatrix} = (\lambda - 1)(\lambda - 5) = 0.$$

If we diagonalize B first, we will get exactly the same result.

8.6   Normal Matrix

A square matrix is said to be normal if and only if it commutes with its hermitian conjugate. That is, A is normal, if and only if

$$AA^{\dagger} = A^{\dagger}A. \tag{8.36}$$

It can be easily shown that all hermitian (or real symmetric), anti-hermitian (or real anti-symmetric), and unitary (or real orthogonal) matrices are normal. All we have to do is to substitute these matrices into (8.36). By virtue of their definition, it is immediately clear that the two sides of the equation are indeed the same.

So far, we have shown that every hermitian matrix is diagonalizable by a unitary similarity transformation. In what follows, we will prove the generalization of this theorem that every normal matrix is diagonalizable.

First, if the square matrix A is given, that means all elements of A are known, so we can take its hermitian conjugate $A^\dagger$. Then let

$$B = \frac{1}{2}(A + A^\dagger),$$

$$C = \frac{1}{2i}(A - A^\dagger),$$

So,

$$A = B + iC. \tag{8.37}$$

Since $(A^\dagger)^\dagger = A$ and $(A + B)^\dagger = A^\dagger + B^\dagger$,

$$B^\dagger = \frac{1}{2}(A + A^\dagger)^\dagger = \frac{1}{2}(A^\dagger + A) = B,$$

$$C^\dagger = \frac{1}{2i^*}(A - A^\dagger)^\dagger = -\frac{1}{2i}(A^\dagger - A) = C.$$

Thus, B and C are both hermitian. In other words, a square matrix can always be decomposed into two hermitian matrices as shown in (8.37). Furthermore,

$$BC = \frac{1}{4i}\left(A^2 - AA^\dagger + A^\dagger A - A^{\dagger 2}\right)$$

$$CB = \frac{1}{4i}\left(A^2 - A^\dagger A + AA^\dagger - A^{\dagger 2}\right).$$

It is clear that if $A^\dagger A = AA^\dagger$, then $BC = CB$. In other words, if A is normal, then B and C commute.

We have shown in the previous section that if B and C commute, then they can be simultaneously diagonalized. That is, we can find a unitary matrix S such that $S^{-1}BS$ and $S^{-1}CS$ are both diagonal. Since

$$S^{-1}AS = S^{-1}BS + iS^{-1}CS,$$

it follows that $S^{-1}AS$ must also be diagonal.

Conversely, if $S^{-1}AS = D$ is diagonal, then

$$\left(S^{-1}AS\right)^\dagger = S^{-1}A^\dagger S = D^\dagger = D^*,$$

since S is unitary and D is diagonal. It follows that

$$S^{-1}AA^\dagger S = (S^{-1}AS)\left(S^{-1}A^\dagger S\right) = DD^*,$$
$$S^{-1}A^\dagger AS = (S^{-1}A^\dagger S)(S^{-1}AS) = D^*D.$$

Since $DD^* = D^*D$, clearly $AA^\dagger = A^\dagger A$. Therefore, we conclude,

- **A matrix can be diagonalized by a unitary similarity transformation if and only if it is normal.**

Thus, both hermitian and unitary matrices are diagonalizable by a unitary similarity transformation.

The eigenvalues of a hermitian matrix are always real. This is the reason why in quantum mechanics observable physical quantities are associated with the eigenvalues of hermitian operators, because the result of any measurement is, of course, a real number. However, the eigenvectors of a hermitian matrix may be complex, therefore the unitary matrix that diagonalizes the hermitian matrix is, in general, complex.

A real symmetric matrix is a hermitian matrix, therefore its eigenvalues must also be real. Since the matrix and the eigenvalues are both real, the eigenvectors can be taken to be real. Therefore, the diagonalizing matrix is a real orthogonal matrix.

Unitary matrices, including real orthogonal matrices, can be diagonalized by a unitary similarity transformation. However, the eigenvalues and eigenvectors of a unitary matrix are, in general, complex. Therefore, the diagonalizing matrix is not a real orthogonal matrix, but a complex unitary matrix. For example, the rotation matrix is a real orthogonal matrix, but, in general, it can only be diagonalized by a complex unitary matrix.

8.7 Functions of a Matrix

8.7.1 Polynomial Functions of a Matrix

Any square matrix A can be multiplied by itself. The associative law of matrix multiplication guarantees that the operation of A times itself n times, which is denoted as A^n, is an unambiguous operation. Thus,

$$A^m A^n = A^{m+n}.$$

Moreover, we have defined the inverse A^{-1} of a non-singular matrix in such a way that $AA^{-1} = A^{-1}A = I$. Therefore, it is natural to define

$$A^0 = A^{1-1} = AA^{-1} = I, \quad \text{and} \quad A^{-n} = \left(A^{-1}\right)^n.$$

With these definitions, we can define polynomial functions of a square matrix in exactly the same way as scalar polynomials.

For example, if $f(x) = x^2 + 5x + 4$, and $A = \begin{pmatrix} 1 & 1 \\ 2 & 3 \end{pmatrix}$, we define $f(A)$ as

$$f(A) = A^2 + 5A + 4.$$

Since

$$A^2 = \begin{pmatrix} 1 & 1 \\ 2 & 3 \end{pmatrix} \begin{pmatrix} 1 & 1 \\ 2 & 3 \end{pmatrix} = \begin{pmatrix} 3 & 4 \\ 8 & 11 \end{pmatrix},$$

$$f(A) = \begin{pmatrix} 3 & 4 \\ 8 & 11 \end{pmatrix} + 5 \begin{pmatrix} 1 & 1 \\ 2 & 3 \end{pmatrix} + 4 \begin{pmatrix} 1 & 0 \\ 0 & 1 \end{pmatrix} = \begin{pmatrix} 12 & 9 \\ 18 & 30 \end{pmatrix}.$$

It is interesting to note that $f(A)$ can also be evaluated by using the factored terms of $f(x)$. For example,

$$f(x) = x^2 + 5x + 4 = (x+1)(x+4),$$

so

$$f(A) = (A+I)(A+4I)$$

$$= \left[\begin{pmatrix} 1 & 1 \\ 2 & 3 \end{pmatrix} + \begin{pmatrix} 1 & 0 \\ 0 & 1 \end{pmatrix} \right] \left[\begin{pmatrix} 1 & 1 \\ 2 & 3 \end{pmatrix} + 4 \begin{pmatrix} 1 & 0 \\ 0 & 1 \end{pmatrix} \right]$$

$$= \begin{pmatrix} 2 & 1 \\ 2 & 4 \end{pmatrix} \begin{pmatrix} 5 & 1 \\ 2 & 7 \end{pmatrix} = \begin{pmatrix} 12 & 9 \\ 18 & 30 \end{pmatrix}.$$

Example 8.7.1 Find $f(A)$, if

$$A = \begin{pmatrix} 1 & 1 \\ 2 & 3 \end{pmatrix}, \quad \text{and} \quad f(x) = \frac{x}{x^2 - 1}.$$

Solution 8.7.1

$$f(A) = \frac{A}{A^2 - I} = A(A^2 - I)^{-1} = \begin{pmatrix} 1 & 1 \\ 2 & 3 \end{pmatrix} \begin{pmatrix} 2 & 4 \\ 8 & 10 \end{pmatrix}^{-1} = \frac{1}{6} \begin{pmatrix} -1 & 1 \\ 2 & 1 \end{pmatrix}.$$

Note that $f(A)$ can also be evaluated by partial fraction. Since

$$f(x) = \frac{x}{x^2 - 1} = \frac{1}{2} \frac{1}{x - 1} + \frac{1}{2} \frac{1}{x + 1},$$

$$f(A) = \frac{1}{2}(A - I)^{-1} + \frac{1}{2}(A + I)^{-1}$$

$$= \frac{1}{2}\begin{pmatrix} 0 & 1 \\ 2 & 2 \end{pmatrix}^{-1} + \frac{1}{2}\begin{pmatrix} 2 & 1 \\ 2 & 4 \end{pmatrix}^{-1} = \frac{1}{6}\begin{pmatrix} -1 & 1 \\ 2 & 1 \end{pmatrix}.$$

8.7.2 Evaluating Matrix Functions by Diagonalization

When the square matrix A is similar to a diagonal matrix, the evaluation of $f(A)$ can be considerably simplified.

If A is diagonalizable, then

$$S^{-1}AS = D,$$

where D is a diagonal matrix. It follows that

$$\dot{D}^2 = S^{-1}ASS^{-1}AS = S^{-1}A^2S,$$
$$D^k = S^{-1}A^{k-1}SS^{-1}AS = S^{-1}A^kS.$$

Thus,

$$A^k = SD^kS^{-1},$$
$$A^n + A^m = SD^nS^{-1} + SD^mS^{-1} = S\left(D^n + D^m\right)S^{-1}.$$

If $f(A)$ is a polynomial, then

$$f(A) = Sf(D)S^{-1}.$$

Furthermore, since D is diagonal and the elements of D are the eigenvalues of A,

$$D^k = \begin{pmatrix} \lambda_1^k & 0 & \cdots & 0 \\ 0 & \lambda_2^k & \cdots & 0 \\ \cdot & \cdot & \cdots & 0 \\ 0 & 0 & 0 & \lambda_n^k \end{pmatrix},$$

$$f(D) = \begin{pmatrix} f(\lambda_1) & 0 & \cdots & 0 \\ 0 & f(\lambda_2) & \cdots & 0 \\ \cdot & & \cdot & \cdots & 0 \\ 0 & 0 & 0 & f(\lambda_n) \end{pmatrix}.$$

Therefore,

$$f(A) = S \begin{pmatrix} f(\lambda_1) & 0 & \cdots & 0 \\ 0 & f(\lambda_2) & \cdots & 0 \\ \cdot & \cdot & \cdots & 0 \\ 0 & 0 & 0 & f(\lambda_n) \end{pmatrix} S^{-1}.$$

Example 8.7.2 Find $f(A)$, if

$$A = \begin{pmatrix} 0 & -2 \\ 1 & 3 \end{pmatrix}, \quad \text{and} \quad f(x) = x^4 - 4x^3 + 6x^2 - x - 3.$$

Solution 8.7.2 First, find the eigenvalues and eigenvectors of A,

$$\begin{vmatrix} 0 - \lambda & -2 \\ 1 & 3 - \lambda \end{vmatrix} = (\lambda - 1)(\lambda - 2) = 0.$$

The eigenvectors corresponding to $\lambda_1 = 1$ and $\lambda_2 = 2$ are, respectively,

$$\mathbf{u}_1 = \begin{pmatrix} 2 \\ -1 \end{pmatrix}, \quad \mathbf{u}_2 = \begin{pmatrix} 1 \\ -1 \end{pmatrix}.$$

Therefore,

$$S = \begin{pmatrix} 2 & 1 \\ -1 & -1 \end{pmatrix}, \quad S^{-1} = \begin{pmatrix} 1 & 1 \\ -1 & -2 \end{pmatrix},$$

and

$$D = S^{-1} A S = \begin{pmatrix} 1 & 0 \\ 0 & 2 \end{pmatrix}.$$

Thus,

$$f(A) = S f(D) S^{-1} = S \begin{pmatrix} f(1) & 0 \\ 0 & f(2) \end{pmatrix} S^{-1}.$$

Since

$$f1) = -1, \quad f(2) = 3,$$

$$f(A) = S f(D) S^{-1} = \begin{pmatrix} 2 & 1 \\ -1 & -1 \end{pmatrix} \begin{pmatrix} -1 & 0 \\ 0 & 3 \end{pmatrix} \begin{pmatrix} 1 & 1 \\ -1 & -2 \end{pmatrix} = \begin{pmatrix} -5 & -8 \\ 4 & 7 \end{pmatrix}.$$

Example 8.7.3 Find the matrix A such that

$$A^2 - 4A + 4I = \begin{pmatrix} 4 & 3 \\ 5 & 6 \end{pmatrix}.$$

Solution 8.7.3 Let us first diagonalize the right-hand side,

$$\begin{vmatrix} 4-\lambda & 3 \\ 5 & 6-\lambda \end{vmatrix} = (\lambda - 1)(\lambda - 9) = 0.$$

The eigenvectors corresponding to $\lambda_1 = 1$ and $\lambda_2 = 9$ are, found to be, respectively,

$$\mathbf{u}_1 = \begin{pmatrix} 1 \\ -1 \end{pmatrix}, \quad \mathbf{u}_2 = \begin{pmatrix} 3 \\ 5 \end{pmatrix}.$$

Thus,

$$S = \begin{pmatrix} 1 & 3 \\ -1 & 5 \end{pmatrix}, \quad S^{-1} = \frac{1}{8}\begin{pmatrix} 5 & -3 \\ 1 & 1 \end{pmatrix},$$

and

$$D = S^{-1}\begin{pmatrix} 4 & 3 \\ 5 & 6 \end{pmatrix} S = \begin{pmatrix} 1 & 0 \\ 0 & 9 \end{pmatrix}.$$

Therefore,

$$S^{-1}(A^2 - 4A + 4I) S = S^{-1}\begin{pmatrix} 4 & 3 \\ 5 & 6 \end{pmatrix} S = \begin{pmatrix} 1 & 0 \\ 0 & 9 \end{pmatrix}.$$

The left-hand side must also be diagonal, since the right-hand side is diagonal. Since we have shown that , as long as $S^{-1}AS$ is diagonal, $S^{-1}A^k S$ will be diagonal, we can assume

$$S^{-1}AS = \begin{pmatrix} x_1 & 0 \\ 0 & x_2 \end{pmatrix}.$$

It follows that

$$S^{-1}(A^2 - 4A + 4I) S = \begin{pmatrix} x_1^2 - 4x_1 + 4 & 0 \\ 0 & x_2^2 - 4x_2 + 4 \end{pmatrix} = \begin{pmatrix} 1 & 0 \\ 0 & 9 \end{pmatrix},$$

which gives

$$x_1^2 - 4x_1 + 4 = 1,$$
$$x_2^2 - 4x_2 + 4 = 9.$$

From the first equation, we get $x_1 = 1,\ 3$, and from the second equation we obtain $x_2 = 5,\ -1$. Therefore, there are four possible combinations for $\begin{pmatrix} x_1 & 0 \\ 0 & x_2 \end{pmatrix}$, namely

$$\Lambda_1 = \begin{pmatrix} 1 & 0 \\ 0 & 5 \end{pmatrix}, \quad \Lambda_2 = \begin{pmatrix} 1 & 0 \\ 0 & -1 \end{pmatrix}, \quad \Lambda_3 = \begin{pmatrix} 3 & 0 \\ 0 & 5 \end{pmatrix}, \quad \Lambda_4 = \begin{pmatrix} 3 & 0 \\ 0 & -1 \end{pmatrix}.$$

Thus, the original equation has four solutions:

$$A_1 = S\Lambda_1 S^{-1} = \begin{pmatrix} 1 & 3 \\ -1 & 5 \end{pmatrix} \begin{pmatrix} 1 & 0 \\ 0 & 5 \end{pmatrix} \frac{1}{8} \begin{pmatrix} 5 & -3 \\ 1 & 1 \end{pmatrix} = \frac{1}{2} \begin{pmatrix} 5 & 3 \\ 5 & 7 \end{pmatrix},$$

and similarly

$$A_2 = \frac{1}{4} \begin{pmatrix} 1 & -3 \\ -5 & -1 \end{pmatrix}, \quad A_3 = \frac{1}{4} \begin{pmatrix} 15 & 3 \\ 5 & 17 \end{pmatrix}, \quad A_4 = \frac{1}{2} \begin{pmatrix} 3 & -3 \\ -5 & 1 \end{pmatrix}.$$

For every scalar function that can be expressed as an infinite series, a corresponding matrix function can be defined. For example, with

$$e^x = 1 + x + \frac{1}{2}x^2 + \frac{1}{3!}x^3 + \cdots,$$

we can define

$$e^A \doteq I + A + \frac{1}{2}A^2 + \frac{1}{3!}A^3 + \cdots.$$

If A is diagonalizable, then

$$S^{-1}AS = D, \quad A^n = SD^n S^{-1},$$

$$e^A = S\left(I + D + \frac{1}{2}\dot{D}^2 + \frac{1}{3!}D^3 + \cdots\right)S^{-1},$$

where

$$D = \begin{pmatrix} \lambda_1 & \cdot & \cdots & 0 \\ 0 & \lambda_2 & \cdots & 0 \\ \cdot & \cdot & \cdots & \cdot \\ \cdot & \cdot & \cdots & \lambda_n \end{pmatrix}.$$

It follows that

$$e^A = S\begin{pmatrix} 1 + \lambda_1 + \frac{1}{2}\lambda_1^2 \cdots & & & 0 \\ 0 & 1 + \lambda_2 + \frac{1}{2}\lambda_2^2 \cdots \cdots & & 0 \\ \cdot & \cdot & \cdots & \cdot \\ 0 & 0 & \cdots 1 + \lambda_n + \frac{1}{2}\lambda_n^2 \cdots \end{pmatrix} S^{-1}$$

$$= S\begin{pmatrix} e^{\lambda_1} & \cdot & \cdots & 0 \\ 0 & e^{\lambda_2} & \cdots & 0 \\ \cdot & \cdot & \cdots & \cdot \\ 0 & 0 & \cdots & e^{\lambda_n} \end{pmatrix} S^{-1}.$$

Example 8.7.4 Evaluate e^A if $A = \begin{pmatrix} 1 & 5 \\ 5 & 1 \end{pmatrix}$.

Solution 8.7.4 Since A is symmetric, it is diagonalizable.

$$\begin{vmatrix} 1-\lambda & 5 \\ 5 & 1-\lambda \end{vmatrix} = \lambda^2 - 2\lambda - 24 = 0,$$

which gives $\lambda = 6, -4$. The corresponding eigenvectors are found to be

$$\mathbf{u}_1 = \begin{pmatrix} 1 \\ 1 \end{pmatrix}, \quad \mathbf{u}_2 = \begin{pmatrix} 1 \\ -1 \end{pmatrix}.$$

Thus,

$$S = \begin{pmatrix} 1 & 1 \\ 1 & -1 \end{pmatrix}, \quad S^{-1} = \frac{1}{2} \begin{pmatrix} 1 & 1 \\ 1 & -1 \end{pmatrix}.$$

Therefore,

$$e^A = S \begin{pmatrix} e^6 & 0 \\ 0 & e^{-4} \end{pmatrix} S^{-1} = \frac{1}{2} \cdot \begin{pmatrix} 1 & 1 \\ 1 & -1 \end{pmatrix} \begin{pmatrix} e^6 & 0 \\ 0 & e^{-4} \end{pmatrix} \begin{pmatrix} 1 & 1 \\ 1 & -1 \end{pmatrix}$$

$$= \frac{1}{2} \cdot \begin{pmatrix} (e^6 + e^{-4}) & (e^6 - e^{-4}) \\ (e^6 - e^{-4}) & (e^6 + e^{-4}) \end{pmatrix}.$$

8.7.3 The Cayley-Hamilton Theorem

The famous Cayley-Hamilton theorem states that every square matrix satisfies its own characteristic equation.

That is, if $P(\lambda)$ is the characteristic polynomial of the nth-order matrix A

$$P(\lambda) = |A - \lambda I| = c_n \lambda^n + c_{n-1} \lambda^{n-1} + \cdots + c_0,$$

then

$$P(A) = c_n A^n + c_{n-1} A^{n-1} + \cdots + c_0 I = 0.$$

To prove this theorem, let $\mathbf{x}_i$ be the eigenvector corresponding to the eigenvalue λ_i. So

$$P(\lambda_i) = 0, \qquad A\mathbf{x}_i = \lambda_i \mathbf{x}_i.$$

Now

$$P(A)\mathbf{x}_i = \left(c_n A^n + c_{n-1} A^{n-1} + \cdots + c_0 I\right) \mathbf{x}_i$$
$$= \left(c_n \lambda_i^n + c_{n-1}\lambda_i^{n-1} + \cdots + c_0\right) \mathbf{x}_i$$
$$= P(\lambda_i)\mathbf{x}_i = 0\mathbf{x}_i.$$

Since this is true for any eigenvector of A, $P(A)$ must be a zero matrix.
For example, if

$$A = \begin{pmatrix} 1 & 2 \\ 2 & 1 \end{pmatrix},$$

$$P(\lambda) = \begin{vmatrix} 1-\lambda & 2 \\ 2 & 1-\lambda \end{vmatrix} = \lambda^2 - 2\lambda - 3.$$

$$P(A) = \begin{pmatrix} 1 & 2 \\ 2 & 1 \end{pmatrix}\begin{pmatrix} 1 & 2 \\ 2 & 1 \end{pmatrix} - 2\begin{pmatrix} 1 & 2 \\ 2 & 1 \end{pmatrix} - 3\begin{pmatrix} 1 & 0 \\ 0 & 1 \end{pmatrix}$$
$$= \begin{pmatrix} 5 & 4 \\ 4 & 5 \end{pmatrix} - \begin{pmatrix} 2 & 4 \\ 4 & 2 \end{pmatrix} - \begin{pmatrix} 3 & 0 \\ 0 & 3 \end{pmatrix} = \begin{pmatrix} 5-3-2 & 4-4 \\ 4-4 & 5-3-2 \end{pmatrix} = \begin{pmatrix} 0 & 0 \\ 0 & 0 \end{pmatrix}.$$

Inverse by Cayley-Hamilton Theorem. The Cayley-Hamilton theorem can be used to find the inverse of a square matrix. Starting with the characteristic equation of A

$$P(\lambda) = |A - \lambda I| = c_n\lambda^n + c_{n-1}\lambda^{n-1} + \cdots + c_1\lambda + c_0 = 0,$$

we have

$$P(A) = c_n A^n + c_{n-1}A^{n-1} + \cdots + c_1 A + c_0 I = 0.$$

Multiplying this equation from the left by A^{-1}, we obtain

$$A^{-1}P(A) = c_n A^{n-1} + c_{n-1}A^{n-2} + \cdots + c_1 I + c_0 A^{-1} = 0.$$

Thus,

$$A^{-1} = -\frac{1}{c_0}\left(c_n A^{n-1} + c_{n-1}A^{n-2} + \cdots + c_1 I\right).$$

Example 8.7.5 Find A^{-1} by Cayley-Hamilton theorem if

$$A = \begin{pmatrix} 5 & 7 & -5 \\ 0 & 4 & -1 \\ 2 & 8 & -3 \end{pmatrix}.$$

Solution 8.7.5

$$P(\lambda) = \begin{pmatrix} 5-\lambda & 7 & -5 \\ 0 & 4-\lambda & -1 \\ 2 & 8 & -3-\lambda \end{pmatrix} = 6 - 11\lambda + 6\lambda^2 - \lambda^3,$$

$$P(A) = 6I - 11A + 6A^2 - A^3 = 0.$$

$$A^{-1}P(A) = 6A^{-1} - 11I + 6A - A^2 = 0.$$

$$A^{-1} = \frac{1}{6}\left(A^2 - 6A + 11I\right).$$

$$A^{-1} = \frac{1}{6}\left[\begin{pmatrix} 5 & 7 & -5 \\ 0 & 4 & -1 \\ 2 & 8 & -3 \end{pmatrix} \begin{pmatrix} 5 & 7 & -5 \\ 0 & 4 & -1 \\ 2 & 8 & -3 \end{pmatrix} - 6 \begin{pmatrix} 5 & 7 & -5 \\ 0 & 4 & -1 \\ 2 & 8 & -3 \end{pmatrix} + 11 \begin{pmatrix} 1 & 0 & 0 \\ 0 & 1 & 0 \\ 0 & 0 & 1 \end{pmatrix} \right]$$

$$= \frac{1}{6}\left[\begin{pmatrix} 15 & 23 & -17 \\ -2 & 8 & -1 \\ 4 & 22 & -9 \end{pmatrix} - \begin{pmatrix} 30 & 42 & -30 \\ 0 & 24 & -6 \\ 12 & 48 & -18 \end{pmatrix} + \begin{pmatrix} 11 & 0 & 0 \\ 0 & 11 & 0 \\ 0 & 0 & 11 \end{pmatrix} \right]$$

$$= \frac{1}{6}\begin{pmatrix} -4 & -19 & 13 \\ -2 & -5 & 5 \\ -8 & -26 & 20 \end{pmatrix}.$$

It can be readily verified that

$$A^{-1}A = \frac{1}{6}\begin{pmatrix} -4 & -19 & 13 \\ -2 & -5 & 5 \\ -8 & -26 & 20 \end{pmatrix} \begin{pmatrix} 5 & 7 & -5 \\ 0 & 4 & -1 \\ 2 & 8 & -3 \end{pmatrix} = \begin{pmatrix} 1 & 0 & 0 \\ 0 & 1 & 0 \\ 0 & 0 & 1 \end{pmatrix}.$$

High Powers of a Matrix. An important application of the Cayley-Hamilton theorem is in the representation of high powers of a matrix. From the equation $P(A) = 0$, we have

$$A^n = -\frac{1}{c_n}\left(c_{n-1}A^{n-1} + c_{n-2}A^{n-2} \cdots + c_1 A + c_0 I\right). \tag{8.38}$$

Multiplying through by A

$$A^{n+1} = -\frac{1}{c_n}(c_{n-1}A^n + c_{n-2}A^{n-1} \cdots + c_1 A^2 + c_0 A), \tag{8.39}$$

and substituting A^n from (8.38) into (8.39), we obtain

$$A^{n+1} = \left(\frac{c_{n-1}^2}{c_n^2} - \frac{c_{n-2}}{c_n}\right) A^{n-1} + \cdots + \left(\frac{c_{n-1}c_1}{c_n^2} - \frac{c_0}{c_n}\right) A + \frac{c_{n-1}c_0}{c_n^2} I. \quad (8.40)$$

Clearly, this process can be continued. Thus any integer power of a matrix of order n can be reduced to a polynomial of the matrix, the highest degree of which is at most $n - 1$. This fact can be used directly to obtain high powers of A.

Example 8.7.6 Find A^{100}, if $A = \begin{pmatrix} 1 & 3 \\ 3 & 1 \end{pmatrix}$

Solution 8.7.6 Since

$$\begin{vmatrix} 1-\lambda & 3 \\ 3 & 1-\lambda \end{vmatrix} = \lambda^2 - 2\lambda - 8 = (\lambda - 4)(\lambda + 2) = 0,$$

the eigenvalues of A are $\lambda_1 = 4$ and $\lambda_2 = -2$. The eigenvalues of A^{100} must be λ_1^{100} and λ_2^{100}, that is

$$A^{100}\mathbf{x}_1 = \lambda_1^{100}\mathbf{x}_1, \qquad A^{100}\mathbf{x}_2 = \lambda_2^{100}\mathbf{x}_2.$$

On the other hand, from the Cayley-Hamilton theorem, we know that A^{100} can be express as a linear combination of A and I, since A is of second-order matrix ($n = 2$).

$$A^{100} = \alpha A + \beta I,$$

thus

$$A^{100}\mathbf{x}_1 = (\alpha A + \beta I)\mathbf{x}_1 = (\alpha\lambda_1 + \beta)\mathbf{x}_1,$$
$$A^{100}\mathbf{x}_2 = (\alpha A + \beta I)\mathbf{x}_2 = (\alpha\lambda_2 + \beta)\mathbf{x}_2.$$

Therefore,

$$\lambda_1^{100} = \alpha\lambda_1 + \beta, \qquad \lambda_2^{100} = \alpha\lambda_2 + \beta.$$

It follows

$$\alpha = \frac{1}{\lambda_1 - \lambda_2}(\lambda_1^{100} - \lambda_2^{100}) = \frac{1}{6}(4^{100} - 2^{100}),$$

$$\beta = \frac{1}{\lambda_1 - \lambda_2}(\lambda_1\lambda_2^{100} - \lambda_2\lambda_1^{100}) = \frac{1}{3}(4^{100} + 2^{101}).$$

Hence,

$$A^{100} = \frac{1}{6}(4^{100} - 2^{100})A + \frac{1}{3}(4^{100} + 2^{101})I.$$

Exercises

1. Find the eigenvalues and eigenvectors of the matrix

$$\begin{pmatrix} 19 & 10 \\ -30 & -16 \end{pmatrix}$$

Ans. $\lambda_1 = 4$, $\mathbf{x}_1 = \begin{pmatrix} 2 \\ -3 \end{pmatrix}$; $\lambda_2 = -1$. $\mathbf{x}_2 = \begin{pmatrix} 1 \\ -2 \end{pmatrix}$.

2. Find the eigenvalues and eigenvectors of the matrix

$$\begin{pmatrix} 6 - 2i & -1 + 3i \\ 9 + 3i & -4 + 3i \end{pmatrix}$$

Ans. $\lambda_1 = 2$, $\mathbf{x}_1 = \begin{pmatrix} 1 - i \\ 2 \end{pmatrix}$; $\lambda_2 = i$. $\mathbf{x}_2 = \begin{pmatrix} 1 - i \\ 3 \end{pmatrix}$.

3. Find the eigenvalues and eigenvectors of the matrix

$$\begin{pmatrix} 2 & -2 & 1 \\ 2 & -4 & 3 \\ 2 & -6 & 5 \end{pmatrix}$$

Ans. $\lambda_1 = 0$, $\mathbf{x}_1 = \begin{pmatrix} 1 \\ 2 \\ 2 \end{pmatrix}$; $\lambda_2 = 1$, $\mathbf{x}_2 = \begin{pmatrix} 1 \\ 1 \\ 1 \end{pmatrix}$; $\lambda_3 = 2$, $\mathbf{x}_3 = \begin{pmatrix} 0 \\ 1 \\ 2 \end{pmatrix}$.

4. If $U^\dagger U = I$, show that (a) the columns of U form an orthonormal set; (b) $UU^\dagger = I$ and the rows of U form an orthonormal set.

5. Show that eigenvalues of anti-hermitian matrix are either zero or pure imaginary.

6. Find the eigenvalues and eigenvectors of the following matrix

$$\frac{1}{5}\begin{pmatrix} 7 & 6i \\ -6i & -2 \end{pmatrix},$$

and show explicitly that the two eigenvectors are orthogonal.

Ans. $\lambda_1 = 2$, $\mathbf{x}_1 = \begin{pmatrix} 2i \\ 1 \end{pmatrix}$; $\lambda_2 = -1$, $\mathbf{x}_2 = \begin{pmatrix} 1 \\ 2i \end{pmatrix}$.

7. Show the two eigenvectors in the previous problem are orthogonal. Construct an unitary matrix U with the two normalized eigenvectors, and show that

$$U^\dagger U = I.$$

Ans. $U = \frac{1}{\sqrt{5}}\begin{pmatrix} 2i & 1 \\ 1 & 2i \end{pmatrix}$.

8. Find the eigenvalues and eigenvectors of the following symmetric matrix

$$A = \frac{1}{5}\begin{pmatrix} 6 & 12 \\ 12 & -1 \end{pmatrix}.$$

Construct an orthogonal matrix U with the two normalized eigenvectors and show that

$$\tilde{U}AU = \Lambda$$

where Λ is a diagonal matrix whose elements are the eigenvalues of A.

Ans. $U = \frac{1}{5}\begin{pmatrix} 4 & -3 \\ 3 & 4 \end{pmatrix}$, $\Lambda = \begin{pmatrix} 3 & 0 \\ 0 & -2 \end{pmatrix}$.

9. Diagonalize the hermitian matrix

$$A = \begin{pmatrix} 1 & 1+i \\ 1-i & 2 \end{pmatrix}$$

with a unitary similarity transformation

$$U^{\dagger}AU = \Lambda.$$

Find the unitary matrix U and the diagonal matrix Λ.

Ans. $U = \begin{pmatrix} -\frac{1+i}{\sqrt{3}} & \frac{1+i}{\sqrt{6}} \\ \frac{1}{\sqrt{3}} & \frac{2}{\sqrt{6}} \end{pmatrix}$, $\Lambda = \begin{pmatrix} 0 & 0 \\ 0 & 3 \end{pmatrix}$.

10. Diagonalize the symmetric matrix

$$A = \begin{pmatrix} 1 & 1 & 0 \\ 1 & 0 & 1 \\ 0 & 1 & 1 \end{pmatrix}$$

with a similarity transformation

$$U^{\dagger}AU = \lambda.$$

Find the orthogonal matrix U and the diagonal matrix Λ.

Ans. $U = \begin{pmatrix} \frac{1}{\sqrt{2}} & \frac{1}{\sqrt{6}} & \frac{1}{\sqrt{3}} \\ 0 & -\frac{2}{\sqrt{6}} & \frac{1}{\sqrt{3}} \\ -\frac{1}{\sqrt{2}} & \frac{1}{\sqrt{6}} & \frac{1}{\sqrt{3}} \end{pmatrix}$, $\Lambda = \begin{pmatrix} 1 & 0 & 0 \\ 0 & -1 & 0 \\ 0 & 0 & 2 \end{pmatrix}$.

11. Diagonalize the symmetric matrix

$$A = \frac{1}{3}\begin{pmatrix} -7 & 2 & 10 \\ 2 & 2 & -8 \\ 10 & -8 & -4 \end{pmatrix}$$

with a similarity transformation

$$\tilde{U} A U = \Lambda.$$

Find the orthogonal matrix U and the diagonal matrix Λ.

Ans. $U = \begin{pmatrix} \frac{2}{3} & \frac{1}{3} & \frac{2}{3} \\ \frac{2}{3} & -\frac{2}{3} & -\frac{1}{3} \\ \frac{1}{3} & \frac{2}{3} & -\frac{2}{3} \end{pmatrix}$, $\Lambda = \begin{pmatrix} 0 & 0 & 0 \\ 0 & 3 & 0 \\ 0 & 0 & -6 \end{pmatrix}$.

12. If A is a symmetric matrix $\left(\text{so } \tilde{A} = A\right)$, S is an orthogonal matrix and $A' = S^{-1} A S$, show that A' is also symmetric.

13. If $\mathbf{u}$ and $\mathbf{v}$ are two column matrices in a two dimensional space, and they are related by the equation

$$\mathbf{v} = C\mathbf{u}$$

where

$$C = \begin{pmatrix} \cos\theta & -\sin\theta \\ \sin\theta & \cos\theta \end{pmatrix},$$

find C^{-1} by the following methods:

(a) By Cramer's rule,

(b) Show that C is orthogonal, so $C^{-1} = \tilde{C}$,

(c) The equation $\mathbf{v} = C\mathbf{u}$ means that C rotates $\mathbf{u}$ to $\mathbf{v}$. Since $\mathbf{u} = C^{-1}\mathbf{v}$, C^{-1} must rotates $\mathbf{v}$ back to $\mathbf{u}$. Therefore, C^{-1} must also be a rotation matrix with an opposite direction of rotation.

14. Find the eigenvalues λ_1, λ_2 and the corresponding eigenvectors of the two-dimensional rotation matrix

$$C = \begin{pmatrix} \cos\theta & -\sin\theta \\ \sin\theta & \cos\theta \end{pmatrix}.$$

Find the unitary matrix U, such that

$$U^{\dagger} C U = \begin{pmatrix} \lambda_1 & 0 \\ 0 & \lambda_2 \end{pmatrix}.$$

Ans. $e^{i\theta}$, $\begin{pmatrix} 1 \\ -i \end{pmatrix}$, $e^{-i\theta}$, $\begin{pmatrix} 1 \\ i \end{pmatrix}$, $U = \frac{1}{\sqrt{2}} \begin{pmatrix} 1 & 1 \\ -i & i \end{pmatrix}$.

15. Show that the canonical form (in which there is no cross-product terms) of the quadratic expression

$$Q(x_1, x_2) = 7x_1^2 + 48x_1 x_2 - 7x_2^2$$

is

$$Q'(x_1', x_2') = 25x_1^2 - 25x_2^2,$$

where

$$\begin{pmatrix} x_1 \\ x_2 \end{pmatrix} = S \begin{pmatrix} x_1' \\ x_2' \end{pmatrix}.$$

Find the orthogonal matrix S.

Ans. $S = \frac{1}{5} \begin{pmatrix} 4 & -3 \\ 3 & 4 \end{pmatrix}$.

16. If $A = \begin{pmatrix} 0 & -2 \\ 1 & 3 \end{pmatrix}$ and $f(x) = x^3 + 3x^2 - 3x - 1$, find $f(A)$.

Ans. $\begin{pmatrix} -13 & -26 \\ 13 & 26 \end{pmatrix}$.

17. If $A = \begin{pmatrix} 1 & 0 \\ 2 & \frac{1}{4} \end{pmatrix}$, find A^n and $\lim_{n \to \infty} A^n$.

Ans. $A^n = \begin{pmatrix} 1 & 0 \\ \frac{8}{3} - \frac{8}{3} \left(\frac{1}{4}\right)^n & \left(\frac{1}{4}\right)^n \end{pmatrix}$, $\lim_{n \to \infty} = \begin{pmatrix} 1 & 0 \\ \frac{8}{3} & 0 \end{pmatrix}$

18. Solve for X, . if $X^3 = \begin{pmatrix} -6 & 14 \\ -7 & 15 \end{pmatrix}$.

Ans. $X = \begin{pmatrix} 0 & 2 \\ -1 & 3 \end{pmatrix}$.

19. Solve the equation

$$M^2 - 5M + 3I = \begin{pmatrix} 1 & -4 \\ 2 & -5 \end{pmatrix}.$$

Ans. $M_1 = \begin{pmatrix} 0 & 2 \\ -1 & 3 \end{pmatrix}$, $M_2 = \begin{pmatrix} -1 & 4 \\ -2 & 5 \end{pmatrix}$, $M_3 = \begin{pmatrix} 6 & -4 \\ 2 & 0 \end{pmatrix}$, $M_4 = \begin{pmatrix} 5 & -2 \\ 1 & 2 \end{pmatrix}$.

20. According to Cayley-Hamilton theorem, every square matrix satisfies its own characteristic equation. Verify this theorem for each of the following matrices:

(a) $\begin{pmatrix} 3 & 4 \\ 5 & 6 \end{pmatrix}$, (b) $\begin{pmatrix} -1 & -2 \\ 3 & 4 \end{pmatrix}$.

21. Find A^{-1} by Cayley-Hamilton theorem if $A = \begin{pmatrix} 2 & 1 & 0 \\ 1 & 2 & 0 \\ 0 & 0 & 3 \end{pmatrix}$.

Ans. $A^{-1} = \frac{1}{3} \begin{pmatrix} 2 & -1 & 0 \\ -1 & 2 & 0 \\ 0 & 0 & 1 \end{pmatrix}$.

References

This bibliograph includes the references cited in the text and a few other books and tables that might be useful.

1. M. Abramowitz, I.A. Stegun, *Handbook of Mathematical Functions* (Dover, New York, 1972)
2. H. Anton, C. Rorres, *Elementary Linear Algegra*, 6th edn. (Wiley, New York, 1991)
3. G.B. Arfken, H.J. Weber, *Mathematical Methods for Physicists*, 5th edn. (Academic Press, San Diego, 2001)
4. M.L. Boas, *Mathematical Methods in the Physical Sciences*, 3rd edn. (Wiley, New York, 2006)
5. T.C. Bradbury, *Mathematical Methods with Applications to Problems in the Physical Sciences* (Wiley, New York, 1984)
6. E. Butkov, *Mathematical Physics* (Addison-Wesley, Reading 1968)
7. F.W. Byron Jr., R.W. Fuller, *Mathematics of Classical and Quantum Physics* (Dover, New York, 1992)
8. T.L. Chow, *Mathematical Methods for Physicists: A Concise Introduction* (Cambridge University Press, Cambridge, 2000)
9. R.V. Churchill, J.W. Brown, *Complex Variable and Applications*, 5th edn. (McGraw-Hill, New York, 1989)
10. H. Cohen, *Mathematics for Scientists and Engineers* (Prentice-Hall, Englewood Cliffs, 1992)
11. R.E. Collins, *Mathematical Methods for Physicists and Engineers* (Reinhold, New York, 1968)
12. R.P. Feynman, R.B. Leighton, M. Sands, *The Feynman Lectures on Physics*, vol. I, Chapter 22 (Addison-Wesley, Reading 1963)
13. M.D. Greenberg, *Advanced Engineering Mathematics*, 2nd edn. (Prentice Hall, Upper Saddle River, 1998)
14. G.H. Golub, C.F. Van Laon, *Matrix Computations* (John Hopkins University Press, Baltimore, 1983)
15. I.S. Gradshteyn, I.M. Ryzhik, *Table of Integrals, Series and Products* (Academic Press, Orlando, 1980)
16. D.W. Hardy, C.L. Walker, *Doing Mathematics with Scientific WorkPlace and Scientific Notebook, Version 5* (MacKichan, Poulsbo, 2003)
17. S. Hasssani, *Mathematical Methods: For Students of Physics and Related Fields* (Springer, New York, 2000)
18. F.B. Hilderbrand, *Advanced Calculus for Applications*, 2nd edn. (Prentice-Hall, Englewood Cliffs, 1976)

© The Editor(s) (if applicable) and The Author(s), under exclusive license to Springer Nature Switzerland AG 2022
K.-T. Tang, *Mathematical Methods for Engineers and Scientists 1*,
https://doi.org/10.1007/978-3-031-05678-9

19. H. Jeffreys, B.S. Jeffreys, *Mathematical Physics* (Cambridge University Press, Cambridge, 1962)
20. D.E. Johnson, J.R. Johnson, *Mathematical Methods in Engineering Physics* (Prentice-Hall, Upper Sadddle River, 1982)
21. D.W. Jordan, P. Smith, *Mathematical Techniques: An Introduction for the Engineering, Physical, and Mathematical Sciences*, 3rd edn. (Oxford University Press, Oxford, 2002)
22. A.K. Kapoor, *Complex Variables* (World Scientific, Singapore, 2011)
23. E. Kreyszig, *Advanced Engineering Mathematics*, 8th edn. (Wiley, New York, 1999)
24. B.R. Kusse, E.A. Westwig, *Mathematical Physics: Applied Mathematics for Scientists and Engineers*, 2nd edn. (Wiley, New York, 2006)
25. Y.K. Kwok, *Applied Complex Variables for Scientists and Engineers* (Cambridge University Press, Cambridge, 2002)
26. H. Kober, *Dictionary of Conformal Representations* (Dover Publications Inc, New York, 1952)
27. S.M. Lea, *Mathematics for Physicists* (Brooks/Cole, Belmont, 2004)
28. E. Maor, *e: The Story of a Number* (Princeton University Press, Princeton, 1994)
29. H. Margenau, G.M. Murphy, *Methods of Mathematical Physics* (Van Nostrand, Princeton, 1956)
30. J. Mathew, R.L. Walker, *Mathematical Methods of Physics*, 2nd edn. (Benjamin, New York, 1970)
31. D.A. McQuarrie, *Mathematical Methods for Scientists and Engineers* (University Science Books, Sausalito, 2003)
32. P.M. Morse, H. Feshbach, *Methods of Theoretical Physics* (McGraw-Hill, New York, 1953)
33. T. Needham, *Visual Complex Analysis* (Clarendon Press, Oxford, 1997)
34. R. Penrose, W. Rindler, *Spinors and Space-Time*, vol. 1 (Cambridge University Press, Cambridge, 1984)
35. G. Polya, *Complex Variables* (Wiley, New York, 1974)
36. E.G. Phillips, *Functions of a Complex Variable* (Oliver & Boyd, Edinburgh, 1957)
37. M.C. Potter, J.L. Goldber, E.F. Aboufadel, *Advanced Engineering Mathematics*, 3rd edn. (Oxford University Press, New York, 2005)
38. W.H. Press, S.A. Teukolsky, W.T. Vettering, B.P. Flannery, *Numerical Recipes*, 2nd edn. (Cambridge University Press, Cambridge, 1992)
39. K.F. Riley, M.P. Hobson, S.J. Bence, *Mathematical Methods for Physics and Engineering*, 2nd edn. (Cambridge University Press, Cambridge, 2002)
40. E.B. Saff, A.D. Snider, *Fundamentals of Complex Analysis with Applications to Engineering and Science*, 3rd edn. (Pearson Education, Upper Saddle River, 2003)
41. B.T. Smith, J.M. Boyle, J. Dongarra, B. Garbow, Y. Ikebe, V.C. Klema, C.B. Moler, *Matrix Eigensystem Routines: EISPACK Guide*, 2nd edn. (Springer-Verlag, Heidelberg, 1976)
42. W.G. Strang, *Linear Algebra and Its Applications*, 2nd edn. (Academic Press, New York, 1980)
43. K.A. Stroud, D.J. Booth, *Advanced Engineering Mathematics*, 4th edn. (Industrial Press, New York, 2003)
44. A. Tucker, *Linear Algebra* (Macmillan, New York, 1993)
45. C.R. Wylie, L.C. Barrett, *Advanced Engineering Mathematics*, 5th edn. (McGraw-Hill, New York, 1982)

Index

Printed in the United States
by Baker & Taylor Publisher Services